Applied and Computational Fluid Mechanics

Scott Post, PhD
Bradley University

JONES AND BARTLETT PUBLISHERS
Sudbury, Massachusetts
BOSTON   TORONTO   LONDON   SINGAPORE

World Headquarters
Jones and Bartlett Publishers
40 Tall Pine Drive
Sudbury, MA 01776
978-443-5000
info@jbpub.com
www.jbpub.com

Jones and Bartlett Publishers
Canada
6339 Ormindale Way
Mississauga, Ontario L5V 1J2
Canada

Jones and Bartlett Publishers
International
Barb House, Barb Mews
London W6 7PA
United Kingdom

Jones and Bartlett's books and products are available through most bookstores and online booksellers. To contact Jones and Bartlett Publishers directly, call 800-832-0034, fax 978-443-8000, or visit our website, www.jbpub.com.

Substantial discounts on bulk quantities of Jones and Bartlett's publications are available to corporations, professional associations, and other qualified organizations. For details and specific discount information, contact the special sales department at Jones and Bartlett via the above contact information or send an email to specialsales@jbpub.com.

Copyright © 2011 by Jones and Bartlett Publishers, LLC

All rights reserved. No part of the material protected by this copyright may be reproduced or utilized in any form, electronic or mechanical, including photocopying, recording, or by any information storage and retrieval system, without written permission from the copyright owner.

MATLAB is a registered trademark of The MathWorks, Inc. in the United States and other countries.

Production Credits
Publisher: David Pallai
Production Director: Amy Rose
Editorial Assistant: Melissa Potter
Associate Production Editor: Tiffany Sliter
Production Assistant: Rebekah Linga
Associate Marketing Manager: Lindsay Ruggiero
V.P., Manufacturing and Inventory Control: Therese Connell
Senior Photo Researcher: Christine Myaskovsky
Composition: Glyph International
Cover and Title Page Design: Kate Ternullo
Cover and Title Page Image: Photo courtesy of NASA Dryden Flight Research Center (NASA-DFRC)
Cover and Title Page Background Photo: © Bram Janssens/Dreamstime.com
Printing and Binding: Malloy, Inc.
Cover Printing: Malloy, Inc.

Library of Congress Cataloging-in-Publication Data
Post, Scott.
 Applied and computational fluid mechanics / Scott Post.
   p. cm.
 Includes bibliographical references and index.
 ISBN-13: 978-1-934015-47-6 (hardcover)
 ISBN-10: 1-934015-47-4 (hardcover)
 1. Fluid mechanics. I. Title.
 TA357.P4865 2010
 620.1'06—dc22
                            2009022415

6048
Printed in the United States of America
14 13 12 11 10   10 9 8 7 6 5 4 3 2 1

Brief Contents

About the Author   v

Preface   vi

Contents   ix

Chapter 1   Introduction   1

Chapter 2   Fluid Statics   25

Chapter 3   Fluid Dynamics   81

Chapter 4   Differential Equations of Fluid Motion   157

Chapter 5   Internal Flow   205

Chapter 6   External Flow   265

Chapter 7   Rotating Machinery   333

Chapter 8   Additional Applications   371

Chapter 9   Fluid Measurement Techniques   427

Appendix A   Properties of Common Fluids   469

Appendix B   Compressible Flow Tables   477

Appendix C   Reynolds Transport Theorem   483

Appendix D   Experimental Uncertainty and Error   485

Appendix E   Additional Resources   493

Appendix F   MATLAB® Codes   497

Appendix G   Glossary   505

Appendix H   Greek Alphabet   509

Appendix I   Conversion Factors   511

Index   515

About the Author

Scott Post received his B.S.M.E. from the University of Missouri-Columbia in 1996, his M.S.M.E. from Purdue University in 1998, and his Ph.D. from Purdue University in 2001. He taught mechanical engineering as an assistant professor at Michigan Technological University from 2001 to 2006, and has been an assistant professor at Bradley University in Peoria, Illinois since 2006. He has also spent three summers working as a summer faculty fellow at NASA's Dryden Flight Research Center in the Mojave Desert. His research interests include computational fluid dynamics, high-speed photography, engines, alternative fuels, fuel sprays, turbulence, and aerodynamics. Scott is currently trying to build a cyclogyro and in his spare time he enjoys cooking and photography.

Preface

Applied and Computational Fluid Mechanics is designed for a traditional undergraduate course in engineering that would use Chapters 1–5 in a one-semester course. The material on open channel flows that is of interest to civil engineering students is included in Chapter 5. For aerospace engineering students, aerodynamics is covered in Chapter 6 and compressible flow in Chapter 8. Most mechanical engineering students will take a course in heat transfer after completing fluid mechanics, so a brief introduction to heat transfer is included in Chapter 8. Additional chapter-specific details are presented in the "Overview of This Book" section.

The fundamental principles of fluid mechanics are conservation of mass, momentum, and energy. Engineering students must understand and be able to apply these basic principles before *designing* fluids engineering systems. Design is the science of applying the fundamental equations of engineering to develop a solution that meets the client's needs for minimum cost. Engineering design is not based on guesswork or gut feelings, but on precise calculations of the relevant physical processes and accurate estimates of costs, including the initial capital costs and ongoing operating costs.

This book emphasizes the use of numerical methods with a computer to solve problems. Although MATLAB® codes are included in Appendix F, the Resource DVD, and on the website, the text presents the numerical methods needed to solve certain problems as general algorithms that can be coded into whatever software package or programming language the student uses. Some of the algorithms in the book can even be used with Microsoft® Excel® spreadsheet software.

Overview of This Book

Chapter 1 covers the fundamentals of dimensions, units, and properties for fluid flows that will be used in the rest of the book. In Chapter 2 we focus on hydrostatics and begin with problems of calculating fluid forces from a fluid at rest. We discuss fluid motion in Chapter 3, in which the student will learn how to calculate the thrust forces and moments generated due to fluids in motion, as well as the usefulness of the energy equation for solving fluid flow problems. To calculate the viscous friction forces that arise when a fluid moves over a solid boundary, it is necessary to examine the fluid flow on very small scales so that the velocity profile in the vicinity of a wall may be calculated (Equation 1.8 may be used to calculate the shear stress at the wall, which can then be used to find the total force). To do this, the governing fluid equation must be expressed in differential form, which we discuss in Chapter 4. Also included in Chapter 4 is a discussion on how computer software programs, such as CFD packages, use discrete/finite approximations to solve the differential equations of fluid flow. In Chapter 5 we analyze the internal flow of fluids through pipes and other ducts and channels. The student will learn how to size pipes for a given application and compute the required pumping power for a given system, or predict the flow rate in an existing system. We discuss external flow in more detail in Chapter 6, with specific emphasis on aerodynamics of planes and ground vehicles as well as the principles of wind-tunnel testing. Most pumps, turbines, and compressors are rotating devices, and we emphasize their analysis in Chapter 7. Chapter 8 contains an overview of various fluid devices, including those used for propulsion and power transmission. The question of how to make measurements of fluid flows is answered in Chapter 9, where we also explore non-dimensionalization techniques for measured data. Finally, there are several useful data tables in the appendices, as well as MATLAB® codes that can be used to solve some of the more complex calculation problems in this book.

For Instructors

Solutions to the exercises, additional code, simulations, suggested projects, and Microsoft® PowerPoint® slides are available online for qualified instructors at http://www.jbpub.com/catalog/9781934015476/. Designated instructors' materials are for qualified instructors only. Jones and Bartlett reserves the right to evaluate all requests.

Acknowledgments

I would like to thank my many colleagues at Michigan Technological University, NASA-Dryden Flight Research Center, and Bradley University, as well as my professors and peers at the University of Missouri and Purdue University, all of whom have helped me in my career in various ways. I want to specifically mention Song-Lin (Jason) Yang at MTU, who encouraged me to pursue the summer faculty fellowship at NASA; Corey Diebler (currently at NASA-Langley) and Tom Bunce at NASA-Dryden for taking the time to work with me during my summers at NASA; and Marty Morris at Bradley University for his unending enthusiasm and support.

I want to thank my dissertation advisor, John Abraham, for never letting me take the easy way out, even if I did not appreciate it at the time. I also want to mention two professors from Mizzou: David "Doc" Wollersheim, who kindled my interest in fluid mechanics, and James Seaba, who advised my first research project and encouraged me to go to graduate school. And of course I must acknowledge my wife Suzanna, who never complained through the two years of my working nights and weekends on this book, and my new baby David, who kept me company on many of those late nights.

Scott L. Post
Peoria, IL

Contents

About the Author v

Preface vi

Chapter 1   **Introduction**   1
- 1.1   Motivation to Study Fluid Mechanics   1
- 1.2   Dimensions and Units   2
  - 1.2.1   Metric and English Unit Systems   2
  - 1.2.2   Unit Systems for Pressure   4
- 1.3   Properties of Fluids   6
  - 1.3.1   Density   6
  - 1.3.2   Viscosity   7
  - 1.3.3   Heat of Vaporization and Vapor Pressure   11
  - 1.3.4   Surface Tension   13
  - 1.3.5   Continuum Assumption   16
  - 1.3.6   Speed of Sound   17
- 1.4   Accuracy and Significant Digits   18
- 1.5   Types of Fluid Flows   19
- References   21
- Problems   22

Chapter 2   **Fluid Statics**   25
- 2.1   Hydrostatic Formula   25
  - 2.1.1   Derivation from a Differential Control Volume   25
  - 2.1.2   Atmospheric Pressure Distribution   29
  - 2.1.3   Application to Manometry   32
- 2.2   Application of the Hydrostatic Formula to Forces on Submerged Surfaces   38
  - 2.2.1   Force on a Flat Plate   39
  - 2.2.2   Force on a Curved Surface   44
  - 2.2.3   Numerical Integration to Obtain Forces   47

　　　　　　　2.2.4　Center of Pressure and Line of Action　49
　　　　　　　2.2.5　Shortcuts to Calculating Forces　55
　　　2.3　Buoyant Forces on Submerged and Floating Objects　58
　　　　　　　2.3.1　Forces on Submerged Objects　59
　　　　　　　2.3.2　Forces on Floating Objects　65
　　　　　　　2.3.3　Stability of Floating Objects　67
　　　2.4　Uniformly Accelerating Systems　69
　　　　　　　2.4.1　Uniform Rectilinear Acceleration　69
　　　　　　　2.4.2　Solid Body Rotation　71
　　　　　　Summary　76
　　　　　　References　76
　　　　　　Problems　76

Chapter 3　Fluid Dynamics　81
　　　3.1　Conservation of Mass　81
　　　　　　　3.1.1　Steady-State Flow　85
　　　　　　　3.1.2　Transient Flow　91
　　　3.2　Conservation of Momentum　94
　　　　　　　3.2.1　Derivation of Momentum Equation from Newton's Second Law　94
　　　　　　　3.2.2　Application of Fluid Thrust Forces　97
　　　　　　　3.2.3　Moving Control Volumes　103
　　　　　　　3.2.4　Accelerating Control Volumes　104
　　　　　　　3.2.5　Numerical Integration of Ordinary Differential Equations　108
　　　　　　　3.2.6　Torques and Angular Momentum　113
　　　　　　　3.2.7　Rotating Machinery　116
　　　3.3　Conservation of Energy　120
　　　　　　　3.3.1　Methods of Extracting or Adding Energy to a Fluid　121
　　　　　　　3.3.2　Friction Losses　121
　　　　　　　3.3.3　Power　121
　　　3.4　The Bernoulli Equation　123
　　　　　　　3.4.1　Derivation of the Bernoulli Equation　123
　　　　　　　3.4.2　Dynamic and Stagnation Pressures　133
　　　　　　　3.4.3　Derivation of Hydrostatic Pressure Distribution from the Bernoulli Equation　141
　　　　　　Summary　142
　　　　　　Problems　143

Chapter 4   **Differential Equations of Fluid Motion**   157
    4.1   Navier–Stokes Equations   157
           4.1.1   Derivation of Navier–Stokes Equations   158
           4.1.2   Turbulent Flow   163
    4.2   Laminar Flow Solutions   167
           4.2.1   Flat Plate Boundary Layers   168
           4.2.2   Streamlines, Streaklines, and Pathlines   174
           4.2.3   Creeping Flow   175
           4.2.4   Lubrication Flow   179
    4.3   Introduction to Computational Fluid Dynamics   188
           4.3.1   Approximating Derivatives   188
           4.3.2   Solving Partial Differential Equations   191
           4.3.3   Example Laminar Flow Solution
                    with Finite Differences   191
           4.3.4   Turbulence Modeling   195
           4.3.5   Limitations of CFD   198
           Summary   199
           References   200
           Problems   200

Chapter 5   **Internal Flow**   205
    5.1   Laminar Flow in Closed Ducts   205
           5.1.1   Laminar Velocity Profile   206
           5.1.2   Wall Shear Stress   210
           5.1.3   Energy Loss Due to Friction   212
    5.2   Turbulent Flow   216
           5.2.1   Critical Reynolds Number   217
           5.2.2   Kinetic Energy Correction Factor   221
           5.2.3   Effects of Surface Roughness   223
           5.2.4   Entrance Length   227
           5.2.5   Noncircular Pipes   229
           5.2.6   Solving Pipe Flow Problems   233
    5.3   Open Channel Flow   240
           5.3.1   Laminar Flow Solutions in Uniform Flow   242
           5.3.2   Gradually Varying Flows   245
           5.3.3   Hydraulic Jumps   249
           5.3.4   Weirs   250
    5.4   Complex Pipe Systems   251
    5.5   Secondary Losses   254
           Summary   257

References 257
Problems 258

Chapter 6 External Flow 265
- 6.1 Introduction to Aerodynamics 265
- 6.2 Viscous Drag 269
- 6.3 Form (Pressure) Drag 273
  - 6.3.1 Drag on a Sphere 273
  - 6.3.2 Strouhal Number 276
  - 6.3.3 Parachute Aerodynamics 280
  - 6.3.4 Sports Balls Aerodynamics 280
  - 6.3.5 Numerical Simulation of Two-Dimensional Trajectories 287
- 6.4 Lift 290
  - 6.4.1 Airfoils 290
  - 6.4.2 Lift-Induced Drag 293
  - 6.4.3 Lift Distribution for Minimum Drag 298
  - 6.4.4 Gliding Flight 300
- 6.5 Vehicle Aerodynamics 302
  - 6.5.1 Aerodynamics of Passenger Cars 302
  - 6.5.2 Aerodynamics of Tractor-Trailers 303
  - 6.5.3 Wave Drag 305
  - 6.5.4 Supersonic Drag 307
  - 6.5.5 Transonic Drag 310
  - 6.5.6 Wind-Tunnel Testing 311
  - 6.5.7 Aerodynamic Stability 317
  - 6.5.8 Lifting Bodies 318
- 6.6 Transient Drag 322

Summary 323
References 323
Problems 325

Chapter 7 Rotating Machinery 333
- 7.1 Conservation of Angular Momentum 333
- 7.2 Pumps, Compressors, Fans, and Propellers 335
  - 7.2.1 Ideal Centrifugal Pump 338
  - 7.2.2 Pump Scaling Laws 342
  - 7.2.3 Net Positive Suction Head (NPSH) and Cavitation 343
  - 7.2.4 Pump Performance Curves 344

        7.2.5   Fans   348
        7.2.6   Propellers   349
  7.3   Turbines   351
        7.3.1   Francis, Pelton, and Kaplan Turbines   351
        7.3.2   Centrifugal Turbines   353
        7.3.3   Applications   355
  7.4   Wind Turbines   358
        Summary   367
        References   368
        Problems   368

Chapter 8   Additional Applications   371
  8.1   Rockets   371
        8.1.1   Rocket Fuels   372
        8.1.2   Transient Rocket Equation   376
        8.1.3   Compressible Flow   378
  8.2   Jet Engines   391
        8.2.1   Turbojets and Turbofans   391
        8.2.2   Ramjets   396
        8.2.3   Scramjets   396
        8.2.4   Pulse Detonation Engines   398
  8.3   Liquid Sprays   398
        8.3.1   Fuel Injectors   398
        8.3.2   Other Applications   409
  8.4   Flow for Electronics Cooling   411
        8.4.1   Air Cooling   413
        8.4.2   Liquid Cooling   415
  8.5   Flow in Biological Systems   415
        8.5.1   Internal Flows   415
        8.5.2   External Flows   417
        Summary   423
        References   423
        Problems   424

Chapter 9   Fluid Measurement Techniques   427
  9.1   Velocity Probes   427
        9.1.1   Pitot Tube   428
        9.1.2   Hot-Wire Anemometer   431
        9.1.3   Laser-Based Measurements   432
        9.1.4   Flow Visualization   436

	9.2	Flow Rate Measurements 443
		9.2.1 Venturi Meter 443
		9.2.2 Rotameter 445
		9.2.3 Orifice Plate Meter 447
		9.2.4 Laminar Flow Element 449
		9.2.5 Turbine Meters 450
		9.2.6 Ultrasonic Flow Meter 451
		9.2.7 Coriolis Meter 451
		9.2.8 Accuracy of Flow Meters 451
	9.3	Pressure Transducers 452
	9.4	Nondimensionalization of Flow Data 453
		9.4.1 Common Nondimensional Numbers 454
		9.4.2 Buckingham Pi Theorem 456
		Summary 463
		References 463
		Problems 463

Appendix A **Properties of Common Fluids** 469

Appendix B **Compressible Flow Tables** 477

Appendix C **Reynolds Transport Theorem** 483

Appendix D **Experimental Uncertainty and Error** 485

Appendix E **Additional Resources** 493

Appendix F **MATLAB® Codes** 497

Appendix G **Glossary** 505

Appendix H **Greek Alphabet** 509

Appendix I **Conversion Factors** 511

Index 515

1   Introduction

In This Chapter
- Motivation to Study Fluid Mechanics
- Dimensions and Units
- Properties of Fluids
- Accuracy and Significant Digits
- Types of Fluid Flows

The objective of this book is to aid in the understanding of fluid dynamics and to develop the tools for solving problems in which fluid flows are important.

■ 1.1   Motivation to Study Fluid Mechanics

Fluid mechanics is important in many engineering applications, including pipe flow, pump sizing, fluid power, aerodynamics, design of fluid storage containers, and many others. Engineers often design complex systems that include fluid components, such as cars, trucks, and other vehicles; commercial and residential buildings; and municipal water management systems.

    With mature knowledge of fluid mechanics, an engineer can calculate the flow rate a given pump can supply, the power generated by a hydroelectric power plant, the efficiency of a pump or turbine, friction losses in fluid bearings, drag force on vehicles, and friction loss in pipes. Fluid mechanics can also be used to analyze the flows inside biological systems; calculate flow rates in canals, rivers, and other open channels; predict the trajectory of projectiles; predict rocket and jet engine performance; and predict wind turbine power generation rates. The principles of fluid mechanics are used to design airplanes, complex pipe systems, convective cooling systems for electrical components, and automobile bodies for drag reduction. Fluid mechanics is even required in some structural engineering problems, such as calculating the forces on flanges and joints in piping systems, or the stresses in pipes, containers, and reservoirs that carry fluids. Thus, a knowledge of fluid mechanics is an essential part of any engineer's education.

1.2 Dimensions and Units

The student of engineering would be well served to memorize the units of commonly used variables. This saves time when working problems, and also helps to ensure accuracy in computations—all proper engineering equations are dimensionally consistent. The four fundamental dimensions used in fluid mechanics—and in engineering in general—are length, mass, time, and temperature. Temperature is a measure of the average internal energy of the molecules.

1.2.1 Metric and English Unit Systems

The primary unit system used worldwide is the m-k-s system. The fundamental units are meters (m) for length, kilograms (kg) for mass, and seconds (s) for time. These fundamental units are combined to form derived units, such as Newtons for force, Joules for energy, and Watts for power, expressed in terms of the fundamental units as follows:

$$\text{Force } 1 \text{ N} = 1 \text{ kg m/s}^2$$
$$\text{Energy } 1 \text{ J} = 1 \text{ kg m}^2/\text{s}^2$$
$$\text{Power } 1 \text{ W} = 1 \text{ kg m}^2/\text{s}^3$$

The most commonly used prefixes in the metric system, which progress in steps of three orders of magnitude, are listed in Table 1.1.

An antiquated unit system that is still encountered in reading and practice is the c-g-s system, in which centimeters (cm), grams (g), and seconds (s) are the fundamental units. Three of the derived units in this system are dyne for force, erg for energy,

Table 1.1 Common Prefixes for Metric Units

Prefix	Symbol	Multiplication factor
Giga	G	10^9
Mega	M	10^6
Kilo	k	10^3
Milli	m	10^{-3}
Micro	μ	10^{-6}
Nano	n	10^{-9}

and poise for viscosity. A chart showing these units expressed in terms of the fundamental units and compared to the m-k-s units follows:

Property	c-g-s system	m-k-s equivalent
Force	$1 \text{ dyn} = 1 \text{ g cm/s}^2$	$10^5 \text{ dyn} = 1 \text{ N}$
Energy	$1 \text{ erg} = 1 \text{ g cm}^2/\text{s}^2$	$10^7 \text{ erg} = 1 \text{ J}$
Power	$1 \text{ erg/s} = 1 \text{ g cm}^2/\text{s}^3$	$10^7 \text{ erg/s} = 1 \text{ W}$
Viscosity	$1 \text{ P} = 1 \text{ g/cm/s}$	$10 \text{ P} = 1 \text{ Pa s}$

1 poise (P) is equal to one gram per centimeter per second $1 \text{ P} = 1 \dfrac{\text{g}}{\text{cm} \cdot \text{s}}$
10 poise is equal to 1 Pascal times 1 second $10 \text{ P} = 1 \text{ Pa} \cdot \text{s}$

In the English unit system still in use in the United States, the fundamental units of length and time are the foot (ft) and second (s), respectively. The fundamental units of mass are the pound mass (lbm) and, less commonly, the slug. The pound force (lbf) is the unit people use when speaking of their weight.

$1 \text{ lbf} = 1 \text{ slug ft/s}^2 = 32.2 \text{ lbm ft/s}^2$

$1 \text{ slug} = 32.2 \text{ lbm}$

The basic unit of pressure is pounds per square inch (psi).

$1 \text{ psi} = 1 \text{ lbf/in}^2 = 144 \text{ lbf/ft}^2$

1 lbf/ft² is sometimes abbreviated as 1 psf. The basic unit of power in the English system is the horsepower.

$1 \text{ hp} = 550 \text{ ft lbf/s}$

Fluid pumps are usually sold by their horsepower rating.

Although volume may be adequately expressed in terms of cubic meters (m³), liters (L) and milliliters (mL) are commonly used for fluid volumes. A milliliter is the volume occupied by 1 cubic centimeter (cc) of a fluid. So 1 mL of a fluid has a volume of 10^{-6} m³. A liter is equal to 1000 mL or 0.001 m³. In the English unit system, no less than eight different units of measure are used for fluid volume; the most common ones in engineering systems are gallons and cubic feet. Conversion factors between common metric and English units are given inside the cover of this book and in the appendices.

1.2.2 Unit Systems for Pressure

The local atmospheric pressure at any locale depends on the altitude above sea level, the geometric latitude, and the local weather conditions. The average atmospheric pressure at sea level is called a *standard atmosphere*, denoted as 1 atm.

A barometer, shown in Figure 1.1, is used to measure atmospheric pressure. A standard atmosphere gives a reading of 760 mm of mercury (mm Hg), or 29.9 inches. One torr is the pressure exerted by 1 mm Hg, so 1 atm = 760 torr. In the metric system, the standard unit of pressure is the Pascal (Pa), which is defined as 1 N of force exerted over 1 square meter of area, so that 1 Pa = 1 N/m^2 = 1 kg/(m·s^2). Another unit of pressure in the metric system is the bar, which is defined as 1 bar = 10^5 Pa, so that 1 bar is very nearly equal to 1 atm. Table 1.2 summarizes these units.

When atmospheric pressure is listed as being 101,325 Pa, that measurement is of the *absolute pressure* of the air. The absolute pressure scale is relative to a reading of zero for a perfect vacuum. Most pressure gages, except barometers, read not the absolute pressure but rather the *gage pressure*. The gage pressure is the pressure *relative to the local surrounding atmosphere*. For example, if the pressure in a car tire is measured as 30 psi (gage) (also denoted psig), then it is pressurized to 30 psi *above* the surrounding atmospheric pressure.

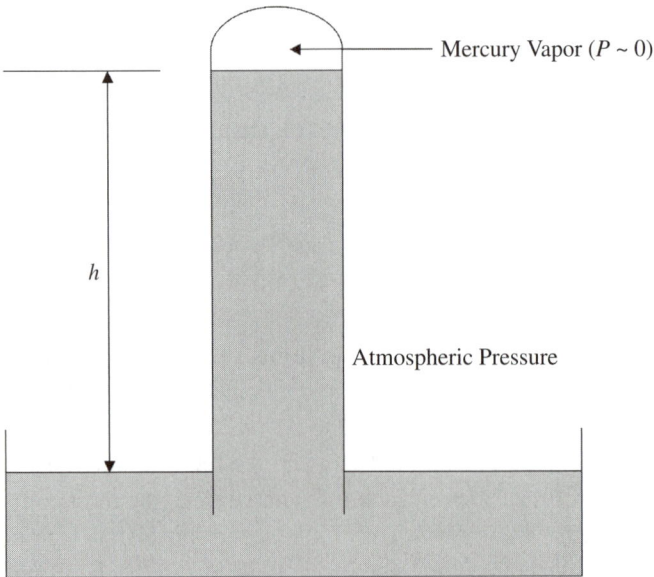

FIGURE 1.1 Diagram of standard mercury barometer for measuring atmospheric pressure.

Table 1.2 Pressure Measurements

Unit	Abbreviation	Relation to standard atmospheric pressure
Atmosphere	atm	1.0
Pascal	Pa	101,325
Psi	psi	14.7
Torr	torr	760
Inch Hg		29.92
Inch H$_2$O		414
mm Hg		760
Bar	bar	1.01325

EXAMPLE 1.1

Calculate the density of the air in your car tire, if the tire pressure is 30 psig and the temperature is 70°F.

SOLUTION When using the ideal gas law to calculate density, you should always use the absolute pressure. The ideal gas law is

$$P = \rho \mathcal{R} T \tag{1.1}$$

where $\mathcal{R} = R/M$, and M is the molecular weight of the gas, in units of kg/kmol. So if the local atmospheric pressure on the day you measure your tire pressure is 14.5 psi (absolute) (also denoted psia), then the absolute pressure inside the tire is 30.0 + 14.5 = 44.5 psia. If the tires are in equilibrium with the surrounding air at 70°F, then after converting the temperature to absolute units you can use the ideal gas law. (Note that if you had just been driving your car, the tires will probably be hotter than the surrounding air.)

$$\rho = \frac{PM}{RT} = \frac{44.5 \frac{\text{lbf}}{\text{in}^2} \, 29.0 \frac{\text{lbm}}{\text{lbmol}}}{1545 \frac{\text{ft} \cdot \text{lbf}}{\text{lbmol}°\text{R}} \, 530°\text{R}} \, 144 \frac{\text{in}^2}{\text{ft}^2} = 0.227 \frac{\text{lbm}}{\text{ft}^3}$$

For simplicity, we use the symbol R to represent the universal gas constant in this book.

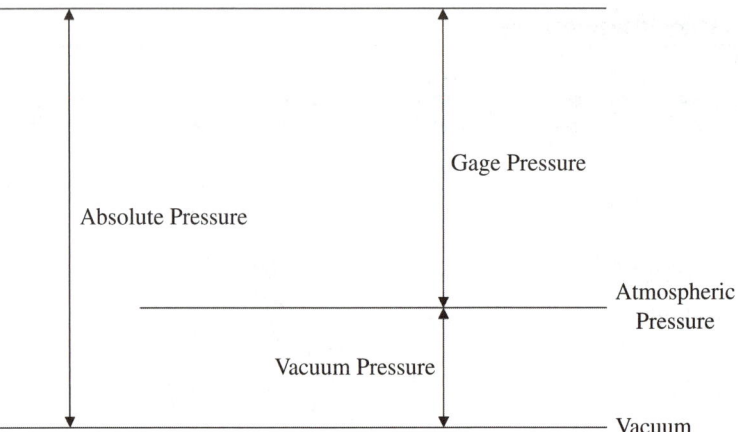

FIGURE 1.2 Pressure systems diagram.

Another pressure system that is sometimes used is vacuum pressure. If the local conditions cause the pressure to decrease below the surrounding atmosphere, then the difference in pressure can be termed the *vacuum pressure*. Sometimes a person will say that a pump can draw so many psi of vacuum. The different pressure systems are illustrated in Figure 1.2.

1.3 Properties of Fluids

The two most frequently used fluid properties in this textbook are density and viscosity. Other fluid properties that are important in specific applications include heat of vaporization, vapor pressure, surface tension, and speed of sound.

1.3.1 Density

The *density* of a fluid is the amount of mass it contains per unit volume:

$$\rho = \frac{m}{\forall} \tag{1.2}$$

The density is the reciprocal of the specific volume, v, which is commonly used in thermodynamics.

$$\rho = \frac{1}{v} \tag{1.3}$$

At room temperature the density of water is a little less than 1000 kg/m³, or 998 kg/m³ at 22°C. For the majority of the problems in this textbook, the density of water will be rounded to a value of 1000 kg/m³. For applications where high precision is required, the density should be found using pressure and temperature and an appropriate software program or reference table (such as Table A.1 in Appendix A).

A quantity related to density is the *specific gravity*, usually denoted by the abbreviation SG:

$$SG = \frac{\rho}{\rho_{ref}} \tag{1.4}$$

where the reference fluid is water for liquids and air for gases, and is taken at a state of 20°C and 1 atm.

We often model liquids as incompressible, meaning that the density is constant regardless of the flow conditions or the state of the fluid. Furthermore, we usually assume the specific heats of liquids to be constant. In reality, liquids are somewhat, compressible. As an example, the density of seawater increases 5% from the ocean's surface to its deepest depth.

Another property related to density is the *specific weight*, γ, which has units of weight (force) per volume:

$$\gamma = \rho g \tag{1.5}$$

So, for example, for water at 4°C at sea level, $\gamma = (1000 \text{ kg/m}^3)(9.8 \text{ m/s}^2) = 9800 \text{ N/m}^3$.

1.3.2 Viscosity

One unique property of a fluid is that it will deform continuously under shearing stresses, no matter how small the stress. On the other hand, a solid will deform by a finite amount proportional to the stress applied, and then come to rest. In a solid the stress is proportional to the strain. A fluid, however, is free to move and so does not have a finite strain but continues to move as long as the shear stress is applied.

The amount that a fluid will deform or move under a given stress is determined by a property called the *viscosity*. For example, suppose there is layer of fluid 1-cm deep in a large container, and a thin flat plate is placed on the surface of the liquid and then moved across the surface, parallel to the bottom of the container. We expect there to be more resistance to the motion if the fluid is oil rather than water because oils are more viscous. Another way to say this is that water flows more readily than oil.

Sir Isaac Newton considered this problem of the resistance force of a fluid to parallel motion. A schematic illustrating the problem is shown in Figure 1.3. Based on his observations he proposed the following equation, which has come to be known as Newton's law of viscosity for the force between two parallel layers of fluid:

$$F = \frac{\mu A U}{y} \tag{1.6}$$

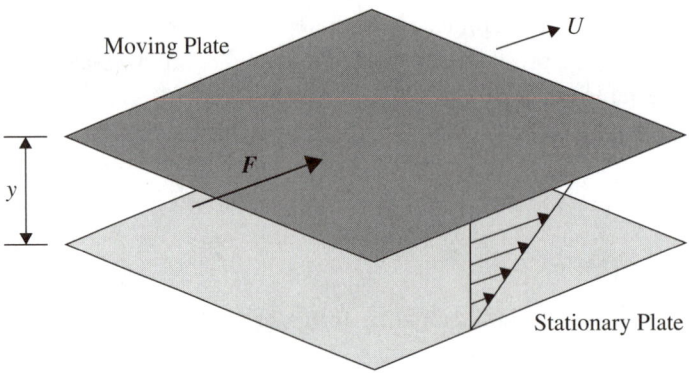

FIGURE 1.3 Diagram illustrating the problem considered in the derivation of Newton's hypothesis. Each plate has area A.

The symbol U is used for the velocity, y for the perpendicular distance between plates or layers, and μ for the coefficient of viscous resistance. In modern engineering, we replace the force, F, and the contact area, A, with the shear stress, τ, since $\tau = F/A$.

Furthermore, taking the limit as the perpendicular distance y between fluid layers (or solid surfaces) goes to zero, we can replace y with dy and U with du, so Equation 1.6 becomes

$$\tau = \mu \frac{du}{dy} \tag{1.7}$$

for the special case of planar two-dimensional shear flow. This is illustrated in Figure 1.4.

Any fluid for which Equation 1.7 holds true with a constant coefficient of viscosity μ is termed a *Newtonian fluid*. Newton's law of viscosity is not derived from any fundamental theory, but it agrees with observations and measurements for many fluids under a wide range of conditions to sufficient accuracy for engineering fluids; thus it is widely used. A Newtonian fluid is also isotropic and has a mechanical pressure equal to the thermodynamic pressure. In fact, when the fluid is at rest, the only stress is the thermodynamic pressure. Those fluids whose viscosity varies according to the flow conditions are referred to as *non-Newtonian fluids*. When the viscosity is constant, the shear is linearly proportional to the strain rate. Most gases and simple liquids like water and oils are Newtonian fluids.

Another way to understand viscosity is to approach it from the molecular level. Viscosity arises from the collisions of individual molecules of a fluid. Higher-velocity molecules tend to transfer some of their kinetic energy when they collide with lower-velocity molecules. Averaged over many collisions the net effect is a diffusion of kinetic energy from fluids of high velocity to fluids of lower velocity. Thus when there

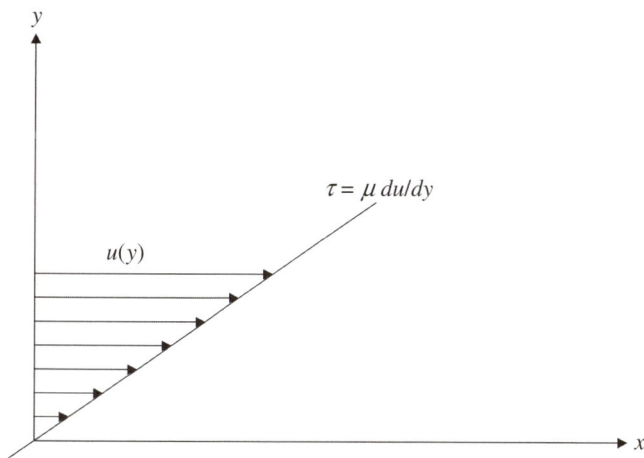

FIGURE 1.4 Illustration of the relationship of shear stress with strain rate.

is a relative shearing motion between two layers of fluid or between a fluid and a solid surface, viscosity acts to resist the motion. (This contributes to the air drag of moving vehicles and projectiles.) The macroscopic viscosity can be derived from molecular properties. For a monatomic ideal gas, the atoms are typically so far apart that intermolecular forces can be safely neglected and only direct collisions need be considered. For such a gas Newton's law may be exactly derived, although the derivation is beyond the scope of this book.

Experiments show that for a wide range of conditions the dependence of viscosity on pressure is small enough that it can be neglected in both gases and liquids. However, viscosity does depend on temperature. In gases, the *viscosity increases with increasing temperature*, as higher temperature leads to higher molecular velocities and increases the effective rate of diffusion. In liquids the trend is quite the opposite. The *viscosity of liquids decreases with increasing temperatures*. The viscosity of a liquid is inversely proportional to the mobility of the molecules. Since the molecules are close together, it is more difficult for them to move in a liquid than in a gas. The viscosity of a liquid is determined by the cohesive forces between closely spaced molecules. These cohesive forces decrease with an increase in temperature. An important implication of this principle is the flow of oil in a car engine on a cold morning.

Non-Newtonian fluids exhibit more complex behavior. The study of non-Newtonian fluids is called *rheology*. Polymers are the most common class of non-Newtonian fluids. Other examples include slurries, cement mix, paint, blood, and ketchup.

The viscosity of liquids can be measured by many means. The device most commonly used for automotive lubricating oils is the Saybolt viscometer (shown in Figure 1.5), which gives a reading in Saybolt seconds. A standardized chamber is filled

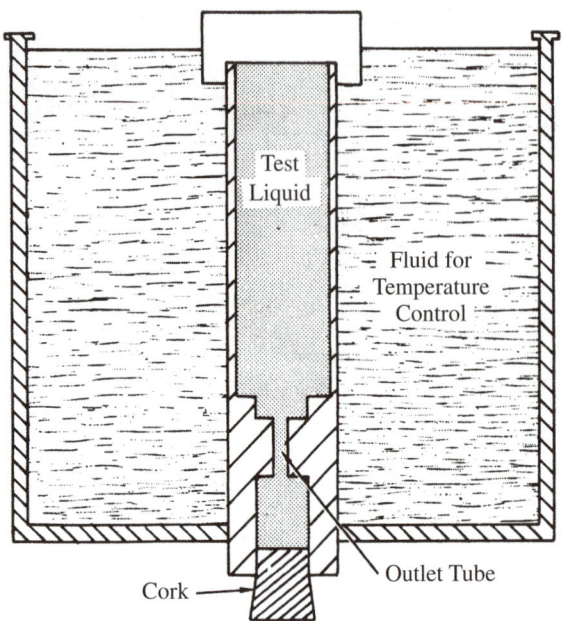

FIGURE 1.5 Diagram of a Saybolt viscometer, from [Audel70]. Used with permission of Wiley.

with 60 mL of oil and placed in the viscometer, which maintains a constant temperature in the oil. A cork plug is then removed from the bottom of the chamber, and the oil is allowed to flow into a beaker beneath the hole. The amount of time it takes for the beaker to be filled up to 60 mL is termed Saybolt universal seconds. The conversion from Saybolt universal seconds (SUS) to viscosity (normally in centistokes) is obtained through an empirical equation:

$$\nu = at - \frac{b}{t} \tag{1.8}$$

where the constants a and b are curve fit for particular temperature ranges. Typical values are $a = 0.0022$ and $b = 1.8$. The use of a Saybolt tube is standard practice in the United States. As an example, SAE 30 oil takes 30 seconds for the 60 mL in the viscometer to flow out of the 1.76-mm-diameter orifice at 100°F. In other countries similar tests are used, though the details vary.

An oil's viscosity at 0°F (−18°C) determines its rating for winter grades of oil, such as 5W, 10W, and 20W, and SAE grade 20, 30, 40, and 50 oils are determined by their viscosity at 212°F (100°C). The higher the grade number, the thicker the oil, and the higher its viscosity and resistance to flow.

The *viscosity index* is a measure of how much the viscosity of an oil changes with temperature. The lower the viscosity index, the larger the decrease in viscosity as

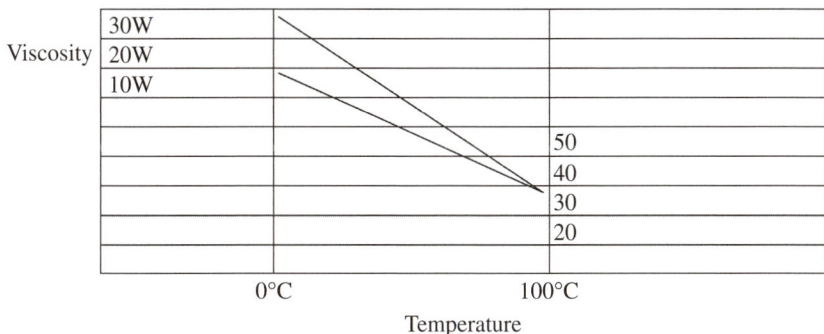

I FIGURE 1.6 Variation of different motor oil grades' viscosities with respect to temperature.

temperature is increased. Hence a higher viscosity index is desirable for automotive applications. To increase the viscosity index, large-molecule additives with molecular weights ranging from 100 to 1000 are used. Multigrade oils have a higher viscosity index than single-grade oils. For multiweight oils the first number indicates the viscosity at −18°C and the second at 100°C. A 5W-30 oil would typically be used in cooler climates, a 10W-40 oil in temperate climates, and a 15W-50 oil in hot climates. See Figure 1.6 for a rough schematic of trends in oil viscosity with temperature.

Other viscometers in use include the Ostwald viscometer, which measures the time for a liquid to flow between two markings in a small capillary tube.

The *kinematic viscosity* is defined as the ratio of a fluid's viscosity to its density.

$$\nu = \frac{\mu}{\rho} \tag{1.9}$$

In most gases, the coefficients of kinematic viscosity, ν, molecular diffusivity, D, and thermal diffusivity, α, are nearly the same. Figure A.1 and Table A.5 in Appendix A list the properties of many common fluids, including viscosity. As can be seen, the viscosity of water is relatively low at 0.001 kg/m-s. The effects of viscosity on fluid flows will be discussed in Chapters 4 through 6.

1.3.3 Heat of Vaporization and Vapor Pressure

The *heat of vaporization* is the amount of energy per unit mass that is needed to vaporize a liquid at a given state. A popular formula for the variation in heat of vaporization, h_{fg}, with temperature, T, is

$$h_{fg}(T) = h_{fg}(T_{bn})\left(\frac{T_{cr} - T}{T_{cr} - T_{bn}}\right)^{0.38} \tag{1.10}$$

where T_{cr} is the critical temperature of the fluid, and T_{bn} is the normal boiling point of the fluid at 1 atmosphere of pressure.

The *vapor pressure*, $P_{vap} = P_{vap}(T)$, is the pressure at which a liquid boils (or at which a gas condenses into a liquid) and is at equilibrium. If $P > P_{vap}$, then boiling will not occur, but *evaporation* can occur if there is another species present (usually air) that is not saturated. *Cavitation* can occur where the hydrodynamic liquid pressure drops below P_{vap} due to changes in pressure in the fluid flow. This normally only happens in areas of high acceleration. An example where this occurs is in the orifice nozzles of diesel fuel injections.

When two or more fluids are mixed, the type of mixture that forms depends on the relative surface tensions between the two fluids, as well as other factors such as the solubility. Once a mixture is made, if it is to be used often it is convenient to be able to calculate the mixture properties easily. Averaged property values work for viscosity and surface tension, but not always for the boiling point and critical point. Furthermore, fluid mixtures typically do not have a fixed boiling point. Instead of listing the boiling point temperature at a fixed pressure, their behavior is described by a *distillation curve*. Figure 1.7 shows distillation curves for various fuels. Note that the distillation curve for ethanol is flat since ethanol is not a mixture but a pure substance.

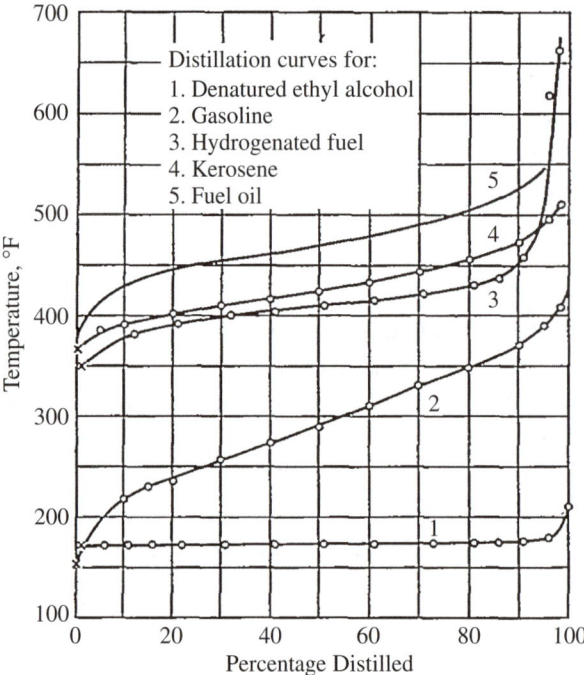

FIGURE 1.7 Typical distillation curves for various fuels. Courtesy of NASA.

1.3.4 Surface Tension

Whenever there is an interface between a liquid and a gas, or between two immiscible liquids, there is a *surface tension* associated with that interface. This force arises because the molecules of a fluid like each other better than they like molecules of another surface. At the microscopic level, the surface tension is related to the van der Waals force of attraction between molecules. The molecules in the middle of a fluid experience attractive forces from all their neighbors equally so that there is no net direction to the sum of these forces. For molecules on the surface, however, there are no neighboring molecules on one side, so they experience a net pull inward, toward the other fluid molecules. This is shown in Figure 1.8.

The surface tension generally attempts to minimize the surface area of the interface, which is why rain drops on a car or water drops on a shower curtain generally take a circular shape. Work must be done to expand the surface, hence increasing its area, and to draw more molecules to the surface layer. Thus, surface tension is usually given in units of force per length, such as N/m, but it is equally correct to think of surface tension as the surface energy per unit interface area and use units of J/m^2. For example, the energy required to rupture a liquid surface is proportional to the increase in surface area thus derived. Fuel injectors work by converting mechanical energy to surface energy to break up a continuous liquid into dispersed liquid drops. This increase in surface area leads to an increased vaporization rate. Table 1.3 lists the surface tension values at standard conditions for some common fluids.

Surface tension generally decreases with increasing temperature and approaches zero as the critical temperature, T_{cr}, of the fluid is approached. The Guggenheim–Katayama formula for surface tension dependence on temperature is

$$\sigma = \sigma_0 \left(1 - \frac{T}{T_{cr}}\right)^n \tag{1.11}$$

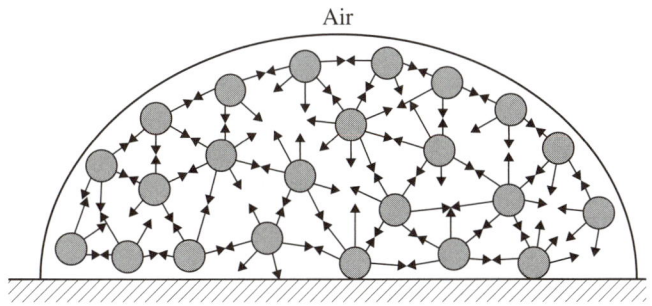

FIGURE 1.8 Diagram illustrating molecular behavior of surface tension. Courtesy of NASA.

Table 1.3 Surface Tensions of Common Liquids at STP in Contact with Air

Liquid	Surface tension (J/m^2)
Water	0.072
Mercury	0.480
Ethanol	0.023
Methanol	0.022
Glycerin	0.063
Olive oil	0.033

The static *contact angle* is the angle at which the liquid–vapor interface meets a solid surface. It depends on not only the surface tension of the fluid but also the characteristics of the surface. The contact angle is measured from the inside of the drop. Generally, the higher the surface tension, the larger the contact angle. A *hydrophilic* surface, one to which water readily adheres, will have a low surface tension and a small contact angle. A *hydrophobic* surface is the opposite: It has a high surface tension and large contact angle, and water does not readily adhere to it. Sometimes the effective surface tension of a substance can be changed by applying a coating or treating the surface. One example of this is Rain-X™. Rain-X is a glass treatment that makes a car's windshield more hydrophobic so that rain drops have a larger contact angle and hence less contact area to which to adhere to the glass surface. The drops then are more easily dislodged by air blowing over the vehicle. Figure 1.9 shows water

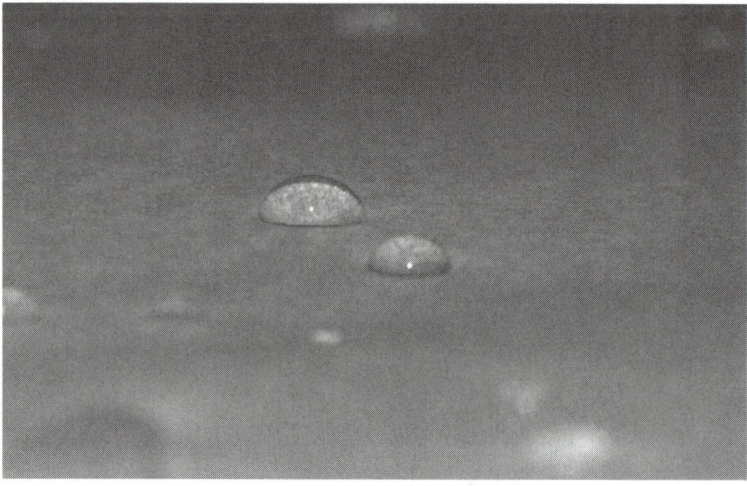

FIGURE 1.9 Water drops on a hydrophobic surface.

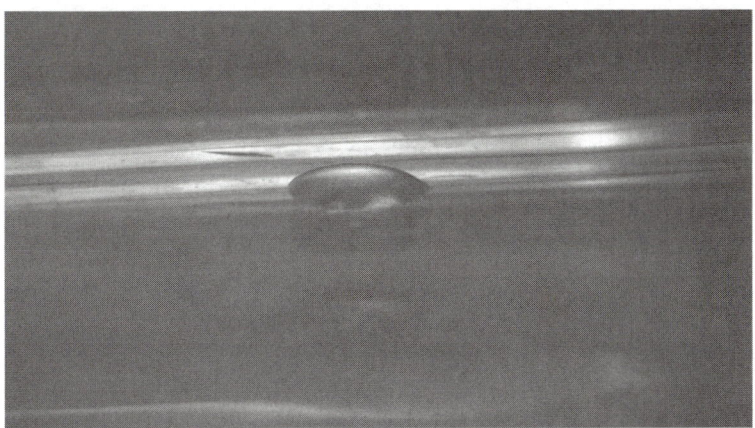

FIGURE 1.10 Water drop on a hydrophilic surface.

drops on a hydrophobic surface, and Figure 1.10 shows a water drop on a hydrophilic surface. Figure 1.11 illustrates how the contact angle is defined.

Although contact angles are important and well-defined for liquid drops in air or on a solid surface, there are many other types of multiphase mixtures, such as liquid–liquid mixtures and liquid–solid mixtures. When one phase is mixed well throughout the other, the resulting mixture is called a suspension. Multiphase suspensions include foams, emulsions, fogs, mists, sprays, gels, and aerosols. When two liquids are miscible, they can be mixed together to form a single homogeneous phase. For example, water and ethanol are miscible, but water and vegetable oil are not. A mixture of two immiscible liquids is an *emulsion*; examples include margarine, photographic film coating, and cutting fluids in machining. Emulsions can be made temporarily by rapidly mixing the two liquids, but eventually the liquids will settle out into two distinct phases unless an emulsifying agent is used to stabilize the emulsion. These agents are usually similar to a surfactant, which is a material added to a liquid to decrease its surface tension.

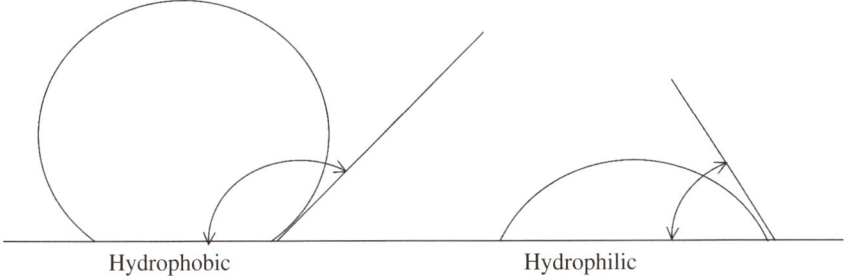

FIGURE 1.11 Definition of contact angle for a liquid drop on a solid surface.

1.3.5 Continuum Assumption

We always assume in this book that we can treat fluids as a continuum. That is, we assume that any fluid or part of a fluid we are looking at is extensive enough to contain such a large number of molecules that its properties are well defined and continuous in space. This assumption may not be valid for a microchannel flow, in which case a different approach must be used to account for the effects of individual molecules on the overall fluid flow. However, such problems are beyond the scope of this book.

Related to the continuum assumption is the no-slip boundary condition: Molecular interactions cause the fluid to be in equilibrium with a contact surface. For a large particle (such as a baseball) traveling through the air, the flow can be treated as a continuum and the no-slip boundary condition can be applied at the surface of the baseball. For a much smaller object (such as a particle of dust settling in quiescent room air), if the mean free path of the gas is comparable to the particle size, then the effects of individual air molecules may become important. The mean free path of a gas, λ, is given by

$$\lambda = \frac{1}{\sqrt{2}\eta \pi d_m^2} \tag{1.12}$$

where η is the number density of gas molecules in terms of number per volume, and d_m is the equivalent hard-sphere diameter of the molecules. (Molecules are not really solid spheres, but it is often useful to model them as such.) For air, the average molecular diameter is about 3.7 Å, or 3.7×10^{-10} m. The number density at STP can be calculated from the ideal gas law and by employing Avogadro's number:

$$\eta = \frac{P}{RT}N_{AV} = \frac{101{,}350 \frac{N}{m^2}}{8314 \frac{N \cdot m}{mol \cdot K} \, 293 \text{ K}} 6.02 \cdot 10^{23} \frac{\text{molecules}}{\text{mol}} = 2.5 \cdot 10^{22} \frac{\text{molecules}}{m^3} \tag{1.13}$$

The mean free path of air at STP is 0.066 μm.

In fluid mechanics, it is common to use nondimensional numbers to characterize a problem. For example, in the case of small spherical particles settling in quiescent air at room temperature and standard atmospheric pressure, the effects of slip at the particle surface will become important when the diameter of the particle is less than about 1 μm. But what about other situations, where the pressure, temperature, gas composition, and geometry of the problem may vary? In these cases, the Knudsen number can be used. The Knudsen number is defined as the ratio of the mean free path to a geometric length scale.

$$\text{Kn} = \frac{2\lambda}{D} = \frac{\lambda}{R} \tag{1.14}$$

For internal flows the length scale R would represent the radius of a pipe or tube, and for external flows it would be the radius of a drop or particle moving through the fluid. A gas may be considered as a continuous fluid from values of Kn < 0.01. Otherwise the statistical effects of individual molecules must be accounted for. For most engineering applications the continuum assumption can be safely invoked. The exceptions include the aerodynamics of space vehicles in the upper atmosphere and flows in microchannels.

1.3.6 Speed of Sound

We also assume that all properties in the fluid are continuous and continuously differentiable, so that there are no discontinuities in the flow. In nature, the only time a discontinuity is present in a macroscopic fluid flow is when a shock wave is present, or at a liquid–gas interface. A shock wave occurs when a fluid moves at the speed of sound, or when an object moves through a fluid at the speed of sound. (Supersonic flows will be discussed in more detail in Chapter 8.)

A common nondimensional number used for aircraft velocity is the Mach number.

$$\text{Ma} = \frac{V}{a} \tag{1.15}$$

where by tradition the symbol a is used for the speed of sound. From thermodynamics it can be shown that the speed of sound is

$$a = \sqrt{\frac{\partial P}{\partial \rho}\bigg|_s} \tag{1.16}$$

For an ideal gas this can be rearranged to

$$a = \sqrt{k \frac{\partial P}{\partial \rho}\bigg|_T} \tag{1.17}$$

where $k = c_P/c_V$ is the specific heat ratio. Since the ideal gas law can be written as $P = \rho RT$, then

$$a = \sqrt{kRT} \tag{1.18}$$

where $R = R_u/M$. For air at STP ($P = 1$ atm, $T = 20°C$), $k = 1.4$ and $R = 287$ J/kg-K. At sea level at STP, $a = 340$ m/s (760 mph) in air. As long as the Mach number is low, the gas may be treated as incompressible. That is, its density will not change as

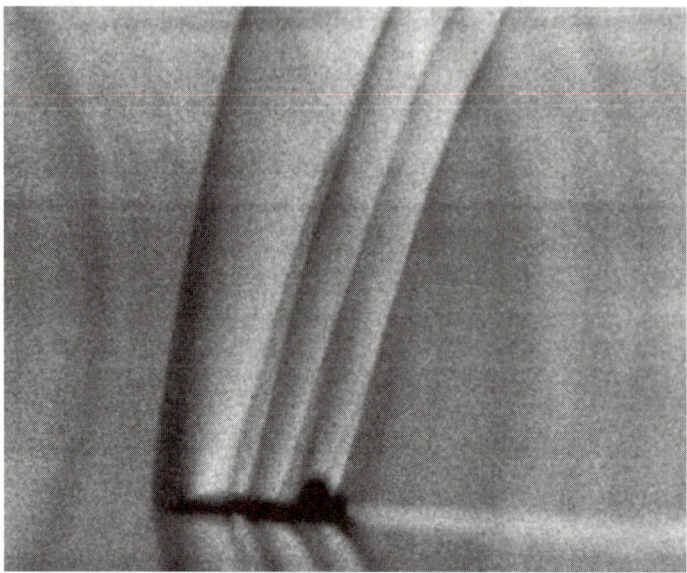

FIGURE 1.12 Schlieren photographs of shock waves at Mach 1.1. Courtesy of NASA Dryden Flight Research Center.

a function of velocity of the flow. If $V \ll a$, then pressure waves travel fast relative to the flow, and the effects of an impulse on the system are felt nearly instantaneously everywhere. So when a jet of fluid (either liquid or gas) exits out into the atmosphere, as long as Ma < 0.3, it is safe to assume that the exit pressure of the jet equals the surrounding atmospheric pressure. As long as the velocities are subsonic everywhere in the flow, it is safe to assume that all properties are continuous.

In nature, the only discontinuity that appears in fluid flows is a shock wave. The pressure, temperature, and density of gas all change instantaneously over a shock wave. Figure 1.12 shows a photograph of a shock wave.

1.4 Accuracy and Significant Digits

Most common measurement devices for fluid flows (thermometers, thermocouples, thermistors, pressure transducers, and flowmeters) only have three digits of accuracy. That is, in real-world situations their accuracy is usually only $\pm 1\%$. Therefore, there is nothing to be gained by carrying more than three digits in calculations. For example, writing 3,287,398.78 Pa gives no more significant information than writing 3.29 MPa, and the extra digits are distracting to the reader.

EXAMPLE 1.2

Say you are calculating the velocity of a water stream by placing a float on the water, such as a small piece of cork, and using a stopwatch to time how long it takes the cork to travel between two marked points whose distance has been measured with a tape measure. Then you can calculate velocity by $V = \Delta x/\Delta t$. How many digits should you include in your reported calculation for velocity?

SOLUTION The accuracy of using a hand-held stopwatch is probably around ± 0.05 s. Accuracy in distance measurements is probably about 0.5 times the smallest division on the scale. Thus, if your recorded time is less than 10 seconds, you will only have 2 reliable digits in your time measurement (that for seconds and for tenths of seconds), and you should include 2 digits in your reported calculations, even if the distance measurement is accurate to 3 or more digits.

EXAMPLE 1.3

With a very accurate barometer you measure the atmospheric pressure to be 101,047 Pa. You use a thermometer to measure the temperature as 22°C. If you use the ideal gas law to calculate density, how many digits should you use in the answer?

SOLUTION Accuracy in thermocouples and thermometers is usually no better than ± 0.5°C. Since the ideal gas law uses absolute temperature to calculate the density, the temperature can be expressed as $T = 22 + 273 = 295$ K ± 0.5 K, so the relative uncertainty in the temperature is about 0.25%. Thus, you should use three or possibly four digits to express the density, assuming the barometer is accurate to the six digits listed.

■ 1.5 Types of Fluid Flows

To know which type of analysis to apply to a fluid problem, we must first identify the type of flow. There are many different ways fluid flows can be classified: Is the fluid a liquid or a gas? Is it single phase or a multiphase mixture? Is it steady or transient? Is it laminar or turbulent? Is it an internal, wall-bounded flow, or an external flow around an object? Is it subsonic or supersonic? Is it on the macroscale, where the fluid can be considered continuous, or on the microscale, where molecular effects have to be considered? In multiphase flow, the stratification of the flow is important and will determine whether it is bubbly, slug flow, two-layer, and so on, as shown in Figure 1.13.

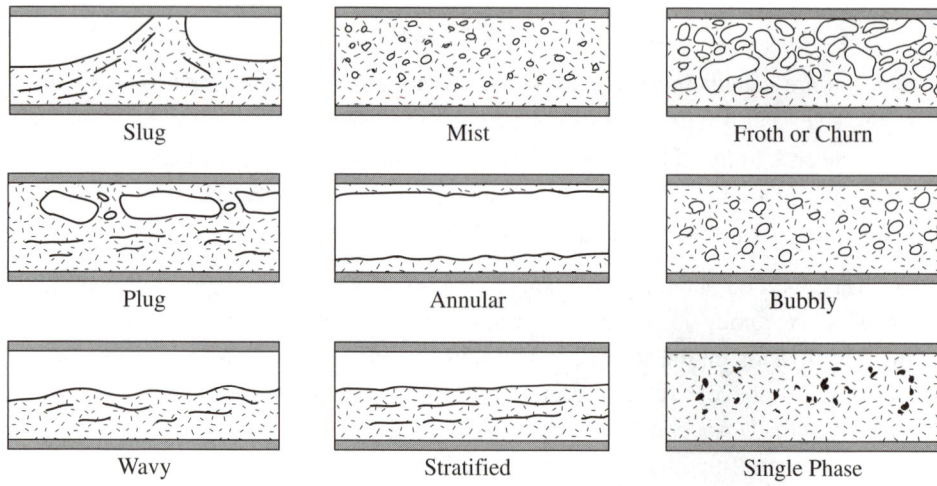

FIGURE 1.13 Schematics of different types of multiphase flows, from [Blevins92]. Used with permission.

One of the most important classifications is whether a given flow is laminar or turbulent. Laminar flows are orderly, with the flow moving smoothly and layers of fluid maintaining their relative positions to each other. Turbulent flows are more chaotic, with seemingly random motion of the fluid in different directions. Examples of turbulent flows include clouds, diesel fuel sprays, cream in coffee, nebulas, and exhaust plumes. Figure 1.14 shows the extremely large scale turbulence seen in a nebula.

FIGURE 1.14 Turbulent flow in a nebula. Courtesy of NASA, ESA, and the Hubble Heritage Team (STScI/AURA).

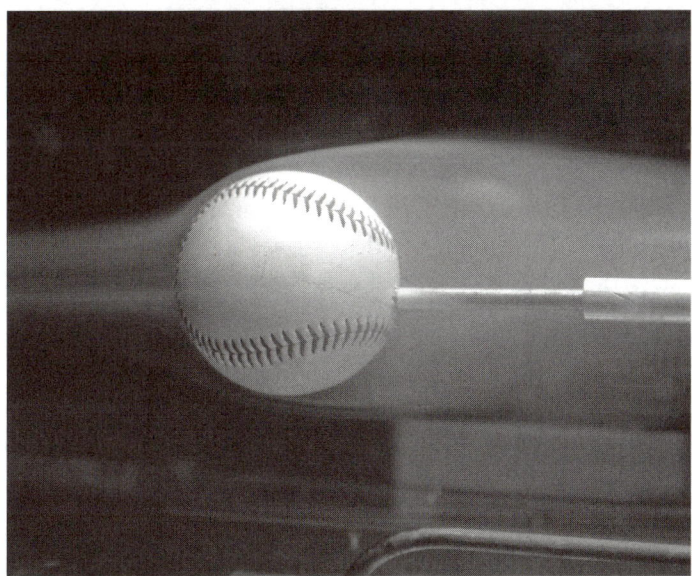

FIGURE 1.15 Picture of laminar flow around a baseball.

This classification is important, for example, in flow cytometry, where cells are sampled by immersing them in a carrier fluid and making them pass through the test section. As long as the velocity of the cells is positive, they are counted only once each. However, increasing the flow velocity to increase the sampling rate will also tend to induce turbulence in the flow, which could cause some of the cells to be transported backward in a turbulent eddy and be counted two or more times. In this case, turbulence is undersirable. But in cases where mixing needs to be enhanced, inducing turbulence is desirable.

Generally speaking, the higher the flow velocity, the more likely the flow is to be turbulent. Baseballs travel over a range of speeds where they see both laminar (such as shown in Figure 1.15) and turbulent flows in a game.

Regardless of what type of flow is present—laminar or turbulent, simple single-phase or a more complex mixture—the three conservation equations always apply. These are conservation of mass, conservation of momentum, and conservation of energy, which will be covered in Chapter 3. Most of the problems presented in this book will deal with steady-state, incompressible flow of single-phase fluids.

References

[Audel70] Black, Perry. 1970. *Audel Pump Handbook*. Howard W. Sams & Co.

[NACA33] Rothrock, A., and C. Waldron. 1933. NACA Report 435. Fuel Vaporization and Its Effect on Combustion in a High-Speed Compression-Ignition Engine.

[Blevins92] Blevins, R. 1992. *Applied Fluid Dynamics Handbook*. Krieger.

Problems

1. There are approximately 384 quadrillion gallons of water in Earth's oceans. Convert this number to liters. What is the mass of this amount of water in kg? In lbm? In slugs?

2. What is the volume in mL contained in a 12-oz beverage can? How many cans are equivalent to a 1-gal milk jug?

3. How many teaspoons are in 1 gal?

4. Convert 1.0 gal to the following units: cubic inches, cubic feet, cubic meters, milliliters, cups, pints, teaspoons, and soda cans.

5. What would be the radius of a sphere that contained 1 picoliter (pL) of water?

6. What is the change in the density of liquid water from 20°C to 80°C at 1 atm? What is the change in density from 1 atm to 10 atm, 100 atm, and 1000 atm at 20°C?

7. If you are traveling in your car at a speed of 60 mph, what is your speed in units of m/s?

8. What is the prefix for 10^{-15}?

9. At the same temperature and pressure, which gas would you expect to have the higher viscosity, helium or argon? Why?

10. Dimethyl ether (DME) is a fuel that has been proposed as an alternative to diesel fuel. What are the viscosities of these two fuels at room temperature? (Assume that DME is compressed so that it is in the liquid state.) What are the implications of this for the fuel handling system?

11. What is the recommended grade of oil for use in the engine of your car? (If you do not own a car, pick a car that you would like to own.) What is the viscosity of this oil at 0°F? At 100°F?

12. If the velocity profile in a pipe is parabolic and follows the equation $U(r) = U_{max}(1 - (r/R)^2)$, what is the shear stress at the wall if $U_{max} = 1.0$ m/s, $R = 5$ cm, and the liquid is water? If it is air? What would be the total force on a 0.1-m-long section of pipe in each case?

13. How much does the viscosity of SAE 30 oil change from 20°C to 50°C?

14. How does the presence of impurities affect the surface tension of liquid water in air?

15. When you have a small amount of cereal left in your bowl, the pieces tend to stick together. Can you explain what effect surface tension has on this phenomenon?

16. Predict the capillary rise for water in a glass tube of inner diameter 0.5 mm. How would this value change if the water was heated from standard conditions of 20°C to a temperature of 50°C?

17. What is the specific gravity of each of the following liquids? ethanol, glycerin, mercury, and kerosene.

18. What is the specific gravity of each of the following gases? hydrogen, helium, argon, and sulfur hexaflouride?

19. How do you think the effects of slip flow would affect the settling velocity of a small particle? That is, would the particle sink faster or slower than what you would predict if you applied the no-slip boundary condition to the surface?

20. Use the ideal gas law to calculate the density of steam at 100 psia and 400°F. ($R = 49{,}700$ ft-lbf/(slugmol-°R), 1 slug = 32.2 lbm, and 1 lbf = 32.2 lbm ft/s^2). Repeat the calculation in metric, at $P = 6.8$ bar and $T = 204.5$°C. ($R = 8314$ J/kmol-K). (Note: normally you should not use the ideal gas law with steam—it would be a good idea to check the accuracy of your answers.)

21. A scuba tank is designed to hold 50 standard cubic feet (SCF) of air when filled to its design pressure of 3000 psig at 80°F. Calculate the required volume of the tank and the interior length of the tank if its inner diameter is 6 inches.

22. On the evening news, the meteorologist states that the local barometric pressure is 29.0 inches of mercury. What is this in units of Pa, psi, torr, and bars?

23. In a certain Michigan city that receives large amounts of lake-effect snow, the local building code calls for the roofs of all new buildings to be able to withstand a pressure of 70 psf from the snow load. What is this pressure in units of psi and Pa?

24. When considering the motion of grain down a conveyor belt, can the grain be treated as a fluid? Why or why not? What about the motion of sand in an hourglass?

25. What is the significance of the nondimensional Reynolds number in fluid mechanics? What is the Reynolds number of a baseball thrown at 100 mph?

26. An oil pipeline is to be constructed in which it is known that the fluid friction between the oil and the pipe walls will cause a loss in pressure of 100 Pa for every meter of pipe (friction losses will be dealt with in Chapter 6). Also, pumps are to be used that provide a pressure rise of 1.0 MPa for each pump. How far apart can the pumps be spaced? The vapor pressure of the oil is 20 kPa, and the surrounding atmospheric pressure is 100 kPa.

27. What are the major contributions to the study of fluid mechanics of each of the following researchers? (a) Archimedes, (b) Eiffel, (c) Ludwig Prandtl, (d) Leonardo Da Vinci, (e) Leonhard Euler, (f) Isaac Newton, (g) Reynolds, (h) Cayley, (i) Daniel Bernoulli, (j) Evanaeslista Torricelli, (k) Blaise Pascal, (l) G. B. Venturi, (m) Clemens Herschel

28. What are the major contributions to the study of aerodynamics of each of the following researchers? (a) Cayley, (b) Prandtl, (c) Dryden, (d) von Karman, (e) Anton Flettner
29. How does the size of its molecule affect the macroscopic properties of a fluid?
30. At what size for a settling dust particle does slip flow become important?
31. What are all the different fluids that are used in your car?
32. If you measure the mass of a fluid contained in a volume of 0.174 m^3 to be 25.0 g, how many digits should you present in your calculation of the fluid density?

2 Fluid Statics

In This Chapter
- Hydrostatic Formula
- Application of the Hydrostatic Formula to Forces on Submerged Surfaces
- Buoyant Forces on Submerged and Floating Objects
- Uniformly Accelerating Systems

The objective of this chapter is to develop proficiency in calculating and using hydrostatic pressure distributions to solve problems in fluid statics.

■ 2.1 Hydrostatic Formula

2.1.1 Derivation from a Differential Control Volume

Consider a cube of fluid inside a large body of fluid at rest, as shown in Figure 2.1. Treating the cube as a control volume, it is possible to draw a free-body diagram and apply Newton's second law,

$$\vec{F} = m\vec{a} \tag{2.1}$$

if all the forces can be identified.

Since the fluid is at rest, there can be no viscous forces, and hence the shear stress is zero at all surfaces of the cube. If the shear stress were nonzero, the fluid would be put into motion. There is a normal stress acting on each surface of the cube, which is the pressure of the surrounding fluid. The fluid contained within the cube also has weight. There are no other forces acting on the cube. Thus there are six pressure forces and one gravity force. Since Equation 2.1 is a vector equation, it can be decomposed into each of the three coordinate axes. In the x-direction the force balance will be

$$P\,dy\,dz - \left(P + \frac{\partial P}{\partial x}\,dx\right)dy\,dz = 0$$

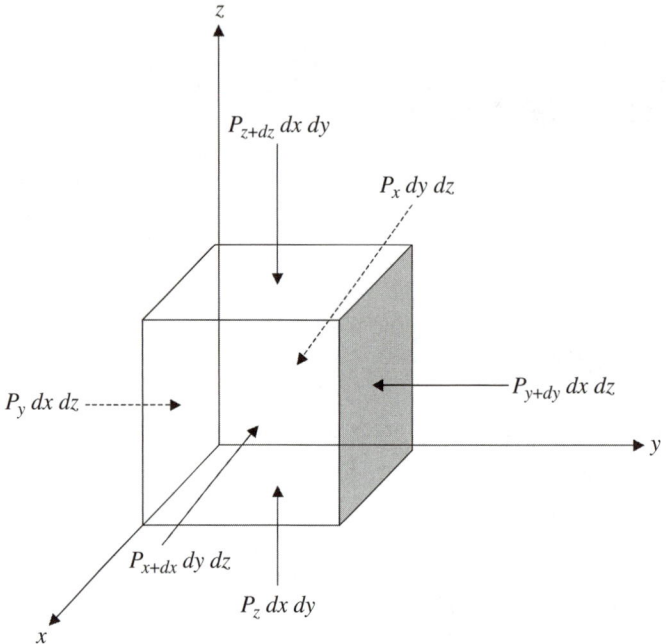

FIGURE 2.1 Cubic control volume of differential size used to derive the hydrostatic pressure distribution.

In the y-direction, it will be

$$P\,dx\,dz - \left(P + \frac{\partial P}{\partial y}dy\right)dx\,dz = 0$$

Simplifying these two equations yields

$$\frac{\partial P}{\partial x} = 0 \quad \text{and} \quad \frac{\partial P}{\partial y} = 0$$

In the z-direction, in addition to the pressure forces there is also the weight of the fluid within the cube. The force balance in the z-direction is

$$P\,dx\,dy - \left(P + \frac{\partial P}{\partial z}dz\right)dx\,dy - \rho g\,dx\,dy\,dz = 0$$

2.1 Hydrostatic Formula

Dividing through by $dx\,dy$ yields

$$P - \left(P + \frac{\partial P}{\partial z}dz\right) - \rho g\,dz = 0$$

Since it has already been demonstrated that P is not a function of x or y, the partial derivative $\partial P/\partial z$ can be replaced by a regular derivative dP/dz. The resulting hydrostatic equation is

$$\frac{dP}{dz} = -\rho g \tag{2.2}$$

This equation can be further simplified if something about the density is known, so that it can be expressed as $\rho = \rho(z)$. The simplest case is that of an incompressible liquid, where the density is constant.

$$P_2 - P_1 = -\rho g(z_2 - z_1) \tag{2.3}$$

It is often more convenient to think in terms of change in pressure and change in height rather than absolute (signed) values. In this case Equation 2.3 can be restated as

$$\Delta P = \rho g\,\Delta z \tag{2.4}$$

where the minus sign has been omitted, and it is understood that pressure increases downward in a fluid. One thing that is evident from the hydrostatic equation is that the pressure in a fluid depends only on the depth within the fluid and not at all on the width of the fluid or the container in which it is held. The water pressure at 1 meter below sea level is the same whether off the coast of Florida or off the coast of Africa.

For a liquid surface under atmospheric conditions and no motion, the pressure just above the interface and just below the interface will be the same. Since the pressure does not change in the horizontal direction, the interface between a liquid at rest and a gas must be a straight line, as illustrated in Figure 2.2. The upside down triangle symbol, ∇, will be used to represent the free surface of a liquid in this textbook. Actually, the interface does curve in the vicinity of the walls of the container due to surface tension forces. This can be significant in small capillary tubes. However, for all the problems in this chapter, assume that the vessel is large enough that the interface can be treated as a flat surface.

Figure 2.3 shows differently shaped containers with the same depth to illustrate the principle that hydrostatic pressure depends only on depth in a static fluid. The pressure depends only on the depth from the surface, not on the shape of the vessel, and the pressure at the bottom is the same for all three vessels.

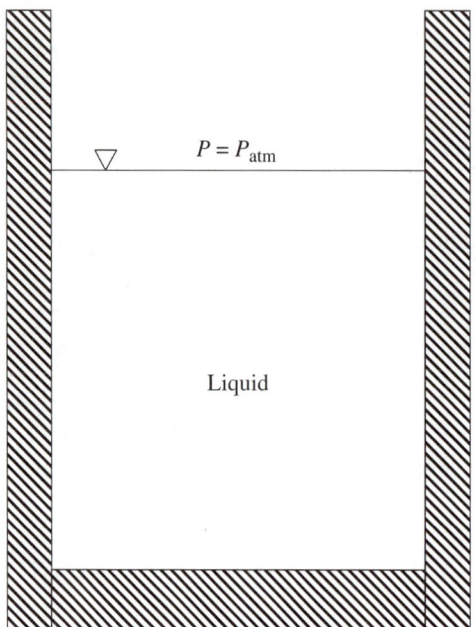

FIGURE 2.2 Illustration that the liquid level in an open vessel with fluid at rest will be horizontal.

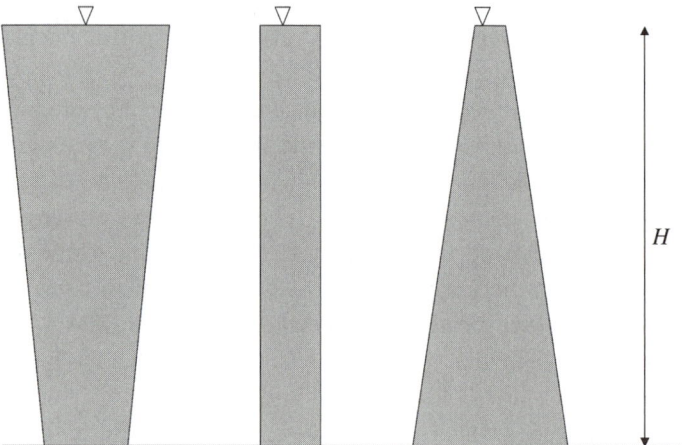

FIGURE 2.3 Three differently shaped containers all of the same height, H, and filled with the same liquid. The pressure at the bottom is the same in all three cases.

2.1.2 Atmospheric Pressure Distribution

Since gases are generally compressible, the integration of Equation 2.2 is more complicated for them than it is for liquids. For most engineering applications with gases, the ideal gas law is valid:

$$\rho = \frac{PM}{RT} \quad (2.5)$$

where R is the universal gas constant and M is the molecular weight. The specific gas constant for air is $R_{air} = R/M = (8314 \text{ J/kmol-K})/(29 \text{ kg/kmol}) = 287 \text{ J/kg-K}$. So for an ideal gas Equation 2.2 becomes

$$\frac{dP}{dz} = -\frac{PM}{RT} g \quad (2.6)$$

This differential equation can be solved by separating variables:

$$\frac{dP}{P} = -\frac{dz}{T} \frac{g}{R/M}$$

To integrate the right-hand side, some information about $T(z)$ must be known. If the gas is isothermal, then T can be treated as a constant, and the integration is trivial:

$$\ln(P_2) - \ln(P_1) = -\frac{g}{(R/M)T}(z_2 - z_1) \quad (2.7)$$

Using the mathematical identity $\ln(a) - \ln(b) = \ln(a/b)$ and taking the exponential of both sides, Equation 2.7 becomes

$$P_2 = P_1 \exp\left[-\frac{g}{(R/M)T}(z_2 - z_1)\right] \quad (2.8)$$

In the troposphere, measurements of temperature with changing altitude are well fit with a linear profile of the form

$$T = T_o - Bz \quad (2.9)$$

where T_o is the temperature at sea level, and the constant B is called the *lapse rate* and has units of °C/m or °F/ft. Substituting Equation 2.9 into Equation 2.6 will give

$$\frac{dP}{P} = -\frac{dz}{(T_o - Bz)} \frac{g}{R/M}$$

Table 2.1 Atmospheric Pressure and Temperature as a Function of Elevation

Elevation (m)	Pressure (bar)	Temperature (K)
0	1.013	288
1,000	0.899	282
2,000	0.795	275
3,000	0.701	269
4,000	0.616	262
5,000	0.540	256
10,000	0.264	223
15,000	0.120	217
20,000	0.055	217
30,000	0.012	227
40,000	0.003	250
50,000	0.001	270

Source: [USAtm76]

which can then be integrated to produce

$$P_2 = P_o\left(1 - \frac{B z_2}{T_o}\right)^{gM/BR} \tag{2.10}$$

where P_o is the pressure at sea level. Measurements give $B = 0.00357°\text{R/ft} = 0.0065 \text{ K/m}$, to be used with a value of $T_o = 288 \text{ K} = 519°\text{R}$. Equation 2.10 is valid up to an elevation of about 36,000 ft.

Table 2.1 lists the profile of pressure and temperature in the U.S. standard atmosphere. More detailed information on atmospheric properties is included in Appendix A.

As illustrated in Figure 2.4, the atmosphere can be roughly divided into four layers: the troposphere, the stratosphere, the mesosphere, and the thermosphere. Most jetliners cruise in the troposphere at an altitude around 30,000 ft. The Concorde flew at a much higher altitude—in the stratosphere—to take advantage of the lower air density and reduced drag, as well as to be above atmospheric turbulence. The stratosphere is the location of the ozone layer that protects humans and animals on Earth's surface from ultraviolet radiation. Like most combustion devices, the Concorde's engines emit nitric oxides, which deplete ozone according to the chemical reaction $NO + O_3 \rightarrow NO_2 + O_2$. While the nitric oxides emitted at lower elevations may eventually diffuse high enough to attack stratospheric ozone, emissions at higher elevations are more dangerous.

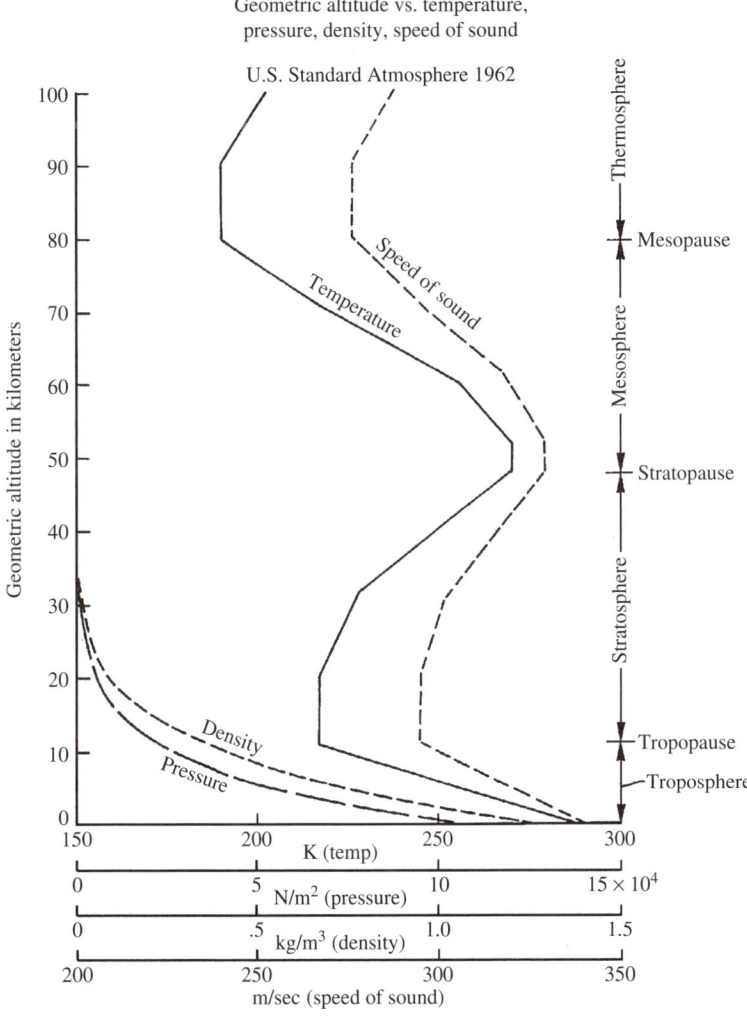

FIGURE 2.4 Architecture of the atmosphere. Reproduced from [NASA75].

The highest altitudes attained by jets with air-breathing engines are around 100,000 ft (by aircraft such as the SR-71). NASA's solar-powered Helios was able to attain an altitude of almost 100,000 ft. Sounding balloons can reach as high as 170,000 ft. The thermosphere starts at 300,000 ft, and satellites typically orbit at 800,000 ft or higher. Cumulus and cirrus clouds are found in the troposphere, and at higher elevations clouds such as noctilucent and nacreous may be found. The aurora occurs in the thermosphere.

2.1.3 Application to Manometry

A *manometer* is a device that uses a static column of liquid to measure pressure differences between two points by employing the hydrostatic formula:

$$P_2 - P_1 = -\rho g(z_2 - z_1)$$

Another useful piece of information is the fact that any two points at the same elevation in a continuous mass of the same fluid will have the same pressure. For a simple manometer to work, the density of the manometer fluid must be greater than that of the fluid being measured because the denser fluid will settle to the bottom. Also, the manometer fluid must be a liquid. If it were a gas it would diffuse away. A pressure stated in so many inches of water or inches (or millimeters) of mercury is a pressure taken with a manometer. For example, to measure air pressure, a small hole can be tapped in the passageway and fitted to some clear tubing that is bent down into a U-shape and filled with water. The air in the passageway will flow down into the tube, pushing the water in the opposite side up until the hydrostatic pressure of the water equals the air pressure, and equilibrium is reached. As long as the tap is flush with the airflow, it can be assumed that the static air pressure (and not the dynamic or stagnation pressure, which will be discussed in the next chapter) is being measured.

EXAMPLE 2.1

Consider air flowing through a 5-cm-diameter pipe, with a 5-mm-diameter hole tapped to measure the static pressure, as shown in Figure 2.5. Find the air pressure at point A.

SOLUTION If the difference in height of the two legs is $H = 8$ cm, the air pressure can be calculated from this measurement. The change in pressure between the two legs can be calculated using Equation 2.4. We first convert the difference in height into meters so that everything is in consistent units: $\Delta z = 8$ cm $= 0.08$ m. The difference in pressure in the water (right leg) is $\Delta P = \rho g \, \Delta z = (1000$ kg/m$^3)$ $(9.8$ m/s$^2)(0.08$ m$) = 784$ kg/m-s$^2 = 784$ (kg m/s$^2)$/m$^2 = 784$ N/m$^2 = 784$ Pa. On the air side the difference in pressure is $\Delta P = \rho g \, \Delta z = (1.2$ kg/m$^3)(9.8$ m/s$^2)$ $(0.08$ m$) = 0.94$ Pa, which is about 0.1% of the pressure change in the other leg, and can be safely neglected.

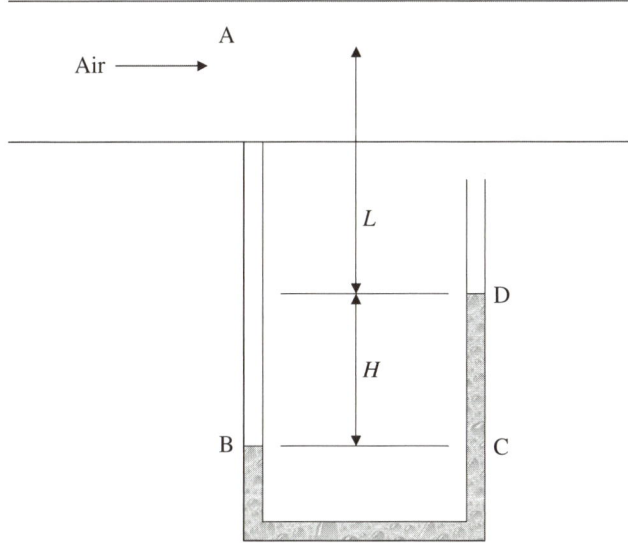

FIGURE 2.5 Water manometer used to measure the air pressure in a pipe at point A.

EXAMPLE 2.2

A U-tube manometer may also be used to measure a differential pressure, or the change in pressure between two locations. Figure 2.6 shows a schematic of such a setup. This U-tube manometer measures *differences* in pressure between two points in a flow. Derive the equation for the pressure difference from the manometer reading. (Note: For measuring air pressures, we often use an inclined manometer because pressure changes are so small.)

SOLUTION In this case the difference in the height of water between the two legs is measured to be 12.4 cm. Then the pressure balance is

$$P_A + \rho_{air}gL + \rho_A gH = P_B + \rho_{air}gL + \rho_{water}gH$$

Solve for the pressure difference, $\Delta P = P_A - P_B$:

$$P_A - P_B = \rho_{water}gH - \rho_{air}gH = (\rho_{water} - \rho_{air})gH$$

Notice that L dropped out of the equation.

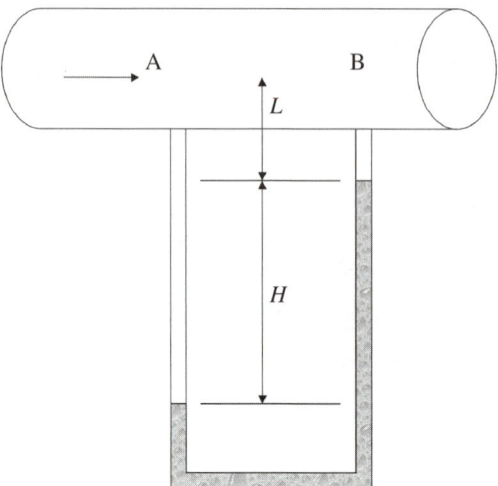

FIGURE 2.6 Differential U-tube manometer used for measuring the difference in pressure between two points A and B in a flow.

EXAMPLE 2.3

Two cylindrical columns are used in the hydraulic jack device shown in Figure 2.7. The area under piston A is 5 in², and the area under piston B is 500 in². What is the required force to be applied on piston A to support a weight at B of 9000 lbf? Piston A is 15 ft higher than piston B, and the device uses a hydraulic fluid of specific gravity 0.8. Neglect the weight of the pistons themselves.

SOLUTION The hydrostatic equation can be used to find the change in pressure between the bottom of piston A and at a location z_L. Since the fluid is continuous in the bottom part of the jack, the pressures at z_L and z_R will be the same because they are at the same elevation. So we can start by finding the required pressure at z_R, knowing that $P = F/A = 9000$ lbf/500 in² = 18 psig. This is a gage pressure, since not only does the fluid support the weight placed on piston B, but it must also push against the atmospheric pressure present above B. So then $z_L = z_R = 18$ psig and $P_A = P_L - \rho g H = P_L - \gamma H$. The pressure at A is less than the pressure at z_L since A is at a higher elevation. $P_A = 18$ psi $- 0.8(62.4$ lbf/ft³$)(15$ ft$)(1$ ft²/144 in²$) = 12.8$ psig. So then the downward force that must be applied at piston A is $F = PA = (12.8$ psi$)(5$ in²$) = 64.0$ lbf. Note that we do not have to include the force of the atmospheric pressure at A since the opposing pressure was expressed as a gage pressure. This shows the usefulness of using gage pressures in fluid mechanics calculations. Unless otherwise stated, in the rest of this chapter all pressures used in calculation problems will be assumed to be gage pressures.

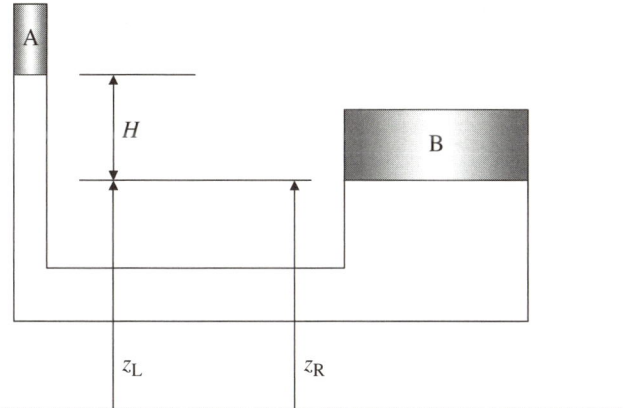

FIGURE 2.7 Diagram for a hydraulic jack.

EXAMPLE 2.4

Consider the mercury manometer shown in Figure 2.8, which is used to measure the pressure in the oil pipe. Approaching this problem from the reverse side, if P_A is known, predict the value of H. (This is called an *inverse problem*.)

SOLUTION Note that since the right leg of the manometer is open to the atmosphere, $P_D = P_{atm}$, and since points B and C are at the same elevation in a continuous fluid, $P_B = P_C$. Applying the hydrostatic equation to each leg yields

$$P_B - P_A = \rho_{oil} g L + \rho_{oil} g H \quad \text{and} \quad P_C - P_D = \rho_{Hg} g H$$

Using the information above, combine the two equations,

$$P_A - P_D = \rho_{Hg} g H - \rho_{oil} g H - \rho_{oil} g L$$

and solve for H:

$$H = \frac{P_A - P_D + \rho_{oil} g L}{\rho_{Hg} g - \rho_{oil} g}$$

Given that $P_A = 20$ psig, $L = 2.5$ ft, $SG_{Hg} = 13.6$, $SG_{oil} = 0.8$, and $\gamma_{H_2O} = (\rho g)_{H_2O} = 62.4$ lbf/ft^3, calculate H.

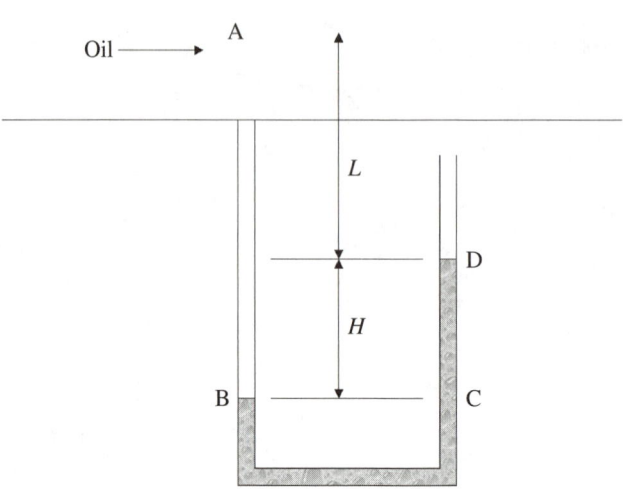

FIGURE 2.8 Mercury manometer used to measure the oil pressure in a pipe at point A.

$$H = \frac{20\,\frac{\text{lbf}}{\text{in}^2}\left(\frac{144\,\text{in}^2}{1\,\text{ft}^2}\right) + 0.8(62.4\,\frac{\text{lbf}}{\text{ft}^3})(2.5\,\text{ft})}{(13.6 - 0.8)(62.4\,\frac{\text{lbf}}{\text{ft}^3})} = 3.76\,\text{ft}$$

Notice that only three significant digits are carried in the final answer.

EXAMPLE 2.5

Consider the simple U-tube manometer shown in Figure 2.9, which is used to measure the pressure at point A in the vessel. Derive an equation for the vapor pressure at point A in terms of the manometer readings.

SOLUTION Here we will denote the height of manometer fluid in the left leg as z_1 and the height in the right leg as z_2, and the elevation of point A as z_A. The pressure at z_1 must be greater than the pressure at A, since it is at a lower point and has some additional weight of fluid above it. Applying the hydrostatic equation to the left leg, and assuming constant density, yields

$$P_A - P_1 = -\rho_A g(z_A - z_1)$$

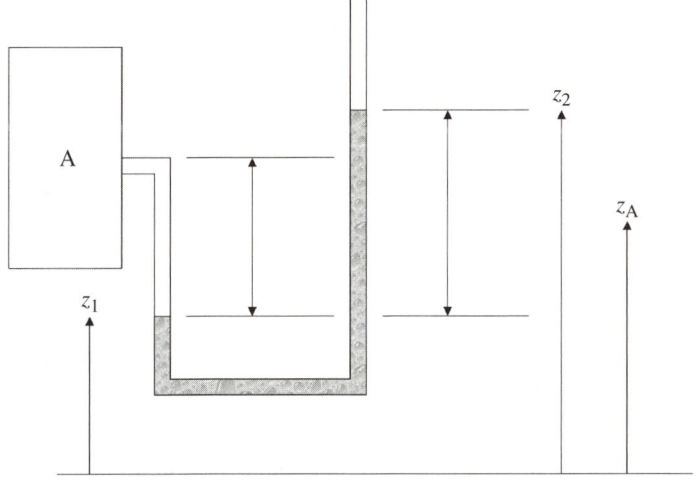

FIGURE 2.9 Manometer used to find the vapor pressure at point A.

The right leg of the manometer is open to the atmosphere. Applying the hydrostatic equation to it gives

$$P_{atm} - P_1 = -\rho_M g(z_2 - z_1)$$

where ρ_M is the density of the manometer liquid. These two equations can be combined to eliminate P_1:

$$P_A + \rho_A g(z_A - z_1) = P_{atm} + \rho_M g(z_2 - z_1)$$

Solving for P_A gives

$$P_A = P_{atm} + \rho_M g(z_2 - z_1) - \rho_A g(z_A - z_1)$$

It is often convenient to use gage pressures in most fluids problems. $P_{A,gage} = P_{A,abs} - P_{atm}$. So

$$P_{A,gage} = \rho_M g(z_2 - z_1) - \rho_A g(z_A - z_1)$$

2.2 Application of the Hydrostatic Formula to Forces on Submerged Surfaces

For a submerged surface, the net force due to fluid pressure can be obtained by integrating Equation 2.2 over the surface of the object:

$$F = \int_A P \, dA = \int_A \rho g h \, dA \tag{2.11}$$

The coordinate axis h is usually defined to be perpendicular downward from the fluid surface, as illustrated in Figure 2.11 in Example 2.6. The differential surface area needs to be expressed in terms of the fluid depth, h, in order to perform the integration of Equation 2.3. Applications of hydrostatic forces include submarines, dams (such as the one pictured in Figure 2.10), floodgates, boats, and aquariums.

The hydrostatic force *always* acts perpendicular to the solid surface. At this point, some specific examples will best serve to demonstrate the use of Equation 2.11.

FIGURE 2.10 Hoover dam. Courtesy of U.S. Bureau of Reclamation.

2.2 Application of the Hydrostatic Formula to Forces on Submerged Surfaces

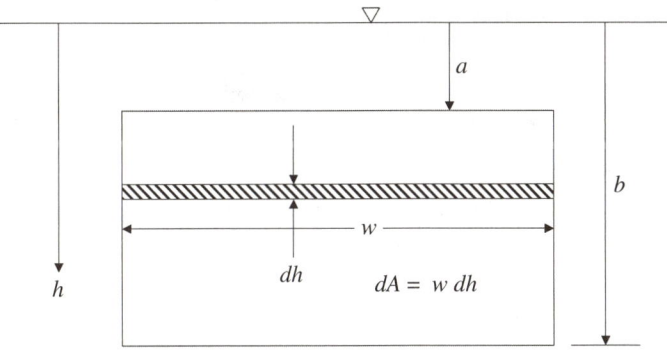

FIGURE 2.11 Schematic for calculating the hydrostatic force on a flat rectangular plate.

2.2.1 Force on a Flat Plate

EXAMPLE 2.6

We first take the relatively easy case of a flat rectangular plate vertically oriented in the fluid (perhaps a door or panel on a container), as illustrated in Figure 2.11. On one side of the plate is water, and on the other side is air. Find the force on the plate. (Note that if the plate were in the middle of a tank with water on both sides, there would be no net force since the water would push on both sides equally.)

SOLUTION For this rectangular plate, $dA = w\,dh$, where w is the width of the plate, and the plate extends between $h = a$ and $h = b$.

$$F = \int_A \rho g h\, dA = \int_a^b \rho g h w\, dh = \rho g w \int_a^b h\, dh = \rho g w \frac{h^2}{2}\bigg|_a^b$$

So for a numerical example, take $b = 3$ m, $a = 1$ m, $w = 4$ m, and the fluid to be water, so that $\rho = 1000$ kg/m³. Then the magnitude of the force is

$$F = 0.5(1000 \text{ kg/m}^3)(9.8 \text{ m/s}^2)(4 \text{ m})[(3 \text{ m})^2 - (1 \text{ m})^2]$$
$$= 156{,}800 \text{ kg-m/s}^2 = 157{,}000 \text{ N}$$

rounding off to three significant digits.

EXAMPLE 2.7

In this example, the plate is perpendicular to the surface, but the width w is not constant. This is illustrated with a trapezoidal plate in Figure 2.12. Find the force on the plate.

SOLUTION We set up this problem exactly as in the previous example, with the only difference being that w is not constant although it is a function of h. At a depth of 1 m, $w = 7$ m, and at a depth of 6 m, $w = 4$ m, so $w = 7 + 1(7 - 4)/(6 - 1) - (7 - 4)/(6 - 1)h = 7.6$ m $- 0.6\,h$. Then

$$F = \int_A \rho g h\, dA = \int_a^b \rho g h w\, dh = \rho g \int_a^b (7.6 - 0.6h)h\, dh$$

$$= \rho g \left(7.6 \frac{h^2}{2} - 0.6 \frac{h^3}{3}\right)\bigg|_a^b$$

Since the limits of integration are $b = 6$ m at the bottom of the plate and $a = 1$ m at the top, the force, F, can be calculated as

$$F = \rho g \left(7.6 \frac{h^2}{2} - 0.6 \frac{h^3}{3}\right)\bigg|_1^6$$

$$= (9800 \text{ N/m}^3)\left[\left(7.6 \frac{6^2}{2} - 0.6 \frac{6^3}{3}\right) - \left(7.6 \frac{1^2}{2} - 0.6 \frac{1^3}{3}\right)\right]$$

where the specific weight of water, $\gamma = \rho g = 9800$ N/m^3, has been substituted for ρg. The value of F is 882,000 N.

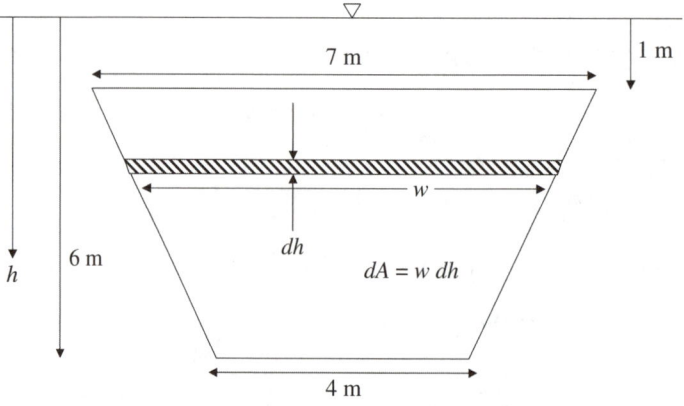

FIGURE 2.12 Schematic for calculating the hydrostatic force on a trapezoidal plate.

2.2 Application of the Hydrostatic Formula to Forces on Submerged Surfaces

EXAMPLE 2.8

A gate has been placed at an incline between two walls to hold back the water in a reservoir. The gate is 6 m long and 3 m wide. The top of the gate is 7 m below the water surface, and the bottom is fixed 10 m below the water surface. Calculate the force on the gate.

SOLUTION In this example $dA = w\, dh/\sin\theta$, where θ is the angle between the plane of the gate and a line perpendicular to the surface. This is illustrated in Figure 2.13.

$$F = \int_A \rho g h\, dA = \int_a^b \rho g h w\, dl = \frac{\rho g w}{\sin\theta} \int_a^b h\, dh$$

The limits of integration are between 7 m and 10 m:

$$F = \frac{\rho g w}{\sin\theta} \int_a^b h\, dh = \frac{(1000\ \text{kg/m}^3)(9.8\ \text{m/s}^2)(3\ \text{m})}{\sin 3} \left.\frac{h^2}{2}\right|_{7\,\text{m}}^{10\,\text{m}}$$

$$= 58{,}800\ \text{Pa}\left(\frac{(10\ \text{m})^2}{2} - \frac{(7\ \text{m})^2}{2}\right) = 1{,}500{,}000\ \text{N}$$

In this case it may have been more convenient to define a coordinate axis l, which is aligned with the length of the plate. Then $dA = w\, dl$. Since the pressure varies with the depth, h, the problem can be solved once a relationship between h and l

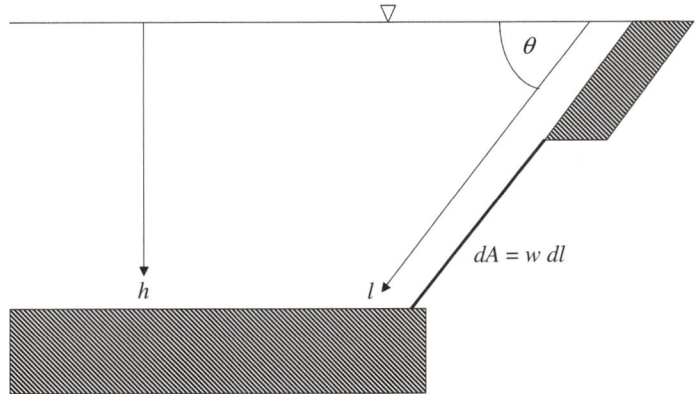

FIGURE 2.13 Schematic for calculating the hydrostatic force on an inclined gate.

is established. Either h or l can be eliminated from the integral; it does not matter which.

$$F = \int_A \rho g h \, dA = \int_a^b \rho g h w \, dl = \rho g w \int_a^b h \, dl$$

EXAMPLE 2.9

Consider a circular panel in a fluid. Calculate the force on the panel.

SOLUTION For this problem w is not constant, but changes with depth, so $dA = w(h) \, dh$. A formula for $w(h)$ must be constructed. For a circle of radius R, the width at any given depth can be calculated by drawing a triangle as shown in Figure 2.14. Using the Pythagorean theorem gives

$$R^2 = \left(\frac{w}{2}\right)^2 + (h - h_{center})^2$$

This can be rearranged to solve for w as a function of h:

$$w = 2\sqrt{R^2 - (h - h_{center})^2}$$

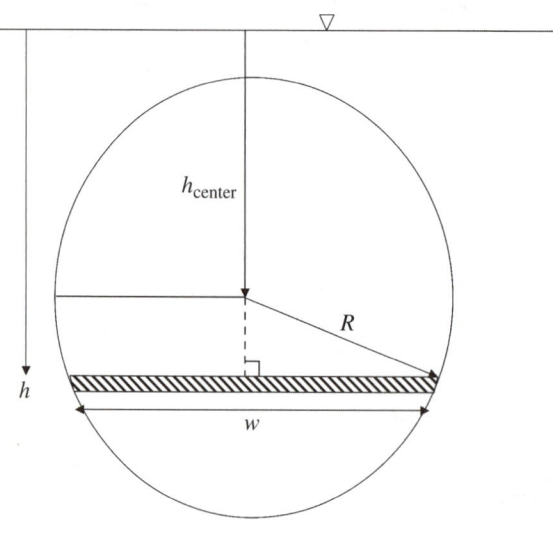

FIGURE 2.14 Schematic for calculating the hydrostatic force on a circular panel.

2.2 Application of the Hydrostatic Formula to Forces on Submerged Surfaces

Substituting in Equation 2.11 yields

$$F = \int_a^b \rho g h w \, dh = \rho g \int_a^b w(h) h \, dh = \rho g \int_a^b 2\sqrt{R^2 - (h - h_{center})^2} \, h \, dh$$

The limits of integration for this case are $a = h_{center} + R$, $b = h_{center} - R$. It turns out in this case, after much work, that an analytical solution to the integral can be found, and $F = \rho g h_{center} \pi R^2$. As an alternative to trying to find integral formulas in a math reference book, the integration can be performed numerically with the use of software packages such as MATLAB, and using Algorithm 2.1, which will be presented later in this chapter. Though performing a computational approximation always introduces some error into the calculation, these errors are usually small enough that they can be neglected for engineering purposes, provided certain precautions are exercised.

EXAMPLE 2.10

Consider the trapezoidal water reservoir, filled to the top with an oil of specific gravity 0.90, shown in Figure 2.15. Calculate the forces on all the sides and the bottom of the reservoir, as well as the total weight of oil in the reservoir.

SOLUTION Starting with the weight, which is equal to the volume of the reservoir times the specific gravity, we use the formula for the area of a trapezoid: $A = \frac{1}{2}(B_1 + B_2)H = 0.5(2 \text{ m} + 4 \text{ m})(3 \text{ m}) = 9 \text{ m}^2$. The volume in the reservoir is $V = AL = (9 \text{ m}^2)(6 \text{ m}) = 54 \text{ m}^3$. So the weight of the oil is $(0.9)(9800 \text{ N/m}^3)(54 \text{ m}^3) = 476{,}000 \text{ N}$. The force on the trapezoidal end plate can be calculated the same way as in Example 2.6:

$$F = \int_0^3 \rho g h \left(4 - \frac{2}{3}h\right) dh = (0.9)\left(9800 \frac{\text{N}}{\text{m}^3}\right) \left[(2 \text{ m})h^2 - \frac{2}{9}h^3\right]_{h=0\text{m}}^{h=3\text{m}} = 106{,}000 \text{ N}$$

The force on the inclined side plates can be calculated the same way as in Example 2.7:

$$F = \int_0^3 \rho g h w \frac{dh}{\sin \theta} = (0.9)\left(9800 \frac{\text{N}}{\text{m}^3}\right)(6 \text{ m}) \left[\frac{h^2}{2(0.949)}\right]_{h=0\text{m}}^{h=3\text{m}} = 251{,}000 \text{ N}$$

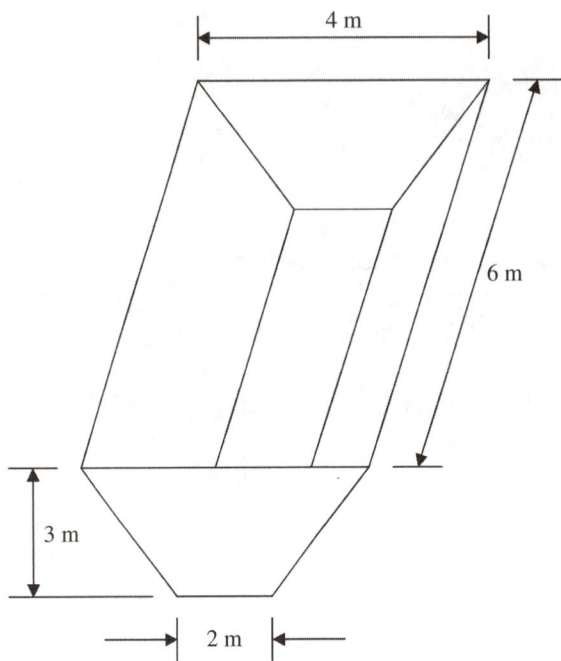

FIGURE 2.15 Diagram of a fluid reservoir containing oil with SG = 0.90.

The force on the bottom plate is equal to the weight of oil directly above it. $F = (0.9)(9800 \text{ N/m}^3)(2 \text{ m})(3 \text{ m})(6 \text{ m}) = 318{,}000 \text{ N}$. Notice that the sums of the forces on the bottom and side plates do not equal the weight of the oil. That is because most of the force on the side plates is directed outward. The portion that is directed downward is $F \cos \theta = 251{,}000 \text{ N}(1/3.16) = 79{,}000 \text{ N}$. So the total downward pressure force on the reservoir is $318{,}000 \text{ N} + 2(79{,}000 \text{ N}) = 476{,}000$ N, which does indeed equal the weight of oil in the reservoir, as it should.

2.2.2 Force on a Curved Surface

EXAMPLE 2.11

Consider a gate of quarter-circle cross section holding back fluid in a reservoir, and as shown in Figure 2.16. The gate has width, w, perpendicular to the page, so it actually has the shape of a quarter cylinder of radius R.

2.2 Application of the Hydrostatic Formula to Forces on Submerged Surfaces

FIGURE 2.16 Diagram of a quarter cylinder gate to illustrate the method for calculating force on a curved surface.

SOLUTION In problems dealing with curved surfaces, it is usually easiest to compute the two (or three) orthogonal components of the force separately. The coordinate system origin could be placed either at the left end of the gate or at its center of curvature, but it will be more helpful to place it at the hinge on the left end of the gate. The equations for the forces in the x- and z-directions are

$$F_x = \int_{z_{min}}^{z_{max}} P\, dA_x \quad \text{and} \quad F_z = \int_{x_{min}}^{x_{max}} P\, dA_z$$

where $dA_x = w\, dz$ and $dA_z = w\, dx$. (The area required for calculating the x-force is the area that is perpendicular to the x-axis—in this case, $w\, dz$). Thus it will be necessary to relate the (x, z) coordinates of the gate to the depth, h, which determines the pressure. The relationship between the fluid depth, h, and the vertical axis, z, is $h = R - z$, which is valid for $0 < z < R$. The equation for the surface of the gate in (x, z) coordinates is

$$(x - R)^2 + z^2 = R^2$$

We solve this to get z in terms of the other variables:

$$z = \sqrt{2Rx - x^2}$$

Substituting, we can write h in terms of x:

$$h = R - z = R - \sqrt{2Rx - x^2}$$

Now we can solve for the forces, beginning with F_x:

$$F_x = \int_{z_{min}}^{z_{max}} P \, dA_x = \int_0^R (\rho g h)(w \, dz) = \rho g w \int_0^R (R - z) \, dz$$

$$= \rho g w \left(Rz - \frac{z^2}{2} \right) \bigg|_{z=0}^{z=R} = \frac{\rho g w R^2}{2}$$

Note that we had to write h in terms of z since we were integrating with respect to z. To solve for the upward force in the z-direction we must write h in terms of x.

$$F_z = \int_{x_{min}}^{x_{max}} P \, dA_z = \int_0^R (\rho g h)(w \, dx) = \rho g w \int_0^R (R - \sqrt{2Rx - x^2}) \, dx$$

$$F_z = \rho g w \left\{ Rx - \left[\frac{x-R}{2}\sqrt{2Rx - x^2} + \frac{R^2}{2} \cos^{-1}\left(\frac{R-x}{R} \right) \right] \right\} \bigg|_{z=0}^{z=R}$$

$$F_z = \rho g w \left(R^2 - \frac{\pi}{4} R^2 \right) = \rho g w R^2 \left(1 - \frac{\pi}{4} \right)$$

Now let us assign values of $R = 4$ m, $w = 5$ m, and say the fluid is an oil with specific gravity of 0.85. Then we can calculate the values of the two forces:

$$F_x = \frac{\rho g w R^2}{2} = \frac{(850 \, \frac{\text{kg}}{\text{m}^3})(9.8 \, \frac{\text{m}}{\text{s}^2})(5 \text{ m})(4 \text{ m})^2}{2} = 333{,}200 \text{ N}$$

and

$$F_z = \rho g w R^2 \left(1 - \frac{\pi}{4} \right) = \left(850 \, \frac{\text{kg}}{\text{m}^3} \right)\left(9.8 \, \frac{\text{m}}{\text{s}^2} \right)(5 \text{ m})(4 \text{ m})^2 \left(1 - \frac{\pi}{4} \right)$$

$$= 143{,}000 \text{ N}$$

Note that the horizontal force is greater than the vertical force. (This should make intuitive sense—the gate area perpendicular to the x-axis increases toward the bottom of the gate, where the pressure is greater.) The total force can be calculated using the Pythagorean theorem:

$$F = \sqrt{F_x^2 + F_z^2} \tag{2.12}$$

So $F = \sqrt{(333{,}200 \text{ N})^2 + (143{,}000 \text{ N})^2} = 362{,}600 \text{ N}$.

2.2 Application of the Hydrostatic Formula to Forces on Submerged Surfaces

2.2.3 Numerical Integration to Obtain Forces

In all of these equations, regardless of the geometry of the problem, the force may be calculated as long as a relationship between h and dA can be deduced. In some cases the integration may be tedious, or, if the surface is complex enough, an analytical integral may not exist. In such cases the use of a short computer program to perform the integration can be useful.

ALGORITHM 2.1

To approximate the integral, I, of a function $f(x)$ taken between the limits $x = a$ and $x = b$, the trapezoid rule may be used. The formula to approximate I with the trapezoid rule is given in Equation 2.13. The trapezoid rule approximates the area under the curve by summing the area of a number of discrete trapezoids that are fit to the function. This is illustrated graphically in Figure 2.17.

$$I = \frac{\Delta x}{2}\left[f(x_1) + 2\sum_{i=2}^{n-1} f(x_i) + f(x_n)\right] \tag{2.13}$$

If the data points are equally spaced along x (so that $x_{i+1} - x_i = \Delta x$ = constant), we can divide the region of integration into n intervals, where $n = (b - a)/\Delta x$. Then we approximate the area under the curve $f(x)$ between a and b as a series of trapezoids of width Δx. This method is $O(\Delta x)^2$, meaning that as the number of approximation intervals is increased, the integration step size, Δx, decreases, and the error, which is proportional to Δx^2, rapidly decreases. (When it is written that the error is $O(\Delta x)^2$, this is read as the error is "order of $(\Delta x)^2$," and mathematically this means the error is proportional to $(\Delta x)^2$.) Thus a more accurate answer

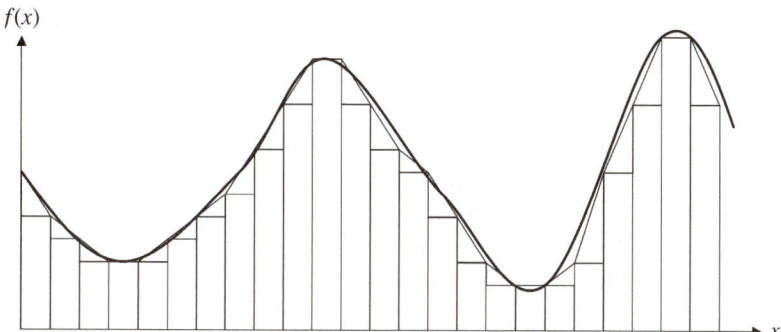

FIGURE 2.17 Illustration of the trapezoidal rule used to numerically approximate the integral of an arbitrary function.

is obtained as Δx is decreased. The same principle applies to Computational Fluid Dynamics (CFD) and finite element computations. There is a limit, however, to how small Δx can be made. Most modern computers use 64 bits to store decimal numbers, which normally results in the equivalent of about 16 decimal digits. Thus if we try to use 10^{16} or more integration intervals, the numerical round-off error due to a finite number of digits being stored in the computer will start to become large, and we could get inaccurate answers.

Algorithm Steps

1. Define a, b, and n.
2. Calculate the step size, Δx.
3. Evaluate the summation in Equation 2.13 by programming a loop:

    ```
    ISUM = 0
    For I = 2 to (N-1)
            ISUM = ISUM + 2*f(x(I))
    END LOOP
    ```
4. Evaluate *I* by using the formula in Equation 2.13.

    ```
    ISUM = 0.5 * Δx * (ISUM + f(x(1)) + f(x(n)))
    ```
5. Check for programming errors by checking to see if the magnitude of the answer makes sense.

This algorithm is implemented in MATLAB® (see Appendix F for MATLAB code) and applied to Example 2.11, with the results presented in Example 2.12. It is only necessary to assume that a sufficiently small integration step size is used for the code to be accurate enough for engineering calculations. This requirement is analogous to using a fine enough grid in finite element calculations or small enough cell sizes in CFD calculations. Here the function to be integrated is $f(h) = \rho g h w \, dh$. It is not necessary for any of these variables to be constants to perform a numerical integration, as long as the function form with respect to the depth, h, is known.

EXAMPLE 2.12

Solve the curved gate problem of Example 2.11 using Algorithm 2.1 and a computer.

SOLUTION The result presented was computed using a MATLAB code (found in Appendix F), but Algorithm 2.1 could be implemented in other software packages, such as MATHCAD®, Maple®, Mathematica®, EES, Fortran, C, or even

2.2 Application of the Hydrostatic Formula to Forces on Submerged Surfaces

Microsoft® Excel®. Many software packages have a built-in integration function, but the function must be entered in the proper format. Using $n = 100$ integration intervals, the step size used is $\Delta x = R/100 = (4 \text{ m})/100 = 0.04 \text{ m}$, and $\Delta z = R/100 = (4 \text{ m})/100 = 0.04 \text{ m}$. For the force in the x-direction, the MATLAB code gives $F_x = 333{,}200$ N, which exactly matches the analytical answer. For the force in the z-direction, MATLAB gives $F_z = 143{,}210$ N, which is about 0.1% off from the analytical answer.

EXAMPLE 2.13

Compute the total compressive force on a submarine at a depth of 2000 m. Model the submarine as a cylinder of diameter 10 m and length 50 m.

SOLUTION Strictly speaking, we should integrate the pressure force over the curved surface of the submarine to find the exact value of the force, but in this case we can make an engineering approximation that simplifies the problem. If the top of the submarine is at a depth of 2000 m, the pressure there is $P = \rho g h = \gamma h = (9800 \text{ N/m}^3)(2000 \text{ m}) = 19.6$ MPa. The pressure at the bottom end of the sub would be $P = (9800 \text{ N/m}^3)(2010 \text{ m}) = 19.7$ MPa, a difference of only 0.5%, which is negligible. So we can approximate the force as the pressure times the total surface area, where the total surface area of a cylinder is $A = 2\pi R^2 + 2\pi R L = 2\pi(5 \text{ m})^2 + 2\pi(5 \text{ m})(50 \text{ m}) = 1730 \text{ m}^2$, and the force is $F = (19.6 \text{ MPa})(1730 \text{ m}^2) = 33.9 \times 10^9$ N.

2.2.4 Center of Pressure and Line of Action

We have demonstrated how to calculate the net force due to hydrostatic pressures, but sometimes it is also necessary to know the line of action of that force to complete static calculations and design a support structure. To get a line of action we must choose a datum around which to sum moments. Just as in the static beam problems you would study in a fundamental statics course, we can replace a load distribution with an equivalent point force that, applied at the proper location, will result in the same moment being felt on the supports.

$$M_o = \int_A lP \, dA = \int_A l\rho g h w \, dh \tag{2.14}$$

where l is the distance from the reference point about which moments are being taken. For a given problem, evaluating the integral in Equation 2.14 is largely a matter of relating l to h. For a simple problem where the solid object is perpendicular to the surface of the liquid, the reference point O can be chosen so that the l and h-axes coincide, as shown in Figure 2.18.

So for this case $l = h$, so this can be substituted directly into Equation 2.14, and w is a constant.

$$M_o = \int_A \rho g h^2 w \, dh = \rho g w \int_0^L h^2 \, dh$$

The location where the force acts is referred to as the *center of pressure*, CP, and can be calculated from the following equation:

$$l_{CP} = \frac{M_o}{F} \tag{2.15}$$

EXAMPLE 2.14

Consider the dam shown in Figure 2.18. To calculate the reaction forces at the base of the dam, including the moment, we need to know not only the magnitude of the hydrostatic force but also exactly where it acts. We need to know this to determine if the dam is wide enough to keep from tipping over, for example.

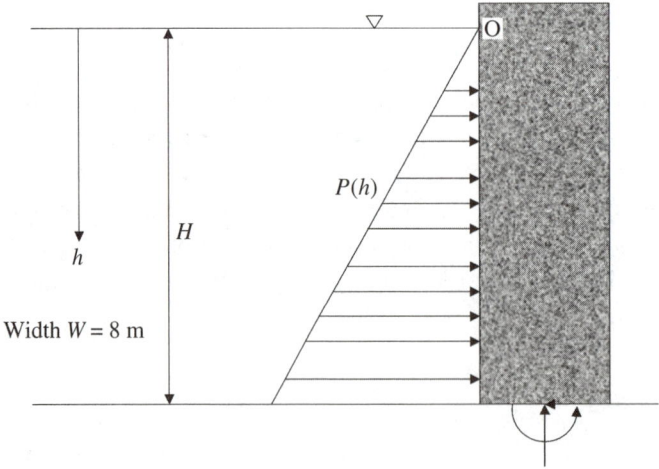

FIGURE 2.18 Schematic showing the forces and moments exerted on a dam.

2.2 Application of the Hydrostatic Formula to Forces on Submerged Surfaces

SOLUTION As usual, it is safe to assume that atmospheric air acts on the backside of the dam, so the pressure distribution shown in Figure 2.18 is for the gage pressure in the water, and the net force on the dam can be calculated using the gage pressures. The first step is to calculate the net fluid force, as has been done in the previous examples.

$$F = \int_A \rho g h \, dA = \int_a^b \rho g h W \, dh = \rho g W \int_0^H h \, dh = \frac{1}{2}\rho g W H^2$$

$$F = \frac{1}{2}\rho g W H^2 = 0.5\left(1000 \frac{\text{kg}}{\text{m}^3}\right)\left(9.8 \frac{\text{m}}{\text{s}^2}\right)(8 \text{ m})(10 \text{ m})^2 = 3{,}920{,}000 \text{ N}$$

To find the line of action for this force, we must calculate the net moment of hydrostatic pressure distribution. Moments must be taken about a point or axis, and for this problem a convenient location is on the upper-left surface of the dam where it intersects the water surface. Denoting this point as O, we can calculate the moments about O using Equation 2.14:

$$M_O = \int_A lP \, dA = \int_0^H h\rho g h W \, dh = \rho g W \int_0^H h^2 \, dh = \frac{1}{3}\rho g W H^3$$

Substituting the values gives

$$M_O = \frac{1}{3}\left(1000 \frac{\text{kg}}{\text{m}^3}\right)\left(9.8 \frac{\text{m}}{\text{s}^2}\right)(8 \text{ m})(10 \text{ m})^3 = 26{,}133{,}000 \text{ N-m}$$

Now, since we have decided that it is expedient to replace the hydrostatic pressure distribution with an equivalent force, we should recognize that this moment, M_O, is equal to the product of this force and its distance from the point O. Using Equation 2.15, the distance to the line of action is h_{CP} = 26,113,000 N-m/3,920,000 N = 6.67 m. So for this problem, where the force $F = \frac{1}{2}\rho g w h^2$, we have $l_{CP} = (\frac{1}{3}\rho g w h^3)/(\frac{1}{2}\rho g w h^2) = \frac{2}{3} h$. That means that the equivalent force acts at the center of pressure, which is two-thirds of the way down from the liquid surface. This is equivalent to the centroid of the triangular pressure distribution shown in Figure 2.18. For objects that do not go all the way to the top of the water, the factor will be different from $\frac{2}{3}$. In fact, the deeper an object gets below the surface, the closer the center of pressure gets to the centroid of the object.

To find the line of action for hydrostatic forces on complex objects, it is probably best to use numerical integration, as discussed in Algorithm 2.1. This can be used as before to calculate the center of pressure. The only difference is that the function to be integrated is $f(h) = l\rho g h w \, dl$, where l is the lever arm, and must be expressed as a function of h. After the integral is calculated, Equation 2.15 is used to calculate the distance to the line of action. The MATLAB code is included in Appendix F.

EXAMPLE 2.15

For the rectangular flat plate in Example 2.6, compute the location of the center of pressure, which is the point where the line of action of the hydrostatic force intersects the plate. (See Figure 2.19.)

SOLUTION Since the plate is oriented vertically, the lever arm for calculating moments is simply the depth from the surface, h. So Equation 2.14 is

$$M_O = \int_A lP \, dA = \int_{h_{min}}^{h_{max}} h\rho g h w \, dh = \rho g w \int_{h_{min}}^{h_{max}} h^2 \, dh = \rho g w \left. \frac{h^3}{3} \right|_{h_{min}}^{h_{max}}$$

The dimensions for this plate were a width of 4 m, with the top 1 m below the water surface and the bottom 3 m below the surface. Substituting these values into the equation for the moment yields

$$M_O = \rho g w \left. \frac{h^3}{3} \right|_{h_{min}}^{h_{max}}$$

$$= \left(1000 \, \frac{\text{kg}}{\text{m}^3}\right)\left(9.8 \, \frac{\text{m}}{\text{s}^2}\right)(4 \text{ m}) \left[\frac{(3 \text{ m})^3}{3} - \frac{(1 \text{ m})^3}{3}\right] = 339{,}700 \text{ N-m}$$

This is the total moment, or torque, about an axis on the water's surface that the hydrostatic force exerts on the plate. Since the force was already calculated to be $F = 157{,}000$ N, Equation 2.15 can be used to calculate the depth to the center of pressure:

$$l_{CP} = \frac{M_O}{F} = \frac{339{,}700 \text{ N-m}}{157{,}000 \text{ N}} = 2.16 \text{ m}$$

Since the center of mass of the plate is at $\frac{1}{2}(1 \text{ m} + 3 \text{ m}) = 2$ m, the center of pressure is below the center of mass. When we replace the continuous loading distribution on a plate with a single force, it is not enough to know the magnitude of the force; we must also know the direction and line of action of the force to calculate the required support forces and moments on the plate or gate or any other object subjected to a hydrostatic force.

2.2 Application of the Hydrostatic Formula to Forces on Submerged Surfaces

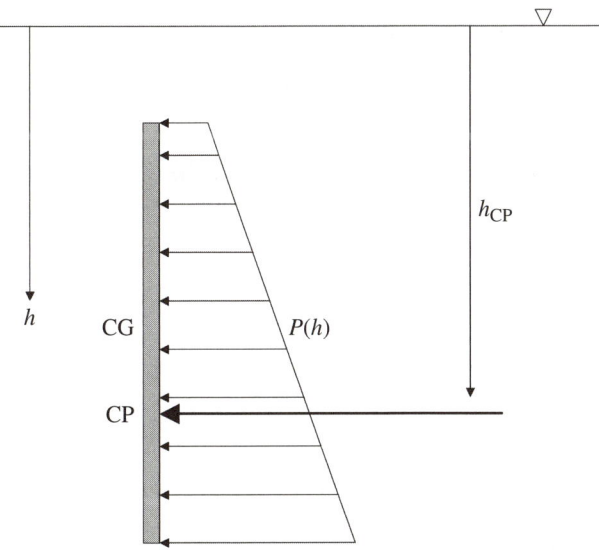

FIGURE 2.19 Schematic showing how to calculate the line of action of a hydrostatic force on a flat plate.

EXAMPLE 2.16

For the trapezoidal plate in Example 2.7, compute the location of the center of pressure.

SOLUTION Once again, since the plate is oriented vertically, the lever arm is equal to the depth, so $l = h$. Remember for this problem $w = 7.6$ m $- 0.6\, h$.

$$M_O = \int_A lP\, dA = \int_{h_{min}}^{h_{max}} h\rho g h w\, dh = \rho g \int_{h_{min}}^{h_{max}} h^2(7.6 - 0.6h)\, dh$$

$$= \rho g \left(7.6 \frac{h^3}{3} - 0.6 \frac{h^4}{4} \right) \Big|_{h_{min}}^{h_{max}}$$

Substituting the values of $h_{min} = 1$ m and $h_{max} = 6$ m, and using the density of water gives a moment of 3,434,000 N-m. Since the force F is 882,000 N, the depth to the center of pressure is $h_{CP} = 3{,}434{,}000$ N-m$/882{,}000$ N $= 3.89$ m.

EXAMPLE 2.17

For the inclined gate in Example 2.8, compute the location of the center of pressure on the gate. (See Figure 2.20.)

SOLUTION Once again, the momentum about a point on the water surface extrapolated from the plane of the gate must be calculated. Using l to represent an axis aligned with the gate, the relationship between the axis l and the depth h is $l = h/\sin\theta$.

$$M_O = \int_A lP\, dA = \int_{h_{min}}^{h_{max}} \frac{h\rho g h w\, dh}{\sin^2\theta} = \frac{\rho g w}{\sin^2\theta} \int_{h_{min}}^{h_{max}} h^2\, dh = \frac{\rho g w}{\sin^2\theta}\left(\frac{h^3}{3}\right)\Big|_{h_{min}}^{h_{max}}$$

Substituting in the numerical values yields a moment of 25,750,000 N-m. Then the distance to the center of pressure along the l-axis is $l_{CP} = M/F = 25{,}720{,}000\text{ N-m}/1{,}500{,}000\text{ N} = 17.2$ m, and the depth at the location of the center of pressure is $h_{CP} = l_{CP}\sin\theta = 8.58$ m, or just below the centroid of the inclined gate.

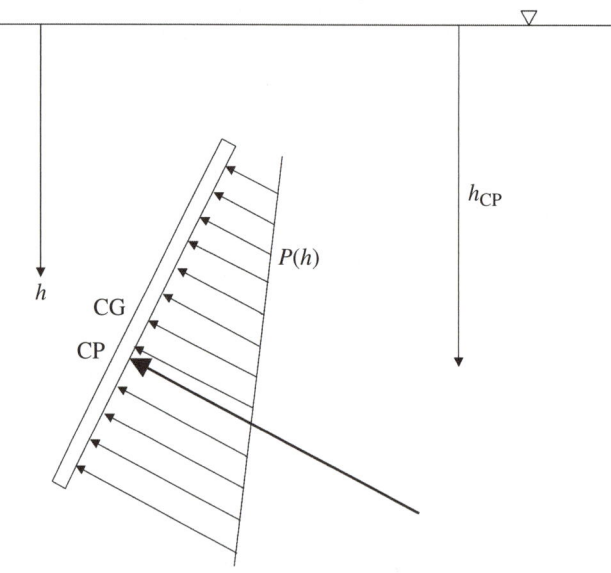

FIGURE 2.20 Diagram showing the force distribution on an arbitrary inclined gate.

2.2 Application of the Hydrostatic Formula to Forces on Submerged Surfaces

EXAMPLE 2.18

For the circular panel in Example 2.9, compute the location of the center of pressure on the vertically oriented panel.

SOLUTION The hydrostatic moment on this panel is

$$M_O = \int_a^b h\rho g h w \, dh = \rho g \int_a^b w(h) h^2 \, dh = \rho g \int_a^b 2\sqrt{R^2 - (h - h_{center})^2}\, h^2 \, dh$$

where $b = h_{center} + R$ and $a = h_{center} - R$. There actually is an analytical solution to this integral, and when the limits of integration are applied the result is

$$M_O = \rho g \left(\frac{\pi R^4}{4} + \pi h_{center}^2 R^2 \right)$$

and the depth to the center of pressure is

$$h_{CP} = \frac{M_O}{F} = h_{center} + \frac{R^2}{4 h_{center}}$$

As a numerical example, if the center of the circular panel is 5 m below the water surface, and the panel has a radius of 1 m, then the force on the panel is 154,000 N, and the center of pressure is at a depth of 5.05 m, or 0.05 m below the center of the panel.

2.2.5 Shortcuts to Calculating Forces

For problems where only the force is needed, or to double-check the calculation of an integral for force, the following observations may be useful.

- The horizontal force on a curved or inclined surface is equal to the horizontal force on a projection of the surface.
- The vertical force on a curved or inclined surface is equal to the weight of fluid above it.
- The total force on any submerged surface is equal to the product of the pressure at the centroid and the submerged area of the object.

Following is the procedure for the centroid method.

1. Find the depth of the centroid of the object from the fluid surface.
2. Calculate the surface area of the object.

3. Calculate the force: $F = P_{CG}A = \rho g h_{CG} A$.
4. Find the moment of inertia about the proper axis for the object
5. Calculate the distance below the centroid for the center of pressure:
   $\Delta l = I \sin \theta / (h_{CG} A)$, where I is the moment of intertia.

Unless the moment of inertia is already known, this method does not save any time. It is a useful way to double-check an answer for force, but the integral method is more general and is recommended, because of the difficulty in finding the line of action and the complex formula with the moment of inertia. In fact, for a complex shape for which the moment of inertia is not known, an integral always will have to be performed.

EXAMPLE 2.19

Use the centroid method to compute the force on the plate in Example 2.6.

SOLUTION The area of the plate is $A = (2 \text{ m})(4 \text{ m}) = 8 \text{ m}^2$, and the centroid is at the center of the rectangle, 2 m below the water's surface. So the pressure at the center of gravity is $P_{CG} = \rho g h_{CG} = (1000 \text{ kg/m}^3)(9.8 \text{ m/s}^2)(2 \text{ m}) = 19{,}600 \text{ Pa}$. The force is $PA = (19{,}600 \text{ Pa})(8 \text{ m}^2) = 157{,}000 \text{ N}$, the same number as calculated before with direct integration.

All of the previous problems in this chapter are of a type in which all the specifications of the problem are given, and only the hydrostatic force and/or the moment had to be calculated. However, most practical engineering problems tend to be design problems. That is, a task is specified, and the engineer must develop a safe, cost-effective way to perform that task. The next example is such a problem.

EXAMPLE 2.20

The water level of a reservoir is to be regulated by a spillway that will open and allow flow when the water level is too high, but stays closed otherwise. The system must be a passive system. That is, no sensors or actuators that require an external electric power source may be used. The specifications are as shown in Figure 2.21, with a maximum allowable height of 4 m above the spillway, which is a square passageway of size by 2 m by 2 m. Design this spillway.

SOLUTION So what options are available to the engineer of this problem? Obviously the principles of hydrostatics must be employed, but beyond that no

2.2 Application of the Hydrostatic Formula to Forces on Submerged Surfaces

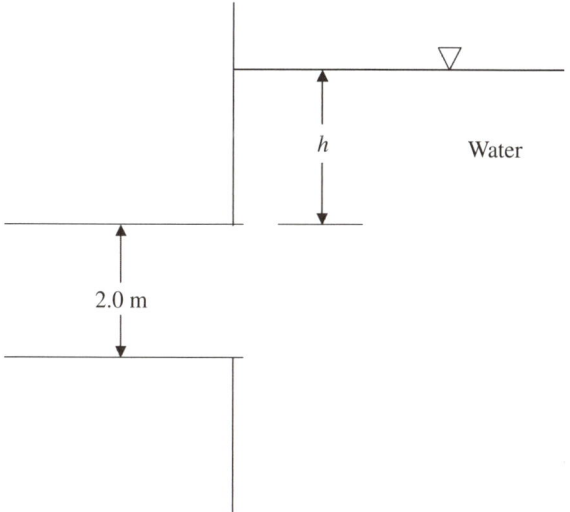

FIGURE 2.21 Geometry of the problem for which a passive drain gate must be designed.

specific design has been specified. Many previous problems in this chapter dealt with calculating forces and moments on plates, so you may think of installing a metal plate in such a way that it will open under the desired conditions. Even with this decision made, there are still many ways to implement the design. The plate could be hinged at the top, the bottom, or somewhere in the middle. A counterweight connected through a rope-and-pulley system could be used to hold the gate shut, or a float could be used on the water side that is connected to the gate. Perhaps a stopper could be positioned on the wall of the spillway near the gate to ensure that it only moves in one direction or to keep it from opening too far. There are many possible solutions; one is sketched in Figure 2.22.

For this solution, a square gate has been placed in the open channel, with a stopper behind it so that the gate can only open clockwise. The gate is hinged exactly at its centroid, which is important since the line of action of a force always acts below the centroid, so that the moment due to hydrostatic force will tend to open the gate clockwise. To prevent the gate from opening prematurely, a counterweight has been attached to the gate 0.5 m *above* the hinge, which will pull the gate counterclockwise and tend to keep it shut. Now the size of the counterweight must be chosen, so the maximum magnitude of the allowed hydrostatic force on the gate should be calculated. The area of the gate is $A = (2 \text{ m})(2 \text{ m}) = 4 \text{ m}^2$. When

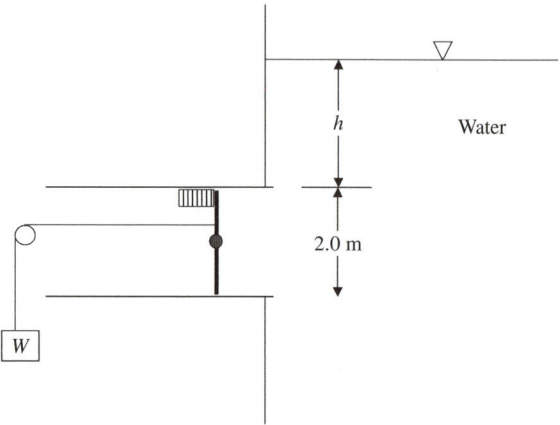

FIGURE 2.22 Solution to the spillway design problem using a gate, stopper, and weight.

the depth above the gate is 4 m, the center of the gate will be 5 m below the water surface, so the pressure there is $P_{CG} = (5 \text{ m})(9800 \text{ N/m}^3) = 49{,}000$ Pa. Using the centroid method, we find the force on the gate: $F = P_{CG} A = 196{,}000$ N. Next, we must calculate the moment the hydrostatic force exerts on the gate so that the center of pressure can be located. From Example 2.14, we find

$$M_O = \rho g w \left.\frac{h^3}{3}\right|_{h_{min}}^{h_{max}} = \left(1000 \frac{\text{kg}}{\text{m}^3}\right)\left(9.8 \frac{\text{m}}{\text{s}^2}\right)(2 \text{ m})\left[\frac{(6 \text{ m})^3}{3} - \frac{(4 \text{ m})^3}{3}\right] = 993{,}100 \text{ N-m}$$

and the depth to the center of pressure is $h_{CP} = 993{,}100$ N-m$/196{,}000$ N $= 5.067$ m, or 0.067 m below the hinge. Summing moments on the hinge to find the magnitude of the counterweight that will just keep the gate closed gives $W(0.5 \text{ m}) = 196{,}000$ N$(0.067$ m$)$, which results in $W = 26{,}200$ N. A slightly smaller value of W should probably be used, to account for friction in the pulley.

2.3 Buoyant Forces on Submerged and Floating Objects

All objects that are either immersed in or floating on a fluid experience a *buoyant force*. The buoyant force is an upward force that results from the pressure on the bottom of an object being higher than the pressure on the top due to the hydrostatic pressure distribution. Whether a body is floating on the liquid surface or completely immersed within the liquid, it experiences a buoyant force equal to the weight of the fluid it displaces. A submerged object displaces a volume of fluid equal to its own volume. A floating body displaces a volume equal to the volume of the object under the waterline. The buoyant force acts through the center of volume, which may not coincide with the center of mass. To be in equilibrium, the weight and the buoyant force must act along the same line, though this equilibrium may not be a *stable* equilibrium.

2.3.1 Forces on Submerged Objects

EXAMPLE 2.21

Consider the submerged cube of material shown in Figure 2.23, with its top and bottom faces oriented parallel to the liquid surface, and having characteristic length S. Find the buoyant force on this cube.

SOLUTION The forces on the four side faces (the faces perpendicular to the x- and y-axes) will all cancel, and the only net force will be in the vertical direction. The downward force on the top surface is equal to the pressure times the area: $F = (\gamma H)(S^2)$. The upward force on the bottom surface is $F = (\gamma(H + S))(S^2)$, so the net upward force on the cube is $F_{net} = (\gamma(H + S))(S^2) - (\gamma H)(S^2) = \gamma S^3$. This upward net force is the buoyant force, F_B. For this cube, the buoyant force is equal to the specific weight of the fluid times the volume of the cube, or the weight of fluid displaced. The term *displaced* is used because when the cube is initially lowered into the fluid, it must displace a volume of fluid S^3, which will cause the fluid level to rise.

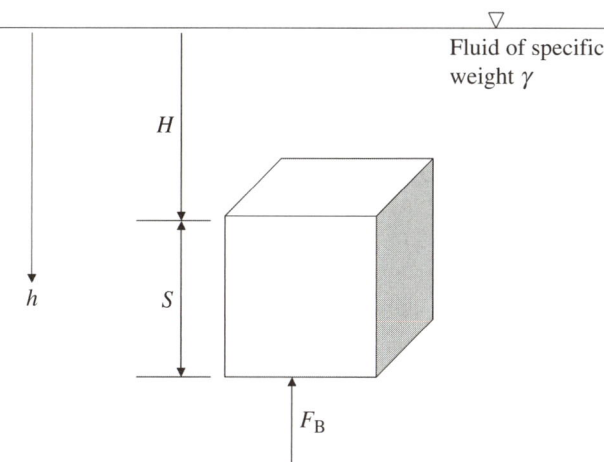

FIGURE 2.23 Illustration of the derivation of the buoyant force from the hydrostatic pressure distribution, using a cube.

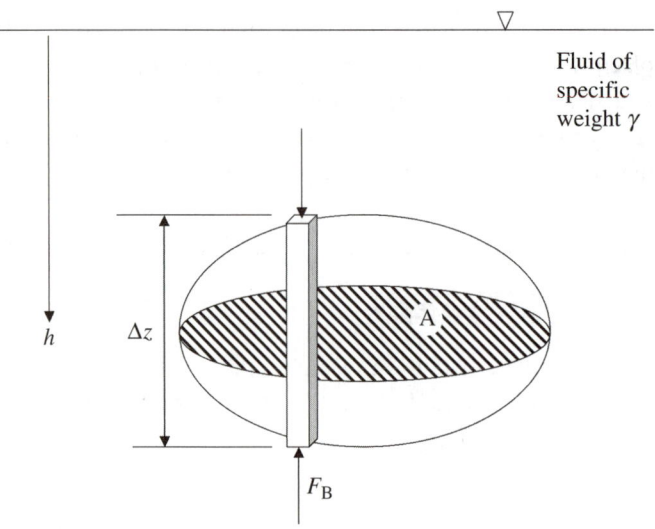

FIGURE 2.24 Illustration of the derivation of the buoyant force for an arbitrary body.

So for a cube, the buoyant force is equal to the weight of liquid that the cube displaces. Can this be generalized to arbitrary shapes? Yes it can, by integrating the hydrostatic force of an object over the surface area. This is illustrated in Figure 2.24 for a generic submerged body.

To perform the integration of the hydrostatic force over the total area, it will be useful to divide the object into a series of differential vertical elements, so that the net sideways force on each element is zero, and only the vertical forces need be considered. The net vertical force on each differential element is $dF = P\,dA = (\gamma \Delta z)(dx\,dy)$. It can be seen that the total volume of the object can be calculated by summing the volume of each vertical element,

$$\mathcal{V} = \int d\mathcal{V} = \int \Delta z\, dx\, dy$$

and the total force is calculated by integrating the differential force on each element:

$$F = \int dF = \int \gamma\, \Delta z\, dx\, dy = \gamma \int \Delta z\, dx\, dy = \gamma \mathcal{V} \tag{2.16}$$

This shows that for any submerged object the buoyant force is equal to the weight of fluid displaced.

2.3 Buoyant Forces on Submerged and Floating Objects

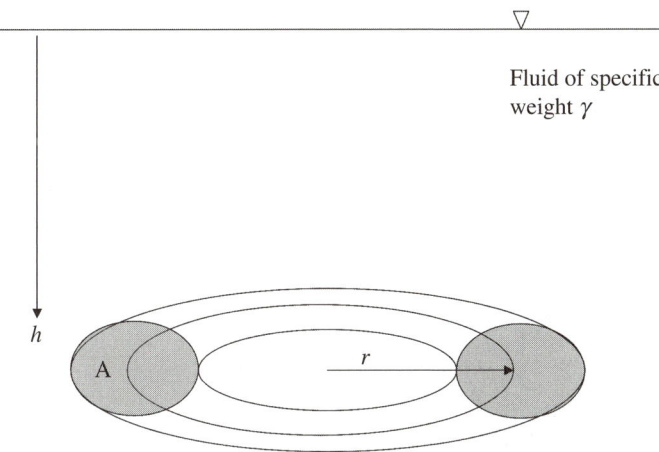

FIGURE 2.25 Schematic of a submerged torus.

The *theorem of Pappus and Guldinus* is useful for calculating the volume of objects of rotation for which it is necessary to find the buoyant force. This theorem states that the volume of an object of revolution is equal to the cross-sectional area of the object times the distance (circumference) that the centroid of the cross section sweeps when the object is formed. These terms are visually defined in Figure 2.25.

EXAMPLE 2.22

Calculate the buoyant force on a donut-shaped object (a torus) submerged in water, with the object having inner diameter 2 cm and outer diameter 10 cm.

SOLUTION So the diameter of the circular cross section of the torus must be (10 cm − 2 cm)/2 = 4 cm, and the radius 4 cm/2 = 2 cm = 0.02 m. The cross-sectional area of the torus is $\pi/4(0.04 \text{ m})^2 = 0.00126 \text{ m}^2$. The centroid of the cross section is 3 cm from the axis of rotation, so the circumference of the circle swept out by the centroid is 2π (3 cm) = 0.188 m, and the volume of the torus is $(0.00126 \text{ m}^2)(0.188 \text{ m}) = 0.000237 \text{ m}^3$. Assuming the torus is submerged in water, the buoyant force is $(9800 \text{ N/m}^3)(0.000237 \text{ m}^3) = 2.32 \text{ N}$.

EXAMPLE 2.23

The story (perhaps apocryphal) is told that the king of Syracuse paid a craftsman to make a gold crown. When the king received the crown, he was suspicious that the craftsman had not made it out of pure gold, but had substituted some other

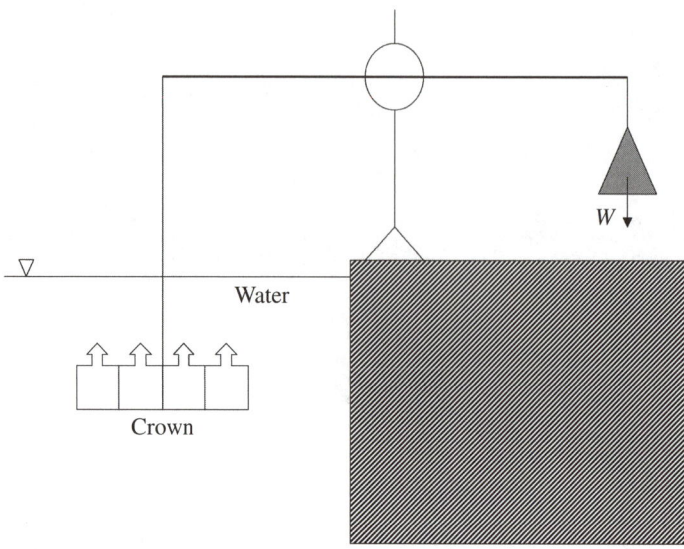

FIGURE 2.26 Schematic for Archimedes' solution.

metal and merely gold-plated the crown and kept the extra gold for himself. But how could the king check without destroying the crown? If he cut it open and it turned out to be pure gold, he would be humiliated and would have destroyed his expensive crown. The crown was of intricate design, so the total volume of the crown could not be easily calculated by geometrical means.

SOLUTION Archimedes was called in to solve the problem. He set up a balance scale, shown in Figure 2.26, whereby he could measure the weight of the crown both in plain air and while the crown was submerged in water. He measured the weight of the crown in air to be 8.5 N. (It is doubtful he actually used units of Newtons, since Newton had not been born yet. Perhaps he used units like "stones." But for the purposes of this problem we will use modern units.) He then submerged the crown in water and measured its weight to be 8.0 N. The weight in water is less due to the buoyant force pushing upward on the crown. Can you guess how he calculated the density of the material in the crown to figure out if it was pure gold or not?

First, the buoyant force will be equal to the difference between the regular weight of the crown and its weight when measured submerged in water. So F_B = 8.5 N − 8.0 N = 0.5 N. Since we know from Equation 2.16 that the buoyant force is equal to the volume of the submerged object times the specific weight of water, the volume of the crown can be calculated by $V_{crown} = F_B/\gamma_{H_2O}$ = 0.5 N/9800 N/m³ = 0.000052 m³. The specific weight of the crown can then be calculated by $\gamma_{crown} = W/V$ = 8.5 N/0.000052 m³ = 163,000 N/m³. Finally, the specific gravity of the crown is SG = 163,000 N/m³/9800 N/m³ = 16.7. Fortunately for Archimedes, gold is a very dense metal, so any filler metal the craftsman may have used would have a lower density than gold. The specific gravity of gold is 19.3 (which Archimedes could have obtained using the same apparatus and procedure described above if he did not already know the value). So the crown obviously was not pure gold, and the craftsman most likely came to an unhappy end.

In case you were wondering, there was a buoyant force on the crown when Archimedes weighed it in plain air, but the specific weight of air at sea level is only 11.5 N/m³, and so the buoyant force of air on the crown would only be $F_B = \gamma_{air}V$ = (11.5 N/m³)(0.000052 m³) = 0.001 N, which can be safely neglected. (Note: In one version of this story, Archimedes found a piece of gold that weighed the same as the crown and put it in a container of water and measured the water height, and then removed the gold and put the crown in the container and measured the water height again to see if they displaced the same amount. This type of measurement would not be very accurate—see Problem 39 at the end of the chapter.)

In the early part of the 20th century, there was great interest in the commercial and military use of lighter-than-air airships, due to their greater carrying capacity and longer range than primitive airplanes. The airship industry largely collapsed after the Hindenburg disaster of 1937. The burning and crash of the Hindenburg has fueled the popular perception that hydrogen is extremely dangerous. However, the cause of the Hindenburg disaster was most likely not the hydrogen that provided it with buoyancy, but a flammable coating on its outer skin. Hydrogen flames are more difficult to see because much of the luminosity in other flames comes from radicals or soot containing carbon. (If you have ever seen footage of a space shuttle launch, you may have noticed that the flames from the solid rocket boosters are bright and luminous, while the flame from the main engines of the space shuttle, which uses hydrogen for the fuel, is a much more faint, almost transparent, light blue.) The fact that the newsreel footage of the Hindenburg shows bright flames strongly suggests something besides the hydrogen is burning. Since hydrogen is the lightest gas, it is very diffusive, and is actually rather difficult to ignite when a hydrogen tank is punctured because the hydrogen is quickly diluted in air to concentrations below the flammability limit. In contrast, gasoline, for example, is more dangerous because the vapors are relatively heavy and tend to stay in the vicinity of a spill.

> **EXAMPLE 2.24**
>
> A weather balloon 2 m in diameter is filled with helium and tethered to the ground at sea level as shown in Figure 2.27. The weight of the balloon skin is 10.5 N. Assume that both the balloon and the surrounding air are at 101,300 Pa and 290 K. Calculate the tension in the string holding the balloon and the final altitude at which the balloon will come to rest if the string is cut.
>
> SOLUTION To find the tension in the string, we must find the net vertical force on the balloon, which will be equal to the buoyant force minus the weight of the balloon. We will need the density of both the helium in the balloon and the surrounding air. The ideal gas law will work for both:
>
> $$\rho_{air} = \frac{PM}{RT} = \frac{(101{,}300 \text{ Pa})(29 \text{ kg/kmol})}{(8314 \text{ J/kmol-K})(290 \text{ K})} = 1.21 \frac{\text{kg}}{\text{m}^3}$$
>
> $$\rho_{He} = \frac{PM}{RT} = \frac{(101{,}300 \text{ Pa})(4 \text{ kg/kmol})}{(8314 \text{ J/kmol-K})(290 \text{ K})} = 0.17 \frac{\text{kg}}{\text{m}^3}$$
>
> So the buoyant force is equal to the weight of air displaced,
>
> $$F_B = \rho_{air} g \forall = \left(1.21 \frac{\text{kg}}{\text{m}^3}\right)\left(9.8 \frac{\text{m}}{\text{s}^2}\right)\left(\frac{\pi}{6}(2 \text{ m})^3\right) = 49.7 \text{ N}$$

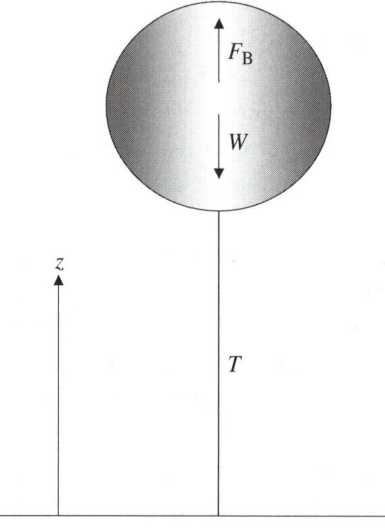

FIGURE 2.27 Schematic for calculating the buoyancy of a weather balloon. T is the tension in the string.

and the total weight of the helium in the balloon plus the skin of the balloon is

$$W = \rho_{He} g V\!\!\!\!/ + W_{skin} = \left(0.17 \frac{kg}{m^3}\right)\left(9.8 \frac{m}{s^2}\right)\left(\frac{\pi}{6}(2\,m)^3\right) + 10.5\,N = 17.4\,N$$

So the net force, and hence the tension, T, in the string, is

$$T = F_B - W = 49.7\,N - 17.4\,N = 32.3\,N$$

After the balloon is released it will come to rest when the buoyant force equals the weight. If we assume no helium leaks out, and thus the weight is constant, we must find the density at which the buoyant force equals the weight, and the altitude can be determined from that.

$$W = \rho_{air} g V\!\!\!\!/ = \rho_{air}\left(9.8 \frac{m}{s^2}\right)\left(\frac{\pi}{6}(2\,m)^3\right) = 17.4\,N$$

Solving for the density yields $\rho_{air} = 0.42$ kg/m³. And Table 2.1 shows that the altitude is approximately 9700 m, or 32,000 ft.

Balloons can be made to float in air either by filling them with a gas that has a lower molecular weight than air, such as hydrogen or helium, as shown in Figure 2.28, or by heating the air in a balloon so that its density decreases. Submarines must be able to maintain a neutral buoyancy to maintain their depth relative to the ocean's surface. They do this with ballast tanks that can fill with air or water to change the overall density of the submarine.

2.3.2 Forces on Floating Objects

For an object floating on a fluid's surface, the buoyant force is equal to the weight of fluid displaced by the object. It should be obvious that the volume of fluid displaced is equal to the portion of the volume of the floating object that is *below the waterline*, as illustrated in Figure 2.29. In static equilibrium the buoyant force is equal to the weight of the floating object. Boats can be made out of metals that are much denser than water, as long as they contain large portions of hollow volumes so that they displace enough water to support their weight and thereby will float. So for a floating object, $F_B = \gamma V\!\!\!\!/_{disp}$, where $V\!\!\!\!/_{disp} < V\!\!\!\!/$, the total volume of the object.

FIGURE 2.28 A helium balloon. Courtesy of NASA.

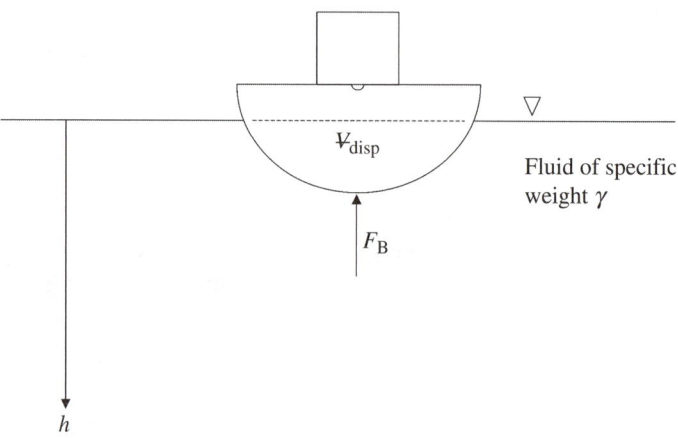

FIGURE 2.29 Schematic for calculating buoyancy of a floating boat.

A classic brain twister students are often asked is as follows: A small boat is floating in an enclosed swimming pool. The boat is then removed, cut into pieces, and placed back into the swimming pool. All the pieces sink to the bottom. Will the water level be higher or lower than what it originally was?

2.3.3 Stability of Floating Objects

EXAMPLE 2.25

Consider two beams of triangular cross section, one floating with its point up, and the other floating with its point down, as shown in Figure 2.30. The triangles are equilateral triangles, and the specific gravity of the beams is 0.8. Which configuration is more stable?

SOLUTION CG marks the center of gravity of each beam, and B the center of buoyancy, or the center of the volume under the waterline. The general test of stability for a system is to perturb or displace it slightly from its initial position or condition and see whether it would return to that condition. This test works for any system, not just for floating objects. The calculations are laborious, but it can be shown that the configuration on the left is not stable, while that on the right is.

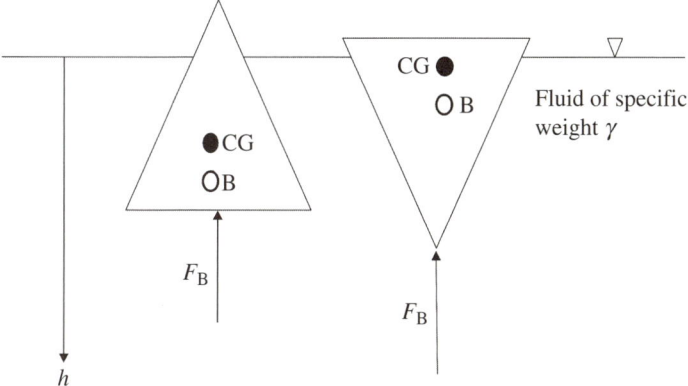

FIGURE 2.30 Schematic for determining which floating object is stable.

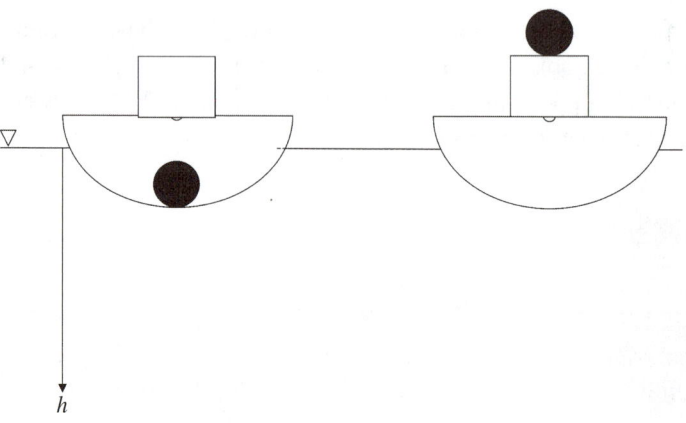

FIGURE 2.31 Comparison of a stable and an unstable boat.

A more obvious example is shown in Figure 2.31. The black dot represents a very heavy weight placed in each boat. If the boat on the left is tipped clockwise, the weight moves to the left, and the center of buoyancy will remain largely unchanged in position at the center of the submersed volume. Now the buoyant force and the weight are still equal in magnitude, but they are no longer collinear. Since the weight is to the left of the buoyant force, there will be a counterclockwise moment of the boat, which acts to restore it to its original position. However, when the boat on the right is tipped clockwise, the center of gravity moves to the right, and the net moment on the boat will be clockwise, which will act to tip over the boat—it is not stable. Thus a general principle in shipbuilding is that the lower the center of gravity is, the more stable the ship is.

There are rare instances when an *unstable* boat is desired. In competitive kayaking, for example, the higher a kayak rests on the water, the less submerged area there is, and the less hydrodynamic drag. So kayakers prefer unstable kayaks, but it takes a great deal of skill to control them. Unstable kayaks also maneuver more quickly.

Buoyancy is also important in dynamic systems with moving fluids. Figure 2.32 shows the effects buoyancy has on a candle flame. In microgravity, without the convective currents generated by buoyancy in normal gravity, the soot generated by the flame diffuses spherically outward in all directions, and the candle burns much cooler.

2.4 Uniformly Accelerating Systems

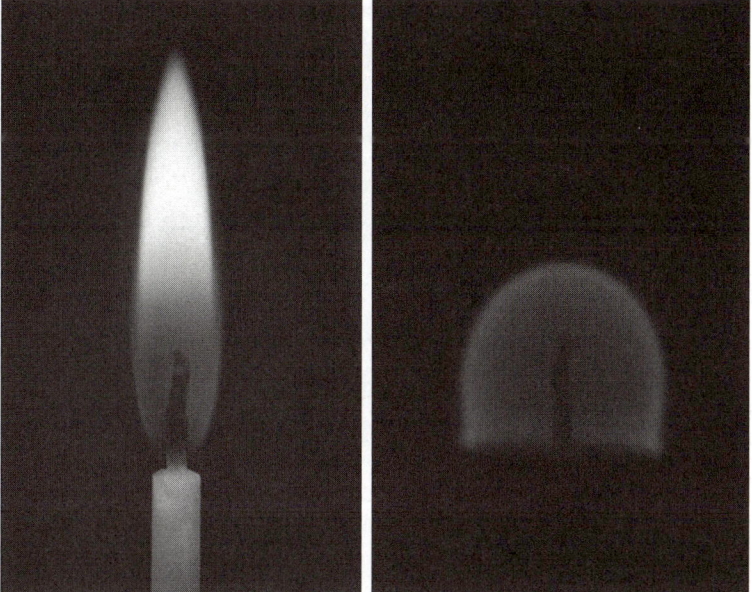

FIGURE 2.32 Candles in regular gravity (left) and microgravity (right). Courtesy of NASA Glenn Research Center (GRC).

■ 2.4 Uniformly Accelerating Systems

Even though uniformly accelerating systems involve motion, they are discussed here because they can be analyzed with the hydrostatic formula, with the addition of an acceleration besides that due to gravity. If a fluid moves in bulk, so that all parts of the fluid move simultaneously together and do not change position with respect to each other, then we refer to it as *rigid body motion*, even though fluids in general are not rigid. An example of this would be moving a bucket of water. If the fluid all moves at the same velocity, then there will be no viscous forces, because there must be a velocity gradient in order for viscosity to generate a shear stress, as shown in Equation 1.6. So reconsider the element of fluid shown in Figure 2.1, which has been redrawn in Figure 2.33 with the addition of a bulk rectilinear acceleration. Now the acceleration of the fluid within the element must also be considered, in addition to the pressure forces on the surfaces and the weight of the fluid within the element.

2.4.1 Uniform Rectilinear Acceleration

Summing forces in the x-direction on the fluid element and applying Newton's second law ($F_x = ma_x$) yields

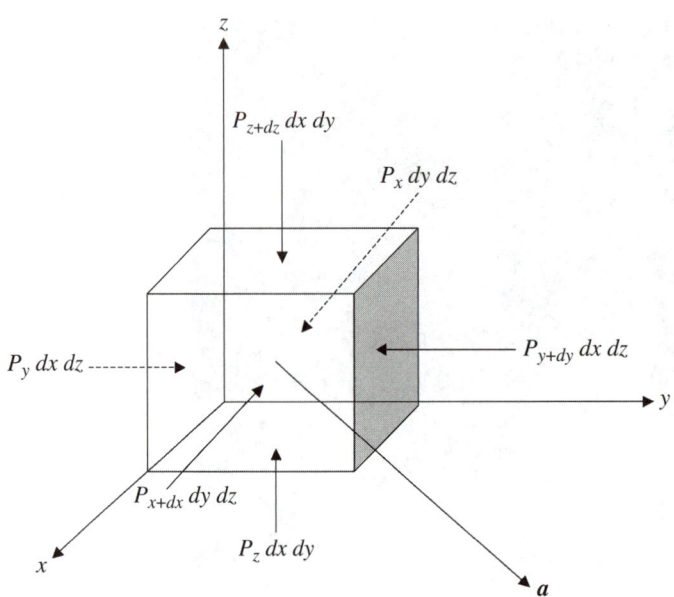

FIGURE 2.33 Schematic for deriving the pressure change in a fluid undergoing rigid body motion.

$$P\, dy\, dz - \left(P + \frac{\partial P}{\partial x} dx\right) dy\, dz = (\rho\, dx\, dy\, dz) a_x$$

Dividing through by the volume of the fluid element, $d\mathcal{V} = dx\, dy\, dz$, gives

$$\frac{P - \left(P + \frac{\partial P}{\partial x} dx\right)}{dx} = \rho a_x$$

Simplifying this equation yields

$$\frac{\partial P}{\partial x} = -\rho a_x$$

Similarly, in the y-direction

$$\frac{\partial P}{\partial y} = -\rho a_y$$

and in the z-direction the weight of the fluid element must also be taken into account, so that

$$\frac{\partial P}{\partial z} = -\rho a_z + \rho g$$

In vector notation:

$$\nabla P = \rho(\vec{g} - \vec{a}) \tag{2.17}$$

EXAMPLE 2.26

One example where this is important is in the liquid fuel tanks of rockets. If a kerosene-fueled rocket accelerates upward with an acceleration of 3 g, what is the pressure gradient in the fuel tank?

SOLUTION The density of kerosene is around 800 kg/m³. The upward acceleration adds to the effect of gravity in increasing the pressure with depth in the tank, so the pressure gradient is

$$dP/dz = \rho(g - a) = (800 \text{ kg/m}^3)(9.8 \text{ m/s}^2 - (-3 \cdot 9.8 \text{ m/s}^2)) = 31{,}360 \text{ Pa/m}$$

2.4.2 Solid Body Rotation

Although it may not seem like it at first, a uniformly rotating fluid also satisfies the condition of zero velocity gradient. If we align the system so that the rotation is about the z-axis, as shown in Figure 2.34, then with a rotation rate of Ω in radians per second, the acceleration vector is

$$\vec{a} = \vec{\Omega} \times (\vec{\Omega} \times \vec{r}) \tag{2.18}$$

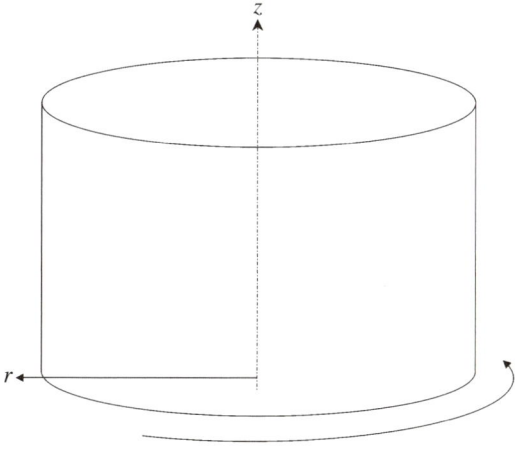

FIGURE 2.34 Schematic for rotational rigid body motion.

This is the centripetal acceleration, which is equal to $a = V^2/r$, if V is the tangential velocity.

As we will see in Chapter 4, if the condition $\mu \nabla^2 \vec{V} = 0$ is met, then the viscous forces are zero. This is obviously the case for linear rigid body acceleration, since the velocity does not vary spatially—both its first and second derivatives in all directions are zero. For the system with rigid body rotation, although the velocity does change across the x-axis, all the molecules do retain their original positions with respect to each other, so there is no relative motion between molecules, and hence no shear stress. So for the fluid system in rigid body rotation, the change in pressure along the radial axis is

$$\frac{dP}{dr} = \rho \omega^2 r \tag{2.19}$$

What can we say about the surface of a fluid undergoing rigid body motion? Just as a driver is thrown back into his seat when he rapidly accelerates his car forward, so when a container of fluid is accelerated will the fluid be pushed toward the back, opposite the direction of acceleration. So the fluid level will no longer be parallel to the ground. But what exactly is the shape of the interface? Remember that at a flat fluid–air interface, the pressure is equal to the surrounding atmospheric pressure, so along the surface we can say that $P = P_{atm} = $ constant. This means that along the surface the slope of the interface must be such that the change in pressure due to acceleration in the x-direction is balanced by the change in pressure in the z-direction. This condition is met when the slope is

$$\frac{dz}{dx} = -\left(\frac{a_x}{g + a_z}\right) \tag{2.20}$$

and the angle of the surface is related to the slope through the relation

$$\theta = \tan^{-1}\left(\frac{a_x}{g + a_z}\right) \tag{2.21}$$

For the fluid undergoing rigid body rotation, the acceleration in the x-direction is not constant, but varies with distance, r. So we do not expect the surface to be perfectly straight. Integrating the expression for pressure yields

$$P = \frac{1}{2}\rho \omega^2 r^2 - \rho g z + C \tag{2.22}$$

Again assuming that the pressure along the surface is constant, Equation 2.22 can be rearranged to solve for the value of z as a function of r that gives 0 gage pressure along the surface:

$$z = \frac{\omega^2 r^2}{2g} \tag{2.23}$$

In a cylindrical coordinate system, Equation 2.23 defines the surface of a paraboloid. Note that the equations for the surface profile were derived neglecting surface tension forces. As long as the length scales are large enough, this approximation is acceptable. However, for microscale flows surface tension forces will be important and must be considered.

One obvious application of rigid body rotation is a centrifuge. Both gas and liquid centrifuges are employed in various applications. The rating of a centrifuge is normally given in terms of the acceleration it generates measured in g. Centrifuges are used in biology for separating cells. Gas centrifuges, which spin in the range of 100,000 rpm, are used in uranium enrichment.

EXAMPLE 2.27

A mirror with parabolic cross section will focus incoming parallel beams of light onto a single point (shown in Figure 2.35), which is useful in telescopes. Spherical mirrors are much easier to make but do not have this property, leading to spherical aberrations. To make a parabolic mirror, molten metal is poured into a rotating mold and allowed to cool. The mirror to be designed should have a diameter of 2 m and a focal point 3 m from the vertex of the mirror. What rotation rate of the mold is necessary to achieve the desired shape?

SOLUTION The equation for a parabola may be written as $r^2 = 4Hz$, where H is the distance to the focal point, as shown in Figure 2.35. Solving this equation for z yields $z = r^2/(4H)$, but from Equation 2.23 we also know $z = r^2 \omega^2/(2g)$. These two equations for z can be combined to solve for the angular velocity, $\omega^2 = g/(2H)$. So then

$$\omega = \sqrt{\frac{g}{2H}} = \sqrt{\frac{9.8 \text{ m/s}^2}{2(3.0 \text{ m})}} = 1.28 \text{ rad/s} = 12.2 \text{ rpm}$$

We could also calculate the change in surface height of the mirror from the center to the outer edge, which we need to know to ensure that the mold is deep enough.

$$z = r^2/(4H) = (1 \text{ m})^2/(4(3 \text{ m})) = 0.083 \text{ m} = 8.3 \text{ cm}$$

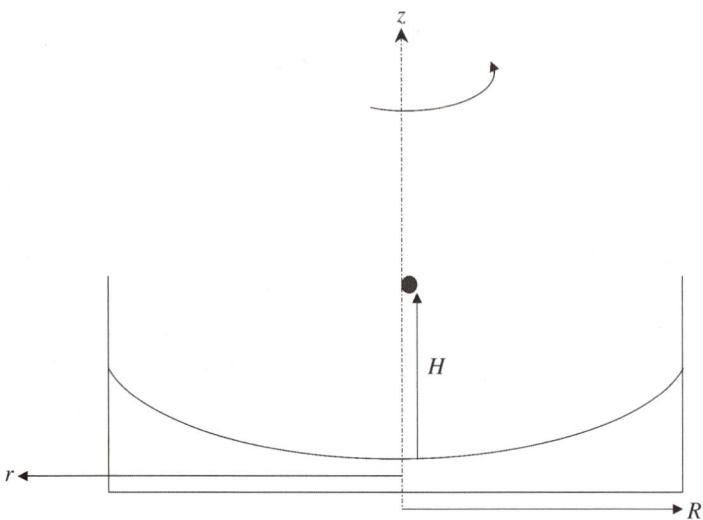

FIGURE 2.35 Schematic of a parabolic mirror.

EXAMPLE 2.28

A centrifuge is being used to sort materials of different densities by generating a high pressure gradient in test tubes. The test tubes are mounted on hinges so that they swing outward as the centrifuge revs up to speed. Find the acceleration, pressure, and force at the bottom end of the test tubes for the following conditions: The centrifuge spins at 1000 rad/s, and tubes are essentially horizontal at that speed. The inner diameter of the test tube is 1 cm, its length is 10 cm, it initially contains fluid to a depth of 8 cm, and the density of the material in the test tube is close to the density of water. The top of the test tube is hinged 3 cm away from the axis of rotation.

SOLUTION Under operating conditions, the bottom of the test tube will be 10 cm + 3 cm = 13 cm away from the axis of rotation. So the acceleration there will be $a = \omega^2 r = (1000 \text{ rad/s})^2 (0.13 \text{ m}) = 130{,}000 \text{ m/s}^2$. This is equivalent to about 13,300 g. Assuming the pressure at the top (inner) end of the tube is atmospheric, the gage pressure at the bottom of the tube can be calculated by integrating the pressure gradient:

$$P_{\text{end}} - P_{\text{top}} = \int_{r_{\text{min}}}^{r_{\text{max}}} \frac{\partial P}{\partial r} dr = \int_{r_{\text{min}}}^{r_{\text{max}}} (\rho \omega^2 r) dr = \rho \omega^2 \left. \frac{r^2}{2} \right|_{r_{\text{min}}}^{r_{\text{max}}}$$

Since $P_{top} = 0$,

$$P_{end} = \left(1000\,\frac{\text{kg}}{\text{m}^3}\right)(1000\,\text{s}^{-1})^2 \frac{1}{2}[(0.13\,\text{m})^2 - (0.03\,\text{m})^2] = 8{,}000{,}000\,\text{Pa}$$

or more compactly, $P_{end} = 8.0$ MPa (gage). The force on the end of the tube is

$$F = PA = (8.0 \times 10^6\,\text{Pa})\frac{\pi}{4}(0.01\,\text{m})^2 = 628\,\text{N}$$

Two types of centrifuges are shown in Figures 2.36 and 2.37.

If a container is small enough compared to its radius of rotation, it may be possible to treat it as being in rectilinear acceleration. For example, consider a cup of water placed at the outer edge of a merry-go-round. If the merry-go-round has a radius of 2 m, the cup has a diameter of 5 cm, and the outer edge of the cup sits at the outer edge of the merry-go-round, then the difference in acceleration at the inner and outer edges of the cup is

$$\% \text{ change} = \frac{a_{outer} - a_{inner}}{a_{inner}} = \frac{\omega^2 r_{outer} - \omega^2 r_{inner}}{\omega^2 r_{inner}}$$

$$= \frac{r_{outer} - r_{inner}}{r_{inner}} = \frac{2\,\text{m} - (2\,\text{m} - 0.05\,\text{m})}{(2\,\text{m} - 0.05\,\text{m})} = 2.6\%$$

which might be small enough to be negligible for some applications.

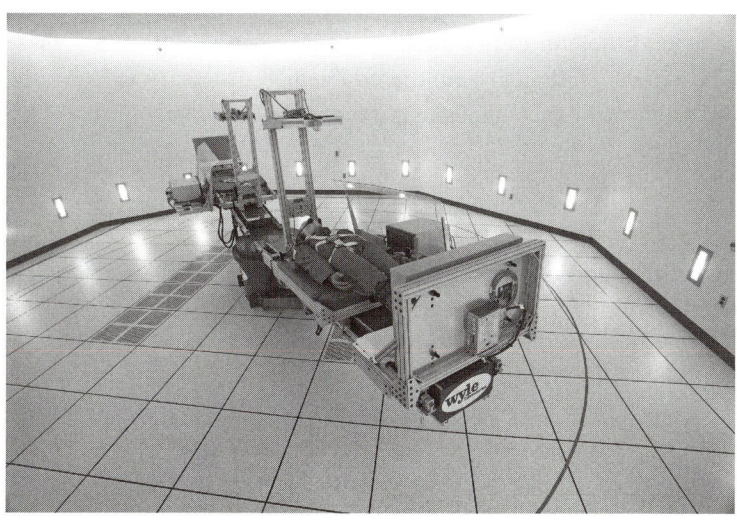

FIGURE 2.36 Centrifuge used for exercise for astronauts in space. Courtesy of NASA/JSC.

FIGURE 2.37 Large human-rated centrifuge. Courtesy of NASA Ames Research Center.

Summary

After reading this chapter and working through the problems, you should be able to apply the hydrostatic formula to calculate the pressure at any point in a fluid at rest, integrate the pressure to calculate the force on a solid object (whether submerged or floating), and solve fluid statics problems using these calculated forces. Additionally, you should understand that it is a pressure difference across a surface, not the pressure itself, that leads to a net force. You should be able to calculate the line of action for a hydrostatic force, which acts through the center of pressure. It is important to work through the end-of-chapter problems in this book. Only by successfully solving these problems will you master the skills and gain the confidence necessary to solve the more complex problems you will encounter in your engineering career.

References

[USAtm] National Oceanic and Atmospheric Administration, National Aeronautics and Space Administration, and U.S. Air Force. 1976. *U.S. Standard Atmosphere.*
[NASA75] Talay, T. 1975. *Introduction to the Aerodynamics of Flight.* NASA SP-367.

Problems

1. What is the depth of the deepest part of the ocean? What is the pressure at that depth? How much has the density of seawater changed at that depth from its value at the surface?

2. What is the value of the acceleration due to gravity at sea level? At a 30,000-ft altitude (typical altitude of jetliners)? At a 60-m altitude (edge of space)?
3. What is the typical local atmospheric pressure at the location where you live?
4. What is the typical local atmospheric pressure in Denver, Colorado?
5. What pure liquids are denser than water?
6. What are the lightest and heaviest gases (at STP)?
7. How much water is contained in Earth's oceans?
8. Who first derived and published the hydrostatic equation?
9. If a human being can generate 3 psi of vacuum pressure, what is the longest length of straw one can use to drink soda?
10. In *Star Wars: Episode I*, the planet Naboo is said to have a water core that connects to the surface oceans. Assuming Naboo is an Earth-sized planet, what is the gage pressure of the water at the center of the planet?
11. A container is filled with oil 4 m deep and water 8 m deep. What is the gage pressure at the bottom of the tank? Take the density of water to be 1000 kg/m^3 and the specific gravity of the oil to be 0.88. Will the oil or the water be on top in the tank? If the bottom of the tank is a circular plate with diameter of 5 m, what is the net fluid force acting on it?
12. If a reasonable height of a manometer is 1 m, what is the dynamic range of air pressures you could read with a water-filled manometer? What is the range of water pressures you could read with a mercury-filled manometer?
13. For a simple U-tube mercury manometer used to measure pressure in water flows, a difference in height between the two legs of 10 cm corresponds to what pressure difference in the water? Repeat the calculation for a water manometer used to measure air pressures. Why are inclined manometers used?
14. Consider an outdoor swimming pool of cylindrical shape, with a height of 5 ft and diameter of 12 ft. What is the weight of water contained in it if it is completely full? What are the possible failure modes for the container?
15. Consider a door 3 ft wide and 7 ft tall with its base at the bottom of a water tank filled to a depth of 16 ft. Calculate the force on the door. If you were on the outside of the tank, would you be able to push the door open (and drain the tank)?
16. A reservoir contains ethanol, filled to a depth of 10 m, and has a square cross section of size 10 m by 10 m. There is a 1 m by 1 m gate on the right wall, with the bottom of the gate at the bottom of the reservoir. Calculate the hydrostatic force acting on the gate and determine its location as well.
17. Consider the cylindrical tank of a water tower supported on four legs so that the bottom surface has atmospheric air underneath it. Derive an expression for the

force on (a) the bottom of the tank and (b) the side walls of the tank in terms of the dimensions of the tank, R and H, and any other relevant properties.

18. Design a passive system (no motors or electronic controllers or human intervention required) to maintain the water level between 24 and 25 ft deep in a reservoir. You may place holes, passageways, gates, and so on in the side of the reservoir.

19. In *Star Trek IV*, the crew of the Enterprise has to build a tank to hold whales. If the dimensions of the tank are 15 ft deep by 30 ft by 60 ft, how thick would standard glass walls have to be, using a factor of safety of 2.0?

20. A tractor-trailer hauling gasoline has a tank in the shape of an elliptic cylinder. Calculate the forces on the sides and endplates if it is completely full. The dimensions of the trailer are as follows: length, 40 ft; major axis, 6 ft; minor axis (vertically oriented), 4 ft. Does the force depend on the length of the tank?

21. A water tank in the shape of a cube of size 1 m by 1 m by 1 m is open at the top and filled to the brim. If the walls are made of steel (4140), and a factor of safety of 2.0 is to be employed in the design, how thick should the walls be?

22. Repeat Problem 21 for a cylindrical container with diameter of 1 m and height of 1 m.

23. Why is the bottom of a soda can domed inwards? What is the pressure inside a typical soda can?

24. If an airplane flying at 30,000 ft is pressurized to 0.7 atm, is the net force on fuselage inward or outward? If we model the airplane as a cylinder 50 m long and 5 m in diameter, what is the net static pressure force on the fuselage?

25. Calculate the fluid force on a regular hexagonal gate, with the top and bottom parallel to the liquid surface, and each side of length 1 m. The top surface is 3.2 m below the surface of the liquid, which is a fluid of specific gravity 0.78. Also calculate the location of the center of pressure.

26. Calculate the fluid force on a semicircular gate of radius 2 m, with the flat end on top and at a depth 5 m below the water surface.

27. Calculate the fluid force on a dam wall that is inclined at an angle of 45 degrees, is 10 m wide and 8 m long, and holds back water.

28. Design a passive gate system for the problem of Example 2.20, using a different solution than the one shown.

29. Design a stopper that plugs a 0.1-m-diameter hole in the bottom of a tank but that will automatically open when the depth of water exceeds 2 m.

30. A swimming pool is 20 ft wide and 50 ft long, inclined on the bottom from a depth of 4 ft at one end to 12 ft at the other. Calculate the forces on the bottom and all four sides of the swimming pool.

31. If you let go your grip on a helium balloon, what will be the ultimate fate of the balloon? If the helium gas leaks out of the balloon, what will happen to it?
32. What is the buoyant force on a submerged torus of outer diameter 2 m and inner diameter 1 m, if the center of mass of the torus is 4 m below the water's surface?
33. Is there any reason you would intentionally design a boat to be unstable?
34. If a weather balloon is 3 ft in diameter, made of 0.001-in-thick polyethylene, and filled with helium at STP, what is the altitude at which the balloon will come to rest, and what is the net upward force on the balloon at sea level?
35. If the shape of a zeppelin can be modeled as an ellipsoid of length 75 m and diameter 18 m, calculate the buoyant force on the zeppelin at sea level.
36. For Example 2.23, if Archimedes placed the crown in a container of cross section 0.6 m by 0.6 m part-way filled with water, how much would the water rise? If he removed the crown (without losing any water) and put in a brick of gold of the same mass as the crown, how much would the water level rise? Would he have been able to detect this difference with the measurement tools available to him?
37. What was the lifting power of the Hindenberg? What is the lifting power of the Goodyear blimp?
38. If a soccer ball accidentally lands in a swimming pool, what percentage by volume of the ball will be under the water's surface? What percentage of the area will be submerged?

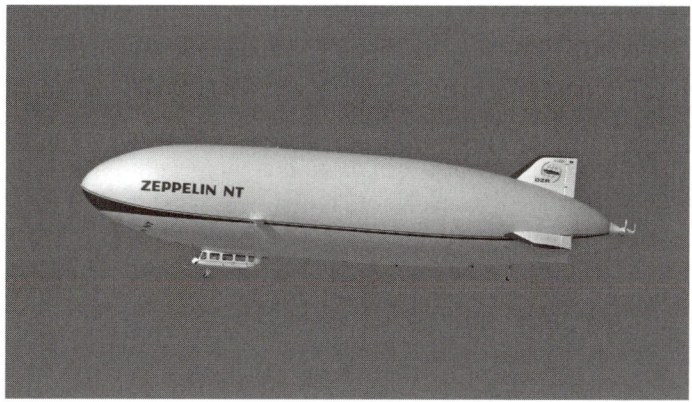

FIGURE 2.38 An ellipsoidal zeppelin. Courtesy of Zeppelin NT.

39. To isolate an experiment from the surroundings, a slab of concrete 1 m by 2 m by 2 m is placed in a large tub of mercury of diameter 3.5 m. How much will the level of mercury rise when the block is placed in it?

40. How thick should the walls of a bathysphere be if it will go down to a depth of 1 mile, has a diameter of 2 m, and is made out of stainless steel?

41. A cup contains water and ice cubes. If the cup is left alone long enough for the ice to melt, will the water level in the cup rise or fall?

42. If a rocket experiences vertical acceleration of 3 g, what is the pressure gradient in its kerosene fuel tank?

43. One method to produce enriched uranium is to use a gas centrifuge. Enriched uranium has a higher ratio of U-235 to U-238 than naturally occurring uranium. Solid uranium hexafluoride (UF6) is heated until it melts and vaporizes. The gas centrifuge is basically a rotating cylinder with lighter-weight gases extracted from the center. If the maximum speed of the outer end of the centrifuge is 500 m/s, compute and plot the pressure distribution in the centrifuge if it is filled with UF6 at a constant temperature of 600 K. The centrifuge is cylinder-shaped. Assume $P = 1$ atm in middle, and a radius of 25 cm. (Note: Tubes up to 10 m long are used in these centrifuges.)

44. Molten metal is poured into a spinning mold. If the centripetal acceleration is 10 g and the dimensions of the tube to be made in the mold are $L = 2$ m, $R_o = 0.15$ m, and $R_i = 0.10$ m, find the required rotational speed (in rpm) of the mold.

45. If a cup containing hot tea is placed at the center of a turntable and spun, what is the maximum rotational speed in rpm before the tea will spill, if the cup is 12 cm tall, 5 cm in diameter, and is initially filled 8 cm deep? What will be the maximum pressure in the cup at this condition? Where is the maximum pressure located?

46. A drag racer puts a half-full coffee cup in the cup holder of his car. The coffee cup is a cylinder 14 cm high and 6 cm in diameter. If she accelerates from rest with a constant acceleration of 15 m/s², will the coffee spill?

47. In the Boston molasses flood of 1919, it is estimated that there were 2.3 million gal of molasses in a 30-ft-high storage tank before it ruptured. What was the pressure in the bottom of the tank before it spilled?

3 Fluid Dynamics

In This Chapter
- Conservation of Mass
- Conservation of Momentum
- Conservation of Energy
- The Bernoulli Equation

The objective of this chapter is to introduce and apply conservation of mass, momentum, and energy to the overall performance of fluid systems.

3.1 Conservation of Mass

One of the fundamental principles of science and engineering is that mass cannot be created or destroyed. Actually, in nuclear reactions mass can be converted to energy according Einstein's famous equation,

$$E = mc^2 \tag{3.1}$$

but we will assume for the time being that none of the fluid flows we will be examining will include nuclear reactions. So a general statement of conservation of mass can be expressed in equation form as

$$\frac{dm_{cv}}{dt} = \sum_{in} \dot{m} - \sum_{out} \dot{m} \tag{3.2}$$

To illustrate the conservation of mass equation, let us draw a control volume (CV) around the system of interest, as shown in Figure 3.1. Since there is no source of mass (because mass cannot be created or destroyed but only be moved from one place to another), the mass in the control volume can only be changed by bringing mass into or out of the control volume.

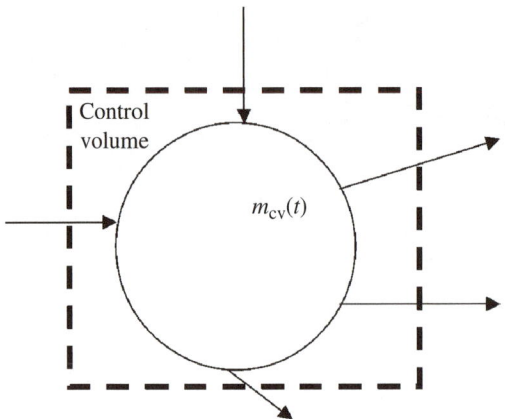

FIGURE 3.1 Schematic of a control volume for conservation of mass analysis.

The mass within the control volume, m_{CV}, is obtained by summing the mass over the entire control volume:

$$m_{CV} = \int_{CV} \rho \, d\forall \tag{3.3}$$

The mass flow rate across a section is

$$\dot{m} = \rho A V \tag{3.4}$$

where V represents an average velocity across the area A. A more general equation for the mass flow rate across a section would be

$$\dot{m} = \int_A \rho V \, dA \tag{3.5}$$

where here the velocity, V, can be a function of position. For a compressible fluid the density could also vary, although we usually neglect this possibility if the cross-sectional area is not too large.

As an example, consider the following velocity profile in a pipe, shown in Figure 3.2:

$$V(r) = V_{max}\left[1 - \left(\frac{r}{R}\right)^2\right] \quad \text{for } 0 < r < R$$

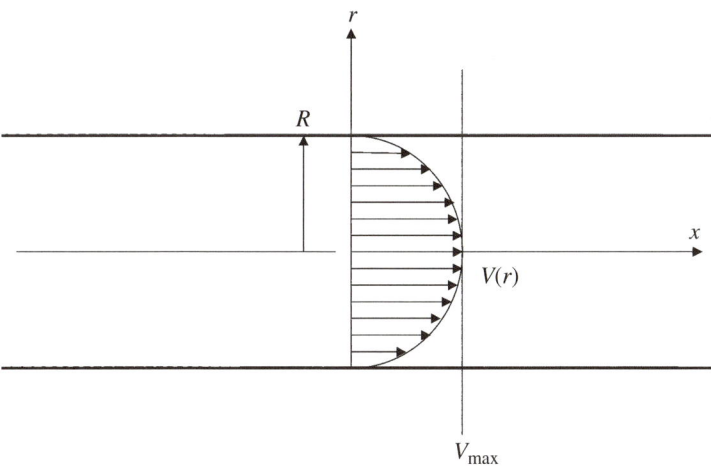

FIGURE 3.2 Velocity profile of a pipe.

We will show later in the chapter that this is the velocity profile of a laminar flow through a circular pipe. Note that the velocity is a function of only the radial location, r. So to perform the integral of Equation 3.5, the differential area, dA, must also be expressed in terms of r. At an arbitrary fixed value of r, an element with thickness of dr must take the shape of a ring, as shown in Figure 3.3. Then the area of the ring can be approximated as the circumference times the thickness, for a ring of negligible thickness. Thus, $dA = 2\pi r\, dr$, and the integral can be expressed as

$$\dot{m} = \rho \int_0^R V(r) 2\pi r\, dr$$

Just as when we calculated hydrostatic forces in Chapter 2, for a single integral we must express all the variables in terms of a single independent variable in order to perform the integration. Here, we substitute the formula for the velocity profile $V(r)$ and perform the integration:

$$\dot{m} = \rho \int_0^R V_{max}\left[1 - \left(\frac{r}{R}\right)^2\right] 2\pi r\, dr = 2\pi\rho V_{max} \int_0^R \left[1 - \left(\frac{r}{R}\right)^2\right] r\, dr$$

$$= 2\pi\rho V_{max} \int_0^R \left(r - \frac{r^3}{R^2}\right) dr$$

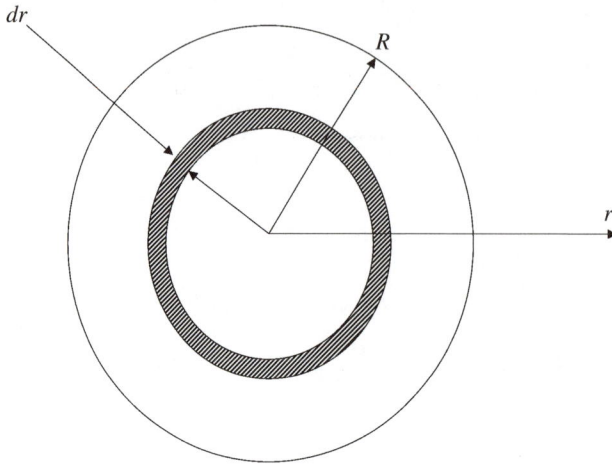

FIGURE 3.3 Schematic of the differential area section used for integrating the velocity profile for a cross section of flow.

Now we are left with just a simple integration of a polynomial, and the mass flow rate is

$$\dot{m} = 2\pi\rho V_{max} \int_0^R \left(r - \frac{r^3}{R^2}\right) dr = 2\pi\rho V_{max} \left(\frac{r^2}{2} - \frac{r^4}{4R^2}\right)\Big|_{r=0}^{r=R}$$

$$= 2\pi\rho V_{max} \left(\frac{R^2}{2} - \frac{R^4}{4R^2} - 0 + 0\right) = \rho \frac{\pi R^2}{2} V_{max}$$

It is often useful to define the mass-averaged velocity for a flow,

$$V_{ave} = \frac{\dot{m}}{\rho A} = \frac{\int_A \rho V \, dA}{\rho A} \tag{3.6}$$

We can also calculate the average velocity for this particular flow as

$$V_{ave} = \frac{\dot{m}}{\rho A} = \frac{\dot{m}}{\rho \pi R^2} = \frac{\rho \pi R^2 V_{max}/2}{\rho \pi R^2} = \frac{V_{max}}{2}$$

Thus for a parabolic flow the average velocity across a cross section is equal to half of the maximum velocity, which is found at the centerline. This statement is only true for parabolic flows; for other shapes of velocity profiles it will not be true.

Most affordable flow rate measurement devices measure volume flow rates rather than mass flow rates. The volume flow rate is commonly denoted by the symbol Q (not to be confused with the Q for a heat transfer in thermodynamics problems) and defined as the volume flow rate of a fluid per unit time:

$$Q = \frac{\dot{m}}{\rho} = AV \tag{3.7}$$

3.1.1 Steady-State Flow

Many engineering devices operate under steady-state conditions most of the time. For an analysis of a system at steady state, we set all the time derivative terms to zero. For conservation of mass, this means the term dm/dt on the left-hand side of Equation 3.2 goes to zero. The resulting equation for conservation of mass for steady-state systems is

$$\sum_{in} \dot{m} = \sum_{out} \dot{m} \tag{3.8}$$

which basically states that in a steady-state system mass cannot accumulate or be depleted within the control volume. Analysis of most devices—pumps, turbines, fans, heat exchangers, and so on—tends to focus on the steady-state case.

EXAMPLE 3.1

Consider oil flowing through the cylindrical passage shown in Figure 3.4. The oil enters through the circular opening in the left and exits through the annular passageway at the right. If the oil inlet flow rate is measured to be 680 mL/hr, find the mass flow rate at the exit in units of kg/s. Also find the average velocity of the oil at the inlet and at the exit. The specific gravity of the oil is 0.89.

SOLUTION: For convenience, we will denote the inlet as point 1, and the exit as point 2. Note that 1 mL of a fluid has a volume of 1 cc = $(0.01 \text{ m})^3 = 10^{-6}$ m³, and 1 hr = 3600 s (you would be surprised how many students have trouble with

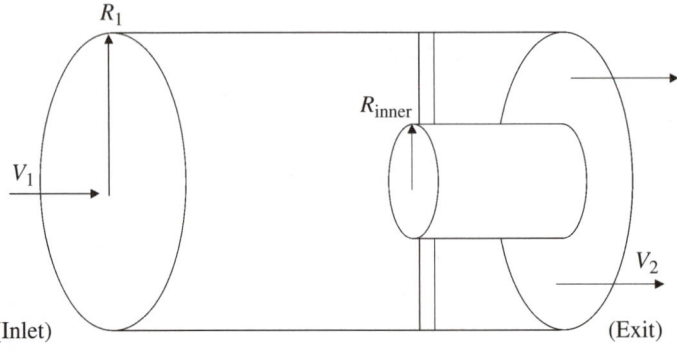

FIGURE 3.4 Schematic of oil flowing through a cylinder.

that conversion factor). Assuming the specific gravity of the oil is 0.89, its density is 890 kg/m³. It is reasonable to assume the device is operating in steady state, so that the outlet mass flow rate must be equal to the inlet mass flow rate, since there is no place inside the device in which oil could accumulate. Since the inlet volume flow rate was already specified and the density is known, the inlet mass flow rate can be calculated as

$$\dot{m}_1 = \rho_1 Q_1 = \left(890 \frac{kg}{m^3}\right)\left(680 \frac{mL}{hr}\right)\left(\frac{1 hr}{3600 s}\right)\left(\frac{10^{-6} m^3}{1 mL}\right) = 1.68 \times 10^{-4} \frac{kg}{s}$$

Since the flow is steady, the outlet flow rate equals the inlet flow rate: $\dot{m}_2 = \dot{m}_1 = 1.68 \times 10^{-4}$ kg/s. To find the inlet and outlet velocities, we must first find the corresponding cross-sectional areas. The inlet is just a simple circle, so the area perpendicular to the flow is $A = (\pi/4)D^2 = (\pi/4)(0.02 \text{ m})^2 = 3.14 \times 10^{-4}$ m². Note that is usually easier to convert nonfundamental units (cm or mm) into fundamental units (m) before proceeding further in the problem. The outlet area is than of an annulus, $A = (\pi/4)(D_o^2 - D_i^2) = (\pi/4)[(0.02 \text{ m})^2 - (0.018 \text{ m})^2] = 5.97 \times 10^{-5}$ m². The average velocity at the inlet is then

$$V_1 = \frac{\dot{m}_1}{\rho_1 A_1} = \frac{1.68 \times 10^{-4} \frac{kg}{s}}{\left(890 \frac{kg}{m^3}\right)(3.14 \times 10^{-4} \text{ m}^2)} = 6.01 \times 10^{-4} \frac{m}{s} = 0.6 \frac{mm}{s}$$

Similarly, the average velocity at the outlet is

$$V_2 = \frac{\dot{m}_2}{\rho_2 A_2} = \frac{1.68 \times 10^{-4} \frac{\text{kg}}{\text{s}}}{(890 \frac{\text{kg}}{\text{m}^3})(5.97 \times 10^{-5} \text{m}^2)} = 3.16 \times 10^{-3} \frac{\text{m}}{\text{s}} = 3.16 \frac{\text{mm}}{\text{s}}$$

For this problem, where the device operates under steady-state conditions and there is a single inlet and a single outlet, the conservation of mass equation simplifies to

$$\rho_1 A_1 V_1 = \rho_2 A_2 V_2 \tag{3.9}$$

Note that for incompressible fluids, such as the oil in this example problem, the density is constant, so $\rho_1 = \rho_2$ and the conservation of mass equation further simplifies to

$$A_1 V_1 = A_2 V_2 \tag{3.10}$$

which is also known as the area-velocity, valid for steady-state flow of incompressible fluids. (This property was first discovered by Leonardo da Vinci.)

EXAMPLE 3.2

Air enters the inlet of a compressor (Figure 3.5) operating at steady state, with intake conditions of atmospheric air at 1 bar, 22°C, and a velocity of 50 m/s. The inlet (point 1) area is 0.1 m², and the outlet (point 2) area is 0.09 m². The outlet pressure and temperature are 700 kPa and 300°C. Find the mass flow rate and the exit velocity.

SOLUTION: The pressures are low enough and the temperature high enough that the ideal gas law can safely be used for air. At the inlet, we can calculate the density by remembering that the average molecular weight of air is approximately 29 (in whatever unit system is being employed: 29 g/mol = 29 kg/kmol = 29 lb/lbmol = 29 slug/slugmol). *The temperature and pressure should always be expressed in absolute units when using the ideal gas law.* Thus, temperature is

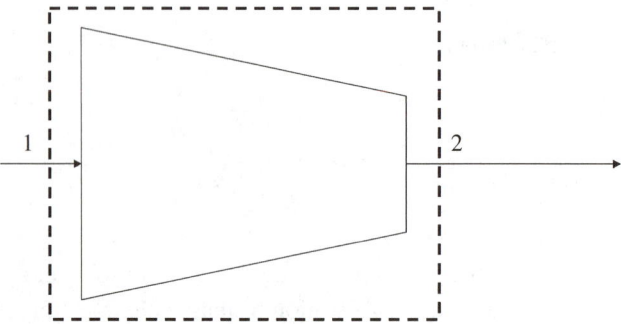

FIGURE 3.5 Diagram of a compressor operating at steady state.

$T_1 = 22°C + 273 = 295$ K. Rearranging the ideal gas law to solve for the density, we have

$$\rho_1 = \frac{P_1 M_1}{RT_1} = \frac{(10^5 \text{ Pa})(29 \frac{\text{kg}}{\text{kmol}})}{(8314 \frac{\text{J}}{\text{kmol} \cdot \text{K}})(295 \text{ K})} = 1.18 \frac{\text{kg}}{\text{m}^3}$$

Note: You should always double-check the units in your calculations. If the units are incorrect, the number probably is incorrect as well. The fundamental unit of pressure, the Pascal, is equal to the force of 1 Newton spread out over 1 square meter, and 1 Joule of energy is equivalent to the work needed to move an object 1 m against 1 N of resisting force, so 1 Pa/1 J = (1 N/m²)/(1 N m) = 1 m^{-3}. In our calculation above, all other units except kg cancel, so we have obtained the correct units of density (mass per volume).

In the same manner, the temperature at the outlet is $T_2 = 300°C + 273 = 573$ K. The density is also calculated using the ideal gas law:

$$\rho_2 = \frac{P_2 M_2}{RT_2} = \frac{(7 \times 10^5 \text{ Pa})(29 \frac{\text{kg}}{\text{kmol}})}{(8314 \frac{\text{J}}{\text{kmol} \cdot \text{K}})(573 \text{ K})} = 4.26 \frac{\text{kg}}{\text{m}^3}$$

Assuming the compressor operates at steady state, the inlet mass flow rate will equal the outlet mass flow rate. With the densities known, there is now enough information to calculate the inlet mass flow rate:

$$\dot{m}_1 = \rho_1 A_1 V_1 = \left(1.18 \frac{\text{kg}}{\text{m}^3}\right)(0.1 \text{ m}^2)\left(50 \frac{\text{m}}{\text{s}}\right) = 5.90 \frac{\text{kg}}{\text{s}}$$

Again, since the device operates at steady state with a single inlet and a single outlet, $\dot{m}_2 = \dot{m}_1 = 5.90$ kg/s. The exit velocity is then

$$V_2 = \frac{\dot{m}_2}{\rho_2 A_2} = \frac{5.90 \,\frac{\text{kg}}{\text{s}}}{(4.26 \,\frac{\text{kg}}{\text{m}^3})(0.09 \text{ m}^2)} = 15.4 \,\frac{\text{m}}{\text{s}}$$

EXAMPLE 3.3

An oil of specific gravity SG = 0.85 flows in the gap between two large parallel flat plates. The top plate moves to the right at a steady velocity of $U_o = 5$ m/s, while the bottom plate is stationary, as sketched in Figure 3.6. The distance between the plates is $H = 2$ mm, and the width of the plates is $W = 50$ cm. Find the mass flow rate of oil between the plates.

SOLUTION: The velocity profile for this particular problem is $u(y) = (U_o y)/H$. The equation for mass flow rate is

$$\dot{m} = \int_A \rho V \, dA = \rho \int_A u \, dA$$

Since $u = u(y)$, dA must also be expressed as a function of y. The geometry is rectangular, so $dA = W \, dy$ and the limits of integration are from $y = 0$ to $y = H$.

$$\dot{m} = \rho \int_0^H uW \, dy = \rho \int_0^H y\frac{U_o}{H}W \, dy = \frac{\rho W U_o}{H}\int_0^H y \, dy = \frac{\rho W U_o}{H}\frac{y^2}{2}\bigg|_{y=0}^{y=H} = \frac{\rho W H U_o}{2}$$

We insert the numerical values to obtain $\dot{m}$ = (850 kg/m³)(0.50 m)(0.002 m) (5 m/s)/2 = 2.125 kg/s.

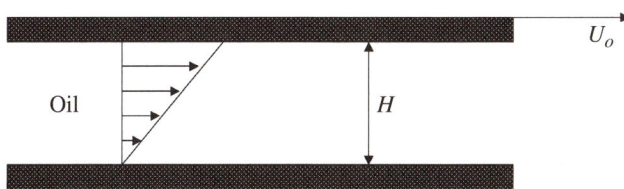

FIGURE 3.6 Diagram of oil flowing between two plates.

EXAMPLE 3.4

Air enters the intake system of an engine at conditions of 1 atm, 22°C, with a measured flow rate of 10,000 L/min. The fuel flow rate into the engine is 0.0136 kg/s. The exhaust exits the tailpipe at 200°C and 1 atm. Find the mass flow rate and the volume flow rate of the exhaust.

SOLUTION: Assume the engine is operating under steady-state conditions. To get the mass flow rate of the air flowing into the engine, we must first calculate the density using the ideal gas law.

$$\rho_{air} = \frac{P_{air} M_{air}}{RT_{air}} = \frac{(101{,}325 \text{ Pa})(29 \frac{\text{kg}}{\text{kmol}})}{(8{,}314 \frac{\text{J}}{\text{kmol} \cdot \text{K}})(295 \text{ K})} = 1.20 \frac{\text{kg}}{\text{m}^3}$$

The mass flow rate of air is equal to density times the volume flow rate:

$$\dot{m} = \rho Q = \left(1.20 \frac{\text{kg}}{\text{m}^3}\right)\left(10{,}000 \frac{\text{L}}{\text{min}}\right)\frac{1 \text{ min}}{60 \text{ s}} \frac{1 \text{ m}^3}{1{,}000 \text{ L}} = 0.20 \frac{\text{kg}}{\text{s}}$$

Conservation of mass for this steady-state system requires that the sum of the incoming mass flow rates of fuel and air must equal the outgoing mass flow rate of the exhaust:

$$\dot{m}_{exhaust} = \dot{m}_{air} + \dot{m}_{fuel} = 0.20 \frac{\text{kg}}{\text{s}} + 0.0136 \frac{\text{kg}}{\text{s}} = 0.214 \frac{\text{kg}}{\text{s}}$$

To get the volume flow rate of the exhaust, the density of the exhaust gases must be known. For this problem we will assume the molecular weight of the exhaust gas is the same as air even though the actual value depends on the composition of the exhaust gases. Thus,

$$\rho_{exh} = \frac{P_{exh} M_{exh}}{RT_{exh}} = \frac{(101{,}325 \text{ Pa})(29 \frac{\text{kg}}{\text{kmol}})}{(8{,}314 \frac{\text{J}}{\text{kmol} \cdot \text{K}})(473 \text{ K})} = 0.747 \frac{\text{kg}}{\text{m}^3}$$

and the mass flow rate of the exhaust gas is

$$Q = \frac{\dot{m}}{\rho} = \frac{0.214 \text{ kg/s}}{0.747 \text{ kg/m}^3} = 0.286 \frac{\text{m}^3}{\text{s}}$$

For naturally aspirated diesel engines, if the inlet flow rate of air into the engine cannot be measured, it can be approximated as the product of the displacement volume of the engine times the engine speed. So a truck engine of 6.0-L displacement operating at 1800 rpm would have an air intake flow rate of approximately

$$Q = (6.0 \text{ L})\left(1800\,\frac{\text{rev}}{\text{min}}\right)\frac{1\,\text{min}}{60\,\text{s}}\frac{2\pi\,\text{rad}}{1\,\text{rev}}\frac{1\,\text{m}^3}{1000\,\text{L}} = 1.13\,\frac{\text{m}^3}{\text{s}}$$

If the surrounding atmospheric air has a density of 1.2 kg/m³, then the intake mass flow rate of air into the engine would be (1.2 kg/m³)(1.13 m³/s) = 1.36 kg/s. For gasoline-fueled engines, the pressure loss over the throttle valve must be accounted for when estimating the air flow rate, and for turbocharged engines the boost pressure created by the compressor also affects the inlet flow rate. The mass flow of rate is much higher than the mass flow rate of fuel into an engine due to the stoichiometry, air-to-fuel ratio requirements for complete combustion for hydrocarbon fuels.

3.1.2 Transient Flow

In transient flow problems, the time derivative term is important. Applications include filling and emptying of tanks, starting and stopping of fluid devices, acceleration or deceleration of an object in a fluid, sudden gas discharges, and oscillatory or pulsing devices such as a pulse-detonation engine.

EXAMPLE 3.5

Water enters the tank shown in Figure 3.7 with a constant inflow rate of 1 gallon per minute (gal/min). The tank has an inner diameter of 1 ft and a height of 2 ft. The tank is initially empty. Determine the height of water in the tank as a function of time, the time it takes to fill the tank, and the mass of water in the tank when it is filled.

SOLUTION: The interior volume of the tank is the product of the cross-sectional area times the height: $V = (\pi/4)D^2 H = (\pi/4)(1\,\text{ft})^2 (2\,\text{ft}) = 1.57\,\text{ft}^3$.

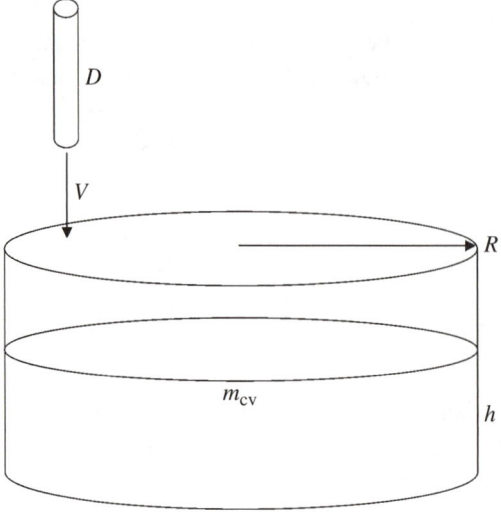

FIGURE 3.7 Diagram of water accumulating in a tank.

The cross-sectional area is $A = (\pi/4) D^2 = (\pi/4)(1 \text{ ft})^2 = 0.785 \text{ ft}^2$. One gallon is equivalent in volume to 231 in^3. To put everything in consistent units, we eliminate the inches:

$$1 \text{ gal} = 231 \text{ in}^3 = 231 \text{ in}^3 \left(\frac{1 \text{ ft}}{12 \text{ in}} \right)^3 = 0.134 \text{ ft}^3$$

Now we apply conservation of mass to a control volume around the tank:

$$\frac{dm_{CV}}{dt} = \sum_{in} \dot{m} - \sum_{out} \dot{m} = \dot{m}_{in}$$

Since the inlet flow rate was given as a volume flow rate rather than a mass flow rate, the relationship between mass and volume flow is needed: $\dot{m} = \rho Q$. The mass in the control volume can be expressed in terms of the geometry of the tank: $m_{CV} = \rho \forall = \rho(\pi/4)D^2 h(t)$, where h is the instantaneous height of water in the tank. Substituting these values into the conservation of mass equation gives

$$\frac{d}{dt}\left(\rho \frac{\pi}{4} D^2 h(t) \right) = \rho Q_{in}$$

The density of water is a constant, so we can divide ρ out of both sides of the equation:

$$\frac{\pi}{4} D^2 \frac{d}{dt}(h(t)) = Q_{in}$$

This simple first-order ordinary differential equation can be solved by separation of variables:

$$dh = \frac{Q_{in}}{\frac{\pi}{4} D^2} dt$$

Integrating both sides yields

$$\int dh = \frac{Q_{in}}{\frac{\pi}{4} D^2} \int dt$$

Since these are indefinite integrals, there will be a leftover constant of integration, C.

$$h(t) = \frac{Q_{in}}{\frac{\pi}{4} D^2} t + C$$

To find the value of C, we must apply an initial condition. For this problem, the initial condition was that the tank was empty at time $t = 0$, so $h(0) = 0$. The only value of C that will satisfy the initial condition is $C = 0$.

To finally solve for $h(t)$, we insert the dimensional values of the flow rate and the cross-sectional area:

$$h(t) = \frac{0.134 \text{ ft}^3/\text{min}}{0.785 \text{ ft}^2}(t) = 0.171 \frac{\text{ft}}{\text{min}} t = 0.00285 \frac{\text{ft}}{\text{s}}(t)$$

The time to fill the tank will be the time when $h = 2$ ft. So $t_{fill} = 2$ ft/(0.00285 ft/s) = 703 s = 11.7 min. The final mass in the tank is the product of density times volume. At standard state conditions water *weighs* 62.4 lbf/ft^3, so the *weight* of the water in the tank when it is full would be (62.4 lbf/ft^3)(1.57 ft^3) = 98 lbf. The *mass* that is in the tank is 98 lbm, or 3.04 slug (1 slug = 32.2 lbm), or 44.5 kg.

3.2 Conservation of Momentum

In this section we will be interested in the *thrust* force caused by a moving fluid. Newton's second law, commonly written as

$$\vec{F} = m\vec{a} \tag{3.11}$$

states that a force $\vec{F}$ applied to an object of mass m will produce an acceleration of magnitude $\vec{a} = F/m$. However, a key assumption must be made to apply Equation 3.11. That assumption is that the force is being applied to a solid object, or an object of constant mass. In fluid engineering problems the mass is obviously not solid, and the mass within a system can change as a function of time. Therefore, we must start with a more general statement of this conservation of momentum principle, namely, that the net sum of all external forces applied on a system will produce a change of *momentum* of that system, where momentum is defined as the product of the mass and the average velocity of the system. The general form of Newton's second law is then

$$\vec{F} = \frac{d(m\vec{V})}{dt} \tag{3.12}$$

For the case of a rigid object, the mass is constant and Equation 3.12 simplifies to the familiar form of Newton's second law:

$$\vec{F} = \frac{d(m\vec{V})}{dt} = m\frac{d\vec{V}}{dt} = m\vec{a} \tag{3.13}$$

However, a fluid is obviously not a rigid object, and the mass in the control volume may also be changing with time, so we need to use the general form of Newton's second law, Equation 3.12.

3.2.1 Derivation of Momentum Equation from Newton's Second Law

The momentum of a control volume is

$$m\vec{V} = \int_\Psi \vec{V} \rho \, d\Psi \tag{3.14}$$

3.2 Conservation of Momentum

In general, the momentum of a fluid can vary *in both time and space*, and the chain rule must be used to find the value of the derivative:

$$\frac{d(m\vec{V})}{dt} = \frac{\partial}{\partial t} \int_\forall \vec{V} \rho \, d\forall + \int_\forall (\nabla \bullet \vec{V}) \rho \, d\forall \tag{3.15}$$

∇ is the gradient operator, which represents the three spatial derivatives in vector form. The divergence theorem relates the volume integral of the divergence (gradient) of a variable to the surface integral of the variable over the boundary of the control volume:

$$\int_\forall (\nabla \bullet \vec{V}) \rho \, d\forall = \int_S \vec{V} \rho (\vec{V} \bullet d\vec{A}) \tag{3.16}$$

Substituting back into Equation 3.15 gives

$$\frac{d(m\vec{V})}{dt} = \frac{\partial}{\partial t} \int_\forall \vec{V} \rho \, d\forall + \int_S \vec{V} \rho (\vec{V} \bullet d\vec{A}) \tag{3.17}$$

Thus the momentum equation for fluid flows is

$$\sum \vec{F} = \frac{\partial}{\partial t} \int_\forall \vec{V} \rho \, d\forall + \int_S \vec{V} \rho (\vec{V} \bullet d\vec{A}) \tag{3.18}$$

The area vector, dA, is positive when it points out of the control volume, and it is normal to the control surface. The dot product of two vectors is equal to the product of their magnitudes times the cosine of the angle between them, as illustrated in Figure 3.8:

$$(\vec{V} \bullet \vec{A}) = VA \cos(\theta) \tag{3.19}$$

To simplify matters, the engineer should always construct the control volume surface so that it is perpendicular to each inlet or outlet flow of fluid, so that the only possible values for θ are 0° and 180°, and $\cos(0°) = 1$ and $\cos(180°) = -1$.

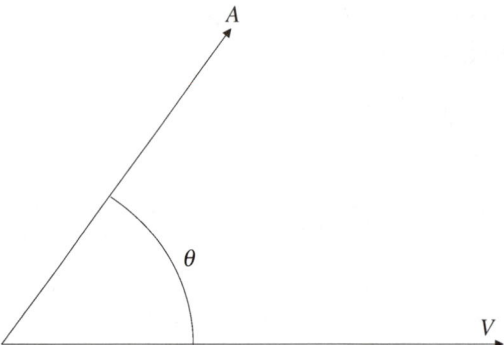

FIGURE 3.8 Schematic of two vectors.

Equation 3.18 may look intimidating, but it usually simplifies for most problems. For example, in the case of uniform velocity profiles at each inlet and outlet, the surface integral simplifies so that

$$\sum \vec{F} = \frac{\partial}{\partial t} \int_{\Psi} \vec{V} \rho \, d\Psi + \sum \pm \dot{m}\vec{V} \qquad (3.20)$$

Further, if the density and velocity are distributed uniformly throughout the control volume, then the first integral simplifies so that

$$\sum \vec{F} = \frac{\partial}{\partial t}(m\vec{V})_{\text{CV}} + \sum \pm \dot{m}\vec{V} \qquad (3.21)$$

If the problem is also for a steady-state system, then the equation further simplifies to

$$\sum \vec{F} = \sum \pm \dot{m}\vec{V} \qquad (3.22)$$

The term on the right-hand side can be thought of as a thrust force. Whenever a moving fluid enters or exits a control volume, it exerts a thrust force on that system. So the mass flow rate at a section of the control surface is equal to ρAV, and the magnitude of the fluid thrust force is equal to ρAV^2.

As with any problem involving forces, whether in solid fluid mechanics, statics or dynamics, the recommended procedure is to define the system by drawing a *control surface* (*S*) around the boundaries of the system. Everything inside the control surface is considered part of the *control volume* ($\forall$). This technique is often called *control volume analysis*. As previously stated, the control surface should be drawn so that it is perpendicular to any inflow or outflow of fluid.

The forces that act on the surfaces of a control volume can include viscous friction forces due to shear stress, pressure forces, and the forces transmitted by solid objects passing through the control volume boundaries. The weight of the fluid in the control volume can also be significant. In this chapter we will focus on frictionless flows to obtain approximate solutions to fluid problems. In Chapters 5 and 6 we will discuss how to include the effects of friction so that more accurate solutions may be obtained.

Even though Equation 3.18 looks complex, for most problems it will greatly simplify. For one thing, a common assumption we make in dealing with fluid problems is that the system operates in steady state. Although this assumption is not valid for accelerating control volumes or systems as they are being started up or shut down, most engineering devices spend most of their operational time in a steady-state condition. The assumption of steady-state operation means that in Equation 3.18 all derivatives with respect to time go to zero. Thus any and all external forces on the control volume must be balanced by the fluid thrust forces under steady-state conditions.

Another assumption that simplifies the equation is that the velocity through each inlet or outlet of the control volume is uniform and can be adequately represented by an average velocity. This assumption works well for turbulent flows through pipes, and except for microscale flows and flows of non-Newtonian fluids, most engineering flows are turbulent.

3.2.2 Application of Fluid Thrust Forces

EXAMPLE 3.6

Consider a jet on a thrust test stand. The inlet area of the jet is 0.10 m², and the outlet area is 0.085 m². The incoming air is at 1.0 bar and 21°C, with an average velocity of 20 m/s. If the outlet velocity is 100 m/s, and the pressure is approximately atmospheric at the exit, calculate the thrust force generated by the engine.

SOLUTION: Assume the jet engine operates at steady state, and neglect the mass flow rate of fuel into the system. The density of the incoming air can be calculated using the ideal gas law:

$$\rho_1 = \frac{P_1 M_1}{RT_1} = \frac{(100{,}000 \text{ Pa})(29 \frac{\text{kg}}{\text{kmol}})}{(8{,}314 \frac{\text{J}}{\text{kmol} \cdot \text{K}})(294 \text{ K})} = 1.19 \frac{\text{kg}}{\text{m}^3}$$

The intake mass flow rate can be calculated as

$$\dot{m}_1 = \rho_1 A_1 V_1 = \left(1.19 \frac{\text{kg}}{\text{m}^3}\right)(0.10 \text{ m}^2)\left(20 \frac{\text{m}}{\text{s}}\right) = 2.38 \frac{\text{kg}}{\text{s}}$$

Under the steady-state assumption and with a negligible fuel flow rate, the exhaust flow rate can be calculated from conservation of mass as equal to the intake flow rate of 2.38 kg/s. The incoming air flow pushes on the test stand to the right, while the exhaust flow produces a much stronger force to the left. Then the net thrust force on the test stand is calculated as

$$F = -\rho_1 A_1 V_1^2 + \rho_2 A_2 V_2^2 = \dot{m}(V_2 - V_1) = 2.38 \frac{\text{kg}}{\text{s}}\left(100 \frac{\text{m}}{\text{s}} - 20 \frac{\text{m}}{\text{s}}\right) = 190 \text{ N}$$

EXAMPLE 3.7

Water is directed through a nozzle and hits a stationary target, as shown in Figure 3.9. After the water hits the plate, it splits into two equal streams moving upward and downward as shown. Calculate the horizontal forces on the target. The cross-sectional area at the nozzle exit is 0.01 m² and the nozzle discharge velocity is 10 m/s.

SOLUTION: Assume steady-state flow and uniform pressure. The momentum equation for this problem will be

$$\sum \vec{F} = \int_S \vec{V} \rho (\vec{V} \bullet d\vec{A})$$

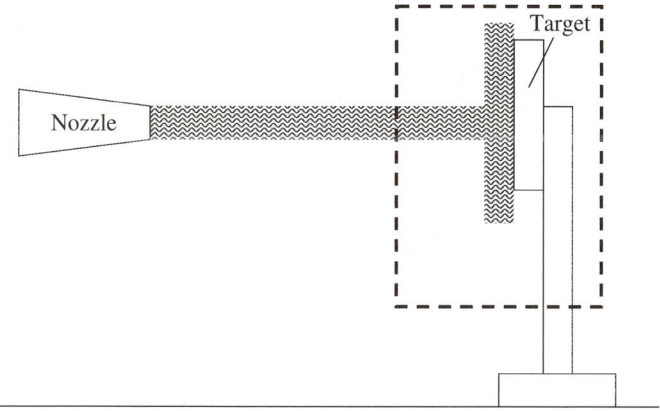

FIGURE 3.9 Water jet impacting flat target.

This vector equation can be split into its scalar x and y components. The only thrust force that acts in the horizontal x-direction is that of the incoming water flow from the left:

$$R_x = \rho V_1(\vec{V}_1 \bullet \vec{A}_1) = -\rho A_1 V_1^2$$

We then substitute the numerical values to compute the magnitude of the support force, R_x:

$$R_x = -\left(1000 \frac{\text{kg}}{\text{m}^3}\right)(0.01 \text{ m}^2)\left(10 \frac{\text{m}}{\text{s}}\right)^2 = -1000 \text{ N}$$

Note that R_x is negative, meaning that the support beam pushes the control volume to the left with a force of 1000 N. Newton's third law states that for every action there is an equal and opposite reaction. In this case, the reaction is the force of the water pushing the plate to the right with a force of 1000 N.

EXAMPLE 3.8

Water is directed through a nozzle, hits a curved stationary target, and is directed back in the direction it came, as shown in Figure 3.10. Find the net force against the target. What would the force on the target in Example 3.7 be if the water struck this curved plate instead of a flat plate?

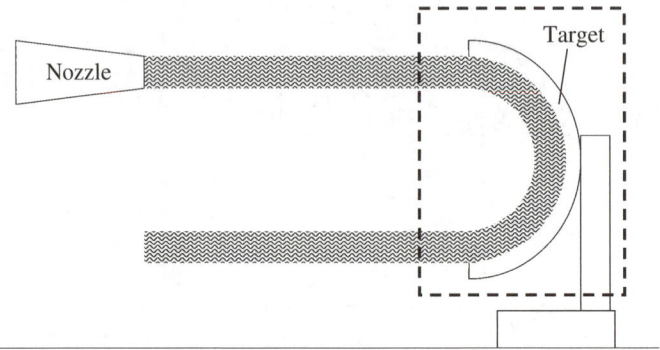

FIGURE 3.10 Water jet striking a curved target.

SOLUTION: The procedure is much the same as the previous example, except that now there are two horizontally moving fluid streams that cross the control surface.

$$R_x = \rho V_1(\vec{V}_1 \bullet \vec{A}_1) + \rho(-V_2)(\vec{V}_2 \bullet \vec{A}_2) = -\rho A_1 V_1^2 - \rho A_2 V_2^2$$

Let us further assume that as the stream of water proceeds around the curved surface, the diameter of the jet remains constant, so that $A_2 = A_1$. Then by conservation of mass $V_2 = V_1$, and the reaction force is

$$R_x = -2\rho A_1 V_1^2 = -2\left(1000 \frac{\text{kg}}{\text{m}^3}\right)(0.01 \text{ m}^2)\left(10 \frac{\text{m}}{\text{s}}\right)^2 = -2000 \text{ N}$$

We see that the thrust force generated by a moving fluid can be doubled by turning the fluid back around in the direction it came from as opposed to merely deflecting it. This is an important principle for the design of devices such as the turbines in hydroelectric power plants.

EXAMPLE 3.9

Water is forced through a nozzle with a gage pressure of 50 kPa as measured at the flange. The diameter at the flange is 10 cm, the diameter at the nozzle exit is 4 cm, and the exit velocity at the nozzle is 15 m/s. What is the net force on the nozzle?

SOLUTION: In this problem, in addition to the fluid thrust forces, there is also a fluid pressure force. It is preferable to use gage pressures in such problems, since the pressure everywhere on the control volume surface except at the flange is atmospheric, and the sum of atmospheric pressure over all the control volume surfaces will cancel no matter how the control volume is drawn. So the net pressure force on the control volume acts to the right and has magnitude

$$F = PA = (50{,}000 \text{ Pa})\left(\frac{\pi}{4}(0.10 \text{ m})^2\right) = 393 \text{ N}$$

To calculate the velocity at the inlet, conservation of mass can be used:

$$V_1 = V_2\left(\frac{D_2}{D_1}\right)^2 = 15\,\frac{\text{m}}{\text{s}}\left(\frac{4 \text{ cm}}{10 \text{ cm}}\right)^2 = 2.4\,\frac{\text{m}}{\text{s}}$$

The mass flow rate of the water is

$$\dot{m} = \rho A V = \left(1000\,\frac{\text{kg}}{\text{m}^3}\right)\frac{\pi}{4}(0.1 \text{ m})^2\left(2.4\,\frac{\text{m}}{\text{s}}\right) = 18.8\,\frac{\text{kg}}{\text{s}}$$

The net fluid thrust force is calculated from the momentum equation:

$$F_{\text{thrust}} = \dot{m}_1 V_1 - \dot{m}_2 V_2 = 18.8\,\frac{\text{kg}}{\text{s}}\left(2.4\,\frac{\text{m}}{\text{s}} - 15\,\frac{\text{m}}{\text{s}}\right) = -237 \text{ N}$$

The thrust force acts to the left. The net force on the flange is

$$F = 393 \text{ N} - 237 \text{ N} = 156 \text{ N}$$

So the net force is to the right, acting to pull the flange apart.

EXAMPLE 3.10

A curved vane turns a jet of fluid through an angle of 20°, as sketched in Figure 3.11. Find the vertical and horizontal forces the fluid exerts on the vane.

SOLUTION: Assume that the magnitude of the velocity of the fluid jet does not change as it moves around the vane, and that the fluid is incompressible. Then, by conservation of mass, $A_1 = A_2$. Now conservation of momentum can be applied to the control volume. The horizontal force due to the vane deflecting the fluid can be calculated as

$$F_x = -\rho A_1 V_1^2 + \rho A_2 V_2^2 \cos\theta = -\rho A_1 V_1^2 (1 - \cos\theta)$$

The vertical force is

$$F_y = -\rho A_2 V_2^2 \sin\theta$$

These are the forces that the vane exerts on the fluid, so by Newton's third law, the force of the fluid on the vane is equal in magnitude but opposite in sign. So the fluid force acts up and to the right.

Example 3.10 is geometrically similar to the turning vanes in a turbine or the effects of the wing of an airplane on the air it flies through. It both cases the goal is to maximize the ratio of the vertical force (which does useful work) to the horizontal force, which is wasted or degrades performance. Aerodynamics will be discussed in more detail in Chapter 6. For now it is sufficient to note that there are many other sources of drag on an airplane besides the change in horizontal momentum of the air due to

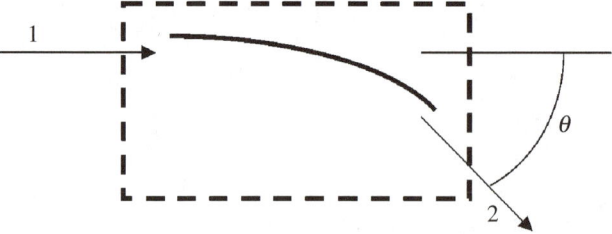

FIGURE 3.11 Fluid jet deflected through an angle θ by the turning vane.

the wings. Note that for airplanes the air is stationary relative to an observer on Earth's surface, and the control volume moves with the plane at the velocity of the plane, and so the velocity of the air impinging on the wing is equal to the forward speed of the plane. Such moving control volumes are discussed next.

3.2.3 Moving Control Volumes

Whenever the control volume is moving, such as when it is attached to a moving vehicle, the relative velocity between the fluid and the control volume must be used.

EXAMPLE 3.11

Consider a wheeled rectangular cart moving to the right at a constant velocity, V, of 10 m/s (Figure 3.12). A water jet of velocity $V_{jet} = 25$ m/s and diameter of 6 cm strikes the cart on the left end. Find the resisting force necessary to keep the cart moving at the constant velocity of 10 m/s.

SOLUTION: Assume the cart moves on frictionless wheels and the jet remains at constant velocity and diameter until it hits the cart. Draw the control volume so that it moves with the cart to the right at a constant velocity, V. The area of the water jet is $(\pi/4)(0.06 \text{ m})^2 = 0.00283 \text{ m}^2$. The momentum equation for the control volume is

$$R_x = -\rho A (V_{jet} - V)^2$$

$$R_x = -\left(1000 \frac{\text{kg}}{\text{m}^3}\right)(0.00283 \text{ m}^2)\left(25 \frac{\text{m}}{\text{s}} - 10 \frac{\text{m}}{\text{s}}\right)^2$$

$$R_x = -637 \text{ N}$$

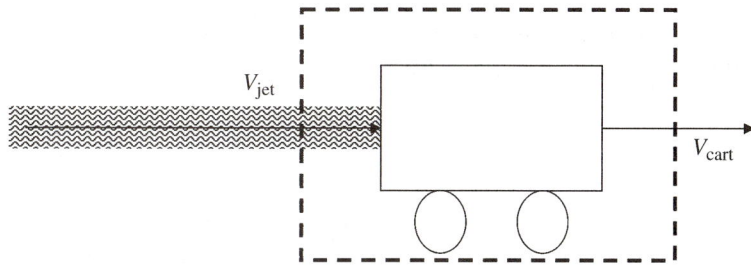

FIGURE 3.12 Diagram to show a control volume moving at constant velocity.

So a resisting force of magnitude 637 N must push on the cart toward the left to keep it moving at the constant velocity of 10 m/s. For moving control volumes, the velocity to use is always the *velocity relative to the control volume*.

3.2.4 Accelerating Control Volumes

When the control volume is accelerating, the problem can no longer be modeled as a steady-state system.

EXAMPLE 3.12

The wheeled cart of Example 3.11 is initially at rest when it is struck by a fluid jet of velocity V_{jet}. If there is no resisting force and friction can be neglected, find an expression for the velocity of the cart $V(t)$ as a function of time.

SOLUTION: For problems with accelerating control volumes, the transient term in the momentum equation (Equation 3.21) must be retained. For this specific problem, there are no external forces, so this momentum equation is

$$F_x = 0 = \frac{d}{dt}(mV) - \rho A(V_{jet} - V)^2$$

The mass of the cart is constant, so m can be taken outside of the derivative:

$$m\frac{dV}{dt} = \rho A(V_{jet} - V)^2$$

This is a first-order ordinary differential equation. Any differential equation can be solved numerically using Algorithm 3.1 (presented later in this chapter), but this particular equation has an analytical solution that can be found using separation of variables. We first put all the terms involving the velocity, V, on the left-hand side, and all the terms involving time, t, on the right-hand side:

$$\frac{dV}{(V_{jet} - V)^2} = \frac{\rho A}{m}dt$$

Now we integrate both sides using indefinite integrals. The integral on the right-hand side is easy enough; the left-hand side is more complex. The left-hand integral can be found either by searching through a table of integrals in a reference book, or by using the transformation $\eta = (V_{jet} - V)$, with $d\eta = -dV$. Thus,

$$\frac{1}{(V_{jet} - V)} = \frac{\rho A}{m} t + C$$

The use of indefinite integrals will result in an arbitrary constant of integration, C, whose value must be determined by applying an initial condition. The initial condition for this problem is that the cart is initially at rest, so $V(t = 0) = 0$. This results in a value of $C = 1/V_{jet}$. Solving for the velocity, V, gives

$$V(t) = V_{jet} \frac{t}{\frac{m}{\rho A V_{jet}} + t}$$

The velocity is plotted as a function of time in Figure 3.13. The velocity of the cart V, asymptotically approaches the velocity of the jet, V_{jet}, as $t \to \infty$. Of course in reality this equation is only valid when the cart is close to the jet outlet. After some distance the jet will start to break up due to aerodynamic forces; it will also fall due to gravity.

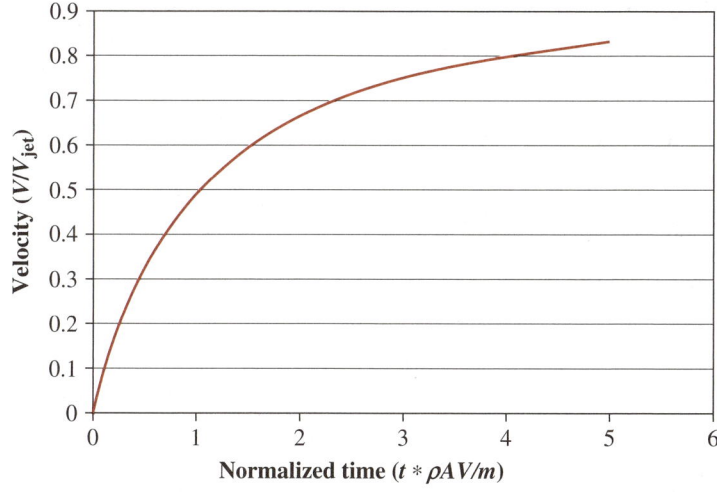

FIGURE 3.13 Velocity as a function of time for an accelerating cart.

EXAMPLE 3.13

Consider a rocket with initial mass M_o, which contains chemical propellants (fuel and oxidizer) that burn at a rate $\dot{m}_e$ and exit at a velocity V_e relative to the rocket body. The rocket is initially at ground level ($y = 0$) and travels vertically upward. The rocket engine is ignited at time $t = 0$. Neglect drag for this problem. Derive a formula for the rocket's height and velocity as a function of time.

SOLUTION: In this problem the control volume needs to move with the rocket as it travels through the atmosphere. There is one fluid thrust force acting on the control volume, and the only other force is the weight of the rocket itself. So the momentum equation for this problem is

$$F = -mg + \dot{m}_e V_e = ma$$

Rearranging this equation to solve for the acceleration a, we can integrate a once to get the velocity, then again to get the height:

$$a = \frac{\dot{m}_e V_e - mg}{m} = \frac{\dot{m}_e V_e}{m} - g$$

It is important to note that the total mass of the rocket is not constant, but decreases with time as the fuel is expended, so $m(t) = M_o - \dot{m}_e t$.

$$a = \frac{\dot{m}_e V_e}{M_o - \dot{m}_e t} - g \qquad (3.23)$$

Now we integrate both sides with respect to time:

$$V = \int a\, dt = \int \left(\frac{\dot{m}_e V_e}{M_o - \dot{m}_e t} - g \right) dt$$

The integral on the right-hand side does have an analytical form, which is

$$V(t) = -V_e \ln\left(\frac{M_o - \dot{m}_e t}{M_o} \right) - gt \qquad (3.24)$$

To get the height, we need to integrate the velocity with respect to time:

$$y(t) = -V_e \left[-t + \left(t - \frac{M_o}{\dot{m}_e} \right) \ln\left(\frac{M_o - \dot{m}_e t}{M_o} \right) \right] - \frac{1}{2} g t^2 \qquad (3.25)$$

assuming the rocket starts at a height of $y = 0$ at time $t = 0$. Of course this formula is only valid up until the time the rocket engine burns out. After that the rocket will still increase in height for a while due to its own inertia.

EXAMPLE 3.14

Work Example 3.13 for the specific values of initial mass $M_o = 5000$ kg, burn rate $\dot{m}_e = 35$ kg/s, and exit velocity $V_e = 2500$ m/s relative to the rocket body for a burn duration of $t = 100$ s. How high will the rocket travel?

SOLUTION: At $t = 100$ s, the elevation of the rocket is

$$y(t) = -2500 \frac{\text{m}}{\text{s}} \left[-100 \text{ s} + \left(100 \text{ s} - \frac{5000 \text{ kg}}{35 \text{ kg/s}}\right) \times \ln\left(\frac{5000 \text{ kg} - (35 \text{ kg/s})(100 \text{ s})}{5000 \text{ kg}}\right) \right] - \frac{1}{2} 9.8 \frac{\text{m}}{\text{s}^2} (100 \text{ s})^2$$

$y(100 \text{ s}) = 72{,}000$ m.

The velocity of the rocket at burnout is

$$V(100 \text{ s}) = -\left(2500 \frac{\text{m}}{\text{s}}\right) \ln\left(\frac{5000 \text{ kg} - (35 \text{ kg/s})(100 \text{ s})}{5000 \text{ kg}}\right) - \left(9.8 \frac{\text{m}}{\text{s}^2}\right)(100 \text{ s}) = 2030 \frac{\text{m}}{\text{s}}$$

After the rocket engine has ceased firing, the rocket will continue to climb under its own inertia for a while. The trajectory after burnout is actually easier to calculate since there is no longer any thrust force acting on the rocket. The only force acting on the rocket is its weight, so Newton's second law for the rocket is $F = ma = -mg$, and the acceleration $a = -g$. So the velocity for the final stage of ascent is $V(t) = V_o - gt$, and the elevation is $y(t) = y_o + V_o t - \frac{1}{2} gt^2$. The maximum altitude will be reached when the rocket loses all its inertia, or $V = 0$. This happens at $t = (2030 \text{ m/s})/(9.8 \text{ m/s}^2) = 207$ s after burnout. The gain in elevation will be $(2{,}030 \text{ m/s})(207 \text{ s}) - \frac{1}{2} (9.8 \text{ m/s}^2)(207 \text{ s})^2 = 210{,}000$ m, for a total height above the earth of 282,000 m.

The height of the rocket as a function of time is plotted in Figure 3.14, and its velocity as a function of time is plotted in Figure 3.15. Of course, these are only approximations, since aerodynamic drag was not accounted for. Drag will be discussed in Chapter 6, and the MATLAB program in Appendix F for calculating trajectories does include the effects of drag. It should also be noted that at these high altitudes the value of the gravitational acceleration is less than it is at sea level and is not a constant.

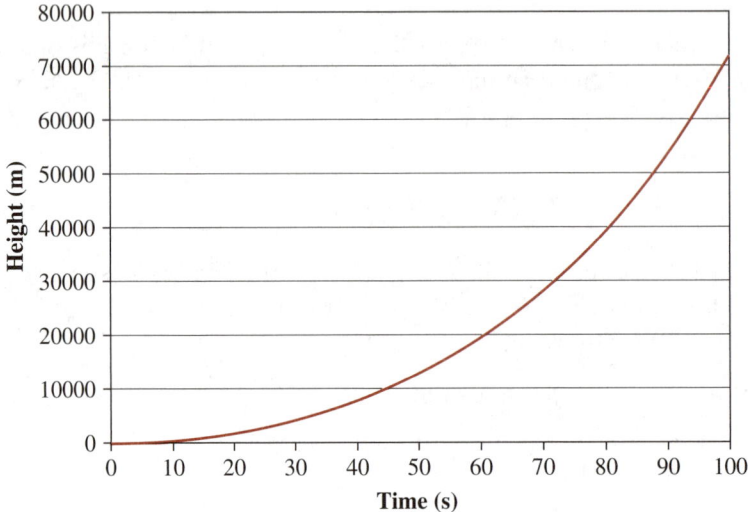

FIGURE 3.14 Height as a function of time for the rocket problem.

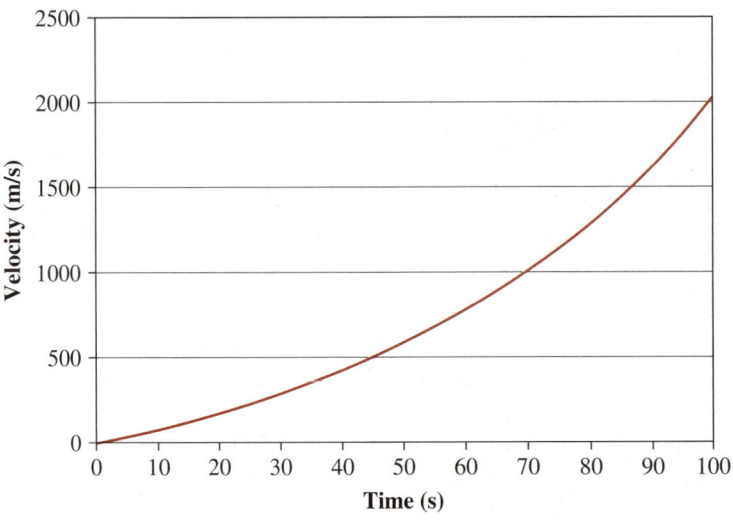

FIGURE 3.15 Velocity as a function of time for the rocket problem.

3.2.5 Numerical Integration of Ordinary Differential Equations

Equation 3.23 is a complex formula that is difficult to derive analytically. Including aerodynamic resistance, multiple stages, or nonconstant thrust forces would make the equation even more formidable. For such problems a numerical solution is often the better choice.

ALGORITHM 3.1 Numerical integration of ordinary differential equations using the Euler method

Many problems can be modeled as ordinary differential equations (ODEs). While some ODEs have analytical solutions, numerical methods can be constructed to approximate the solution to *any* ODE.

We will start by analyzing problems of the form

$$\frac{dy}{dt} = f(y, t) \qquad \text{with an initial condition } y(t_o) = y_o$$

In the case of the rocket problem, we are solving for the velocity as a function of time, t, with the initial velocity set to zero. The simplest method of solving first-order ODEs, and the easiest to code, is the Euler method. The Euler method is formed by constructing a Taylor series about a point t_i. (Note that the numerical solution to the ODE will yield a series of discrete values $y_i(t_i)$.)

$$y(t_{i+1}) = y(t_i) + y'(t_i)\Delta t + \frac{\Delta t^2}{2}y''(\xi)$$

where $\Delta t = y(t_{i+1}) - y(t_i)$. So if $y(t_i)$ represents the approximated value of y that is known, then $y(t_{i+1})$ is the unknown value of y at a future point in time. Now $y'(t) = f(y, t)$, a known function, so a simple approximation can be made:

$$y(t_{i+1}) \approx y(t_i) + \Delta t \cdot f(y, t)$$

This is the Euler method, which is used successively to generate the approximate solution $y(t)$ to an ODE.

At first glance it appears that the error in this method is second order, but keep in mind that we will make a series of approximations over a finite time interval, so the total error in using the Euler method over a length of time is usually first order. Also note that the error grows larger the longer we use the Euler method to march forward. There are higher-order methods that are more accurate than the Euler method, such as the Runge-Kutta method, but they are also more complex and more difficult to program. (In MATLAB the function ODE45 performs Runge-Kutta integration. See the MATLAB help in the drop-down menu for more information.)

For the rocket problem, integrating $F = ma$ will solve for the velocity. To get the height as a function of time, integrate again. Also note that with this numerical method we could also include the drag on the rocket, which is a function of the velocity.

STEPS
1. Create equation in the form `dx/dt = f(x,t)`, `x(0) = x`$_0$. Select time step, Δt.
2. LOOP

    `x(t+`Δ`t) = x(t) + `Δ`t*f(x(t),t)`
3. END LOOP

EXAMPLE 3.15

Repeat Example 3.14, starting with the expression for acceleration, and use Algorithm 3.1 to numerically integrate the acceleration to approximate the velocity and height. Compare the answer to the analytical solution for the first 10 s of flight.

SOLUTION: To use Algorithm 3.1 to find velocity, three things are required: (1) an expression for *dV/dt* solely in terms of *V* and *t*, (2) an initial condition for the value of *V* at time *t* = 0, and (3) the value of the numerical time step, Δt. For this problem an expression for the acceleration, *a*, is known, and since *a* = *dV/dt*, we can use this:

$$\frac{dV}{dt} = \frac{\dot{m}_e V_e}{M_o - \dot{m}_e t} - g$$

Now we substitute numerical values for all parameters except the time, *t*, which is the independent variable.

$$\frac{dV}{dt} = \frac{(35 \text{ kg/s})(2{,}500 \text{ m/s})}{5{,}000 \text{ kg} - (35 \text{ kg/s})t} - 9.8 \frac{\text{m}}{\text{s}^2} = \frac{87{,}500 \text{ m/s}^2}{5{,}000 - (35 \text{ s}^{-1})t} - 9.8 \frac{\text{m}}{\text{s}^2}$$

The initial condition is *V*(0) = 0, since the rocket starts from rest. For this problem we will try a value of the time step $\Delta t = 0.1$ s. At least 1000 time steps should be used, and preferably many more. For a total time of 100 s, 1000 time steps results in a time step size of 0.1 s. So the numerical formula to approximate velocity is

$$V(t + \Delta t) = V(t) + \Delta t \frac{dV}{dt}$$

The height as a function of time can similarly be approximated. Knowing that the velocity is the derivative of distance with respect to time and the initial height is $y(0) = 0$, we have

$$y(t + \Delta t) = y(t) + \Delta t \frac{dy}{dt}$$

At time $t = 0$ the initial acceleration is $a = 17.5$ m/s² $- 9.8$ m/s² $= 7.7$ m/s². So the value of velocity at $t = 0.1$ s is

$$V(0.1 \text{ s}) = 0 \frac{m}{s} + (0.1 \text{ s})\left(7.7 \frac{m}{s^2}\right) = 0.77 \frac{m}{s}$$

The altitude obtained is

$$y(0.1 \text{ s}) = 0 \text{ m} + (0.1 \text{ s})\left(0 \frac{m}{s}\right) = 0 \text{ m}$$

We continue marching forward to the next time step, $t = 0.1$ s $+ 0.1$ s $= 0.2$ s:

$$V(0.2 \text{ s}) = 0.77 \frac{m}{s} + (0.1 \text{ s})\left(7.702 \frac{m}{s^2}\right) = 1.5402 \frac{m}{s}$$

$$y(0.2 \text{ s}) = 0 \text{ m} + (0.1 \text{ s})\left(0.77 \frac{m}{s}\right) = 0.077 \text{ m}$$

We repeat this process until the time has been marched forward far enough to cover the period of interest, in this case for 100 s. Table 3.1 shows the first 10 time steps of the process. Note that for numerical solutions to differential equations, you should carry through as many digits as the computer will handle (usually 8 digits for most decimal numbers, or 16 digits using double precision, which is the default in MATLAB).

Table 3.1 Numerical Solution to the Rocket Problem (First 10 Time Steps)

Time (s)	Height (m)	Velocity (m/s)	Acceleration (m/s²)
0	0	0	7.7
0.1	0	0.77	7.712258581
0.2	0.077	1.541225858	7.724534348
0.3	0.231122586	2.313679293	7.736827337
0.4	0.462490515	3.087362027	7.749137585
0.5	0.771226718	3.862275785	7.761465128
0.6	1.157454296	4.638422298	7.773810002
0.7	1.621296526	5.415803298	7.786172244
0.8	2.162876856	6.194420523	7.798551891
0.9	2.782318908	6.974275712	7.810948979
1	3.479746479	7.755370609	7.823363545

Although the data in Table 3.1 is qualitatively reasonable, it turns out that by the end of 100 s the error the numerical solution is 0.2% in the prediction of the height and 0.1% in the prediction of the velocity. If higher accuracy is required, a smaller time step size must be used. If only 100 time steps had been used in this problem—so that $\Delta t = 1.0$ s—the error would have been 2.3% in the prediction of the height and 1.0% in the prediction of the velocity.

What is *minimum velocity* to reach orbit? Low-earth orbits (LEOs) are generally defined to be in the range of 200 to 2000 km in altitude. Below 200 km there is too much aerodynamic resistance—the vehicle will rapidly fall out of orbit. For a typical space shuttle mission at an orbit of 300 km above Earth's surface, the required orbital velocity is the velocity such that the centripetal acceleration, V^2/r, matches the gravitational acceleration, g, where r is the distance from the center of Earth. At an altitude of 300 km (185 miles), the orbital velocity is about 7,700 m/s (17,300 mph). In a geosynchronous orbit, the object orbits at the same rate that Earth spins, so it stays fixed over one point on Earth's surface. The altitude for geosynchronous orbit (GEO) is very high, around 36,000 km, with a correspondingly lower orbital speed of around 3,000 m/s.

The *escape velocity* of a rocket is the minimum speed needed by the projectile after it has burned its fuel to escape from the influence of Earth's gravity and travel onward. The escape velocity is calculated on the basis of comparing the kinetic energy of the rocket to the potential energy it must overcome to leave Earth's gravitational field. Usually around 10 km/s, the actual value depends on the altitude at burnout and whether aerodynamic drag is a significant factor.

Mathematically, and ignoring aerodynamic forces, the escape velocity for a rocket at a starting point on Earth's surface can be calculated by comparing the initial kinetic energy, $\frac{1}{2}mV^2$, to the change in potential energy required to leave Earth:

$$\Delta PE = \int_R^\infty mg\,dz = m\int_R^\infty g_o\left(\frac{R}{z}\right)^2 dz = mg_o R$$

where R is the radius of Earth, z is the elevation measured relative to the center of Earth, and g_o is the value of gravitational acceleration at Earth's surface. Thus a balance of kinetic and potential energies will yield

$$\frac{1}{2}mV^2 = mg_o R$$

The theoretical escape velocity is then calculated by solving the equation for V:

$$V_{escape} = \sqrt{g_o R} \tag{3.26}$$

The value of g_o is 9.8 m/s², and the average value of Earth's radius is about $R = 6400$ km, so the escape velocity can be estimated as

$$V_{escape} = \sqrt{\left(9.8\,\frac{m}{s^2}\right)(6{,}400{,}000\text{ m})} = 11{,}200\,\frac{m}{s}$$

Again, this calculation neglects air resistance or the possibility of the rocket adding additional thrust as it travels. If a rocket has a velocity less than the escape velocity, it will either fall into orbit or fall to Earth's surface.

For a rocket outside of Earth's atmosphere (so that there is no drag), is V_e the limiting velocity for the rocket? No, it is not. V_e is the velocity of the exiting gases *relative to the rocket*, not relative to Earth. As long as it has fuel, the rocket can keep accelerating. (Of course no object can accelerate up to or past the speed of light.)

3.2.6 Torques and Angular Momentum

When the fluid thrust force does not act through the center of gravity of a system, it can generate a *torque* on the system. A torque acts to put a system into rotational motion, unless there is another opposing torque or moment. Torque is usually expressed in units of N-m or ft-lbf, or occasionally in-lbf, and a torque is equivalent to a moment.

The angular momentum of a system must be defined relative to a coordinate origin. Whereas the linear momentum of a rigid object is the mass times the velocity, the angular momentum of a rigid object is the product of the linear momentum times the length of a lever arm from the object to the coordinate origin. In vector terms this is expressed as

$$\vec{H} = \vec{r} \times (m\vec{V}) \tag{3.27}$$

where $\vec{H}$ is the angular momentum vector. For a fluid system defined within a control volume V, the total angular momentum is

$$\vec{H} = \int_V \rho(\vec{r} \times \vec{V}) \tag{3.28}$$

The time rate of change of the angular momentum of a fluid system is equal to the rate of change of fluid angular momentum within the control volume, plus any changes due to convective fluxes of angular momentum into or out of the control volume:

$$\frac{d\vec{H}}{dt} = \frac{\partial}{\partial t}\int_V \rho(\vec{r} \times \vec{V}) + \int_S \rho(\vec{r} \times \vec{V})(\vec{V} \bullet d\vec{A}) \tag{3.29}$$

Just as a force produces a change in linear momentum, a torque produces a change in angular momentum. For a right-hand coordinate system defined about an origin point, O (see Figure 3.16), a general equation for the conservation of angular momentum is

$$\sum \vec{M}_O = \frac{d\vec{H}}{dt} = \frac{\partial}{\partial t}\int_V \rho(\vec{r} \times \vec{V}) + \int_S \rho(\vec{r} \times \vec{V})(\vec{V} \bullet d\vec{A}) \tag{3.30}$$

So the sum of any external moments or torques on the system is equal to the total change in angular momentum.

For the special case of steady-state flow with uniform velocity profiles at inlets and outlets, we have

$$\sum \vec{M}_O = \sum \dot{m}_{\text{out}}(\vec{r} \times \vec{V}) - \sum \dot{m}_{\text{in}}(\vec{r} \times \vec{V}) \tag{3.31}$$

Equation 3.31 is the form of the conservation of angular momentum equation we will attempt to apply to most problems.

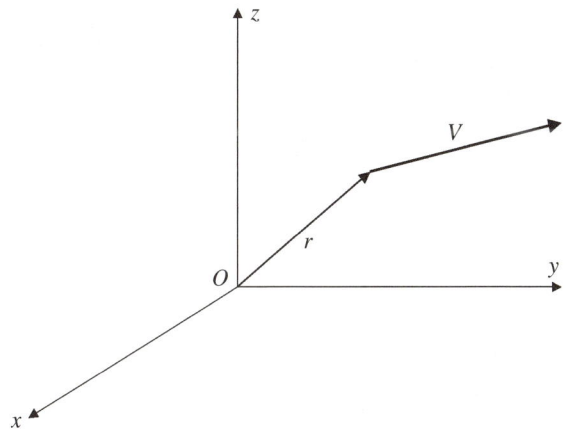

FIGURE 3.16 A right-hand coordinate system for angular momentum.

EXAMPLE 3.16

Crude oil flows through a 2-m length of horizontal pipe of diameter 30 cm, through a 90° bend that directs it downward and out the open end of the pipe and into the reservoir below, as shown in Figure 3.17. The 90° elbow is connected to the straight pipe by a flange. Find all the forces and torques on the flange. The flow rate of oil is 5000 L/min, and the weight of the pipe can be neglected.

SOLUTION: First we convert the flow rate to standard units:

$$Q = \frac{5000 \text{ L}}{\text{min}} \frac{1 \text{ min}}{60 \text{ s}} \frac{1 \text{ m}^3}{1000 \text{ L}} = 0.0833 \frac{\text{m}^3}{\text{s}}$$

The velocity in the pipe is

$$V = \frac{Q}{A} = \frac{0.0833 \text{ m}^3/\text{s}}{\pi (0.15 \text{ m})^2} = 1.18 \frac{\text{m}}{\text{s}}$$

The density of crude oil varies depending on the source. We will use a value of SG = 0.875, representative of crude from the Gulf of Mexico region. The magnitude of the thrust force of the oil moving through a section of the pipe is

$$F = \rho A V^2 = \rho Q V = \left(875 \frac{\text{kg}}{\text{m}^3}\right)\left(0.0833 \frac{\text{m}^3}{\text{s}}\right)\left(1.18 \frac{\text{m}}{\text{s}}\right) = 86.0 \text{ N}$$

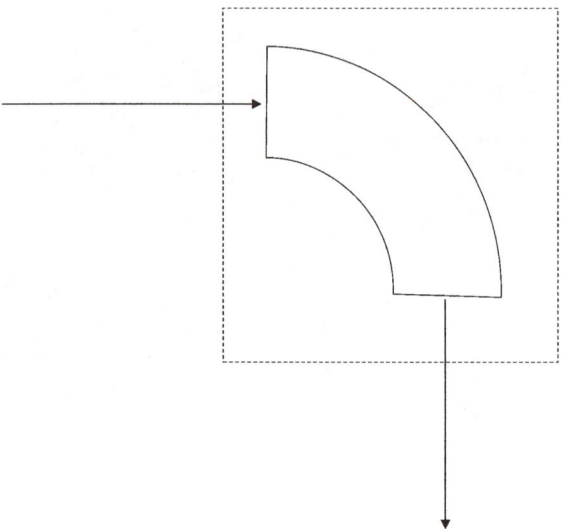

FIGURE 3.17 Sketch of a pipe elbow.

The torque on the flange is equal to the force at the end of the elbow times the perpendicular distance to the flange. Since the angle of the elbow is 90°, $(\vec{r} \times \vec{F}) = rF$. Then,

$$\vec{T} = \vec{r} \times \vec{F} = rF \sin \theta = rF = (2 \text{ m})(86 \text{ N}) = 172 \text{ N} \cdot \text{m}$$

3.2.7 Rotating Machinery

Depending on the problem, it may be more convenient to have the control volume rotate with the device. Consider a simple symmetric two-arm water sprinkler, with 90° bends before the exit, as shown in Figure 3.18.

What prevents the sprinkler from continuously accelerating toward an infinite rotational speed? Obviously the rotational inertia of the system limits the acceleration rate, and the friction will also resist the motion. But even without friction there is a limit to the maximum rotational speed the sprinkler can achieve. The sprinkler can only accelerate as long as a thrust force is being generated. When the sprinkler is moving as fast as the water spraying out of it, no more thrust or acceleration can be achieved. So for a given mass flow rate of water, that produces an exit velocity equal to $V = \dot{m}/(\rho A)$. The relative velocity of the water spray relative to the moving control volume is

$$V_{\text{rel}} = V - R\omega \tag{3.32}$$

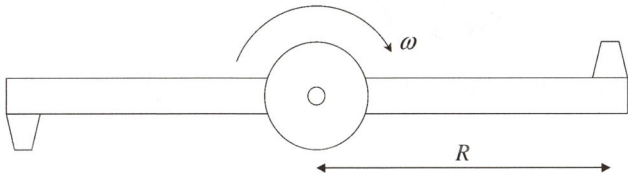

FIGURE 3.18 Schematic of a simple two-arm water sprinkler.

This relative velocity is defined relative to the control volume, which moves with the sprinkler arm. So the maximum possible rotational speed for a frictionless sprinkler is

$$\omega_{max} = V/R \tag{3.33}$$

If the rotational speed and water velocity are measured, the frictional torque on the sprinkler can be deduced from

$$T = \dot{m}RV_{rel} \tag{3.34}$$

Conversely, if the torque and flow rate are known, the predicted rotational speed will be

$$\omega = \frac{V}{R} - \frac{T}{\dot{m}R^2} \tag{3.35}$$

Also note the power associated with a torque is

$$\dot{W} = T\omega \tag{3.36}$$

It is common engineering practice in situations involving rotating shafts to write the equation for shaft power as

$$\dot{W} = 2\pi NT \tag{3.37}$$

where the factor of 2π is included to account for the conversion of revolutions to radians. Note that since rotational speed is usually expressed in rpm, the cycle time in minutes must be converted to seconds to put the power in standard units such as watts. Systems where angular momentum is important include turbomachinery such as compressors and turbines, water sprinklers, dishwasher arms, automotive torque converters, and so on.

EXAMPLE 3.17

A water sprinkler sends 2 gal/min of water through a pipe with $\frac{1}{4}$-in. inner diameter along a 1-ft length from the center hub, and then through a 90° bend, before it sprays the water out, as shown in Figure 3.18. If the sprinkler is held in place so that it does not rotate, what is the torque on the fixture?

SOLUTION: Since there are two sprinkler arms, the flow rate through each arm is 1 gal/min. We convert the flow rate and area to fundamental units so that the outlet velocity can be calculated:

$$Q = \left(1 \frac{\text{gal}}{\text{min}}\right)\left(\frac{1 \text{ min}}{60 \text{ s}}\right)\left(\frac{0.134 \text{ ft}^3}{1 \text{ gal}}\right) = 2.23 \times 10^{-3} \frac{\text{ft}^3}{\text{s}}$$

$$A = \frac{\pi}{4}D^2 = \frac{\pi}{4}\left(0.25 \text{ in } \frac{1 \text{ ft}}{12 \text{ in}}\right)^2 = 3.41 \times 10^{-4} \text{ ft}^2$$

$$V = \frac{Q}{A} = \frac{2.33 \times 10^{-3} \text{ ft}^3/\text{s}}{3.41 \times 10^{-4} \text{ ft}^2} = 6.83 \frac{\text{ft}}{\text{s}}$$

The density of water in standard English units is 1.94 slug/ft^3. 1 lbf = 1 slug-ft/s^2. The torque can be calculated as

$$T = \dot{m}RV = \rho QRV = \left(1.94 \frac{\text{slug}}{\text{ft}^3}\right)\left(2.33 \times 10^{-3} \frac{\text{ft}^3}{\text{s}}\right)(1 \text{ ft})\left(6.83 \frac{\text{ft}}{\text{s}}\right) = 0.031 \text{ ft} \cdot \text{lbf}$$

Since there are two sprinkler arms, the total torque is $2 \times 0.031 = 0.062$ ft-lbf.

EXAMPLE 3.18

The same water sprinkler as in Example 3.17 is now released so that it is free to rotate. What is the maximum rotational speed the sprinkler arm can obtain if friction between it and the support is negligible?

SOLUTION: If there is no torque, all of the angular momentum of the fluid goes into rotating the sprinkler. Once the sprinkler is turned on, it will continue to accelerate until the water can no longer push it faster. In the absence of friction and other resisting forces, this will occur when the sprinkler outlet on the end of the arm moves at the same translational velocity as the water running through it:

$$\omega_{\text{max}} = \frac{V}{R} = \frac{6.83 \text{ ft/s}}{1.0 \text{ ft}} = 6.83 \frac{\text{rad}}{\text{s}} \times \frac{1 \text{ rev}}{2\pi \text{ rad}} \times \frac{60 \text{ s}}{1 \text{ min}} = 65 \text{ rpm}$$

FIGURE 3.19 Picture of a Pelton waterwheel. Courtesy of Voith.

> If the sprinkler arm rotated any faster than this, it would be dragging the water along with it, rather than being propelled forward by the exiting thrust. Of course the actual rotational speed of the sprinkler will be less than this value, since there is friction between the sprinkler arm and its support. But this calculation provides an upper limit on the maximum possible rotational speed.

Another type of rotating machinery is the waterwheel. Lester Pelton reported in the 1870s that the waterwheels he tested were only about 40% efficient. In seeking to improve their design, he found that a bucket that deflected all the water to one side was more efficient than one that deflected it to both sides. To make the arrangement symmetric and eliminate any possible side forces, he put two buckets side by side to split the flow equally, as shown in Figure 3.19. Thus the Pelton waterwheel was born and was much superior to the previous technology. Its design is very similar to modern water turbines.

The power generated by a Pelton waterwheel is given by

$$\dot{W} = \rho \omega R Q (V_{jet} - \omega R)(1 - \cos \beta) \tag{3.38}$$

where V_{jet} is the velocity of the jet striking the wheel. Thus $(V_{jet} - \omega R)$ is the *relative velocity* of water hitting the wheel.

> **EXAMPLE 3.19**
>
> Find the maximum power that can be extracted from a Pelton waterwheel. Will the maximum power occur at the maximum torque or at the maximum rotational speed?
>
> SOLUTION: Assuming that the rotational speed of the waterwheel can be varied by the designer if not the user, the maximum power can be found by taking the derivative of the power (Equation 3.38) with respect to the rotational speed and setting it equal to zero. The derivative is
>
> $$\frac{d\dot{W}}{d\omega} = \rho R Q(1 - \cos \beta) \cdot [V_{jet} - 2\omega R]$$
>
> Setting the derivative equal to zero and solving for ω yields
>
> $$\omega_{MaxPower} = \frac{V_{jet}}{2R}$$
>
> Substituting this value for ω back into the equation for power, we find the maximum power:
>
> $$\dot{W} = \rho \omega R Q \left[V_{jet} - \left(\frac{V_{jet}}{2R}\right) R \right](1 - \cos \beta) = \frac{1}{2}\rho \omega R Q V_{jet}(1 - \cos \beta)$$
>
> So the maximum power occurs at half the maximum rotational speed.

A more detailed analysis of angular momentum for more complex turbomachinery such as pumps, compressors, and turbines is presented in Chapter 7.

3.3 Conservation of Energy

The **first law of thermodynamics**, or the conservation of energy principle, is one of the most fundamental principles of engineering. It states that energy cannot be created or destroyed, it can only be transferred from one form to another. The most general form of the conservation of energy equation for a control volume applied to an open system is

$$\frac{dE}{dt} = \pm \dot{Q} \pm \dot{W} + \sum_{in} \dot{m}\left(h + \frac{V^2}{2} + gz\right) - \sum_{out} \dot{m}\left(h + \frac{V^2}{2} + gz\right) \qquad (3.39)$$

where *h* is the *enthalpy*, which is defined as the sum of internal energy and specific flow work, so that $h = u + P/\rho$, where u is the specific internal energy of the fluid in units of kJ/kg. The internal energy is the thermal energy associated with the random motion of the molecules that make up the fluid. In traditional thermodynamics notation, $\dot{Q}$ represents a transfer of heat into or from the system, and $\dot{W}$ represents mechanical work. E is the total energy of the fluid in the control volume and can include thermal energy, kinetic energy, and potential energy.

3.3.1 Methods of Extracting or Adding Energy to a Fluid

For a stationary gas, energy can be added to the gas by compressing it. Energy can also be added to a fluid by adding heat to it—such as in a heat exchanger—or via chemical reaction—such as combustion of a fuel in the presence of an oxidized gas (usually air) in an engine. For moving liquids, which are essentially incompressible, engineers have created many devices to add energy to those fluids. The most common examples of mechanical devices that add energy to fluids are pumps and compressors. Both perform essentially the same function, but when the working fluid is a liquid the term *pump* is used, and when the working fluid is a gas the term *compressor* is used. Most common mechanical devices that extract work from a fluid are called *turbines*, regardless of whether the fluid is liquid or gas. There are a variety of turbine designs, which will be discussed in more detail in Chapter 7.

3.3.2 Friction Losses

In this chapter the effects of friction on fluid motion have been ignored. In reality, we know that for the steady flow of fluid through a pipe, friction between the fluid and the walls of the pipes resists the fluid motion, resulting in a loss of pressure. Methods of calculating friction losses for internal flows in pipes will be presented in Chapter 5, and the friction loss for external flows over vehicles, termed *drag*, will be discussed in Chapter 6.

3.3.3 Power

How much power is required to move a fluid? In Section 3.2 we introduced the *force* required to move a fluid, but how does the force relate to the power needed to move or pressurize a fluid? Since the fluid thrust force is proportional to $\rho A V^2$, and the power is equal to the force times velocity, it stands to reason that the power required to accelerate a fluid, as in a jet engine, will be proportional to $\rho A V^3$. More exactly, from the conservation of energy equation, the power associated with a change of velocity across a control volume is

$$\dot{W} = \dot{m}\left(\frac{V_2^2}{2} - \frac{V_1^2}{2}\right) = \frac{1}{2}\rho_2 A_2 V_2(V_2^2 - V_1^2)$$

So, for example, if a fan sucks air in through a large opening and then pushes it out through a small nozzle so that V_1 is small compared to V_2, then the power required by the fan is approximately

$$\dot{W} = \frac{1}{2}\rho_2 A_2 V_2^3$$

The power required to pressurize a liquid can be calculated from the conservation of energy equation as

$$\dot{W} = \dot{m}\left(\frac{P_2}{\rho} - \frac{P_1}{\rho}\right)$$

Since liquids are usually incompressible, the density is constant, and the previous equation can be written as

$$\dot{W} = Q\Delta P \tag{3.40}$$

EXAMPLE 3.20

The turbine in a hydroelectric power plant is 20 m below the top of the water in the reservoir behind the dam. If the pressure at the turbine inlet is 95 kPa, gage; the diameter of the feed pipe into the turbine is 2.0 m; the water velocity going into the turbine is 14 m/s; and there are eight identical turbines in the power plant, how much power could the plant generate?

SOLUTION: The volume flow rate of the water is

$$Q = AV = 8\left(\frac{\pi}{4}(2.0\text{ m})^2\right)\left(14\,\frac{\text{m}}{\text{s}}\right) = 352\,\frac{\text{m}^3}{\text{s}}$$

So the power is

$$\dot{W} = Q\Delta P = \left(352\,\frac{\text{m}^3}{\text{s}}\right)\left(95{,}000\,\frac{\text{N}}{\text{m}^2}\right) = 33.4\text{ MW}$$

This is the maximum power that could be generated if the turbine were 100% efficient. Of course the actual power will be less.

3.4 The Bernoulli Equation

The Bernoulli equation is a simplified form of the energy equation. The assumptions used to derive the Bernoulli equation are

- Steady state
- Frictionless flow
- No work
- No heat transfer
- Single inlet, single outlet

3.4.1 Derivation of the Bernoulli Equation

Starting with the first law of thermodynamics (Equation 3.39) and setting dE/dt, $\dot{Q}$, and $\dot{W}$ equal to zero gives

$$0 = \sum_{in} \dot{m}\left(h + \frac{V^2}{2} + gz\right) - \sum_{out} \dot{m}\left(h + \frac{V^2}{2} + gz\right) \tag{3.41}$$

Furthermore, with the assumption of a single inlet and a single outlet, applying the conservation of mass equation shows that $\dot{m}_{in} = \dot{m}_{out}$, so the mass flow rate can be divided out of the equation:

$$\left(h + \frac{V^2}{2} + gz\right)_{in} = \left(h + \frac{V^2}{2} + gz\right)_{out} \tag{3.42}$$

At this point it is convenient to expand the enthalpy, $h = u + P/\rho$. For most fluids the internal energy is a function of temperature only to a good approximation. If there is no work, heat transfer, or friction, it is reasonable to assume that the temperature is constant from inlet to outlet. So the internal energy is constant and can be canceled from both sides of the equation, yielding

$$\left(\frac{P}{\rho} + \frac{V^2}{2} + gz\right)_{in} = \left(\frac{P}{\rho} + \frac{V^2}{2} + gz\right)_{out} \tag{3.43}$$

Rather than labeling the two points in the flow as inlet and outlet, they could be labeled more generally as points 1 and 2.

$$\frac{P_1}{\rho} + \frac{V_1^2}{2} + gz_1 = \frac{P_2}{\rho} + \frac{V_2^2}{2} + gz_2 \tag{3.44}$$

Alternatively, the Bernoulli equation can be multiplied through by the density, ρ, to put all the terms in units of pressures:

$$P_1 + \frac{1}{2}\rho V_1^2 + \rho g z_1 = P_2 + \frac{1}{2}\rho V_2^2 + \rho g z_2 \tag{3.45}$$

Civil engineers like to express the Bernoulli equation in units of length or elevation:

$$\frac{P_1}{\rho g} + \frac{V_1^2}{2g} + z_1 = \frac{P_2}{\rho g} + \frac{V_2^2}{2g} + z_2 \tag{3.46}$$

In civil engineering terminology, these terms in units of length are called "heads" and represent the energy stored in each different term, as in the common phrase "build up a head of steam." The height above a datum, z, is the *elevation head* and represents the potential energy of the fluid. The term $P/\rho g$ is the *pressure head*, and $V^2/2g$ is the *velocity head*. The Bernoulli equation can also be modified to include work terms from the original energy equation, from which the Bernoulli equation is derived. For work added to the fluid, as with a pump, the *pump head* is defined as

$$h_{\text{pump}} = \frac{\dot{W}_{\text{pump}}}{\dot{m}g} \tag{3.47}$$

where $\dot{W}$ is the actual power input to the pump in units of W or kW. Dividing by the mass flow rate gives the work per unit mass of fluid. Then further dividing by g puts the pump work term in units of length, consistent with the other terms in Equation 3.46. Similarly, the work extracted from the fluid through a device such as a turbine is called the *turbine head*, defined as

$$h_{\text{turbine}} = \frac{\dot{W}_{\text{turbine}}}{\dot{m}g} \tag{3.48}$$

The head loss due to friction, h_f, can also be defined in a similar manner. Calculating the head loss will be covered in Chapter 5.

All the various forms of the Bernoulli equation are equivalent, and any of them may be used. The Bernoulli equation is a useful approximation to the energy equation in many cases, especially cases in which the effects of friction are small. Even when it is not quantitatively accurate, it is often qualitatively accurate enough to reveal important trends. Some examples illustrating the use of the Bernoulli equation follow.

EXAMPLE 3.21

Although Venturi is credited with inventing the Venturi tube, Hershel found an application for it. He discovered that the flow rate of fluid through a Venturi tube could be related to the pressure drop measured from the main section of pipe to the narrowest part of the Venturi. A picture of a basic Venturi flow meter is shown in Figure 3.20. This particular flow meter has a diameter ratio of 2, with the diameter of the inlet pipe denoted by D_1 and the diameter at the throat of the Venturi by D_2. Derive an equation for the volume flow rate, Q, in terms of pressure drop and any other relevant parameters.

SOLUTION: Conservation of mass can be applied between points 1 (inlet) and 2 (outlet) in the Venturi tube.

$$\rho_1 A_1 V_1 = \rho_2 A_2 V_2$$

If we assume that the density is constant (the flow is incompressible), then the density can be canceled out, leaving us with

$$A_1 V_1 = A_2 V_2$$

Since $D_1 = 2D_2$, then $A_1 = 4A_2$, and from conservation of mass $V_2 = 4V_1$. The flow rate is $Q = A_1 V_1$. Next velocity needs to be related to the pressure drop. This can be done through the Bernoulli equation:

$$P_1 + \frac{1}{2}\rho V_1^2 + \rho g z_1 = P_2 + \frac{1}{2}\rho V_2^2 + \rho g z_2$$

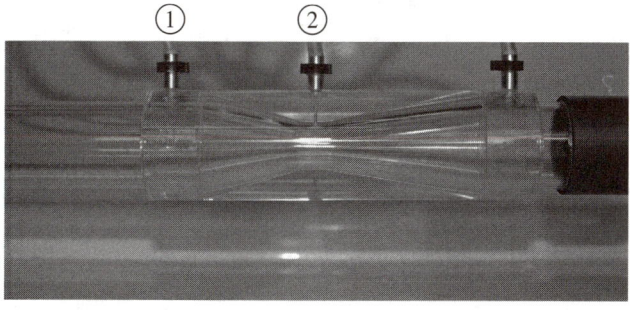

FIGURE 3.20 Picture of a Venturi flowmeter. Courtesy of Armfield.

Since the Venturi is horizontal, $z_1 = z_2$. Now we rearrange the Bernoulli equation to get all the pressure terms on the left-hand side and all the kinetic energy terms on the right-hand side:

$$P_1 - P_2 = \frac{1}{2}\rho(V_2^2 - V_1^2)$$

The measured pressure drop is $\Delta P = P_1 - P_2$. We can also substitute $V_2 = 4V_1$.

$$\Delta P = \frac{1}{2}\rho(16V_1^2 - V_1^2) = \frac{15}{2}\rho V_1^2$$

To introduce the flow rate, we substitute $V_1 = Q/A_1$ and solve the previous equation for Q:

$$Q = A_1\sqrt{\frac{2}{15}\frac{\Delta P}{\rho}}$$

This shows that the flow rate depends on the measured pressure drop, the density of the fluid, and the cross-sectional area of the pipe. If air flows through the Venturi tube, the pressure drop could be measured with a U-tube manometer. Then pressure drop is related to the measured difference in height as

$$\Delta P = \rho_m g\, \Delta h$$

where ρ_m is the density of the fluid in the manometer, and the flow rate is:

$$Q = A_1\sqrt{\frac{2}{15}\frac{\rho_m}{\rho}g\, \Delta h}$$

EXAMPLE 3.22

Consider water flowing through a horizontal converging–diverging nozzle at a rate of 0.0027 m³/s. The diameter of the throat is $\frac{1}{3}$ the size of the diameter of the inlet pipe, and the diameter of the inlet pipe is 8 cm. Assuming frictionless flow, use the Bernoulli equation to estimate the pressure in the throat and the exit, if the inlet pressure is 50 kPa, gage.

SOLUTION: The Bernoulli equation is

$$\frac{P_1}{\rho g} + \frac{V_1^2}{2g} + z_1 = \frac{P_2}{\rho g} + \frac{V_2^2}{2g} + z_2$$

Since the pipe is horizontal, $z_1 = z_2$. We rearrange the equation to solve for the pressure at the throat, P_2:

$$P_2 = P_1 + 1/2\,\rho(V_1^2 - V_2^2)$$

To find the value of the pressure, P_2, we must know the velocities V_1 and V_2. The velocity at the inlet is equal to the flow rate divided by the area, $V_1 = 0.0027/[\pi(0.04\text{ m})^2] = 0.537$ m/s. By conservation of mass, the velocity through the throat must be equal to $3^2 = 9$ times the velocity at the inlet, or 4.83 m/s. The change in pressure can be calculated by inserting these values:

$$P_2 = 50{,}000\text{ Pa} + 0.5\left(1{,}000\,\frac{\text{kg}}{\text{m}^3}\right)\left[\left(0.537\,\frac{\text{m}}{\text{s}}\right)^2 - \left(4.83\,\frac{\text{m}}{\text{s}}\right)^2\right] = 38{,}480\text{ Pa}$$

Note that it is possible for the pressure in the throat to fall below atmospheric pressure, so that it has a negative gage pressure (though the absolute pressure can never be below zero). Also note that the inlet pressure is completely recovered at the exit. In reality, there will be flow losses in the system and the exit pressure will be slightly less than the inlet pressure. This type of flow configuration is known as a Venturi. For a well-designed Venturi, the pressure at the outlet is almost exactly equal to the pressure at the inlet.

Venturis are popular as flowmeters because of their relatively low total pressure loss. Their use is discussed in Chapter 9. Other types of flowmeters are also discussed in Chapter 9, along with the relative advantages and disadvantages of each. Consult a reference such as Blevin's *Applied Fluid Dynamics Handbook* (listed in Appendix E) for further information on the different types of Venturi design and performance.

EXAMPLE 3.23

Water is held in the large container shown in Figure 3.21 and flows out through the small short pipe at the bottom. Use the Bernoulli equation to estimate the velocity of the water exiting the pipe.

SOLUTION: Draw a streamline from a point on the surface of the tank (label it point 1) to the exit of the pipe (label that point 2). Assume that there is negligible friction between the two points so that the Bernoulli equation can be applied. Further assume that the water level in the tank is constant or nearly constant,

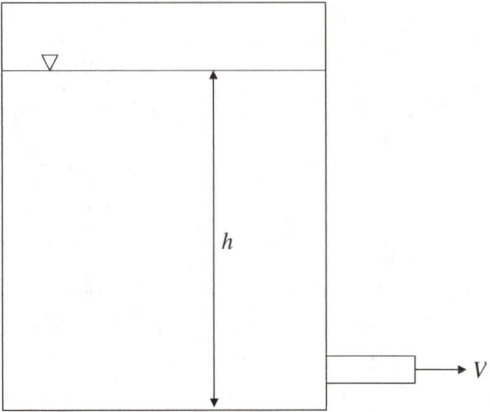

FIGURE 3.21 Sketch of a water container with a small exit pipe.

so that the velocity at point 1 is zero. The pressure at point 1 will be equal to atmospheric pressure since it is at a flat liquid–air interface. The use of gage pressures is convenient in these problems, since then $P_1 = 0$ (gage). The pressure at the pipe exit can also be assumed to be equal to atmospheric pressure as a result of the pressure-jet boundary condition. As long as the velocity at the outlet of a pipe or orifice into a jet is sufficiently subsonic, it can be assumed that the surrounding atmospheric pressure is quickly felt at the jet exit and the out-flowing fluid very quickly reaches the surrounding atmospheric pressure level in the jet. The criterion for this is that the Mach number is less than about 0.3. The Mach number is the ratio of the velocity to the speed of sound, and pressure waves travel at the speed of sound. The speed of sound in water is very high, so it is always safe to use the pressure-jet boundary condition for water jets. So then for this problem $P_2 = 0$ (gage). The remaining nonzero terms in the Bernoulli equation are

$$z_1 = \frac{V_2^2}{2g} + z_2$$

The change in height between the two points is $h = z_1 - z_2$, so the exit velocity is

$$V_2 = \sqrt{2gh}$$

The Bernoulli equation is useful in these cases because it tells us what the maximum velocity could be if the friction losses were negligible.

EXAMPLE 3.24

Suppose you turn on the faucet at your kitchen sink to a moderate flow rate—not so low that the water dribbles, and not so high that the water atomizes into small drops, but at an intermediate flow rate in which you get a nice smooth, continuous liquid surface. (See Figure 3.22.) You will notice that the diameter of the liquid stream gets smaller the farther away from the faucet it gets, until it hits the sink. Explain this behavior.

SOLUTION: Once you turn the faucet on and leave it at a given setting, it only takes a few seconds for the flow to reach steady state, so the steady-state conservation of mass equation (Equation 3.9) can be applied between point 1 at the faucet exit and point 2 some distance below. The density of the water is constant, so it can be divided out of both sides of the equation, leaving $A_1 V_1 = A_2 V_2$. The Bernoulli equation applied between the points 1 and 2 is

$$P_1 + \frac{1}{2}\rho V_1^2 + \rho g z_1 = P_2 + \frac{1}{2}\rho V_2^2 + \rho g z_2$$

The pressure will be a constant, equal to atmospheric pressure throughout the free stream. Since water is incompressible, the density will also be constant. Thus,

$$\frac{1}{2}V_1^2 + g z_1 = \frac{1}{2}V_2^2 + g z_2$$

So as the stream of water flows downward, it loses potential energy, which is converted to kinetic energy. Another way to think of it is that the force due to gravity causes the fluid velocity to accelerate. Conservation of mass in a steady-

FIGURE 3.22 Picture of a smooth, continuous water stream from a kitchen faucet.

state flow of an incompressible fluid requires the product of area and velocity to be a constant. Since the velocity increases, the area must decrease as the water travels farther from the faucet outlet.

EXAMPLE 3.25

To go around an obstacle, a piping system is constructed from two identical long straight pipes of constant diameter, as shown in Figure 3.23. If the friction losses within the pipe can be ignored, find expressions for the pressure at the far end of the pipe system (point 3) and at the location of the highest elevation (point 2).

SOLUTION: Assume that the flow through the pipes is steady-state flow and the fluid is incompressible. Since the diameter of the pipes is constant and the density is also constant, then the velocity must be constant at each location in the pipe system. If there are no friction losses, the Bernoulli equation can be employed. The Bernoulli equation between points 1 and 3 is

$$P_1 + \frac{1}{2}\rho V_1^2 + \rho g z_1 = P_3 + \frac{1}{2}\rho V_3^2 + \rho g z_3$$

From the geometry of the problem $z_3 = z_1$, and we have already verified that $V_3 = V_1$. Therefore, $P_3 = P_1$. All the pressure energy that was lost to potential energy to raise the fluid to higher elevation is regained when the fluid comes back down to its original elevation. (Of course in reality there will be some pressure loss to friction between the fluid and the pipes.) Now, to find the pressure at the apex, we again recognize that $V_2 = V_1$. Thus the kinetic energy terms will fall out of the Bernoulli equation, which can be rearranged to solve for the pressure at the peak, P_2:

$$P_2 = P_1 + \rho g(z_1 - z_2)$$

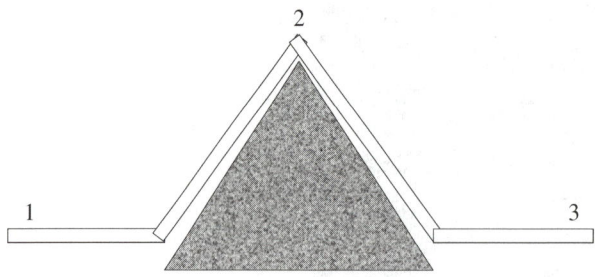

FIGURE 3.23 Side view of a piping system constructed around an obstacle.

3.4 The Bernoulli Equation

EXAMPLE 3.26

For the situation described in Example 3.25, what is the minimum pressure that would have to be supplied to water flowing through point 1 if the height of the obstacle at point 2 is 100 m?

SOLUTION: The pressure can never fall below absolute 0 for any fluid, so setting $P_2 = 0$ and solving for P_1 gives

$$P_1 = \rho g (z_2 - z_1) = (1{,}000 \text{ kg/m}^3)(9.8 \text{ m/s}^2)(100 \text{ m}) = 980{,}000 \text{ Pa (absolute)}$$

For a gas this might be an acceptable answer, but for liquids there is another factor that must be considered. If the pressure in a liquid falls below the vapor pressure, then the liquid will boil and form a gas. If a gas pocket forms that completely fills the cross section of the pipe at a point, this leads to *vapor lock*, and the fluid will cease moving. For water at a temperature of 20°C, the vapor pressure is 2,339 Pa, and so the pressure at point 1 would then have to be at least 982,339 Pa.

One further consideration: it is usually not a good idea to let the pressure in a pipe fall below atmospheric pressure (0 Pa, gage). If it does, a leak in the pipe will cause flow to be drawn inward from the surrounding atmosphere, and the flow within the pipe could become contaminated with air bubbles or other matter brought in with the air.

EXAMPLE 3.27

Determine the minimum pressure required in the pressurized reservoir of a Super-soaker™ water gun to have a firing range of 10 m.

SOLUTION: The Super-soaker water gun, first introduced in 1989, was a considerable improvement over traditional water guns. Once primed, water flows out through the nozzle of a traditional water gun when the depressed trigger forces the piston into the firing chamber. By conservation of mass for an incompressible fluid, the velocity of the exiting spray is

$$V_{spray} = V_{trigger} \frac{A_{piston}}{A_{nozzle}}$$

Thus the maximum firing velocity, and hence the range of the water gun, is limited by the maximum velocity at which the user's finger can push the trigger. And the volume of water in one squirt is limited by the stroke length of the trigger and the area of the piston. But if the piston is made too large, little children are not able to generate enough force to push the trigger.

The improved design of the Super-soaker works because air is a compressible substance but water is not. When the pump is compressed, it pushes water into the pressure chamber, which compresses the air. As the volume of air is reduced, the pressure rises according to the ideal gas law. The water is in equilibrium with the air, so its pressure also rises. When the pump is expanded, it draws water from the water reservoir into the holding chamber for the next stroke. In this design, all the trigger does is hold back the pressurized water in the pressure chamber. When the trigger is pressed, the passageway is opened, allowing water to flow freely through the nozzle.

We can apply the Bernoulli equation between the pressure chamber (point 1) and the nozzle exit (point 2).

$$V_2 = \sqrt{\frac{2P_1}{\rho}}$$

To relate firing distance (range) to the initial spray velocity, we have to make an assumption about the orientation of the gun when it is fired. If the nozzle of the gun is held 1 m above the ground and titled upward at a 45° angle, then (if air resistance is ignored) we can calculate a trajectory. The horizontal distance covered, x, depends on the initial horizontal velocity and the time the spray is in the air:

$$x = V_{o,x}t = (V_{spray} \cos 45°)t$$

We compute the time the water is in the air based on how long it takes the water to fall due to gravity:

$$z = -V_{o,z}t + \frac{1}{2}gt^2 = -(V_{spray} \sin 45°)t + \frac{1}{2}gt^2$$

Setting $z = 1$ m and $x = 10$ m, we solve the two previous equations simultaneously to find t, the time a particle of water spends in the air, and V_{spray}, the minimum velocity required to cover a range of 10 m. The unique solution is $t = 1.50$ s and $V_{spray} = 9.44$ m/s. With the velocity known, we now calculate the required pressure in the gun using the Bernoulli equation:

$$P = \frac{1}{2}\rho V^2 = 0.5\left(1,000 \frac{kg}{m^3}\right)\left(9.44 \frac{m}{s}\right)^2 = 44,600 \text{ Pa, gage}$$

So the required pressure is nearly half an atmosphere.

3.4.2 Dynamic and Stagnation Pressures

The pressure on the front of an object rises as that object moves through a fluid. This pressure is called the *stagnation pressure* and is equal to the pressure at the *stagnation point*. The stagnation pressure is the pressure obtained when all the kinetic energy of the fluid is transformed into pressure energy. From the Bernoulli equation it can be shown that the stagnation pressure, P_{stag}, is

$$P_{stag} = P + \frac{1}{2}\rho V^2 \qquad (3.49)$$

The difference between the stagnation pressure, P_{stag}, and the static pressure, P, is called the *dynamic pressure* and is equal to $\frac{1}{2}\rho V^2$. It is a historical convention to denote the dynamic pressure by the symbol q, so that $q = \frac{1}{2}\rho V^2$. In NASA terminology, the phrase "max q" refers to the point in a space shuttle's ascent when it reaches the maximum dynamic pressure and the aerodynamic loading on the structure is the highest. Although the space shuttle continues to accelerate until it reaches orbit, so that the velocity continues to increase from 0 at liftoff, the density of the atmosphere decreases with increasing elevation, with the result that there is an altitude at which max q is reached, and above that the atmosphere is so sparse that the dynamic pressure decreases even though the space shuttle is accelerating.

Henri Pitot invented a device to measure the dynamic pressure, which can then be used to calculate the velocity in a fluid. The Pitot tube is an ingenious device that combines two tubes, one of which is used to measure the stagnation pressure and the other to measure the static pressure. By connecting a U-tube manometer between the two tubes, the difference in pressure—the dynamic pressure—can be calculated:

$$\Delta P = \frac{1}{2}\rho V^2 \qquad (3.50)$$

If the density of the fluid is known, then the velocity can be calculated. Figure 3.24 shows a research vehicle with a Pitot tube on the front of it to measure velocity. Pitot tubes are still in wide use today on airplanes to measure the relative wind, which is the relative velocity between the airplane and the wind it flies through. Athough GPS can be used to calculate the velocity of the airplane relative to the surface of Earth, the velocity relative to the *air* is needed for calculating the aerodynamic force on the plane. This information is then used to set the responsiveness of the controls and calculate fuel consumption.

FIGURE 3.24 Picture of a NASA ground research vehicle (GRV). The large flat region in front gives rise to a large region of stagnation pressure. Also note the Pitot tube on the front of the vehicle. Courtesy of NASA Dryden Flight Research Center (NASA-DFRC).

EXAMPLE 3.28

A Pitot tube placed in an air duct measures a pressure difference of 245 Pa. The static pressure is also measured to be 110 kPa, absolute, and the temperature is 21°C. Calculate the velocity.

SOLUTION: First, we calculate the density of the air using the ideal gas law:

$$\rho = \frac{PM}{RT} = \frac{(110{,}000 \text{ Pa})(29 \text{ kg/kmol})}{(8{,}314 \frac{\text{J}}{\text{kmol} \cdot \text{K}})(294 \text{ K})} = 1.31 \frac{\text{kg}}{\text{m}^3}$$

Now with the density known, the Pitot tube equation (Equation 3.50), rearranged to solve for the velocity in terms of pressure drop, can be used:

$$V = \sqrt{\frac{2 \, \Delta P}{\rho}}$$

Now we insert all the known values and solve:

$$V = \sqrt{\frac{2(245 \text{ Pa})}{1.31 \frac{\text{kg}}{\text{m}^3}}} = 19.3 \, \frac{\text{m}}{\text{s}}$$

As we have mentioned, you should always verify that the units in your calculations work out correctly. If you have the wrong units, you probably have the wrong number as well. Here, 1 Pa = 1 N/m² = 1 (kg m/s²)/m² = 1 kg/(m s²). So a Pascal divided by kg/m³ will result in units of m²/s², and when the square root is taken the proper units for velocity of m/s will emerge. Note that this would not have been true if other units for pressure, such as kPa or bars, had been used.

EXAMPLE 3.29

Reconsider the tank-filling problem of Example 3.5. Suppose a small circular hole of diameter $d_o = \frac{1}{2}$ in. has been created by a puncture of the bottom of the tank, as shown in Figure 3.25. Determine whether the tank can be filled to the top with this hole present. If so, then determine the time to fill the tank. If not, find the maximum height to which the tank can be filled.

SOLUTION: Conservation of mass can obviously be applied to this problem to yield

$$\frac{dm_{CV}}{dt} = \dot{m}_{in} - \dot{m}_{out}$$

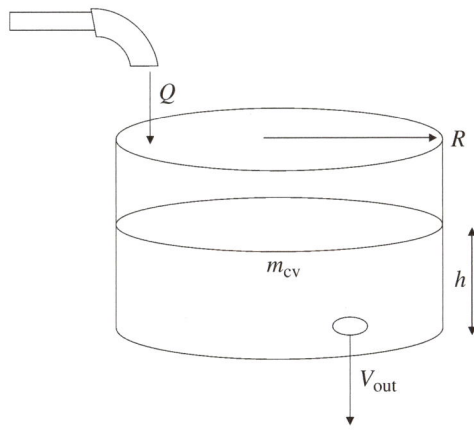

FIGURE 3.25 Sketch of a water tank with a hole in the bottom.

Now the problem is that the outlet mass flow rate is not known. The area of the hole and the density of the water exiting the hole are known, so if we could find the velocity then the mass flow rate exiting the tank would be known. Let us apply the Bernoulli equation in an attempt to calculate the velocity. The outlet is at the hole itself, so that can be our point 2, but where should point 1 be located? A convenient place is at the top of the fluid level of the tank. This is a convenient location because we know that the pressure at the liquid–air interface is equal to atmospheric, and the velocity at that point should be reasonably small enough that it can be neglected. So the general Bernoulli equation is

$$\frac{P_1}{\rho g} + \frac{V_1^2}{2g} + z_1 = \frac{P_2}{\rho g} + \frac{V_2^2}{2g} + z_2$$

The simplifications will be that $P_1 = 0$ (gage pressure) and $V_1 = 0$ (approximately). Further, the pressure at the hole exit should also be equal to atmospheric, so $P_2 = 0$ (gage). Thus the Bernoulli equation simplifies to

$$z_1 = \frac{V_2^2}{2g} + z_2$$

Solving for the exit velocity, V_2, we have

$$V_2 = \sqrt{2g(z_1 - z_2)} = \sqrt{2gh(t)}$$

So the Bernoulli equation tells us that the velocity will vary as the depth of the water in the tank varies. We can now calculate the mass flow rate through the exit:

$$\dot{m}_{out} = \rho_2 A_2 V_2 = \rho \frac{\pi}{4} d_o^2 \sqrt{2gh}$$

Substituting into the conservation of mass equation yields

$$\frac{d}{dt}\left(\rho \frac{\pi}{4} D^2 h\right) = \dot{m}_{in} - \rho \frac{\pi}{4} d_o^2 \sqrt{2gh}$$

The density and the diameter of the tank are constants, so they can be removed from the derivative on the left-hand side and divided out through the equation:

$$\frac{dh}{dt} = \frac{\dot{m}_{in}}{\frac{\pi}{4}\rho D^2} - \left(\frac{d_o}{D}\right)^2 \sqrt{2gh}$$

This is a nonlinear differential equation that is rather difficult to solve. Separation of variables will not work on a differential equation of this type. The best course of action is to get an approximate answer using numerical integration, as described in Algorithm 3.1. This numerical method will create an array of values for h at different times t. To start the algorithm, we need the initial condition, which was given as $h(t = 0) = 0$. Then we will calculate each subsequent value of h using the formula

$$h(t + \Delta t) = h(t) + \Delta t \frac{dh}{dt}$$

Now we need the equation for dh/dt expressed as a function only of h and t, so we substitute numerical values for all other parameters.

$$\frac{dh}{dt} = \frac{0.134 \text{ ft}^3/\text{min}}{(\pi/4) 1 \text{ ft}^2} - \left(\frac{0.5 \text{ in}}{12 \text{ in}}\right)^2 \sqrt{2\left(32.2 \frac{\text{ft}}{\text{s}^2}\right)h}$$

$$= 0.00285 \frac{\text{ft}}{\text{s}} - 0.000278 \sqrt{64.4 \frac{\text{ft}}{\text{s}^2} h}$$

The time step, Δt, must be selected. Since the tank with no hole required 703 s to fill up, a reasonable choice of time step might be $\Delta t = 703 \text{ s}/1000 = 0.7$ s. The results of the numerical integration are plotted in Figure 3.26. It can be seen that the tank will not fill up, as the water reaches an asymptotic height around 1.6 ft.

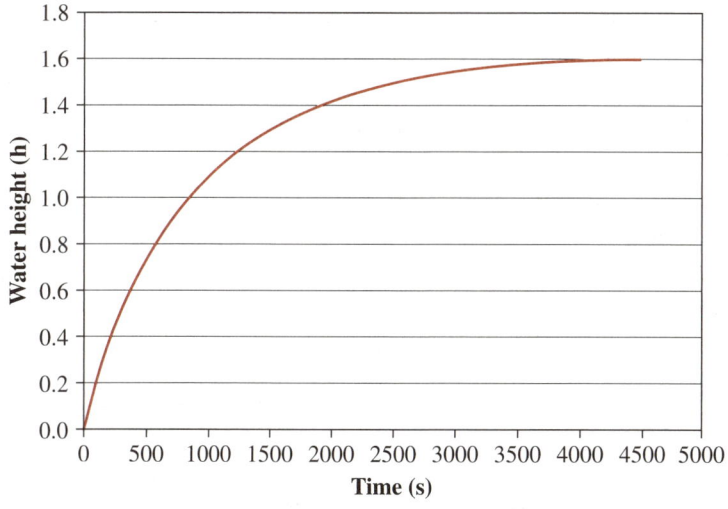

FIGURE 3.26 Computed profile of the height of the water in the tank vs time.

There are also situations in which both the Bernoulli equation and conservation of mass are needed to define the inlet and outlet states of a problem, so that the momentum equation can then be used to find the net force on the system. For flow situations with a single inlet and single outlet, operating under steady-state conditions, the conservation of mass equation is

$$\rho_1 A_1 V_1 = \rho_2 A_2 V_2$$

So while the Bernoulli equation provides a relationship between the flow variations, P, z, V, and ρ, the conservation of mass equation also relates ρ and V. Some examples of this type of problems follow, in which all three equations (mass, momentum, energy) must be used simultaneously.

EXAMPLE 3.30

A toy hovercraft weighs 20 N. It is open at the top to suck in air to the propeller with an inlet diameter of 26 cm. After going through the propeller, the air exits through an annular outlet on the bottom with outer diameter 26 cm and inner diameter 22 cm. Calculate the outlet velocity and the necessary power to be supplied to the propeller.

SOLUTION: We first use conservation of mass to find a relationship between the inlet and exit velocities. Assuming that the density of the air does not change significantly as it goes through the propeller, then $A_1 V_1 = A_2 V_2$. Here, $A_1 = \pi (0.13 \text{ m})^2 = 0.0531 \text{ m}^2$, and $A_2 = \pi (0.13 \text{ m})^2 - \pi (0.11 \text{ m})^2 = 0.0151 \text{ m}^2$. Also, $V_2 = V_1(A_1/A_2) = 3.52 V_1$. Now we can apply the momentum equation to find the necessary thrust force and relate that to the airstream exit velocity. In steady hovering flight above the ground, the momentum equation is

$$W = \rho_2 A_2 V_2^2 - \rho_1 A_1 V_1^2$$

The weight, W, is 20 N and under standard atmospheric conditions the density of air is about 1.2 kg/m³. Thus,

$$20 \text{ N} = \left(1.2 \frac{\text{kg}}{\text{m}^3}\right)(0.0151 \text{ m}^2)(3.52 \, V_1)^2 - \left(1.2 \frac{\text{kg}}{\text{m}^3}\right)(0.0531 \text{ m}^2)(V_1)^2$$

This leaves us with one equation in one unknown, V_1. Solving this equation gives $V_1 = 11.2$ m/s, and then $V_2 = 3.52 V_1 = 39.3$ m/s.

To calculate the required power, we use the energy equation. The power provided by the electric motor is used to accelerate the air to a higher kinetic energy, and so the relevant terms in the energy equation are

$$\dot{W} = \dot{m}\left(\frac{V_2^2}{2} - \frac{V_1^2}{2}\right)$$

The mass flow rate is (1.2 kg/m³)(0.0531 m²)(11.2 m/s) = 0.714 kg/s. The required power is then

$$\dot{W} = 0.714 \frac{\text{kg}}{\text{s}}\left(\frac{(39.2 \frac{\text{m}}{\text{s}})^2}{2} - \frac{(11.2 \frac{\text{m}}{\text{s}})^2}{2}\right) = 504 \text{ W}$$

There are many situations in which both conservation of energy (or the Bernoulli equation) and conservation of momentum can be used together to solve a fluid problem. In some, a force must be calculated, and the Bernoulli equation can be used to estimate the required property data that was not provided by measurements.

EXAMPLE 3.31

The nozzle of a fire hose has an exit diameter of 4 cm, and the base of the nozzle has a diameter of 12 cm, which is the same as the diameter of the hose. If the volume flow rate through the hose is 24 L/s, find the force required to hold the nozzle stationary when the hose is spraying horizontally.

SOLUTION: Assume a firefighter is holding the nozzle at the flange where the nozzle connects to the hose, and draw a control volume that cuts through the hose at the flange and passes just in front of the nozzle on the other side. The pressure of the exiting water jet will be atmospheric, but the pressure of the water at the flange need not be. To calculate the pressure there, we use the Bernoulli equation between point 1, the flange, and point 2, the nozzle exit:

$$\frac{P_1}{\rho} + \frac{V_1^2}{2} + gz_1 = \frac{P_2}{\rho} + \frac{V_2^2}{2} + gz_2$$

Since the hose is horizontal, $z_1 = z_2$, and $P_2 = 0$ if gage pressures are used. Then with those terms eliminated, we can rearrange the Bernoulli equation to solve for P_1:

$$P_1 = \frac{\rho}{2}(V_2^2 - V_1^2)$$

To calculate P_1, we must first calculate the velocities from the flow rate and the diameters. First we put the flow rate in standard units: $Q = 24$ L/s $= 0.024$ m³/s. Then, the velocities are

$$V_1 = \frac{Q}{A_1} = \frac{0.024 \text{ m}^3/\text{s}}{\pi(0.06 \text{ m})^2} = 2.12 \ \frac{\text{m}}{\text{s}}$$

$$V_2 = \frac{Q}{A_2} = \frac{0.024 \text{ m}^3/\text{s}}{\pi(0.02 \text{ m})^2} = 19.1 \ \frac{\text{m}}{\text{s}}$$

Then the pressure at the flange is calculated as

$$P_1 = \frac{1}{2}\left(1{,}000 \ \frac{\text{kg}}{\text{m}^3}\right)\left[\left(19.1 \ \frac{\text{m}}{\text{s}}\right)^2 - \left(2.1 \ \frac{\text{m}}{\text{s}}\right)^2\right] = 180{,}000 \text{ Pa}$$

There will be a pressure force pushing to the right, and a thrust force pushing to the left on the control volume. The momentum equation is

$$R_x + P_1 A_1 = -\dot{m} V_1 + \dot{m} V_2$$

Thus the force that must be applied by the firefighter is

$$R_x = -(180{,}000 \text{ Pa})[\pi(0.06 \text{ m})^2] + \left(1{,}000 \ \frac{\text{kg}}{\text{m}^3}\right)\left(0.024 \ \frac{\text{m}^3}{\text{s}}\right)\left(19.1 \ \frac{\text{m}}{\text{s}} - 2.1 \ \frac{\text{m}}{\text{s}}\right)$$

$$R_x = -2036 \text{ N} + 408 \text{ N} = -1628 \text{ N} \quad \text{(force acts to the left)}$$

In this case the pressure force is more significant than the thrust force.

EXAMPLE 3.32

An air fan sucks in air at atmospheric pressure and 20°C through an inlet of diameter 2 m and expels the air at atmospheric pressure through an outlet of diameter 0.5 m at a velocity of 50 m/s. Find the fluid thrust force generated and the power required to run the fan.

SOLUTION: Assuming the temperature is constant, then the density will also be constant, and the inlet velocity can be calculated from conservation of mass as

$$V_1 = V_2 \left(\frac{D_2}{D_1}\right)^2 = 50\,\frac{m}{s}\left(\frac{0.5\text{ m}}{2.0\text{ m}}\right)^2 = 3.125\,\frac{m}{s}$$

The density can be calculated using the ideal gas law:

$$\rho_1 = \frac{P_1 M_1}{RT_1} = \frac{(101{,}300\text{ Pa})(29\,\frac{kg}{kmol})}{(8{,}314\,\frac{J}{kmol\cdot K})(293\text{ K})} = 1.20\,\frac{kg}{m^3}$$

The net thrust force on the control volume is

$$F = \dot{m}_1 V_1 - \dot{m}_2 V_2 = \rho A_1 V_1 (V_1 - V_2)$$

$$= \left(1.2\,\frac{kg}{m^3}\right)\left(\frac{\pi}{4}(2\text{ m})^2\right)\left(3.125\,\frac{m}{s}\right)\left(3.125\,\frac{m}{s} - 50\,\frac{m}{s}\right) = 552\text{ N}$$

The mass flow rate by itself is 11.8 kg/s. The power required to run the fan is

$$\dot{W} = \dot{m}\left(\frac{V_2^2}{2} - \frac{V_1^2}{2}\right) = 11.8\,\frac{kg}{s}\left\{\frac{1}{2}\left[\left(50\,\frac{m}{s}\right)^2 - \left(3.125\,\frac{m}{s}\right)^2\right]\right\}$$

$$= 14{,}700\text{ W} = 14.7\text{ kW}$$

3.4.3 Derivation of Hydrostatic Pressure Distribution from the Bernoulli Equation

In Chapter 2 we derived the hydrostatic pressure distribution from a *force* balance on a differential element of fluid. The hydrostatic pressure distribution can also be derived from the *energy* equation. Taking the Bernoulli equation and setting the velocity equal to zero gives

$$P_1 + \rho g z_1 = P_2 + \rho g z_2 \tag{3.51}$$

Rearranging this to put similar terms on each side gives

$$P_1 - P_2 = \rho g (z_2 - z_1) \tag{3.52}$$

Limitations of the Bernoulli Equation

A very important fact to remember is that whenever a fluid flows over a solid surface, there will be friction losses. So the Bernoulli equation will give only an approximation. In Chapter 4 methods for including the friction losses in external flow, such as the flow of air over an airplane, will be shown. In Chapter 5 the friction losses for internal flows through pipes and ducts will be presented. For Newtonian fluids the friction losses between adjacent layers of fluid are truly negligible.

■ Summary

After reading this chapter and working through the problems, you should be able to apply the conservation of mass, momentum, and energy principles to fluid flows. In fluid systems, mass is always conserved—it is never created or destroyed but can be moved from one place to another. Energy also is conserved—it is not created or destroyed but can be transferred from one form to another. The rate of change of linear momentum of a system is equal to the net force acting on the system, and the rate of change of angular momentum of a system is equal to the net torque applied to that system. This chapter dealt with integral control volume analysis of fluid systems—only the incoming and outgoing average velocities of the fluid were considered. We also presented the Bernoulli equation, which approximates fluid flow by neglecting friction. Table 3.2 summarizes the equations used in this chapter.

Table 3.2 When to Use Each Conservation Equation

When you need a value for a:	Use the following conservation equation:
Flow rate	Mass
Force	Momentum
Torque	Angular momentum
Work of a machine	Energy
Heat transfer	Energy
Velocity or pressure in frictionless flow	Bernoulli

Problems

1. What is the rate at which water flows from the Mississippi River into the Gulf of Mexico?

2. Approximately how much mass is set in motion with one flap of an eagle's wings?

3. What is the force associated with a fluid cutting jet of diameter 0.5 mm and pressure 50,000 psi? Assume the fluid has the properties of water.

4. Water flow goes through a 90° reducing elbow. The inlet diameter is 30 cm and the outlet diameter is 10 cm, and the flow rate of the water is 21.5 kg/s. The pressure at the outlet is atmospheric. Calculate the net force in both the x- and y-directions to hold the elbow in place.

5. If a proposed submarine design ingested water through a 1-m-diameter hole and used a nuclear reactor to heat it to steam and exit it through another 1-m hole, what is the maximum thrust force? Assume both the inlet and outlet are at 1 atm, and the exiting steam is at the saturated state. You can take the inlet speed of water to be equal to the forward speed of the submarine, and assume a forward speed of 10 m/s.

6. Calculate the thrust force on a ramjet engine with inlet and outlet diameters of 10 cm, inlet velocity of 300 m/s, using hydrogen and air mixed at stoichiometric conditions. Assume that the exhaust gas is at the adiabatic flame temperature and atmospheric pressure. (See Chapter 8 for more information about balanced combustion reactions.)

7. Explain the fluid mechanics behind how a vertical axis wind turbine (VAWT) works. Why is this design less popular than the horizontal axis wind turbine?

8. What is the advantage of using an ion thruster rather than a traditional chemical rocket to position satellites in orbit?

9. Water flows into a nozzle of diameter 10 cm at a rate of 50 L/min. If the nozzle outlet has a diameter of 3 cm, what is the exit velocity, in m/s?

10. Explain how a wind turbine is able to generate power, since conservation of mass requires the air flow rates into and out of the turbine to be equal but conservation of energy requires a change in kinetic energy to generate power?

11. A certain hovercraft design consists of a round intake shaft at the top of inlet area 10 square meters, and the fan exhaust exit at the bottom has an area of 1.5 square meters. If the mass of the vehicle is 600 kg, what is the required minimum intake velocity to make the craft hover above the ground? How much power would the engine have to supply at this condition?

12. A single-engine turbojet flies at 530 mph at 40,000 ft. The inlet and exit areas of the jet engine are 13 ft^2 and 10 ft^2, respectively. The exhaust is at 1,500 ft/s relative to the plane and at 450 psf, gage. Find the thrust of the turbojet.

13. Water flows through a 30° elbow at 0.1 m³/s, the diameter of the pipe is constant at 12 cm, and you can neglect any pressure losses in the flow. What is the force of water on the pipe bend section?

14. A rocket has propellant mass of 7200 kg, structural mass of 800 kg, and payload mass of 60 kg. The specific impulse of the engine is 275 s. The mass flow rate of propellants is 72 kg/s. What is the burnout velocity of the rocket? Plot height and velocity as a function of time, up until burnout. Do you think this rocket will make orbit?

15. A simplified lawn sprinkler is designed so that all the parts and motion are in a horizontal plane. If it has two arms on opposite sides that end with 90° bends, find the maximum possible rotational speed of the sprinkler if it sprays 4.0 gal/min on the lawn, the pipes have inner diameter 6 mm, and the distance from the pivot to the 90° bend is 150 mm.

16. Water is sprayed downward at 1 m/s through a hole of diameter 10 mm. What would be the force on a plate perpendicular to the jet just outside the nozzle? If the plate is placed 4 m below the jet opening, what will be the force on the plate?

17. A space capsule is orbiting Earth at a speed of 8 km/s. To achieve reentry, it needs to be slowed down to 5 km/s. If the mass of the capsule is 1500 kg, the retro-rocket fuel consumption rate is 10 kg/s, and the exhaust velocity is 3000 m/s relative to the capsule, for how long do the retro-rockets need to fire?

18. A water turbine is supplied with 730 L/s of water through a 0.3-m-diameter inlet pipe, and has a 0.4-m-diameter exit pipe. If the water turbine generates 62 kW of power, what is the minimum pressure drop across the turbine?

19. A simple flow-through wind tunnel sucks in atmospheric air through a large opening and accelerates it through the test section. If the maximum velocity in the test section is 110 mph, calculate the pressure and density of the air in the test section. Assume the air in the room is at 20°C and 1.0 bar. The intake is 35 in. by 39 in., and the test section is 10 in. by 14 in.

20. A 7-hp pump is used to move water up an elevation change of 15 m. If the mechanical efficiency of the pump is 85%, what is the maximum flow rate it can handle? (Neglect frictional effects.)

21. A wind turbine has an 88-m blade span diameter and is in an area that has 15 mph average winds. If the efficiency of the turbine is 35%, calculate the power generated by the turbine and the force on the supporting structure. How would these numbers change if the wind speed were 25 mph?

22. If a water main at 75 psig broke open, what is the maximum height the water could reach?

23. What is the stagnation pressure on the nose of a 747 flying at 30,000 ft at 600 mph?

24. A certain decorative fountain can shoot water up to 85 ft high. What is the minimum required water pressure to do this?

25. What is the stagnation pressure on the front of your car when driving at 60 mph?

26. A jet of fluid of velocity V_{jet} and density ρ and area A_{jet} hits a cart with a flat shield so that the jet flow splits and goes off to the side at a 90° angle. If the cart is initially at rest, plot the motion of the cart as a function of time. (You will need to solve a differential equation.) What is the maximum velocity the cart can obtain?

27. A water fountain consists of a vertical pipe 5 m tall, with holes drilled at 1, 2, 3, and 4 m above the bottom. Water flows through the pipe in steady state and any excess dribbles out a hole in the top. Which of the four jets will shoot out the farthest?

28. One of the historical oddities of fluid mechanics is that there is no evidence that Bernoulli ever wrote or used the equation that now bears his name. Who was the person to publish what we today call the Bernoulli equation?

29. Explain the phenomenon of vapor lock. Why is vapor lock in automobile fuel lines a bigger problem in New Mexico than it is in Illinois?

30. What is the equation for the water hammer pressure? When does this pressure arise in fluid mechanics applications?

31. A Harrier VTOL jet (Figure 3.27) is holding stationary in hovering flight. The gross weight of the jet is 20,000 lbf, and it has its exhaust nozzles directed through an angle of 95°. If the plane is not moving in the vertical direction, and it is moving slow enough in the horizontal direction that drag can be neglected, what is the initial horizontal acceleration of the aircraft?

FIGURE 3.27 Picture of a VTOL jump jet. Courtesy of NASA Langley Research Center (NASA-LaRC).

32. An AV-8B Harrier flies in level forward flight at a constant velocity. The exhaust vents are deflected down at an angle of 80°. The drag force on the plane is $F = KV^2$, where $K = 8$ kg/m. Neglect the thrust force associated with the air intake. If the weight of the plane is 60,000 N and the plane is moving slowly enough that the lift force of the air flowing over the wings can be neglected, find the steady forward velocity of the plane. Also find the necessary engine thrust force.

33. Consider a rocket plane such as the one shown in Figure 3.28. When fully loaded with fuel the plane weighs 13,000 lbf (58,500 N). If the drag and lift coefficients are approximately constant, calculate (a) the velocity of the plane, V, needed to maintain the plane in steady level flight (0° angle of attack) at constant speed, and (b) the corresponding thrust force, T, generated by the rocket engine. (c) If the engine orifice diameter D_e is 20 cm and the density of the exhaust gases is 0.1 kg/m³, calculate the velocity of exhaust gas, V_e, needed to generate this thrust force. Neglect the change in weight of the plane as the fuel exits. You may also neglect any moments on the plane due to the fact that the lift and weight vectors are not perfectly aligned—in reality the tail is designed to generate negative lift and trim tabs are used to cancel out this moment. Lift force: $F_L = k_L v^2$, where $k_L = 0.65$ kg/m. Drag force: $F_D = k_D v^2$, where $k_D = 0.30$ kg/m.

34. A pipe system carrying water consists of two smaller pipes of different diameters connected to a larger tube. Pipe 1 has a diameter of 10 cm and a velocity of 1 m/s, and pipe 2 has a diameter of 5 cm and a velocity of 2 m/s. Pipes 1 and 2 merge at

FIGURE 3.28 Picture of an X-1 rocket plane. Courtesy of NASA Dryden Flight Research Center. USAF photo by Lt. Robert A. Hoover.

a junction and flow into pipe 3, which has a diameter of 20 cm. Neglecting all losses (friction, sudden expansion, etc.) and assuming uniform velocity profiles, calculate (a) the velocity in pipe 3, u_3, and (b) the difference in pressure between pipe 1 and pipe 3.

35. A hydroelectric power plant generates power from a water turbine under a dam. If the turbine is placed in the middle of a duct of constant area, A, and the height of water behind the dam is H, (a) derive an expression for the maximum possible power the turbine can generate in terms of the relevant variables. Neglect any friction losses. (Hint: Maximize the turbine power with respect to one of the other variables.) (b) Calculate the maximum possible theoretical power in watts for conditions of $H = 100$ m and $A = 1$ m^2.

36. In a Pelton waterwheel of radius R, the flow can be deflected though an angle as high as 165°. A water jet of mass flow rate m and velocity V strikes the buckets on the wheel. (a) If the wheel is rotating very slowly, find the torque that the jet exerts on the wheel. (b) Alternatively, if there is no torque at all on the wheel and it rests on frictionless bearings, what is the maximum possible rotational speed of the wheel? Under what conditions would you expect to get the maximum power of the wheel: at low speed, maximum speed, or some intermediate speed?

37. A flow of water comes through a nozzle and hits a target, delivering a force of magnitude F. If the diameter of the nozzle is decreased by a factor of 2 and the mass flow rate of water remains constant, what will the new force on the target be?

38. If a pressure vessel contains dry air at 50 psig, what is the maximum velocity that could be obtained by releasing the air into the surrounding atmosphere? What is the corresponding Mach number? Is the assumption of incompressible flow a valid one?

39. Water flows through a 30° elbow at a flow rate of 0.1 m^3/s. The diameter of the pipe is constant at 12 cm, and any pressure losses in the elbow can be neglected. What is the force exerted by the water on the pipe elbow section?

40. Water flows through a pipe at low speed, but with a high pressure of 25 psig. A small hole is drilled in the top of the pipe to make a decorative waterspout as part of a fountain. What is the velocity of the water exiting through the hole, and how high will the water spray?

41. A children's water sprinkler toy is designed as a conical clown's hat, which is levitated by a water jet. Water is shot upward from the base station, striking the hat in the middle and then spraying out downward at an angle onto the children playing below. The diameter at the spray nozzle on the ground is 1 in. and the initial vertical velocity of the water is 20 ft/s. The clown hat is designed to be supported at a

steady distance of 5 ft above the nozzle, and the internal cone angle of the hat is 60°. It can be assumed that the speed of the water entering the hat is the same as the speed of the water leaving the hat. Calculate the maximum weight of a hat that can be supported under these design conditions. Also, for safety concerns, calculate the force children would feel if they put their hands directly over the water jet nozzle.

42. A toy rocket is constructed using a 2-L bottle, which is attached to a swinging hitch by a solid rod. A bicycle pump is used to pump up the emptied bottle with air to a pressure P_o. The empty weight of the bottle is M_o, the length of the connecting rod is R, and the diameter of the bottle top is $D = 2$ cm. If the bottle top is suddenly opened at time $t = 0$, derive an equation for the angular speed, ω, of the bottle rocket about the hinge.

43. A toy water rocket is a cheap plastic container that holds a mixture of water and air. The air is pressurized using a pump, and then the rocket is released, squirting water out the bottom to propel it upward. Explain why this type of two-fluid rocket is used instead of one that simply uses compressed air with no water.

44. A toy water rocket of empty mass 500 g has an internal volume of 400 cc. If half the volume is filled with water and air is pumped in until the pressure reaches 90 kPa, gage, determine (a) the initial exit velocity of the water when the rocket is released, (b) the initial thrust force on the rocket if the exit is a hole of diameter 2 mm, (c) the time it takes for all the water to be expelled from the rocket, and (d) the height obtained when all the water is spent. (Neglect the drag force.)

45. A kite is being flown in a wind of 8 m/s. If the kite is 50 cm by 100 cm and weighs 1 kg, what is the angle it must deflect the wind through to stay aloft?

46. Explain how a sailboat can travel upwind without any engine power or rowing. (This is called tacking into the wind.)

47. Explain how the use of dynamic soaring allows birds and remote-control model gliders to travel faster than the wind speed.

48. A parasailer has a mass of 90 kg, including parachute and cords, and her rectangular parachute has an area of 24 m². If the parachute deflects the air through a 50° angle downward, calculate the speed at which the boat towing the parasailer needs to move to keep her aloft.

49. A water jet of mass flow rate $\dot{m}$ and velocity V strikes the back of a wheeled cart of mass M and is deflected sideways. If the cart is initially at rest, derive an equation for the velocity of the cart as a function of time.

50. For the space shuttle, what is the initial acceleration at liftoff with all three main engines and both solid rocket boosters firing? (See www.howstuffworks.com for shuttle specs.)

51. A small rocket has an initial gross mass of 1000 kg, a fuel plus oxidizer mass of 500 kg, an exhaust gas flow rate of 5 kg/s, and an exhaust gas velocity of 2000 m/s. Find the maximum height the rocket reaches.

52. A turning vane turns a horizontal fluid jet down through an angle θ. This results in an upward lift force and a rearward drag force. Plot the lift to drag ratio as a function of θ, for values of θ between 0° and 90°. How well do these results represent reality?

53. The Coanda effect occurs when a stream of fluid passes over a round object and a low-pressure area is created between the rear half of the object and the fluid stream, causing the fluid to be directed inward rather than outward as would normally be expected. You can demonstrate this at home by putting your finger under the stream of water coming out of the kitchen faucet. When the edge of your finger just penetrates the edge of the water stream, it will deflect toward your finger. For this problem consider a rectangular water sheet of width 1 cm and thickness 2 mm flowing downward with a velocity of 3 m/s. The stream flows over the edge of a cylinder and is turned inward through an angle of 15°. Find the magnitude of the force on the cylinder in the horizontal direction.

54. A Super-soaker water gun with its nozzle 3 ft above ground level and angled upwards at 15° is able to shoot a stream over a horizontal distance of 40 ft. Find the velocity of water when it leaves the nozzle and the minimum required pressure in the firing chamber.

55. For the gun in the previous problem, if the nozzle is 1 mm in diameter and the volume of the firing chamber is 200 mL, for how long can the gun be fired?

56. A watering system consists of a large reservoir of water supported above the ground draining straight down into nozzles that spray outward at an angle, and the firing angle can be adjusted. If the water level in the reservoir is 10 ft above ground level, what is the farthest horizontal distance outward from the base that the nozzles can spray? (The nozzles can be located anywhere from $z = 0$ to $z = 10$ ft above ground level.)

57. A water gun that was popular before Super-soakers were invented was the water bazooka (see Figure 3.29). This consisted of a large piston–cylinder device that pressed water through a small nozzle. If the nozzle was 2 mm in diameter and the cylinder was 5 cm in diameter, what is the force necessary to push on the piston to get a spray with velocity 10 m/s out the nozzle?

58. What is the pumping power required to pump 100 gal/min of water up an elevation of 100 ft? Look at the available pumps in a pump catalog and make a suggestion for which pump size to use.

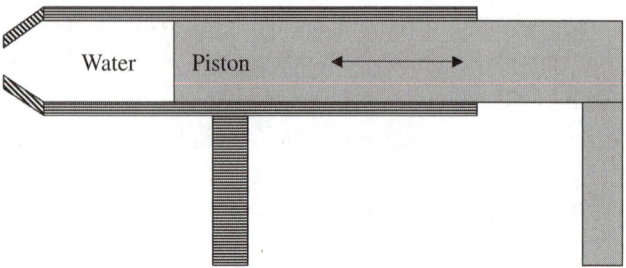

FIGURE 3.29 Sketch of a water bazooka.

59. A boat contains a large reservoir of compressed air that it uses for propulsion. If the compressed air exits through a hole of diameter 15 cm with a velocity of 110 m/s, the boat has a mass of 260 kg, and the total resistance force on the boat due to air and water drag can be modeled as $F_D = kV^2$, where $k = 0.12$ kg/m, plot the speed of the boat as a function of time if it starts from rest, and calculate the top speed of the boat.

60. How much pumping power is required to pump 10 gal/min to the bathrooms on the top floor of the Empire State building from a reservoir at atmospheric pressure at ground level?

61. A hole is accidentally punctured in a pipe containing crude oil at 220 kPa, gage. What is the maximum height to which the oil could spray, and what is the maximum velocity of oil at the hole?

62. An XV-15 tilt-rotor plane (see Figure 3.30) is maintaining position in hovering flight. If the plane has mass of 5000 kg and the rotors have diameter of 7.6 m,

FIGURE 3.30 Picture of an XV-15 experimental tilt-rotor aircraft. Courtesy of NASA Dryden Flight Research Center.

estimate the downwash velocity of the air from the rotors. State any assumptions you make.

63. A water spigot outside a house has a supply pressure of 30 psig. The hose is not long enough to reach to every part of the yard. What is the farthest distance from the end of the hose in the yard that the water can be sprayed? (Neglect friction losses in the hose.)

64. A garden hose has a supply pressure of 30 psig and an outlet diameter of 1.5 cm. How long will it take to fill up an initially empty swimming pool of depth 5 ft, width 10 ft, and length 25 ft?

65. The pump used in the cooling system of an engine in a research test cell has failed, and the antifreeze has collected in the reservoir, leaving the pipes empty. To start up the repaired pump, it must have fluid in it. Someone suggests that the lab technician siphon the antifreeze from the reservoir to get it flowing, but he does not want to risk accidentally swallowing some of the antifreeze. Then he gets the idea to use the building's compressed air supply to spray a jet of air over the open end of the pipe from the reservoir. Will this idea work? If so, what velocity of air is needed if the antifreeze has to be raised up 20 cm out of the reservoir before it flows to the lower elevation of the pump?

66. What would be the pressure required in the firing chamber for a Super-soaker gun to shoot a spray of water 100 ft?

67. A swimming pool of width 10 ft and length 25 ft is filled to a depth of 5 ft. To drain the pool a valve is opened beneath the drain gate on the bottom, which has an open area of 170 cm². How long will it take for 90% of the pool water to be drained?

68. If the flow rate of a river is 10 million kg/s, and a 30-m-high dam is built, what is the maximum amount of hydroelectric power that can be generated, in kW?

69. The turbulent flow of fluid through a pipe is approximately given by

$$V(r) = V_{max}\left[1 - \left(\frac{r}{R}\right)\right]^{1/n}$$

for $0 < r < R$. For a value of $n = 10$, find the mass-averaged velocity of the flow.

70. Define the *hydraulic grade line* and the *energy grade line*. What is the difference between the two? What are the uses of these grade lines?

71. Air flows into a system at 1 MPa (absolute), 300°C, and 10 m/s, and leaves at 135°C and 25 m/s. Find the pressure of the exiting air, assuming steady-state operation of the system and equal-sized inlet and outlet.

72. Water enters a heat exchanger at 15°C and 5 m/s through an entrance of diameter 5 cm and leaves at 85°C through an exit of diameter 8 cm. Calculate the velocity at the exit (a) assuming constant density of the water and (b) accounting for the change in density.

73. You are filling an initially empty barrel of diameter 0.7 m and height 1.5 m with water at a rate of 0.2 L/s. When the barrel is about halfway full, you notice it is leaking through a hole of diameter 0.5 cm near the bottom. You plan is to fill up the barrel, close the top, then quickly flip it over and then seal the hole. Will you be able to fill up the barrel under these conditions?

74. Water is sprayed upward from a hose with supply pressure 50 psig. Using the Bernoulli equation, find the maximum height the water spray can obtain. List all the reasons why the actual height in reality would be less. Which of these reasons do you think has the most significant effect on the actual height of the fluid?

75. What is Toricelli's equation?

76. If the pressure drop in a hurricane can be as high as 0.25 atm, how high could it raise the seawater beneath it over the standard sea level? (This effect is called swell or storm surge.)

77. How would the Bernoulli equation have to be modified for a compressible flow of a gas?

78. Water at 15°C in a large reservoir is sucked up through a pump and raised to an elevation of 10 m above the reservoir level, where it is discharged at a rate of 0.75 kg/s. The pipe diameter is 6 cm. If the mechanical efficiency of the pump is 72%, what is the rate at which electrical power must be supplied to the pump?

79. An electric motor provides 6 kW to a water pump. Water enters the pump at 1500 L/m and 95 kPa through a pipe of diameter 20 cm and exits at 250 kPa through a pipe of diameter 14 cm. Estimate the mechanical efficiency of the pump from these measurements. What happens to the energy that is not converted to useful pressure or kinetic energy?

80. A hydroelectric power plant has a dam that holds the water back 100 m above the downstream water level. If the flowrate of water through the plant is 80 m^3/s, plot the power generated as a function of the efficiency of the genset (turbine plus electric generator combination). Discuss the economic analysis that would have to be done to determine the optimum genset to buy.

81. What are the limitations of the Bernoulli equation?

82. What precautions must be exercised when using a Pitot tube to measure velocity? (In other words, what are the limitations of using a Pitot tube?)

83. What is the value of the temperature rise of water going through a pump of 70% efficiency at a flow rate of 0.3 gal/s with a pump power input of 33 hp?

84. An office water cooler contains water at a height of 0.6 m above the spigot. What will be the velocity of water coming out of the spigot? If the diameter of the spigot is 3 mm, how long will it take to fill a cup of volume 180 mL?

85. An office water cooler in the shape of a cylinder has internal height of 0.7 m and diameter 0.4 m. How many 150-mL cups can be filled from the water cooler?

86. What is meant by the term *convective acceleration*? How is it possible to have fluid acceleration in a steady-state flow? Give an example.

87. An airplane is flying at an altitude of 10,500 m at a speed of Ma = 0.85. What would be the differential pressure (in Pa) measured by the Pitot tube used for the airspeed indicator? (Neglect compressibility effects.)

88. In the previous problem, would the compressibility of the air cause the velocity reading in the Pitot tube to be higher or lower than what is calculated using the incompressible Bernoulli equation?

89. A Pitot tube used to measure air velocity is connected to a U-tube manometer filled with water. The change in height between the two legs of the manometer is 5.1 cm. What is the velocity of the air, assuming it is at standard conditions?

90. Explain the fluid mechanics behind how a nebulizer works.

91. Diesel fuel is pumped vertically from an underground storage tank along a 3-m section of pipe, through a 90° bend, horizontally over 2 m, and then through another 90° bend and down into a tanker truck. The diameter of the pipe is constant at 10 cm. Find all the fluid forces and moments exerted at the ground support of the pipe where it exits the main tank.

92. In a particular experimental apparatus, a water jet hits a target and is deflected back through an angle. It is possible to measure the force on the target and the volume flow rate of water, Q. If you plot $\log(F)$ vs $\log(Q)$, what do you expect the slope of the graph to be?

93. A scale model of a Pelton waterwheel has a radius of 8 cm, and the bucket deflection angle is 160°. With no torque applied, a water flow rate of 0.5 kg/s produces a rotational speed of 600 rpm. What is the maximum power that could be obtained from the wheel?

94. A fan of diameter 50 cm produces a motion of 1 m³/s of air at STP. Determine the fluid thrust force on the fan and the power required to run the fan.

95. An engineer is observing a lawn water sprinkler that has two opposing arms that rotate around the center base and spray out the water. The engineer reasons that it would spray the lawn just as well if it only rotated at a fraction of the speed it does

now, and he figures he could extract some power from it in the process. He hooks up a small electric generator to the top of the center of rotation of the sprinkler. The sprinkler originally rotated at a rate of 240 rpm with a total water supply rate of 75 gal/m to both arms combined. The diameter of the outlets is $\frac{3}{8}$ in., the distance from the centerline of the sprinkler to the outlets is 20 in., and the outlet jets are at 90° angles to the sprinkler arms in the horizontal plane. What is the maximum power that could be produced by the generator, if the minimum allowable sprinkler speed is 20 rpm?

96. Plot the relative efficiency of a Pelton waterwheel as a function of the bucket angle, with the relative efficiency defined as the power generated at a given bucket angle divided by the power generated for a bucket angle of 180°.

97. A diesel fuel injector has an orifice diameter of 200 microns and an injection velocity of 500 m/s. If the average size of the fuel drops formed from the spray is 10 microns, how many drops are formed from one hole of the injector for an injection event lasting 1 millisecond?

98. A hovercraft is designed as shown in Figure 3.31. To improve stability, the two fans are tilted up at an angle of 5° from the horizontal. What percentage of the upward thrust is lost relative to a design with the fans horizontal? How does the tilted design improve stability?

99. An air flow meter reads 100 SCFM of air flowing into an engine. If the measured properties of the air at the meter are 0.998 bar and 77°F, what is the actual volumetric flow rate of the air, in cubic feet per minute, and what is the mass flow rate of the air?

100. In a long straight rectangular channel the depth of the water is measured to be 5 cm and the velocity of the water on the surface of the open channel flow is 10 cm/s. Just before the exit of the channel, the water level rises to 6.5 cm. Estimate the velocity on the surface at that point.

101. An astronaut of mass 65 kg sprays liquid water to hydrate the food. If 10 cubic centimeters are sprayed through a 3-mm-diameter hole over a time of 2 s, what is the acceleration of the astronaut?

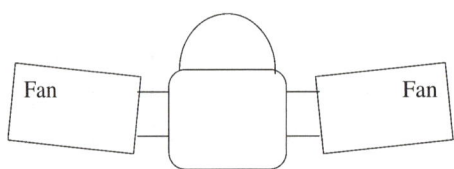

FIGURE 3.31 Hovercraft design.

FIGURE 3.32 Picture of a jet engine on a test stand. Courtesy of NASA.

102. A toy helicopter has mass of 5 kg. What is the minimum power that must be supplied by the motor to keep it in hovering flight, if the rotor diameter is 35 cm?

103. A jet engine sits on a test stand (Figure 3.32) in a wind tunnel and provides 10,000 lbf of thrust. If the incoming velocity is 100 mph and the inlet diameter is 28 cm, what is the exit velocity?

4 Differential Equations of Fluid Motion

In This Chapter
- Navier–Stokes Equations
- Laminar Flow Solutions
- Computational Fluid Dynamics

The objective of this chapter is to introduce the differential equations that govern fluid flows, show how to simplify those equations to solvable form for particular flow situations, and present the basic workings and limitations of computational fluid dynamics software tools.

■ 4.1 Navier–Stokes Equations

In 1857 Thomson wrote, "Now I think hydrodynamics is to be the root of all physical science, and is at present second to none in the beauty of its mathematics." The mathematics of fluid dynamics, as hydrodynamics is now called, is governed by the Navier–Stokes equations. These equations describe how the velocities and fluid properties within a flow field vary in space and time.

The analysis techniques presented in Chapter 3 work well when one-dimensional flows can be assumed or the inner details of the flow are relatively unimportant for engineering purposes. However, for many problems we need to know not only the average flow quantities but also the details of a flow, such as whether flow over a wing will separate and cause the plane to stall. We also need to be able to calculate the velocity profiles in order to estimate the viscous resistance to flow.

The Navier–Stokes equations were originally derived by Navier, but it was not until Stokes independently rederived them about a century later that they became popular as a tool for fluid analysis. Furthermore, it was not until Osborne Reynolds explained how turbulence alters a flow field that it was understood that the Navier–Stokes equations truly do govern all fluid flows.

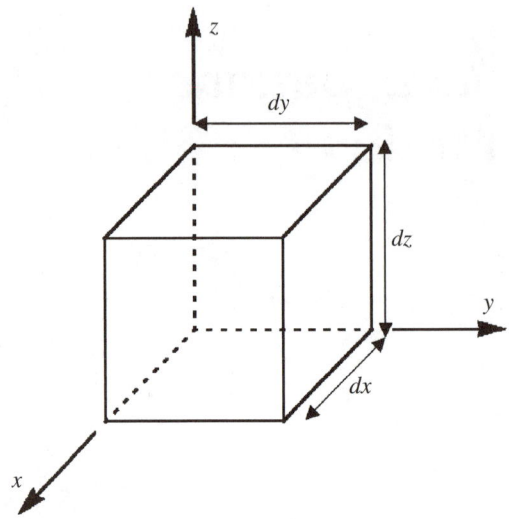

FIGURE 4.1 Control volume used for the derivation of differential equations of fluid flow.

4.1.1 Derivation of Navier–Stokes Equations

Continuity

To derive these equations, we start by defining an infinitesimally small control volume, as shown in Figure 4.1. For the control volume shown in the figure in Cartesian space, the position vector is (x, y, z) and the velocity vector is (u, v, w). Applying conservation of mass to the control volume yields

$$\frac{dM_{CV}}{dt} = \rho u_x dy\, dz - \rho u_{x+dx} dy\, dz + \rho v_y dx\, dz - \rho v_{y+dy} dx\, dz$$
$$+ \rho w_z dx\, dy - \rho w_{z+dz} dx\, dy$$

where $M_{CV} = \rho\, dx\, dy\, dz$. We divide by $(dx\, dy\, dz)$, take the limit as $dx, dy, dz \to 0$, and, using the definition of a derivative, we get

$$\frac{d\rho}{dt} = -\left(\rho \frac{du}{dx} + \rho \frac{dv}{dy} + \rho \frac{dw}{dz}\right) \tag{4.1}$$

For an incompressible fluid in steady-state flow, the conservation of mass equation (also called the continuity equation) is

$$\left(\frac{du}{dx} + \frac{dv}{dy} + \frac{dw}{dz}\right) = 0 \tag{4.2}$$

Principle of Linear Momentum for a Differential Control Volume

The principle of linear momentum for a macroscopic control volume was obtained from Newton's second law in Chapter 3 as

$$\sum \vec{F} = \frac{\partial}{\partial t}(m\vec{V}) + \sum (\dot{m}\vec{V})_{\text{out}} - \sum (\dot{m}\vec{V})_{\text{in}}$$

For the differential control volume,

$$(\dot{m}\vec{V})_{\text{out}} - (\dot{m}\vec{V})_{\text{in}} = \{\rho u \vec{V}|_{x+\Delta x} - \rho u \vec{V}|_x\}\Delta y\, \Delta z + \{\rho v \vec{V}|_{y+\Delta y} - \rho v \vec{V}|_y\}\Delta x\, \Delta z + \{\rho w \vec{V}|_{z+\Delta z} - \rho w \vec{V}|_z\}\Delta x\, \Delta y$$

In the limit of $\Delta x \to 0$, $\Delta y \to 0$, $\Delta z \to 0$, we have

$$(\dot{m}\vec{V})_{\text{out}} - (\dot{m}\vec{V})_{\text{in}} = \left\{\frac{\partial}{\partial x}\rho u \vec{V} + \frac{\partial}{\partial y}\rho v \vec{V} + \frac{\partial}{\partial z}\rho w \vec{V}\right\}\Delta x\, \Delta y\, \Delta z$$

or, using the product rule of the derivative, we obtain

$$(\dot{m}\vec{V})_{\text{out}} - (\dot{m}\vec{V})_{\text{in}} = \rho\left\{u\frac{\partial}{\partial x}\vec{V} + v\frac{\partial}{\partial y}\vec{V} + w\frac{\partial}{\partial z}\vec{V}\right\}\Delta x\, \Delta y\, \Delta z$$

$$+ \left\{\frac{\partial}{\partial x}\rho u + \frac{\partial}{\partial y}\rho v + \frac{\partial}{\partial z}\rho w\right\}\vec{V}\Delta x\, \Delta y\, \Delta z$$

On the other hand, since $m = \rho\, \Delta x\, \Delta y\, \Delta z$, then

$$\frac{\partial}{\partial t}(m\vec{V}) = \left\{\frac{\partial \rho}{\partial t}\vec{V} + \rho\frac{\partial \vec{V}}{\partial t}\right\}\Delta x\, \Delta y\, \Delta z$$

Therefore, the principle of linear momentum for a differential control volume takes the following form:

$$\sum \vec{F} = \rho\left\{u\frac{\partial}{\partial x}\vec{V} + v\frac{\partial}{\partial y}\vec{V} + w\frac{\partial}{\partial z}\vec{V}\right\}\Delta x\, \Delta y\, \Delta z$$

$$+ \left\{\frac{\partial}{\partial x}\rho u + \frac{\partial}{\partial y}\rho v + \frac{\partial}{\partial z}\rho w\right\}\vec{V}\Delta x\, \Delta y\, \Delta z$$

Since by the chain rule (note that the three terms on the far right are convective accelerations) we get

$$\frac{d\vec{V}}{dt} = \frac{\partial \vec{V}}{\partial t} + u\frac{\partial \vec{V}}{\partial x} + v\frac{\partial \vec{V}}{\partial y} + w\frac{\partial \vec{V}}{\partial z} \tag{4.3}$$

and rearranging the continuity equation gives

$$\frac{\partial \rho}{\partial t} + \frac{\partial}{\partial x}(\rho u) + \frac{\partial}{\partial y}(\rho v) + \frac{\partial}{\partial z}(\rho w) = 0$$

We arrive at

$$\sum \vec{F} = \rho \left\{ \frac{d}{dt}\vec{V} \right\} \Delta x\, \Delta y\, \Delta z$$

We will consider three types of forces: gravity, F_g; pressure, F_p; and viscous forces, F_v.

Gravity: $\quad F_g = \rho g\, \Delta x\, \Delta y\, \Delta z$
Pressure: $\quad F_p = (-\nabla P)\Delta x\, \Delta y\, \Delta z$
Viscous forces: $\quad F_v = \mu\, \nabla^2 V$

Finally, combining all these terms yields differential momentum equations in the three coordinate axes:

$$\rho\left(\frac{\partial u}{\partial t} + u\frac{\partial u}{\partial x} + v\frac{\partial u}{\partial y} + w\frac{\partial u}{\partial z}\right) = \rho g_x - \frac{\partial P}{\partial x} + \mu\left(\frac{\partial^2 u}{\partial x^2} + \frac{\partial^2 u}{\partial y^2} + \frac{\partial^2 u}{\partial z^2}\right)$$

$$\rho\left(\frac{\partial v}{\partial t} + u\frac{\partial v}{\partial x} + v\frac{\partial v}{\partial y} + w\frac{\partial v}{\partial z}\right) = \rho g_y - \frac{\partial P}{\partial y} + \mu\left(\frac{\partial^2 v}{\partial x^2} + \frac{\partial^2 v}{\partial y^2} + \frac{\partial^2 v}{\partial z^2}\right) \tag{4.4}$$

$$\rho\left(\frac{\partial w}{\partial t} + u\frac{\partial w}{\partial x} + v\frac{\partial w}{\partial y} + w\frac{\partial w}{\partial z}\right) = \rho g_z - \frac{\partial P}{\partial z} + \mu\left(\frac{\partial^2 w}{\partial x^2} + \frac{\partial^2 w}{\partial y^2} + \frac{\partial^2 w}{\partial z^2}\right)$$

where $V = (u, v, w)$ are the rectangular components of the spatial velocity field. These four equations (conservation of mass [Equation 4.2], plus the three components of the momentum equation [Equations 4.4]) are referred to as the *Navier–Stokes equations*. These are a system of partial differential equations, and there is no general analytical solution to them. However, as will be explored in the rest of the chapter, for certain geometries these equations do simplify and there are a few cases where exact solutions to the differential equations can be obtained. For complex geometries, the equations can only be solved approximately, as is done by CFD computer codes.

In vector form the Navier–Stokes equations can be written as

$$\frac{\partial \rho}{\partial t} + \nabla \bullet (\rho \vec{V}) = 0$$

$$\frac{\partial}{\partial t}(\rho \vec{V}) + \nabla \bullet (\rho \vec{V})\vec{V} = -\nabla P + \nabla \bullet \tau + \rho \vec{f}$$

In cases where there is chemical reaction, heat transfer, or compressible flow so that the temperature changes, a differential form of the energy equation can also be used:

$$\frac{\partial}{\partial t}(\rho e) + \nabla \bullet (\rho e \vec{V}) = -\nabla \bullet \vec{q} + \nabla \bullet (\tau \bullet \vec{V}) + \rho \vec{V} \bullet \vec{f} \tag{4.5}$$

EXAMPLE 4.1

Consider air flowing in a one-dimensional flow along the x-axis at steady state. If the viscosity can be neglected, what does the momentum equation simplify to?

SOLUTION: As with any fluids problem, we start with the full Navier–Stokes equations and see what terms can be thrown away either because they are identically zero or they are very small compared to the other terms. We will assume that the flow is incompressible, so that changes in density can be neglected. For this one-dimensional flow only the x-momentum equation need be considered.

$$\rho\left(\frac{\partial u}{\partial t} + u\frac{\partial u}{\partial x} + v\frac{\partial u}{\partial y} + w\frac{\partial u}{\partial z}\right) = \rho g_x - \frac{\partial P}{\partial x} + \mu\left(\frac{\partial^2 u}{\partial x^2} + \frac{\partial^2 u}{\partial y^2} + \frac{\partial^2 u}{\partial z^2}\right)$$

Since the flow is at steady state, the first term, the transient term du/dt, goes to zero. Since it is one-dimensional flow, the v and w velocities will be zero. Setting the viscosity to zero eliminates the final term on the right-hand side. The remaining terms in the momentum equation are

$$\rho u \frac{du}{dx} = \rho g_x - \frac{dP}{dx}$$

The partial derivatives have been replaced with regular derivatives since x is the only independent variable of the flow. To solve this equation, we multiply all the terms by dx:

$$\rho u\, du = \rho g_x\, dx - dP$$

Now we integrate both sides of the equation between two end states, 1 and 2.

$$\rho \frac{u^2}{2}\bigg|_1^2 = \rho g_x x \big|_1^2 - P\big|_1^2$$

Applying the limits of integration and rearranging yields

$$P_2 + \frac{1}{2}\rho u_2^2 - \rho g_x x_2 = P_1 + \frac{1}{2}\rho u_1^2 - \rho g_x x_1$$

This looks very similar to the Bernoulli equation. In fact, recall that for the Navier–Stokes equations, by convention g is defined positive along the axis. So, for example, if the x-direction points upward, then g acts downward. Replacing g with $-g$ yields

$$P_2 + \frac{1}{2}\rho u_2^2 + \rho g_x x_2 = P_1 + \frac{1}{2}\rho u_1^2 + \rho g_x x_1$$

So the Bernoulli equation is in fact obtained. Note that in this example it was derived from the momentum equation, whereas in Chapter 3 it was derived from the energy equation.

EXAMPLE 4.2

What form do the Navier–Stokes equations take in the limit of a very high Reynolds number?

SOLUTION: Although a high Reynolds number implies turbulent flow, it also implies that the ratio of inertial forces to viscous forces is high. The terms on the left-hand side of the momentum equation, not counting the transient terms, are referred to as the *convective* terms. These terms represent the momentum of the fluid in each of the three directions. They can also be thought of as representing the fluid inertia. So if the viscous forces are small compared to the other terms in the Navier–Stokes equations, then the viscous terms can be eliminated to good approximation. Dropping the viscous terms gives the following equations:

$$\frac{\partial \rho}{\partial t} + \frac{\partial}{\partial x}(\rho u) + \frac{\partial}{\partial y}(\rho v) + \frac{\partial}{\partial z}(\rho w) = 0$$

$$\rho\left(\frac{\partial u}{\partial t} + u\frac{\partial u}{\partial x} + v\frac{\partial u}{\partial y} + w\frac{\partial u}{\partial z}\right) = \rho g_x - \frac{\partial P}{\partial x} \quad (4.6)$$

$$\rho\left(\frac{\partial v}{\partial t} + u\frac{\partial v}{\partial x} + v\frac{\partial v}{\partial y} + w\frac{\partial v}{\partial z}\right) = \rho g_y - \frac{\partial P}{\partial y}$$

$$\rho\left(\frac{\partial w}{\partial t} + u\frac{\partial w}{\partial x} + v\frac{\partial w}{\partial y} + w\frac{\partial w}{\partial z}\right) = \rho g_z - \frac{\partial P}{\partial z}$$

This set of equations is known as the *Euler equations.* Leonhard Euler derived them before Navier for what hydrodynamicists called a "perfect fluid"—one with no viscosity. Obviously the applicability of these equations to engineering problems is limited since all real fluids have viscosity. However, up through the 1950s the Euler equations were used to predict the flow over an airfoil, where the viscous effects of the fluid are restricted to a relatively small *boundary layer* near the surface of the wing. Also, for the special case of steady-state two-dimensional flow of an incompressible fluid, which is approximately valid for subsonic flows over wings, the Euler equations simplify to

$$\frac{\partial u}{\partial x} + \frac{\partial v}{\partial y} = 0$$

$$u\frac{\partial u}{\partial x} + v\frac{\partial u}{\partial y} = g_x - \frac{1}{\rho}\frac{\partial P}{\partial x}$$

$$u\frac{\partial v}{\partial x} + v\frac{\partial v}{\partial y} = g_y - \frac{1}{\rho}\frac{\partial P}{\partial y}$$

4.1.2 Turbulent Flow

An interesting phenomenon that occurs in nature is that when a critical velocity of a flow is exceeded, the flow ceases to be orderly; it transitions to a chaotic, turbulent state. Even for a flow that can be considered steady state as a whole, there will be internal variations in velocity with respect to time due to this turbulence. The Navier–Stokes equations apply to turbulent flows just as they do to the more orderly laminar flows, but they are much more difficult to solve, in part because steady state can never be assumed. Osborne Reynolds sought to deal with this problem by rewriting the Navier–Stokes equations in an averaged sense, rather than their traditional instantaneous form. His ideas frequently have been used as a basis for turbulent modeling in computational fluid dynamics (CFD) codes used in computer simulations of turbulent fluid flows.

It is challenging to describe *turbulence* with words, but it is easy to identify a turbulent flow when one sees it. Turbulent flows are characterized by strong velocity fluctuations and disordered, or chaotic, motions. Figures 4.2 through 4.4 show some examples of turbulent motion.

The criterion for transition from an ordered laminar flow to a chaotic turbulent flow depends on the ratio of inertial forces to viscous forces. This ratio is characterized by a Reynolds number, as originally proposed by Osborne Reynolds and named in his honor:

$$\mathrm{Re} = \frac{\rho V L}{\mu} \qquad (4.7)$$

where L is a characteristic length scale of the flow. The Reynolds number is an example of a nondimensional number. (Nondimensionalization of flow variables is discussed in greater detail in Chapter 9.) There is a critical value of the Reynolds number, Re_{crit}, below which the flow remains laminar and above which it becomes turbulent. The value of Re_{crit} depends on the type of flow. For internal flows through pipes and ducts, Re_{crit} is around 2,000, while for external flows such as the flow over an airplane's wings, Re_{crit} is around 500,000. In reality the transition between laminar and turbulent flow is not so abrupt, and there is a range of Reynolds numbers in the transition region between the two that contains some intermittent flows, but for engineering purposes it is convenient to assume a critical Reynolds number for transition. This approximation

I FIGURE 4.2 Turbulent flow on Jupiter. Courtesy of NASA Jet Propulsion Laboratory.

FIGURE 4.3 Turbulent flow in a nebula. Courtesy of NASA Marshall Space Flight Center Collection.

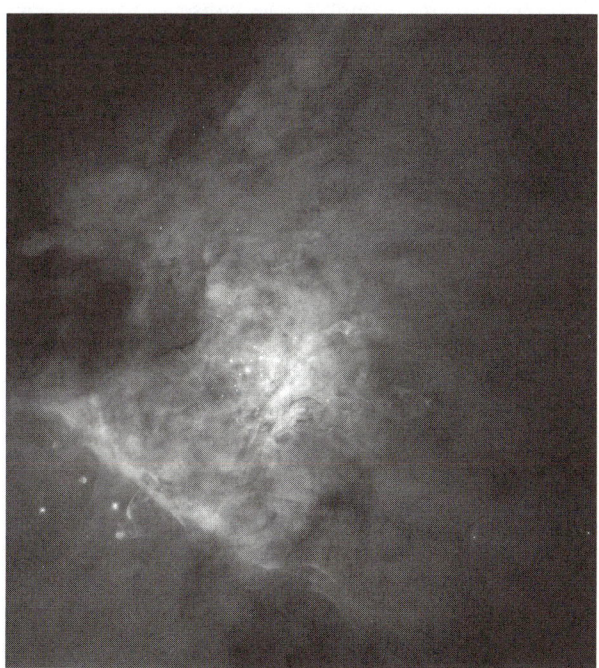

FIGURE 4.4 Turbulent flow in a nebula. Courtesy of NASA/STScI/Rice Univ./C.O'Dell et al.

is sufficiently accurate for most applications, though more exact predictions of the conditions that lead to transition in the flow over wings is an area of open research.

It is important to remember again that the Navier–Stokes equations are valid in general for any fluid. It is possible to solve these equations directly for laminar flow cases of simple (single-phase, Newtonian) fluids. That is, it is possible to solve them numerically. It is only possible to get analytical solutions for the simplest of geometries.

The Reynolds-averaged Navier–Stokes (RANS) equations are the oldest framework for turbulence modeling. The concept of Reynolds-averaging was developed by Osborne Reynolds in the 19th century, though it did not see use until the last half of the 20th century. The idea is that for a turbulent flow, even in steady state, all of the variables are fluctuating in time, so that some statistical approach must be developed to describe the turbulence. In Reynolds' simple model, any fluctuating variable can be expressed as the sum of the mean and variable components:

$$u(t) = \bar{u} + u'(t) \tag{4.8}$$

where the overbar represents the time-averaged quantity.

This same decomposition can be done for all the variables in the flow (u, v, w, P, ρ). Then the Reynolds-decomposed variables are substituted back into the Navier–Stokes equations. It is common practice to neglect density fluctuations because they are difficult both to model and to measure. If the flow being simulated is incompressible, then this assumption is exactly correct. So, for example, the conservation of mass equation becomes

$$\frac{d\bar{\rho}}{dt} = -\bar{\rho}\left(\frac{d(\bar{u}+u')}{dx} + \frac{d(\bar{v}+v')}{dy} + \frac{d(\bar{w}+w')}{dz}\right)$$

The next step is to take a time average of the entire equation:

$$\overline{\frac{d\bar{\rho}}{dt}} = -\bar{\rho}\left(\overline{\frac{d(\bar{u}+u')}{dx}} + \overline{\frac{d(\bar{v}+v')}{dy}} + \overline{\frac{d(\bar{w}+w')}{dz}}\right)$$

The following rules apply:

- The time average of the mean component is the same as the mean $\bar{\bar{u}} = \bar{u}$
- The time average of the fluctuating component is zero, by definition $\overline{u'} = 0$

So the Reynolds-averaged conservation of mass equation becomes

$$\overline{\frac{d\bar{\rho}}{dt}} = -\bar{\rho}\left(\frac{d(\bar{u})}{dx} + \frac{d(\bar{v})}{dy} + \frac{d(\bar{w})}{dz}\right)$$

which is nearly identical to the original equation.

Where things get interesting is in the nonlinear convection terms of the momentum equations. Consider the product of two velocity components:

$$uv = (\bar{u} + u')(\bar{v} + v') = \bar{u}\,\bar{v} + \bar{u}v' + \bar{v}u' + u'v'$$

When we take the time average of this product we are left with the following:

$$\overline{uv} = \overline{\bar{u}\,\bar{v} + \bar{u}v' + \bar{v}u' + u'v'} = \bar{u}\,\bar{v} + \overline{u'v'}$$

Now an important point about turbulence must be made. If the fluctuating velocity components, u' and v', were truly independent random variables, then the average of their product would be zero. In reality it is not, because in real turbulent flows the different velocity components are *correlated* to each other. Turbulent flows exhibit randomness in their behavior, but they do not meet the mathematical definition for a statistical random process. In other words, there is order in the chaos. The mathematical definition of chaos is that a small, even infinitesimal, change in the initial conditions of the system (in this case the Navier–Stokes equations governing the fluid flow) can produce significant large-scale differences downstream in the flow.

So terms of the form $u'v'$ appear in the Reynolds-averaged Navier–Stokes equations that do not appear in the original (or laminar) equations. These terms are called *Reynolds stresses*, and are included as follows:

$$\bar{\rho}\left(\frac{\partial \bar{u}}{\partial t} + \bar{u}\frac{\partial \bar{u}}{\partial x} + \bar{v}\frac{\partial \bar{u}}{\partial y} + \bar{w}\frac{\partial \bar{u}}{\partial z}\right) = \bar{\rho}g_x - \frac{\partial \bar{P}}{\partial x} + \mu\left(\frac{\partial^2 \bar{u}}{\partial x^2} + \frac{\partial^2 \bar{u}}{\partial y^2} + \frac{\partial^2 \bar{u}}{\partial z^2}\right) + \bar{\rho}\left(\frac{\partial \overline{u'^2}}{\partial x} + \frac{\partial \overline{u'v'}}{\partial y} + \frac{\partial \overline{u'w'}}{\partial z}\right)$$

$$\bar{\rho}\left(\frac{\partial \bar{v}}{\partial t} + \bar{u}\frac{\partial \bar{v}}{\partial x} + \bar{v}\frac{\partial \bar{v}}{\partial y} + \bar{w}\frac{\partial \bar{v}}{\partial z}\right) = \bar{\rho}g_y - \frac{\partial \bar{P}}{\partial y} + \mu\left(\frac{\partial^2 \bar{v}}{\partial x^2} + \frac{\partial^2 \bar{v}}{\partial y^2} + \frac{\partial^2 \bar{v}}{\partial z^2}\right) + \bar{\rho}\left(\frac{\partial \overline{u'v'}}{\partial x} + \frac{\partial \overline{v'^2}}{\partial y} + \frac{\partial \overline{v'w'}}{\partial z}\right) \quad (4.9)$$

$$\bar{\rho}\left(\frac{\partial \bar{w}}{\partial t} + \bar{u}\frac{\partial \bar{w}}{\partial x} + \bar{v}\frac{\partial \bar{w}}{\partial y} + \bar{w}\frac{\partial \bar{w}}{\partial z}\right) = \bar{\rho}g_z - \frac{\partial \bar{P}}{\partial z} + \mu\left(\frac{\partial^2 \bar{w}}{\partial x^2} + \frac{\partial^2 \bar{w}}{\partial y^2} + \frac{\partial^2 \bar{w}}{\partial z^2}\right) + \bar{\rho}\left(\frac{\partial \overline{u'w'}}{\partial x} + \frac{\partial \overline{v'w'}}{\partial y} + \frac{\partial \overline{w'^2}}{\partial z}\right)$$

The net effect of these Reynolds stresses is to increase the diffusivity of the flow, relative to a laminar flow in a comparable geometry.

4.2 Laminar Flow Solutions

When the geometry is sufficiently simple, in some cases exact solutions can be found to the Navier–Stokes equations when the flow is laminar. In the next section we examine such a problem, the two-dimensional flat plate boundary layer.

4.2.1 Flat Plate Boundary Layers

One use of the Navier-Stokes equations is to obtain the magnitude of the viscous drag that objects experience aerodynamically when they move through a fluid. Before examining more complex cases in Chapter 7, we will start with the simple case of fluid moving parallel to a flat plate with a sharp leading edge, as shown in Figure 4.5. When a fluid moves over a flat plate with a velocity parallel to the surface of the plate, directly on the object's surface the velocity must go to zero due to the no-slip boundary condition. At sufficiently far enough distances from the solid object, its effects are not felt and the fluid moves unimpeded with its original velocity.

In between the outer region of constant velocity and the surface of the plate with zero velocity, there exists a *boundary layer* of fluid that provides smooth transition and a continuous velocity profile. The discovery of the boundary layer concept was one of many accomplishments of the German researcher Ludwig Prandtl in the early part of the 20th century. Practical examples of plane boundary layers include the flow of air over the wings and fuselage of an airplane and over the top and sides of a car or truck, the flow of water over a boat or submarine, and the flow of wind over solid ground. Boundary layers over curved surfaces are of course more complex than boundary layers

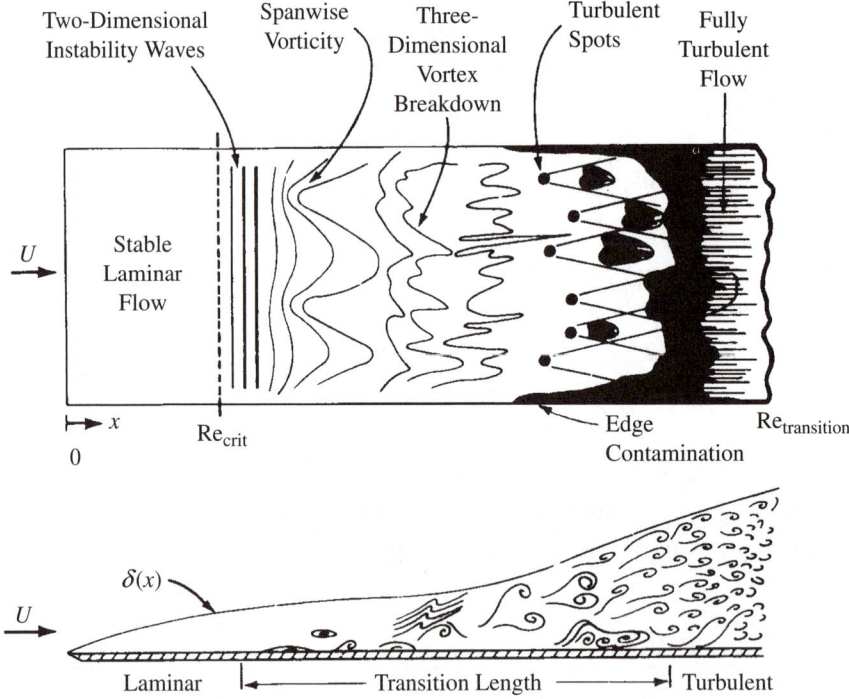

FIGURE 4.5 Schematic of flat plate boundary layers. From F. White, *Viscous Fluid Flow*, 1974, McGraw-Hill Education. Reproduced with permission of The McGraw-Hill Companies.

over perfectly flat surfaces, but they do bear qualitative similarities, and the flat plate boundary layer is much easier to deal with mathematically.

It is common practice in boundary layer flows to define the x-axis parallel with the flow direction and the surface of the solid plate, with its origin at the leading edge of the plate. The y-axis is perpendicular to the solid surface, and the z-axis is along the width of the plate. Recall that in Cartesian coordinates the full three-dimensional transient Navier-Stokes equations are

$$\frac{d\rho}{dt} = -\left(\frac{d(\rho u)}{dx} + \frac{d(\rho v)}{dy} + \frac{d(\rho w)}{dz}\right)$$

$$\rho\left(\frac{\partial u}{\partial t} + u\frac{\partial u}{\partial x} + v\frac{\partial u}{\partial y} + w\frac{\partial u}{\partial z}\right) = \rho g_x - \frac{\partial P}{\partial x} + \mu\left(\frac{\partial^2 u}{\partial x^2} + \frac{\partial^2 u}{\partial y^2} + \frac{\partial^2 u}{\partial z^2}\right)$$

$$\rho\left(\frac{\partial v}{\partial t} + u\frac{\partial v}{\partial x} + v\frac{\partial v}{\partial y} + w\frac{\partial v}{\partial z}\right) = \rho g_y - \frac{\partial P}{\partial y} + \mu\left(\frac{\partial^2 v}{\partial x^2} + \frac{\partial^2 v}{\partial y^2} + \frac{\partial^2 v}{\partial z^2}\right)$$

$$\rho\left(\frac{\partial w}{\partial t} + u\frac{\partial w}{\partial x} + v\frac{\partial w}{\partial y} + w\frac{\partial w}{\partial z}\right) = \rho g_z - \frac{\partial P}{\partial z} + \mu\left(\frac{\partial^2 w}{\partial x^2} + \frac{\partial^2 w}{\partial y^2} + \frac{\partial^2 w}{\partial z^2}\right)$$

Assuming that the flow along the width is uniform, the problem can be reduced to two dimensions in the xy-plane. Further assuming steady-state conditions and incompressible flow we can simplify the Navier–Stokes equations to

$$0 = \frac{du}{dx} + \frac{dv}{dy}$$

$$\rho\left(u\frac{\partial u}{\partial x} + v\frac{\partial u}{\partial y}\right) = -\frac{\partial P}{\partial x} + \mu\left(\frac{\partial^2 u}{\partial x^2} + \frac{\partial^2 u}{\partial y^2}\right)$$

$$\rho\left(u\frac{\partial v}{\partial x} + v\frac{\partial v}{\partial y}\right) = -\frac{\partial P}{\partial y} + \mu\left(\frac{\partial^2 v}{\partial x^2} + \frac{\partial^2 v}{\partial y^2}\right)$$

The boundary conditions are the no-slip condition at the solid surface, $u(y = 0) = 0$, and that the solid surface is impermeable (no fluid can go through it): $v(y = 0) = 0$. Additionally, at distances far enough from the plate the velocity approaches the free-stream fluid velocity: $u(y = \infty) = U_o$. But we still need to make some more assumptions to simplify this enough to be solved. First, we will assume the boundary layer has some finite thickness $\delta(x)$, so that for values of $y > \delta$, $u = U_o$. The boundary layer thickness usually ends up being quite small compared to other length scales of the flow. Second, we will assume a zero pressure-gradient boundary layer, so that $dP/dx = 0$.

This is reasonable for a flat plate. (In reality for airfoils, the pressure does change along x, which is what gives the lift force.) We also assume that things change more rapidly in the y-direction than in the x-direction, so that the second derivative can be neglected in x.

It also useful to define the *kinematic viscosity*, denoted by the Greek letter ν, where $\nu = \mu/\rho$.

$$\frac{du}{dx} + \frac{dv}{dy} = 0 \tag{4.10}$$

$$u\frac{\partial u}{\partial x} + v\frac{\partial u}{\partial y} = \nu\left(\frac{\partial^2 u}{\partial y^2}\right) \tag{4.11}$$

(Note: v is not zero for boundary layer flow, but it will be small compared to u.) We can approximate a solution by assuming a polynomial function of the form

$$u(x,y) = A + B\left(\frac{y}{\delta(x)}\right) + C\left(\frac{y}{\delta(x)}\right)^2$$

The no-slip boundary condition is $u(x, 0) = 0$, so $A = 0$. At the edge of the boundary layer, $u(x, \delta) = U_o = B + C$. We need one more condition, so we will assume the boundary layer merges smoothly with the outer flow, so that

$$\left.\frac{du}{dy}\right|_{y=\delta} = 0$$

That is, not only is the velocity profile continuous but the derivative of the velocity profile is also continuous. The result of this condition is $B + 2C = 0$. Combining this with the condition that $B + C = U_o$, we have two equations and two unknowns, so we can solve to get $B = 2U_o$ and $C = -U_o$. Then the approximate velocity profile is

$$u(x,y) = U_o\left(2\frac{y}{\delta} - \frac{y^2}{\delta^2}\right) \tag{4.12}$$

We still need to find values for $\delta(x)$. Unfortunately there is not a simple theory to find this, so we must rely on empirical measurements. Experiments for laminar flow show that

$$\frac{\delta}{x} \approx \frac{5}{\sqrt{Re_x}} \tag{4.13}$$

where the Reynolds number is always defined as the ratio of inertial forces to viscous forces. For flat plate boundary layers, the length scale used in defining the Reynolds number is the axial distance from the leading edge of the plate:

$$\text{Re}_x = \frac{\rho V x}{\mu} \tag{4.14}$$

So then the scaling is $\delta \sim \sqrt{x}$. This is only an approximate solution.

An exact solution to Equations 4.10 and 4.11 can be found without having to resort to approximations or empirical trends. In 1908 while working under Prandtl, Blasius derived the solution through the use of a variable transformation. He introduced the variable $\eta = y/\delta(x)$. This transformation implies that at each axial location x in the laminar boundary layer, there is some similarity in the shape of the velocity profile to the shape at other values of x. Through this he was able to transform the system of partial differential equations into a single third-order ordinary differential equation in terms of a function $f(\eta)$. The full derivation of Blasius's solution is beyond the scope of this book, but it is often included in graduate fluid mechanics classes. Blasius was not able to find an analytical solution to his differential equation, so he solved it numerically. His solution agrees with the experimental trends that the boundary layer thickness increases with the square root of the axial distance, x, indicated in Equation 4.13.

One of the chief advantages of knowing the velocity profile is that it allows us to calculate the shear stress at the wall since the derivative of the velocity profile at the wall can also be calculated:

$$\tau = \mu \left.\frac{\partial u}{\partial y}\right|_{y=0} = \frac{2\mu U_o}{5x}\sqrt{\frac{\rho U_o x}{\mu}} = 0.4\sqrt{\frac{\rho \mu U_o^3}{x}} \tag{4.15}$$

This is valid when the flow is laminar. In reality not all flows are laminar, because when disturbances in the flow grow and the viscosity is not strong enough to damp them out, turbulent chaotic motion ensues. So when is the flow laminar? For internal flow through a pipe, it is laminar up to a Reynolds number of about 2300, but for *external* flow, the flow can stay laminar up to a *critical Reynolds number* of

$$\text{Re}_{\text{crit}} = 500{,}000 \tag{4.16}$$

Thus for long lengths of plates and/or flows of high velocities, the flow tends to become turbulent. For example, for air at STP and a velocity of 20 m/s (about 45 mph), the critical length for transition to turbulence would be

$$x_{\text{crit}} = \text{Re}_{\text{crit}} \frac{\nu}{V} = 500{,}000 \frac{1.5 \times 10^{-5} \text{ m}^2/\text{s}}{20 \text{ m/s}} = 0.37 \text{ m} = 37 \text{ cm}$$

So the above boundary layer equations give reasonable answers for laminar flow, but what about turbulent flow? Unfortunately, there is no general solution for turbulent flow (which is the chief limitation of the theory of fluid dynamics). An empirical velocity profile is given by the 1/7 power law, which has been shown to be reasonably accurate for turbulent flows along a flat plate:

$$u(x, y) = U_o \left(\frac{y}{\delta}\right)^{1/7} \tag{4.17}$$

The corresponding relation for the boundary layer height is

$$\frac{\delta}{x} \approx \frac{0.16}{\text{Re}_x^{1/7}} \tag{4.18}$$

If you were to plot these with EXCEL or MATLAB, you would see that the turbulent boundary layer is thicker than the laminar one (it grows more quickly), and it has a steeper velocity profile. (See Figure 4.6.) The wall shear stress in turbulent flow is

$$\tau = \mu \frac{\partial u}{\partial y}\bigg|_{y=0} \approx \frac{\rho U_o^2}{\text{Re}_x^{1/7}} \tag{4.19}$$

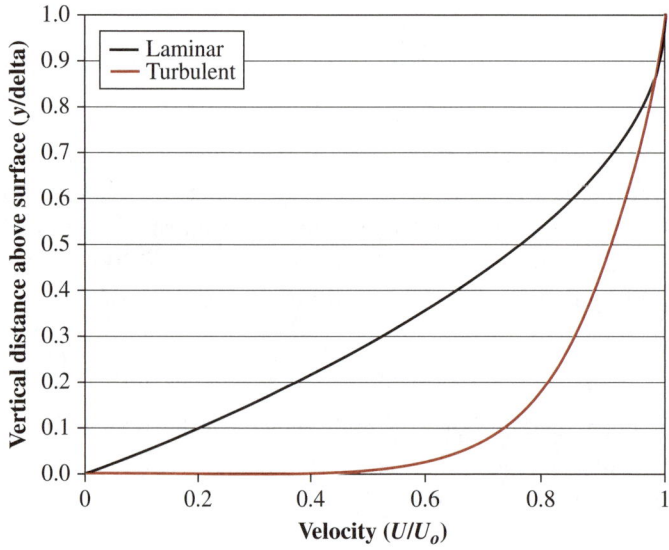

FIGURE 4.6 Comparison of laminar and turbulent velocity profiles in a boundary layer.

4.2 Laminar Flow Solutions

So then, given these equations (Equations 4.15 and 4.19) for shear stress, how would you calculate the friction drag on a wing or vehicle? The frictional drag force is equal to the integral of the shear stress over the area, but the shear stress does not stay constant in external flow:

$$F_D = \int_A \tau \, dA = \int_0^L \tau(x) W \, dx$$

Also note that the nondimensional drag coefficient is defined as

$$C_D = \frac{F_D}{0.5 \rho A U_o^2} \tag{4.20}$$

A popular correlation for the drag coefficient in turbulent boundary layers is

$$C_D = \frac{0.031}{\text{Re}_L^{1/7}} \tag{4.21}$$

This correlation is accurate for a range of Reynolds numbers from $500{,}000 < \text{Re}_x < 10^7$. For cases of mixed flow, where the front part of the boundary layer is laminar and the rear part is turbulent, the overall drag coefficient is

$$C_D = \frac{0.031}{\text{Re}_L^{1/7}} - \frac{1440}{\text{Re}_L} \tag{4.22}$$

The local nondimensional friction coefficient is

$$c_f = \frac{\tau}{0.5 \rho U_o^2} \tag{4.23}$$

A popular correlation for the skin friction coefficient in turbulent boundary layers is

$$c_f = \frac{0.027}{\text{Re}_x^{1/7}} \tag{4.24}$$

This correlation is accurate for a range of Reynolds numbers from $500{,}000 < \text{Re}_x < 10^7$. The difference between the skin friction coefficient and the drag coefficient is that the skin friction coefficient is a nondimensional measure of the local shear stress, whereas the drag coefficient is a nondimensional measure of the drag total force over the entire surface, obtained by integrating the local shear stress over the whole area.

Note the drag coefficient is defined for *external* flow. For *internal* flows, like pipe flow, a friction factor is defined instead. The nondimensional friction factor for *internal* flow is

$$f = \frac{8\tau}{\rho U_o^2} \tag{4.25}$$

where $f = 64/\text{Re}$ for laminar flow. Internal flows will be discussed in Chapter 5, and external flows in Chapter 6.

4.2.2 Streamlines, Streaklines, and Pathlines

A *streamline* (Figure 4.8) is defined as a set of points that are everywhere parallel to the local fluid motion. A *pathline* is the line in space traced out by a single particle of fluid released into the flow at some initial time, t_o. A *streakline* is the locus of points that all passed through a particular point of the flow at some time earlier. Physically, a streakline is created when a small dye injector is placed into a flow for visualization purposes and dye is continually injected into the flow. Figure 4.7 shows streamlines created in water by dye injection, and Figure 4.8 shows streamlines created by smoke in air. All three types of lines become the same for a steady-state

FIGURE 4.7 Flow near the intake of an F-18 model in a water tunnel. The boundary layer splitter plate in front of the inlet can be seen, which serves to keep the inlet out of the fuselage boundary layer so it will reach higher velocities. Courtesy of NASA Dryden Archives.

FIGURE 4.8 Streamlines around a baseball, made visible by smoke.

flow. Longtime exposures with photographic equipment can be used to capture pathlines, and from them to get estimates of the velocity. Pathlines are also called *timelines*.

4.2.3 Creeping Flow

A *creeping flow* is a flow where the fluid motion is very slow; the flow is said to creep along its path. Very low velocities imply very low Reynolds numbers and laminar flows. Further, they also imply that the viscous forces will be strong compared to the inertial or momentum forces. Thus in the limit of creeping flow the governing equations can be reduced in complexity. In Cartesian coordinates, the Navier–Stokes equations can be simplified to give the governing equations for creeping flow. Assuming steady-state flow and neglecting the convective terms, the resulting equations are

$$\rho g_x - \frac{\partial P}{\partial x} + \mu\left(\frac{\partial^2 u}{\partial x^2} + \frac{\partial^2 u}{\partial y^2} + \frac{\partial^2 u}{\partial z^2}\right) = 0$$

$$\rho g_y - \frac{\partial P}{\partial y} + \mu\left(\frac{\partial^2 v}{\partial x^2} + \frac{\partial^2 v}{\partial y^2} + \frac{\partial^2 v}{\partial z^2}\right) = 0 \qquad (4.26)$$

$$\rho g_z - \frac{\partial P}{\partial z} + \mu\left(\frac{\partial^2 w}{\partial x^2} + \frac{\partial^2 w}{\partial y^2} + \frac{\partial^2 w}{\partial z^2}\right) = 0$$

(Note that these are the opposite of the Euler equations, which were formulated for the limit of high Reynolds number.) The extremely low velocities also imply that the flow is incompressible.

Probably the earliest example of an analytical solution for creeping flow is Stokes' solution for flow over a sphere. Stokes formulated the problem of flow past a sphere of radius a at low Reynolds number in a spherical coordinate system, with the flow being symmetric about the axis of the sphere aligned with the flow direction. The velocity of the fluid relative to the sphere is U. The application of this geometry is the fall of a solid sphere through a very viscous liquid. The boundary conditions are (1) the no-slip condition at the surface of the sphere:

$$u_\theta = 0 \quad \text{at } r = a$$

(2) that the surface of the sphere is impermeable:

$$u_r = 0 \quad \text{at } r = a$$

and finally (3) that at distances sufficiently far enough away from the sphere the velocity matches the free-stream velocity:

$$u_\theta = U \sin \theta \quad \text{as } r \to \infty$$

$$u_r = U \cos \theta \quad \text{as } r \to \infty$$

Stokes was able to find a solution to this two-dimensional system for the two velocity components and the pressure:

$$u_r = U \cos \theta \left(1 - \frac{3a}{2r} + \frac{a^3}{2r^3} \right)$$

$$u_\theta = U \sin \theta \left(1 - \frac{3a}{4r} + \frac{a^3}{4r^3} \right)$$

$$P - P_o = -\frac{3}{2} \frac{\mu U a}{r^2} \cos \theta$$

Figure 4.9 shows the streamlines around a body of revolution in creeping flow.

From Stokes' solution, the viscous shear stress and the normal pressure stress on the surface of the sphere can be calculated. Once these stresses are integrated over the entire surface, the drag force can be calculated:

$$F_D = 6\pi \mu U a \tag{4.27}$$

FIGURE 4.9 Streamlines around a cylinder in creeping flow. Note the symmetry of the flow field. Adapted from *Visualized Flow*, Japan Society of Mechanical Engineers, Pergamon Press, 1988.

From this drag force, the nondimensional drag coefficient for a sphere is

$$C_D = \frac{24}{\text{Re}} \tag{4.28}$$

Of course, this equation is valid only for laminar flow. Well before transition to turbulence is reached, however, the wake region begins to grow and vortices are shed off the back of the sphere. This will be discussed in more detail in Chapter 6. Equation 4.28 is quantitatively accurate up to a Reynolds number of $\text{Re} = 1$.

From the drag force, a *settling velocity* can be calculated. The settling velocity is the steady-state velocity a falling particle reaches if it is left free to fall only under the forces of gravity and air resistance. For larger objects, the phrase *terminal velocity* is used. The settling velocities for small particles tend to be small, and hence the Reynolds numbers are small and Stokes' law applies.

For particles for which Stokes' law is valid, the settling velocity is

$$V_{\text{TS}} = \frac{2}{9} \frac{(\rho_{\text{part}} - \rho_{\text{fluid}}) g a^2}{\mu} \tag{4.29}$$

This equation accounts for the buoyant force, as well as the weight of the particle and viscous resistance. The ratio of the response time of a particle to a characteristic time scale of the fluid flow is called the *Stokes number*. The characteristic flow time is $\tau = U/D$.

$$\text{St} = \frac{\tau_{\text{particle}}}{\tau_{\text{flow}}} \tag{4.30}$$

For values of St $\gg$ 1, the particle's trajectory is governed by its own inertia; for St $\ll$ 1 the particle will follow the flow as a tracer particle.

EXAMPLE 4.3

A ball bearing of diameter 1 cm falls through glycerin contained in a cylinder. Estimate the terminal velocity.

SOLUTION: The density of the steel in the ball bearing will be about 7800 kg/m³, and the density of glycerin is about 1200 kg/m³. The viscosity of glycerin is about 1.2 kg/m-s at 23°C. (Note that the viscosity of glycerin varies rapidly with temperature changes.) Using the equation of the settling velocity, we have

$$V_{\text{TS}} = \frac{2}{9} \frac{\left(7800 \frac{\text{kg}}{\text{m}^3} - 1200 \frac{\text{kg}}{\text{m}^3}\right)\left(9.8 \frac{\text{m}}{\text{s}^2}\right)(0.005 \text{ m})^2}{1.2 \frac{\text{kg}}{\text{m} \cdot \text{s}}} = 0.299 \frac{\text{m}}{\text{s}}$$

Now we check the Reynolds number to verify that Stokes' law is valid:

$$\text{Re} = \frac{\rho V D}{\mu} = \frac{(1200 \text{ kg/m}^3)(0.3 \text{ m/s})(0.01 \text{ m})}{1.2 \text{ kg/m-s}} = 3$$

This is slightly greater than 1, so the drag force is probably slightly underpredicted and the terminal velocity slightly overpredicted.

Another application of creeping flow is flow through porous media. Porosity is usually defined as the ratio of the volume of the pores to the total volume of the material, and is denoted by the symbol ϕ. Examples of creeping or seeping flow through porous media include the flow of water through soil or porous rocks and the flow of oil through a filter bed. The average overall velocity or flow rate of the liquid is of more importance than the local velocity through an individual pore. A detailed discussion of such flow is beyond the scope of this book, but you may consult further references listed in Appendix E.

4.2.4 Lubrication Flow

In this section we will present three solutions for the Navier–Stokes equations, for two-dimensional cases of planar flow, pipe flow, and flow in a cylindrical journal bearing. In many practical examples of lubrication flows, viscosity dominates the flow. Two common examples are that of a bearing in a journal or slipper and a guide, or thrust block. If there is no motion, the journal will sink to the bottom of the bearing and contact it. However, if there is sufficient motion of the liquid lubricant, the journal will lift off and will not touch the bearing surface. Thus the lubricant prevents the two solid bodies from touching at all.

Flow Between Two Parallel Plates

In Cartesian coordinates the full three-dimensional transient Navier-Stokes equations are:

$$\frac{d\rho}{dt} = -\left(\frac{d(\rho u)}{dx} + \frac{d(\rho v)}{dy} + \frac{d(\rho w)}{dz}\right)$$

$$\rho\left(\frac{\partial u}{\partial t} + u\frac{\partial u}{\partial x} + v\frac{\partial u}{\partial y} + w\frac{\partial u}{\partial z}\right) = \rho g_x - \frac{\partial P}{\partial x} + \mu\left(\frac{\partial^2 u}{\partial x^2} + \frac{\partial^2 u}{\partial y^2} + \frac{\partial^2 u}{\partial z^2}\right)$$

$$\rho\left(\frac{\partial v}{\partial t} + u\frac{\partial v}{\partial x} + v\frac{\partial v}{\partial y} + w\frac{\partial v}{\partial z}\right) = \rho g_y - \frac{\partial P}{\partial y} + \mu\left(\frac{\partial^2 v}{\partial x^2} + \frac{\partial^2 v}{\partial y^2} + \frac{\partial^2 v}{\partial z^2}\right)$$

$$\rho\left(\frac{\partial w}{\partial t} + u\frac{\partial w}{\partial x} + v\frac{\partial w}{\partial y} + w\frac{\partial w}{\partial z}\right) = \rho g_z - \frac{\partial P}{\partial z} + \mu\left(\frac{\partial^2 w}{\partial x^2} + \frac{\partial^2 w}{\partial y^2} + \frac{\partial^2 w}{\partial z^2}\right)$$

The first approximation we will make is to limit the problem to a two-dimensional case, so any terms involving z or w can be eliminated (assuming g acts in the z-direction):

$$\frac{d\rho}{dt} = -\left(\frac{d(\rho u)}{dx} + \frac{d(\rho v)}{dy}\right)$$

$$\rho\left(\frac{\partial u}{\partial t} + u\frac{\partial u}{\partial x} + v\frac{\partial u}{\partial y}\right) = -\frac{\partial P}{\partial x} + \mu\left(\frac{\partial^2 u}{\partial x^2} + \frac{\partial^2 u}{\partial y^2}\right)$$

$$\rho\left(\frac{\partial v}{\partial t} + u\frac{\partial v}{\partial x} + v\frac{\partial v}{\partial y}\right) = -\frac{\partial P}{\partial y} + \mu\left(\frac{\partial^2 v}{\partial x^2} + \frac{\partial^2 v}{\partial y^2}\right)$$

The z-momentum equation drops out in its entirety. Let us further simplify the problem by assuming we are looking at a steady, incompressible flow. Thus,

$$0 = \frac{du}{dx} + \frac{dv}{dy}$$

$$\rho\left(u\frac{\partial u}{\partial x} + v\frac{\partial u}{\partial y}\right) = -\frac{\partial P}{\partial x} + \mu\left(\frac{\partial^2 u}{\partial x^2} + \frac{\partial^2 u}{\partial y^2}\right)$$

$$\rho\left(u\frac{\partial v}{\partial x} + v\frac{\partial v}{\partial y}\right) = -\frac{\partial P}{\partial y} + \mu\left(\frac{\partial^2 v}{\partial x^2} + \frac{\partial^2 v}{\partial y^2}\right)$$

To simplify this any further we need to know the geometry of the flow so that we can apply boundary conditions and see if any further assumptions can be made. Let us consider the flow between two large parallel plates, at a distance apart of $2H$, with the flow going in the x-direction. Since the plates are solid walls, no fluid can pass through them, so $v(x, H) = v(x, -H) = 0$. The no-slip boundary condition also applies for almost all macroscopic applications, so $u(x, H) = u(x, -H) = 0$. Also note that in a fully developed, steady-state flow, not only can fluid not move through the walls, it cannot accumulate at the centerline either, so $v = 0$ at all values of x and y. This realization simplifies things a great deal, because if $v = 0$, then $dv/dx = 0$ and $dv/dy = 0$. Then we have

$$\frac{du}{dx} = 0$$

$$\rho\left(u\frac{\partial u}{\partial x}\right) = -\frac{\partial P}{\partial x} + \mu\left(\frac{\partial^2 u}{\partial x^2}\right)$$

$$0 = -\frac{\partial P}{\partial y}$$

Since $v = 0$ everywhere, the result in the conservation of y-momentum is that $dP/dy = 0$. The result of the continuity equation is that $du/dx = 0$, or $u = u(y)$ only. We can now substitute this into the x-momentum equation to simplify it.

$$0 = -\frac{\partial P}{\partial x} + \mu\left(\frac{\partial^2 u}{\partial y^2}\right) \quad \text{or} \quad \frac{\partial P}{\partial x} = \mu\left(\frac{\partial^2 u}{\partial y^2}\right)$$

Since the left-hand side is a function of x only, and the right-hand side is a function of y only, the only way the equation can be true in general is if both sides are equal to a constant. So dP/dx = constant, which tells us that the pressure decreases linearly with distance along the flow:

$$\mu\left(\frac{\partial^2 u}{\partial y^2}\right) = C$$

Integrating both sides with respect to y yields

$$\mu\left(\frac{\partial u}{\partial y}\right) = Cy + C_2$$

and integrating again with respect to y gives

$$u = \frac{1}{\mu}\left(\frac{C}{2}y^2 + C_2 y + C_3\right)$$

Now the boundary conditions for u can be applied. We have the two conditions $u(x, H) = u(x, -H) = 0$, but we have three unknown constants. So we need one more condition, which we get by assuming symmetry at the centerline (or center plane): $du/dy = 0$ at $y = 0$. Applying this last condition first yields $\mu(0) = C(0) + C_2$. So $C_2 = 0$. The other two boundary conditions are

$$0 = \frac{1}{\mu}\left(\frac{C}{2}H^2 + C_3\right) \quad \text{so} \quad C_3 = -\frac{C}{2\mu}H^2 \quad \text{and} \quad u = \frac{1}{\mu}\left(\frac{C}{2}y^2 - \frac{C}{2}H^2\right) = \frac{C}{2\mu}(y^2 - H^2)$$

Unfortunately, plugging in $y = -H$ gives the same formula, so we still need one more condition. At this point, can you guess what the sign of the constant C will be? If the total mass flow rate, $\dot{m}$, is known over a width W between the two plates, then we can use the relation

$$\dot{m} = \int_A \rho V\, dA = \int_{-H}^{+H} \rho\left[\frac{C}{2\mu}(y^2 - H^2)\right](W\, dy) = \frac{C\rho W}{2\mu}\int_{-H}^{+H}(y^2 - H^2)\, dy$$

Evaluation of the integral yields

$$\dot{m} = \frac{C\rho W}{2\mu}\left(\frac{y^3}{3} - H^2 y\right)\bigg|_{-H}^{+H} = \frac{C\rho W}{2\mu}\left(\frac{H^3}{3} - H^3 + \frac{H^3}{3} - H^3\right) = -\frac{2C\rho W H^3}{3\mu}$$

So then $C = -\dfrac{3\mu \dot{m}}{2\rho W H^3}$ and

$$u(y) = \dfrac{3\dot{m}}{4\rho W H^3}(H^2 - y^2) \tag{4.31}$$

is the velocity profile. Additionally, the pressure drop between the plates is

$$\dfrac{dP}{dx} = \dfrac{3\dot{m}\mu}{2\rho W H^3} \tag{4.32}$$

Also, the shear stress at the wall can be calculated using Newton's formula, $\tau = \mu\, du/dy$:

$$\tau = \mu \dfrac{3\dot{m}}{4\rho W H^3}(2H) = \dfrac{3\dot{m}\mu}{2\rho W H^2} \tag{4.33}$$

Let us now consider a rectangular control volume of length L along x, and perform a macroscopic force analysis. Since the pressure decreases in x, there is an unbalanced pressure force along x of magnitude $(dP/dx)\,AL$, where $A = WH$, that must be resisted by the shear stress at the wall. The total shear force on this control volume is $F = \tau LW$. Comparing the two formulas above, we note that they differ by a factor of H, so the forces do in fact balance. Note also the following trends: the pressure drop increases with increasing viscosity or flow rate, but decreases as the distance between the plates is increased. This type of planar flow is termed *Poiseuille flow*. This flow is usually stable (laminar) up to a Reynolds number of about 1000.

Cylindrical Pipe Flow

For problems of pipe flow and journal bearing flow, we must write the three-dimensional Navier-Stokes equations in cylindrical coordinates. For expediency, only the final forms of the equations are shown here:

$$\dfrac{d\rho}{dt} = -\left(\dfrac{1}{r}\dfrac{\partial}{\partial r}\rho r v_r + \dfrac{1}{r}\dfrac{\partial}{\partial \theta}\rho v_\theta + \dfrac{\partial}{\partial z}\rho v_z\right) \tag{4.34}$$

$$\rho\left(\dfrac{\partial v_r}{\partial t} + v_r\dfrac{\partial v_r}{\partial r} + \dfrac{v_\theta}{r}\dfrac{\partial v_r}{\partial \theta} + v_z\dfrac{\partial v_r}{\partial z}\right) = \rho g_r - \dfrac{\partial P}{\partial r} + \mu\left[\dfrac{\partial}{\partial r}\left(\dfrac{1}{r}\dfrac{\partial}{\partial r}(r v_r)\right) + \dfrac{1}{r^2}\dfrac{\partial^2 v_r}{\partial \theta^2} + \dfrac{\partial^2 v_r}{\partial z^2} - \dfrac{2}{r^2}\dfrac{\partial v_\theta}{\partial \theta}\right]$$

$$\rho\left(\dfrac{\partial v_\theta}{\partial t} + v_r\dfrac{\partial v_\theta}{\partial r} + \dfrac{v_\theta}{r}\dfrac{\partial v_\theta}{\partial \theta} + v_z\dfrac{\partial v_\theta}{\partial z}\right) = \rho g_\theta - \dfrac{1}{r}\dfrac{\partial P}{\partial \theta} + \mu\left[\dfrac{\partial}{\partial r}\left(\dfrac{1}{r}\dfrac{\partial}{\partial r}(r v_\theta)\right) + \dfrac{1}{r^2}\dfrac{\partial^2 v_\theta}{\partial \theta^2} + \dfrac{\partial^2 v_\theta}{\partial z^2} - \dfrac{2}{r^2}\dfrac{\partial v_r}{\partial \theta}\right] \tag{4.35}$$

$$\rho\left(\dfrac{\partial v_z}{\partial t} + v_r\dfrac{\partial v_z}{\partial r} + \dfrac{v_\theta}{r}\dfrac{\partial v_z}{\partial \theta} + v_z\dfrac{\partial v_z}{\partial z}\right) = \rho g_z - \dfrac{\partial P}{\partial z} + \mu\left[\dfrac{1}{r}\dfrac{\partial}{\partial r}\left(r\dfrac{\partial v_z}{\partial r}\right) + \dfrac{1}{r^2}\dfrac{\partial^2 v_z}{\partial \theta^2} + \dfrac{\partial^2 v_z}{\partial z^2}\right]$$

Now let us simplify the problem again by assuming two-dimensional, incompressible, steady-state flow, with the azimuthal (θ) direction being the one with no significant motion, and neglecting gravity. We have

$$0 = \left(\frac{1}{r}\frac{\partial}{\partial r}rv_r + \frac{\partial}{\partial z}v_z\right)$$

$$\rho\left(v_r\frac{\partial v_r}{\partial r} + v_z\frac{\partial v_r}{\partial z}\right) = -\frac{\partial P}{\partial r} + \mu\left[\frac{\partial}{\partial r}\left(\frac{1}{r}\frac{\partial}{\partial r}(rv_r)\right) + \frac{\partial^2 v_r}{\partial z^2}\right]$$

$$\rho\left(v_r\frac{\partial v_z}{\partial r} + v_z\frac{\partial v_z}{\partial z}\right) = -\frac{\partial P}{\partial z} + \mu\left[\frac{1}{r}\frac{\partial}{\partial r}\left(r\frac{\partial v_z}{\partial r}\right) + \frac{\partial^2 v_z}{\partial z^2}\right]$$

Let us now look at flow through a circular pipe of radius R. It seems logical that $v_r = 0$, since in steady state the flow cannot go through the wall or accumulate in the middle, so the equations will simplify some more:

$$0 = \frac{\partial v_z}{\partial z} \quad \text{and} \quad 0 = -\frac{\partial P}{\partial r} \quad \text{and} \quad \rho\left(v_z\frac{\partial v_z}{\partial z}\right) = -\frac{\partial P}{\partial z} + \mu\left[\frac{1}{r}\frac{\partial}{\partial r}\left(r\frac{\partial v_z}{\partial r}\right) + \frac{\partial^2 v_z}{\partial z^2}\right]$$

Now we can substitute from continuity into the z-momentum:

$$\frac{\partial P}{\partial z} = \mu\left[\frac{1}{r}\frac{\partial}{\partial r}\left(r\frac{\partial v_z}{\partial r}\right)\right] = C_1$$

Then we integrate to find the velocity profile, v_z, as a function of r:

$$\left(r\frac{\partial v_z}{\partial r}\right) = \frac{C_1}{\mu}\frac{r^2}{2} + C_2 \quad \text{and} \quad v_z = \frac{C_1}{\mu}\frac{r^2}{4} + C_2 \ln(r) + C_3$$

At this point, we have to apply boundary conditions to find the three constants. The range of possible values for r is $0 < r < R$, but $\ln(0)$ is undefined, so the only way the formula can work is if $C_2 = 0$. Next, $(R) = 0$ by the no-slip boundary condition, so

$$0 = \frac{C_1}{\mu}\frac{R^2}{4} + C_3 \quad \text{or} \quad C_3 = -\frac{C_1}{\mu}\frac{R^2}{4} \quad \text{so} \quad v_z(r) = \frac{C_1}{4\mu}(r^2 - R^2)$$

And now we can assume that the mass flow rate will be known, so that

$$\dot{m} = \int_A \rho V \, dA = \int_0^R \rho\left[\frac{C}{4\mu}(r^2 - R^2)\right](2\pi r \, dr) = \frac{\pi C \rho}{2\mu}\int_0^R (r^2 - R^2) r \, dr$$

and upon integrating we have

$$\dot{m} = \frac{\pi C \rho}{2\mu} \int_0^R (r^2 - R^2) r \, dr = \frac{\pi C \rho}{2\mu} \left(\frac{r^4}{4} - \frac{R^2 r^2}{2} \right) \Big|_0^R = -\frac{\pi C \rho R^4}{8\mu}$$

so $C = -\dfrac{8\mu \dot{m}}{\pi \rho R^4}$ and

$$v_z(r) = \frac{2\dot{m}}{\pi \rho R^4}(R^2 - r^2) = \frac{2\dot{m}}{\pi \rho R^2}\left[1 - \left(\frac{r}{R}\right)^2\right] \tag{4.36}$$

This shows that the axial velocity profile in a laminar, steady, fully developed pipe flow is parabolic. As with the planar case, we can also calculate the shear stress and pressure drop. The maximum velocity will be at the pipe centerline, where $r = 0$. So

$$V_{max} = \frac{2\dot{m}}{\pi \rho R^2}$$

And so the velocity profile can also be expressed relative to the maximum velocity as

$$V(r) = V_{max}\left[1 - \left(\frac{r}{R}\right)^2\right] \tag{4.37}$$

Journal Bearings

The last type of problem employs a slightly different geometry, that of a journal bearing, which is cylindrical. For this problem $v_z = 0$, but v_r and v_θ will vary with r (still assuming that the problem is axisymmetric so that the velocities do not change with θ). So the governing equations are

$$0 = -\left(\frac{1}{r}\frac{\partial}{\partial r} r v_r\right)$$

$$\rho\left(v_r \frac{\partial v_r}{\partial r}\right) = -\frac{\partial P}{\partial r} + \mu\left[\frac{\partial}{\partial r}\left(\frac{1}{r}\frac{\partial}{\partial r}(r v_r)\right)\right]$$

$$\rho\left(v_r \frac{\partial v_\theta}{\partial r}\right) = \mu\left[\frac{\partial}{\partial r}\left(\frac{1}{r}\frac{\partial}{\partial r}(r v_\theta)\right)\right]$$

4.2 Laminar Flow Solutions

Upon integrating the continuity equation, we get $rv_r = C$. Substituting from continuity that $v_r = C/r$ into the momentum equation gives

$$\rho\left[v_r \frac{\partial}{\partial r}\left(\frac{C}{r}\right)\right] = -\frac{\partial P}{\partial r} + \mu\left[\frac{\partial}{\partial r}\left(\frac{1}{r}\right)\right]$$

This simplifies the momentum equation to

$$-\rho\frac{C^2}{r^3} = -\frac{\partial P}{\partial r} - \frac{\mu}{r^2}$$

The boundary conditions are $v_\theta(R_2) = 0$ and $v_\theta(R_1) = \omega R_1$. If $v_r = 0$, then

$$\frac{1}{2}C_1 r^2 + C_2 = rv_\theta \quad \text{and} \quad v_\theta = \frac{C_1}{2}r + \frac{C_2}{r}$$

C_1 and C_2 can be computed using the boundary conditions, $v_\theta(R_2) = 0$ and $v_\theta(R_1) = \omega R_1$.

$$0 = \frac{C_1}{2}R_2 + \frac{C_2}{R_2} \quad \text{and} \quad \omega R_1 = \frac{C_1}{2}R_1 + \frac{C_2}{R_1}$$

Solving for the unknown constants yields

$$C_1 = -\frac{2\omega R_1/R_2}{\left(1 - \frac{R_1}{R_2}\right)} \quad \text{and} \quad C_2 = -\frac{\omega R_1 R_2}{\left(1 - \frac{R_1}{R_2}\right)}$$

Thus the velocity profile can be expressed as

$$v_\theta = \frac{\omega R_1}{\left(1 - \frac{R_1}{R_2}\right)}\left(\frac{R_2}{r} - \frac{r}{R_2}\right) \tag{4.38}$$

This calculation is for the case of the inner cylinder rotating and the outer cylinder fixed. It is just as easy to find the velocity profile for the case of the inner cylinder fixed and outer cylinder moving, or both cylinders being allowed to move. The most general equation for either or both cylinders rotating is

$$v = \frac{\omega_2 R_2^2(r^2 - R_1^2) + \omega_1 R_1^2(R_2^2 - r^2)}{r(R_2^2 - R_1^2)} \tag{4.39}$$

With the velocity profile known, the shear stress can be calculated, as can the resisting torque and power required to turn the bearing. Note that if the gap thickness in the bearing is small compared to the diameter of the bearing, the viscous liquid film can be approximated as having the same velocity profile as a liquid between two flat plates.

Implicit in the three types of lubrication flow problem we have discussed is the assumption of laminar flow. Most lubrication applications operate at extremely low Reynolds numbers, so that not only is the flow laminar but the viscous terms dominate over the inertial or convective terms in the Navier–Stokes equations. However, there are a few applications at high speed. One example is the seals for the liquid hydrogen turbopumps used in the main engine of the space shuttle. The Reynolds number there is on the order of 10^5.

EXAMPLE 4.4

An automobile crankshaft rotates at the engine speed of 2500 rpm. A journal bearing along the crankshaft is lubricated with engine oil (SAE 5W-30) at the oil temperature of 95°C. The geometry of the bearing is that it has a length of 3 cm and an inner diameter of 7 cm with a gap thickness of 0.075 mm. If there are four journal bearings along the crankshaft, what is the total torque and power required to spin them all?

SOLUTION: The rotational speed of 2500 rpm = 2500 rev/min(1 min/60 s) (2 π rad/1 rev) = 262 rad/s. The linear velocity at the inner bearing surface is (262 rad/s)(0.035 m) = 9.17 m/s. The shear stress on the inner cylinder is

$$\tau = \mu \frac{du}{dy} = (0.009 \text{ kg/m} \cdot \text{s}) \frac{9.17 \text{ m/s}}{0.000075 \text{ m}} = 1100 \text{ Pa}$$

And the total torque is equal to the force times the radial distance from the shaft centerline to the surface:

$$T = FR_1 = (\tau A) R_1 = 2\pi \tau R_1^2 L = \frac{\pi}{2} \tau D_1^2 L$$

Substituting in the values, we have

$$T = \frac{\pi}{2}(1100 \text{ Pa})(0.07 \text{ m})^2(0.03 \text{ m}) = 0.25 \text{ N} \cdot \text{m}$$

The power is the product of the torque times the rotational speed, so

$$\dot{W} = T\omega = (0.25 \text{ N} \cdot \text{m})(262 \text{ rad/s}) = 66 \text{ W}$$

The length of the bearing is (30 mm/0.075 mm) = 400 times the gap thickness, so the assumption of two-dimensional flow is a good one.

One practical aspect of journal bearings is that they heat up when they are used as the viscosity changes. Another is that the shaft initially rests at the bottom of the journal until enough force is generated to push the fluid into the gap. If the gap thickness is small compared to the diameter of the bearing, the flow can be approximated as that between two infinite parallel plates, which simplifies the equations and calculations. The critical Reynolds number for transition to turbulence in journal bearing flows is around 1500. The length scale to be used in the Reynolds numbers for lubrication problems is the gap thickness.

Gas Bearings

In some applications bearings and seals are lubricated with gases instead of liquids. The flows in these devices follow the same basic principles as the lubrication flow of liquids. Since the viscosity of gases is usually two to three orders of magnitude smaller than that of common lubricating liquids, the friction losses will be smaller in gas-lubricated bearings. This is particularly important in applications where very high speeds are desired, such as over 1,000,000 rpm. Other advantages for gases relative to liquids include insensitivity to contamination and high temperature. The disadvantages include the dry friction during starting and a tendency towards instability due to the lack of damping. The gas compressibility also plays a role.

4.3 Introduction to Computational Fluid Dynamics

Computational fluid dynamics (CFD) is widely used in engineering to obtain approximation solutions to fluid flow problems. CFD software takes the Navier–Stokes equations and solves them approximately. To do this, approximations must be developed for each of the derivatives contained within the Navier–Stokes equations. These approximations can be derived using the Taylor series. (For those not interested in the derivation, skip ahead to Equation 4.42 to see the formula used to replace the derivative of a function with an algebraic approximation.)

4.3.1 Approximating Derivatives

If the value of a function and its derivatives are known at a particular location in space, x_o, but the value of the function in general is not known, the general value can be estimated using a *Taylor series*. Even if the function is known, the Taylor series has other uses, such as in the derivation of finite difference approximations to derivatives, which is how we use it here. Let $\Delta x = x - x_o$. (Later on Δx will represent the spacing between grid points in a CFD simulation.) Then the Taylor series about the point x_o can be expressed in terms of Δx:

$$f(x) = f(x_o) + f'(x_o)\Delta x + f''(x_o)\Delta x^2/2 + f'''(x_o)\Delta x^3/6 + \cdots + f^{(n)}(x_o)\Delta x^n/n \tag{4.40}$$

or in compact form as

$$f(x) = \sum_{k=0}^{n} \frac{f^{(k)}(x_o)}{k!}(x - x_o)^k \quad \text{for } n \to \infty \tag{4.41}$$

Of course it is not possible to compute all the derivatives of a function, so to make use of this in practical applications, we must truncate the series after a finite number of terms, n. In order notation, the truncation error of an $(n + 1)$-term Taylor series is $O(\Delta x^{n+1})$. For a finite number of terms, n, we can write $f(x) = P_n(x) + R_n(x)$, where $P_n(x)$ is the Taylor polynomial that approximates f in the vicinity of x_o, and R is the remainder or truncation error. So how big is R? For every x there exists $\xi(x)$ between x and x_o such that

$$R_n(x) = \frac{f^{(n+1)}(\xi(x))}{(n + 1)!}(x - x_o)^{n+1}$$

The term $\xi(x)$ represents the point where the derivative is highest between x and x_o, so that the worst possible error is obtained. To reduce the error, we make n larger or

Δx smaller. Both options incur increased computational costs (time). In many cases the computational engineer is willing to sacrifice accuracy to obtain results in a specified amount of time, if those results are qualitatively accurate enough to serve as a guide for the engineer's design of the system. While in research settings in universities and government labs, researchers may be willing to let a computation run for weeks or even months to obtain a highly accurate solution, in the business world of engineers who work on developing products on a deadline, a day is often the maximum allowable time to run a simulation. If a parametric study of different design alternatives must be done, the computational code may need to run on a time scale of an hour.

Thus far, we have used the Taylor series to show how we can approximate the value of a function. But with a little manipulation, we can also use the Taylor series to approximate the function's derivatives. We will start with the first derivative. We have the general Taylor series:

$$f(x) = f(x_o) + \frac{f'(x_o)}{1!}(x - x_o) + \frac{f''(x_o)}{2!}(x - x_o)^2 + \frac{f'''(x_o)}{3!}(x - x_o)^3 + \frac{f^{(4)}(x_o)}{4!}(x - x_o)^4 + \cdots$$

We rearrange this to solve for $f'(x_o)$:

$$f'(x_o) = \frac{f(x) - f(x_o)}{x - x_o} + \frac{f''(x_o)}{2!}(x - x_o) + \frac{f'''(x_o)}{3!}(x - x_o)^2 + \cdots$$

Using Δx notation, we can write this as

$$f'(x_o) = \frac{f(x_o + \Delta x) - f(x_o)}{\Delta x} + \frac{f''(x_o)}{2!}\Delta x + \frac{f'''(x_o)}{3!}\Delta x^2 + \cdots$$

So now we have a *finite difference* approximation to the derivative $f'(x)$. This will be useful later on. What if we needed to approximate the second derivative? With a little bit of trial and error we can show it is impossible to approximate the second derivative with only the data we have so far—it cannot be done with just two data points; we must have at least one more point. So all of the higher derivatives are assumed to be unknown, and the finite difference approximation is written as

$$f'(x_o) = \frac{f(x_o + \Delta x) - f(x_o)}{\Delta x} + O(\Delta x) \tag{4.42}$$

The mathematical notation $O(\Delta x)$ represents a term that is proportional to Δx. Using three reference points, $(x_o, x_o + \Delta x, x_o - \Delta x)$, we can construct two different Taylor series and combine them to eliminate the first derivative. Then we solve for the second derivative:

$$f''(x_o) = \frac{f(x_o + \Delta x) - 2f(x_o) + f(x_o - \Delta x)}{\Delta x^2} + \frac{1}{12}f^{(4)}(x_o)\Delta x^2 + \cdots \quad (4.43)$$

We could also construct a centered-difference approximation for the first derivative. Proceeding similarly to the above procedure would yield

$$f'(x_o) = \frac{f(x_o + \Delta x) - f(x_o - \Delta x)}{2\Delta x} + \frac{f'''(x_o)}{3}\Delta x^2 + \cdots \quad (4.44)$$

Now we have two different methods to approximate $f'(x)$, so let us see if we can determine which is better. The second method is more accurate, and its error decreases quadratically as the step size, Δx, is decreased. The first method has a linear trend, which we could have predicted by looking at the error term. For the first approximation the error term is $O(\Delta x)$, but for the second it is $O(\Delta x^2)$. Both these methods have the same computational cost (number of operations required). The first method is a one-sided difference, and the second is a centered difference. In general, centered differences give superior approximations for the same number of terms. This approximation is order Δx^2, and is superior the first-order approximation derived previously and also shown here.

$$\frac{df}{dx} \approx \frac{f(x + \Delta x) - f(x)}{\Delta x} \quad (4.45)$$

This one-sided difference formula is useful near boundary points at walls. In the limit of $\Delta x \to 0$, the exact derivative is obtained. For any finite value of Δx, there is an error in the approximation. For the above approximation, this error is linearly proportional to Δx. With these forms for approximations of the derivatives, the Navier–Stokes equations can be attacked *numerically*.

The Navier–Stokes equations represent the *physics* of the fluid problem to be solved. The term *numerics* refers to the mathematical technique used to solve those equations. Such techniques are often classified as either finite difference, finite element, or finite volume. In finite difference techniques, all of the derivatives in the governing equations are replaced with finite difference approximations similar to Equation 4.45. The *computational domain,* or the physical extent of the flow of interest in three-dimensional space, is mapped with grid points. The use of the finite difference approximations changes the system of differential equations into a system of algebraic equations, in which the value of a function f at a grid point is related to the value of the function f at the neighboring grid points, which are related to the values of f at their neighboring grid points, and so on. The function f could represent any of the three components of velocity or pressure or temperature. So for a total number of grid points, n, there will be n unknown values of f and n algebraic equations involving the values of f at the various grid points. These n equations have to be solved simultaneously in a matrix system. This is where the bulk of the computational effort is employed.

4.3.2 Solving Partial Differential Equations

Differential equations that employ two (or more) independent variables are termed *partial* differential equations, as opposed to *ordinary* differential equations, which only have one independent variable. The Navier-Stokes equations are partial differential equations, since the dependent variables (u, v, w, P) are functions of space and time (x, y, z, t). The *order* of a differential equation is equivalent to the highest derivative (first, second, etc.) that appears in the partial differential equation. A differential equation is classified as *linear* if there are no multiplications among dependent variables and their derivatives, so that all coefficients are functions of independent variables. The Navier–Stokes equations are nonlinear because of the convective terms.

The traditional three classes of partial differential equations (PDEs) are *parabolic*, *elliptic*, and *hyperbolic*. In general, different numerical techniques must be applied to each class. A parabolic PDE has a term with a first-order derivative in time, and terms with second-order derivatives in space. An elliptic PDE has multiple second-order spatial derivatives. Elliptic PDEs are seen in steady-state heat conduction problems. An example of an elliptic PDE is Laplace's equation,

$$\frac{\partial^2 f}{\partial x^2} + \frac{\partial^2 f}{\partial y^2} = 0 \tag{4.46}$$

Hyperbolic PDEs have second-order derivatives in both space and time.

4.3.3 Example Laminar Flow Solution with Finite Differences

We now look at an example solution of the Navier–Stokes equations using finite differences to approximate the derivatives for transient two-dimensional flow between two large parallel flat plates. Recall that the flow between two very wide parallel flat plates is known as Poiseuille flow. The height between the two plates is $2H$. The initial velocity of the fluid between the two plates is zero, and at time $t = 0$, a pressure gradient dP/dx is suddenly applied. For two-dimensional flow of an incompressible fluid in Cartesian coordinates with negligible buoyancy effects, the Navier-Stokes equations are

$$\frac{\partial u}{\partial x} + \frac{\partial v}{\partial y} = 0$$

$$\frac{\partial u}{\partial t} + u\frac{\partial u}{\partial x} + v\frac{\partial u}{\partial y} = -\frac{1}{\rho}\frac{\partial P}{\partial x} + \nu\left(\frac{\partial^2 u}{\partial x^2} + \frac{\partial^2 u}{\partial y^2}\right)$$

$$\frac{\partial v}{\partial t} + u\frac{\partial v}{\partial x} + v\frac{\partial v}{\partial y} = -\frac{1}{\rho}\frac{\partial P}{\partial y} + \nu\left(\frac{\partial^2 v}{\partial x^2} + \frac{\partial^2 v}{\partial y^2}\right)$$

These equations can be further simplified by recognizing that the flow is constrained by the geometry to be effectively horizontal for laminar flow, so that the vertical velocity, v, is zero. If $v = 0$, then by the conservation of mass equation $\partial u/\partial x = 0$, and $\partial^2 u/\partial x^2 = 0$. Eliminating the negligible terms from the axial (x) momentum equation gives the final equation to be solved:

$$\frac{\partial u}{\partial t} = -\frac{1}{\rho}\frac{\partial P}{\partial x} + \nu\frac{\partial^2 u}{\partial y^2} \qquad (4.47)$$

The driving pressure gradient, dP/dx, is specified to be constant (not a function of x). This is a parabolic type of partial differential equation. A solution can be arranged by replacing the derivatives with finite differences. A first-order forward difference is used for the time derivative, and a second-order centered difference for the spatial derivative.

$$\frac{\partial u}{\partial t} = \frac{u(y, t + \Delta t) - u(y, t)}{\Delta t} + O(\Delta t)$$

$$\frac{\partial^2 u}{\partial y^2} = \frac{u(y + \Delta y, t) - 2u(y, t) + u(y - \Delta y, t)}{(\Delta y)^2} + O(\Delta y)^2$$

Substituting these finite difference approximations back into the original partial differential equation and dropping the error terms gives

$$\frac{u(y, t + \Delta t) - u(y, t)}{\Delta t} = -\frac{1}{\rho}\frac{dP}{dx} + \nu\left[\frac{u(y + \Delta y, t) - 2u(y, t) + u(y - \Delta y, t)}{(\Delta y)^2}\right]$$

This equation can be rearranged to solve for the velocity at a future time in terms of the currently known velocities. Thus,

$$u(y, t + \Delta t) = u(y, t) + \Delta t\frac{-1}{\rho}\frac{dP}{dx} + \frac{\nu \Delta t}{(\Delta y)^2}[u(y + \Delta y, t) - 2u(y, t)$$

$$+ u(y - \Delta y, t)] \qquad (4.48)$$

Such a formulation is an *explicit* technique. This explicit method is also referred to as a *forward difference* method, and it is *conditionally stable*. Note that the accuracy of this formula would be expressed as $O(\Delta t + \Delta y^2)$. To examine the relative time step effects and help select an optimum time step, we can define the dimensionless *Fourier number* for the computational mesh (sometimes this is also called a *Courant number*):

$$\text{Fo} = \frac{\alpha \Delta t}{\Delta y^2} \tag{4.49}$$

The explicit method will be stable when all the coefficients for velocity are greater than or equal to 0. For a one-dimensional problem, that will be when $\text{Fo} < \frac{1}{2}$; for a two-dimensional problem, when $\text{Fo} < \frac{1}{4}$; and for a three-dimensional problem, when $\text{Fo} < \frac{1}{6}$. For this specific problem the stability criterion is that

$$\frac{\nu \Delta t}{(\Delta y)^2} < \frac{1}{2}$$

Implicit formulations are more stable, but more computationally intensive to solve.

EXAMPLE 4.5

We began this section by considering the transient flow between two large parallel flat plates. Suppose the distance between the plates is $2H = 2$ cm $= 0.02$ m, with an applied pressure gradient of 1.0 Pa/m. The fluid used is water ($\rho = 1000$ kg/m^3, $\nu = 1 \times 10^{-6}$ m^2/s). The spatial grid size is $\Delta y = 0.2$ cm $= 0.002$ m, and the time step is $\Delta t = 0.04$ s. Plot the steady-state spatial distribution of velocity and the transient centerline velocity.

SOLUTION: Figures 4.10 and 4.11 show the results for a stable solution, with a value of the Fourier number of 0.01. The spatial distribution is roughly parabolic, as expected from the analytical theoretical solution, though a smoother curve would be obtained if more grid points were used. From Figure 4.11, it can be seen that the steady-state velocity profile is reached after about 200 s.

Figure 4.12 shows the results for an unstable choice of time step. We kept the spatial grid size the same at 0.002 m, but increased the time step to 2.13 s, resulting in a Fourier number of 0.533, just above the cutoff value of 0.5 for a stable solution. The results are obviously in error. (The MATLAB code used to generate these results can be found in Appendix F.)

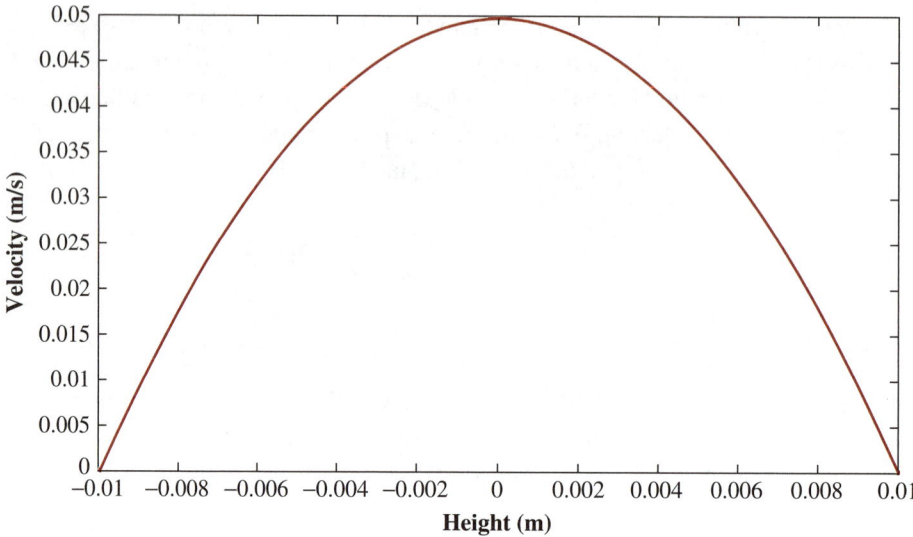

FIGURE 4.10 Computed steady-state spatial distribution of velocity across the channel in Poiseuille flow.

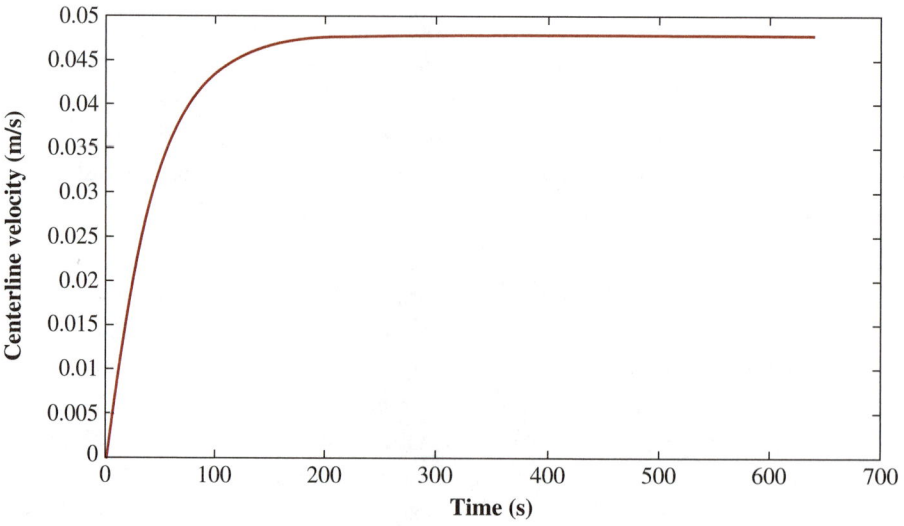

FIGURE 4.11 Computed transient centerline velocity for transient Poiseuille flow.

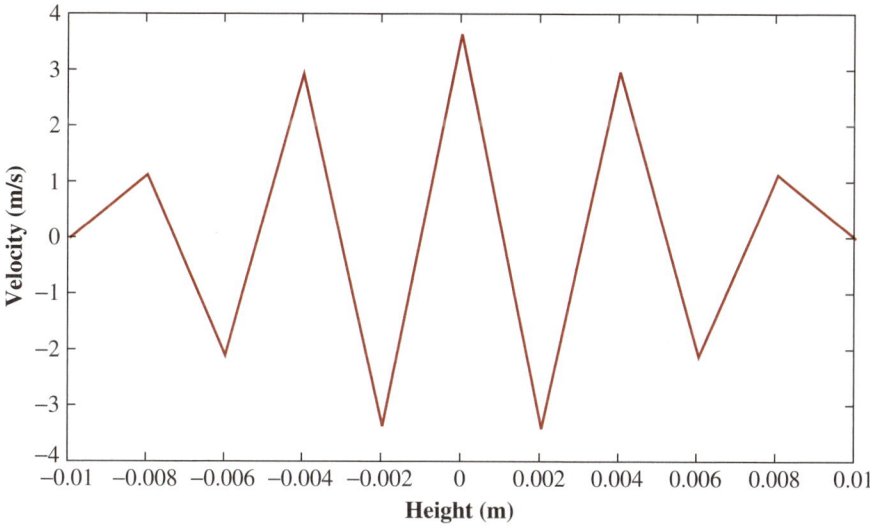

FIGURE 4.12 Spatial distribution of velocity in computed transient Poiseuille flow for an unstable choice of time step, Δt.

4.3.4 Turbulence Modeling

Computational simulations of laminar flows of Newtonian fluids are usually very accurate. However, additional challenges arise for turbulent flows. As discussed previously, the Reynolds stresses that arise in the Reynolds-averaged Navier–Stokes equations have to be modeled. The two most commonly used types of turbulence models are

- RANS (Reynolds-averaged Navier–Stokes equations). Simplest, quickest, least accurate, and provides the least information. Does not work well for oscillatory transient flows.
- LES (Large eddy simulations). A good compromise between accuracy and detail of a model and computational runtime.

Both of these are modeled in an Eulerian reference system. There have been some attempts to model turbulence in a Lagrangian framework, but such methods never really caught on.

Turbulence Characteristics

Integral length scale, l_I, is the largest scale of turbulence motion. It can be determined experimentally by making two-point velocity measurements simultaneously and performing a correlation analysis. The smallest scale is the *Kolmogorov scale*, l_k, which is the scale at which the viscous forces equal the inertial/kinetic forces. At scales smaller than the Kolmogorov scale viscosity damps out any turbulent motion. The ratio of Kolmogorov scale to integral scale, scales with the turbulence Reynolds number to the $-3/4$ power,

$$\frac{l_k}{l_I} \approx \text{Re}_t^{-3/4} \tag{4.50}$$

where Re_t is defined as

$$\text{Re}_t = \frac{u_{\text{RMS}} l_I}{\nu} \tag{4.51}$$

In engines, it has been found that the root mean square (RMS) turbulence intensity inside the cylinder scales with the mean piston speed: $u_{\text{RMS}} \sim \frac{1}{2} U_{\text{piston}}$. In a direct numerical simulation (DNS) *all* of the turbulent length scales must be resolved. In cases where DNS is not a viable solution (pretty much all practical problems), turbulent flow simulations require the introduction of a *turbulence model*. Large eddy simulations (LES) and the Reynolds-averaged Navier–Stokes equations (RANS) formulation, with the *k-ε* model or the Reynolds stress model, are two techniques for dealing with turbulence without resolving all the length scales.

In the *k-ε* model, *k* represents the average turbulence kinetic energy per unit mass (units of m^2/s^2), and ε is the dissipation rate of that turbulent kinetic energy (units of m^2/s^3). The effective turbulent viscosity or diffusivity is

$$\nu_t = C_\mu \frac{k^2}{\varepsilon} \tag{4.52}$$

The average turbulent length scale is

$$l \approx \frac{k^{3/2}}{\varepsilon} \tag{4.53}$$

The time scale is

$$\tau \approx \frac{k}{\varepsilon} \tag{4.54}$$

and the velocity scale is

$$u_{\text{RMS}} \approx \sqrt{k} \tag{4.55}$$

Variations on *k-ε* include re-normalization group (RNG) *k-ε*, realizable *k-ε*, and *k-ω*, among others.

RANS is not a model, but a framework that presents the transient Navier–Stokes equations in such a way that the Reynolds stresses are the only terms that have to be modeled. All the other terms can be solved directly. Thus an *ensemble* version of the governing equations is solved, which introduces these new apparent stresses known as Reynolds stresses. This adds a second-order tensor of unknowns for which various models can provide different levels of closure. The turbulence models used to close the equations are valid only as long as the time over which these changes in the mean occur is large compared to the time scales of the turbulent motion containing most of the energy.

Boussinesq Hypothesis

In 1877 Boussinesq postulated that the momentum transfer caused by turbulent eddies can be modeled with an eddy viscosity. This is in analogy with how the momentum transfer caused by the molecular motion in a gas can be described by a molecular viscosity. The Boussinesq assumption states that the Reynolds stress tensor, τ_{ij}, is proportional to the mean strain rate tensor, S_{ij}. More simply stated, this assumes that the rate of turbulence generation is proportional to the mean velocity gradients in the flow field. Thus,

$$-\rho \overline{u'_i u'_j} = \mu_t \left(\frac{\partial \overline{u}_i}{\partial x_j} + \frac{\partial \overline{u}_j}{\partial x_i} \right) \tag{4.56}$$

Here we have used tensor notation, where the subscript $i = 1, 2, 3$ corresponds to the three coordinate axes x, y, z. This method involves using an *algebraic* equation for the Reynolds stresses, which include determining the turbulent viscosity and solving transport equations for determining the turbulent kinetic energy and dissipation. The most popular model of this type is the k-ε model originally developed by Spalding, but other historical models include the mixing length model developed by Prandtl. The mixing length model is a zero-equation model because no transport equations are solved. On the other hand, the k-ε is a two-equation model because two transport equations are solved (one for k and one for ε).

A large eddy simulation (LES) is a technique in which the smaller eddies are filtered and are modeled using a sub-grid scale model, while the larger energy-carrying eddies are simulated. This method generally requires a more refined mesh than a RANS model, but a far coarser mesh than a DNS solution. Thus, in LES the large-scale motions of the flow are calculated, while the effect of the smaller universal scales (the so called sub-grid scales) are modeled using a sub-grid scale (SGS) model. The most commonly used SGS model is the Smagorinsky model. LES is able to predict instantaneous flow characteristics and resolve turbulent flow structures. Detached eddy simulation (DES) is a modification of LES.

Other complicating factors in CFD include phase change, chemical reaction, non-Newtonian liquids, and supersonic flows. Validation of codes usually refers to wind tunnel or other experiments used to check the accuracy of the results. Although validation using a specific test case does not imply that a CFD method or code is valid for all cases, it is necessary in order to use a code with confidence in the applicability of the solutions.

There are other advanced techniques for modeling fluid flows that will only be mentioned in passing here. These include Lattice Boltzmann methods, which are mesoscale methods—neither macroscopic or microscopic. Molecular Dynamics (MD) simulations of molecules using the Monte Carlo (DSMC) technique are commonly used for microscopic molecular-scale flows. A list of CFD vendors, including open-source codes, is included in Appendix F. All of these codes will include some variation of the k-ε model for turbulence, and more and more are coming with options for LES simulations. Other software packages serve as add-ons or complementary products to CFD tools, including EnSight or FIELDVIEW or Tecplot for postprocessing (visualizing three-dimensional vector fields).

4.3.5 Limitations of CFD

It is important to remember that CFD codes are an engineering tool, and as with any tool they have optimum uses as well as limitations. It is also important to note that CFD codes give approximate answers, not exact solutions, to the Navier–Stokes equations.

Major limitations of CFD:

- Numerical errors
- Modeling errors
- Application of boundary conditions
- Complex geometries
- Interpretation of results

Numerical errors include errors that arise from the fact that grid cells of finite size are used to map a continuous function. The smaller the grid cells are, the more accurate the numerical solution is, but there will always be some numerical error. If the problem is highly dynamic and occupies a wide range of scales, use of adaptive mesh refinement methods is recommended. Figure 4.13 shows an example in the change of a computed spray parameter when the grid size in the simulation is changed.

Modeling errors often arise from instances where our understanding of the basic fluid mechanics is incomplete. Examples include turbulence, the viscosity behavior of complex fluids like polymers and plastics, and the interaction between multiple fluids as in a spray.

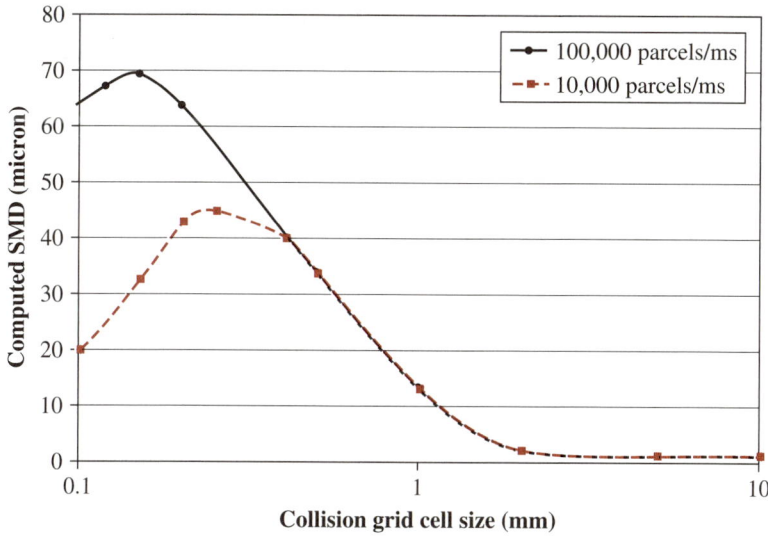

FIGURE 4.13 Example of grid dependency in a CFD simulation. As the grid size gets smaller, the computed solution changes.

Boundary conditions can also be difficult to apply. In the cases of solid boundaries, this is pretty straightforward. No-slip and no leakage can often be assumed. Open boundaries are more problematic. For example, if you were modeling the flow through the exhaust system of a sports bike, what condition do you apply at the end of the pipe where the exhaust exits into the atmosphere? If you specify either the pressure or the velocity profile, you presuppose the solution. For any model, the boundary conditions must be correctly applied. These can be closed or open, natural, specified pressure, or zero gradient.

A final limitation of CFD is the sheer amount of data a modern code can produce on a fast computer. Sometimes the amount of data can be overwhelming. The engineer must choose which parts of the flow field to examine, and how to present that data.

Summary

This chapter presented the derivation of the Navier–Stokes equations from conservation of mass and momentum on a differential control volume. The Navier–Stokes equations were applied to calculate the velocity profile for parallel flow over a flat plate to calculate the resistance force. The Navier–Stokes equations were also applied to lubrication flows. Then we discussed how computer codes arrive at approximate solutions to the Navier–Stokes equations for complex problems where the equations cannot be solved directly, and some caveats about the use of computational fluid dynamics (CFD) software codes were given. After reading this chapter and working through the problems, you should be able to apply the differential equations of conservation of mass, momentum, and energy principles to fluid flows, simplify those equations as appropriate for certain geometries, and determine which terms are significant and which are negligible.

■ References

[White] White, F. *Viscous Fluid Flow*. 1974. McGraw-Hill.

[JSME88] Japan Society of Mechanical Engineers. *Visualized Flow*. 1988. Pergamon Press.

■ Problems

1. What is Couette flow? Give the governing equations for Couette flow.

2. Define RMS fluctuation. If the velocity at a point in a flow is measured at discrete times as 4.4, 4.7, 5.1, 4.8, 5.2, 4.3, 5.1, 4.7, 5.5, 4.9, 4.4, 4.7, and 5.2 m/s, what are the mean and RMS velocities for this data set?

3. Write down the simplified Navier–Stokes equations for the case of steady-state two-dimensional flow of an incompressible fluid, with negligible viscosity and negligible body force (gravity). How many boundary conditions do you need to solve the resulting equations?

4. For the CFD software available at your school, how many different turbulence modeling options are available? What are the options?

5. For two parallel flat plates 1 cm apart, how long does it take for the boundary layers on the top and bottom plates to merge?

6. What is the shear force on a pizza delivery sign parallel to the direction of travel at 40 mph?

7. A journal bearing has inner and outer radii of 29 mm and 30 mm, respectively. The journal width is 50 mm. At a torque of 0.25 N-m the bearing turns at a speed of 3000 rpm Calculate the viscosity of the oil in the bearing.

8. A journal bearing is filled with SAE 50 oil. It has an inner radius of 25 mm and an outer radius of 26 mm. What will the torque be at a rotational speed of 2500 rpm?

9. A journal bearing is filled with SAE 30 oil. It has an inner radius of 24 mm and an outer radius of 25 mm. At a torque of 0.5 N-m, how fast can it spin?

10. How thick is the boundary layer over the wing of an airplane flying at 100 m/s at sea level if the wing is 1 m long?

11. How thick is the boundary layer at the end of a 1-m-wide wing for a plane traveling at sea level at 100 m/s, if the flow is entirely laminar?

12. As a journal bearing started up from rest proceeds to a steady-state condition, the temperature of the lubricant will rise due to frictional effects. How does this affect the torque needed to maintain a steady speed at the warmed-up condition, compared to the initial starting?

13. A journal bearing of length 28 cm is filled with SAE 10W-30 oil. The diameter of the inner shaft is 7 cm, and the thickness of the gap between the inner and outer surfaces is 0.5 mm. What is the torque required to achieve a rotational speed of 600 rpm?

14. A journal bearing has a length of 26 cm, and inner and outer radii of 31.2 mm and 32.0 mm, respectively. The bearing is initially filled with SAE 10W-30 oil at 20°C, and then when it reaches steady-state operation the temperature is 75°C. Find the torque required both at startup and at steady state to maintain a speed of 750 rpm.

15. What is the terminal velocity of a particle of alumina powder of diameter 2 microns?

16. For occupational safety, it is important to consider whether solid particles of dust in the air can be inhaled into the lungs and lodge there, where they can cause damage. The range of sizes of particles that are capable of lodging in the lungs is referred to as the *respirable range*. Why is the respirable range defined as particles having diameter between 0.5 and 2.5 microns?

17. What is the largest size of a water drop for which the Reynolds number is less than 1 and Stokes' law is applicable?

18. A chemical lab fume hood induces an upward velocity of 5 m/s in the air in the hood. What is the largest diameter of water drop that can be carried away by the hood?

19. At what size (diameter) do particles become so small that the no-slip boundary condition is no longer applicable? (Use a criterion that the settling velocity is underpredicted by 5% or more by using Stokes' law.)

20. What is the Cunningham correction factor?

21. A modern technique for turbulence modeling that is becoming more popular as computers become more powerful is the large eddy simulation (LES). What are the requirements on the size of the grid cells for accurate simulations?

22. In LES simulations, what is meant by the term sub-grid scale (SGS) model?

23. What particle density is required for a particle of diameter 10 microns to have the same settling velocity as 1-micron sphere of aluminum falling in air at standard conditions?

24. Discuss qualitatively how you would expect the pressure profile on a golf ball to differ from the Stokes' pressure profile on a sphere in creeping motion. Why are golf balls dimpled?

25. To illustrate how the computational effort increases nonlinearly with the number of grid points used to solve a CFD problem, use a computer language or programming environment (such as MATLAB) to create an n by n matrix A, and fill all the entries with random numbers. Create an n by 1 vector called b, and fill it with random numbers as well. Solve the matrix system $Ax = b$, for the unknown vector x. Plot the time it takes to solve the problem vs the size of the matrix, n, for values of n from 1 to 1000.

26. How does parallel processing reduce the computational time to perform a CFD simulation?

27. A hydraulic system uses a hydraulic fluid with properties similar to SAE 10 motor oil. The pressurized fluid is at conditions of 25,000 kPa and 60°C. The gap between the piston compressing the fluid and the walls of the cylinder is 6 microns thick, and the cylinder diameter is 30 mm. The piston itself is 10 cm long. Determine the leakage rate of hydraulic fluid if the pressure on the other side of the piston is atmospheric.

28. A vertical wall is covered with an adhesive Newtonian fluid (viscosity is constant). The fluid flows down the wall under the force of gravity, with a flow such that the thickness of the film is constant. You can assume that the pressure everywhere in the liquid film is equal to atmospheric pressure. Starting with the full Navier–Stokes equations in Cartesian coordinates, derive an expression for the velocity profile $V(y)$, where V is the downward velocity and y is the perpendicular distance away from the wall.

29. Find the torque and power required to turn a journal bearing using SAE 10W-30 motor oil at 100°C in a bearing of size 4 cm length and 8 cm diameter with a gap thickness of 0.05 mm. Assume the shaft rotates at 3000 rpm and the outer housing of the journal is fixed so that it does not rotate.

30. How is the momentum thickness of a boundary layer defined? Calculate the momentum thickness for a water boundary layer at a Reynolds number of 100,000. What is the ratio of the momentum thickness to the boundary layer thickness?

31. How is the displacement thickness of a boundary layer defined? Calculate the displacement thickness for a water boundary layer at a Reynolds number of 100,000. What is the ratio of the displacement thickness to the boundary layer thickness?

32. Two flat plates are 100 mm apart, with air flowing between them at 10 m/s. Make an estimate for how long (in distance from the leading edge) it will take for the

two boundary layers on top and bottom to merge with each other. Qualitatively, how do you expect the merging distance in reality to differ from your estimation?

33. For the approximate laminar boundary layer profile given in Equation 4.12, calculate the ratio of the distance where $u = 0.95U_o$ to the boundary layer thickness.

34. Repeat the previous problem using the turbulent velocity profile given in Equation 4.17.

35. Derive an expression for the drag coefficient in terms of the Reynolds number for laminar boundary layer flow.

36. Derive an expression for the skin friction coefficient in terms of the Reynolds number for laminar boundary layer flow.

37. A scale model of a landscape is put in a wind tunnel to simulate wind effects on the buildings. Because the wind tunnel is not capable of producing velocities high enough to match the Reynolds number for the full-scale flow, the turbulent boundary layer is "tripped" by adding roughness to the front edge of the base plate. If the length of the plate is $L = 5$ m, the air speed is 30 m/s, and the boundary layer is fully turbulent from the leading edge, estimate the boundary layer thickness at the trailing edge of the base plate ($x = L$). How thick would the boundary layer be if it were laminar?

38. For the Wright brothers' early gliders, the length of the wings was around $L = 1$ m, and flying velocities were on the order of 10 m/s. Approximating the wing as a flat plate, what percentage of the flow was laminar?

39. Water flows over a flat plate of width 2 m and length 5 m at a speed of 15 m/s. Estimate the drag force on one side of the plate.

40. Air flows over a flat plate of width 2 m and length 5 m at a speed of 15 m/s. Estimate the drag force on one side of the plate.

41. Air flows through a cylinder of inner diameter 0.7 m and length 5 m at a flow rate of 1 m³/s. Estimate the drag force on the cylinder by modeling it as a flat plate. Is this a good approximation?

42. For the laminar pipe flow velocity profile, plot the error in estimating the first derivative with a finite difference formula as a function of the step size, Δr.

43. Solve the transient Poiseuille flow problem for a case of zero pressure gradient and an initial uniform velocity of 1.0 m/s. How long does it take the velocity to decay to zero?

5 Internal Flow

In This Chapter
- Laminar Flow in Closed Ducts
- Turbulent Flow
- Open Channel Flow
- Complex Pipe Systems
- Secondary Losses

The objective of this chapter is to provide the tools needed for calculating friction losses in closed pipe systems and open channel flows, and for designing basic flow systems.

■ 5.1 Laminar Flow in Closed Ducts

At low velocities in a flow, viscosity tends to dominate the flow structure and keep the streamlines parallel, as shown in Figure 5.1. As the velocity of the flow increases, it reaches a point where the instabilities in the flow have sufficient inertia so that the viscosity is no longer able to damp them out. At that point irregular flow patterns emerge, leading to turbulence.

The Reynolds number was previously defined in Chapters 1 and 4 as the ratio of the inertial forces to the viscous forces, given in equation form by

$$\text{Re} = \frac{\rho V L}{\mu}$$

For pipe flow the length scale L of most importance is obviously the internal diameter D. The velocity scale V can be logically defined as the *mass-averaged velocity*:

$$V = \frac{\dot{m}}{\rho A}$$

CHAPTER 5 INTERNAL FLOW

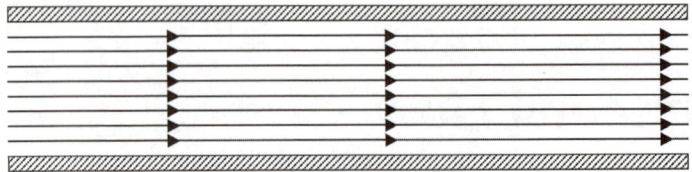

FIGURE 5.1 Schematic of streamlines in laminar pipe flow.

When the Reynolds number is less than about 2,000 or so, the flow is likely to be laminar. When the Reynolds number is above 2,000, the flow begins to transition to turbulence, becoming fully turbulent by a Reynolds number of 10,000.

5.1.1 Laminar Velocity Profile

Figure 5.2 shows the geometry for a typical pipe flow problem. In the case of laminar flow, it is actually possible to obtain a solution from the Navier-Stokes equations (Chapter 4) for steady-state flow through a round pipe. The general three-dimensional Navier-Stokes equations in cylindrical coordinates are

$$\frac{d\rho}{dt} = -\left(\frac{1}{r}\frac{\partial}{\partial r}\rho r v_r + \frac{1}{r}\frac{\partial}{\partial \theta}\rho v_\theta + \frac{\partial}{\partial x}\rho u\right)$$

$$\rho\left(\frac{\partial v_r}{\partial t} + v_r\frac{\partial v_r}{\partial r} + \frac{v_\theta}{r}\frac{\partial v_r}{\partial \theta} + u\frac{\partial v_r}{\partial x}\right) = \rho g_r - \frac{\partial P}{\partial r} + \mu\left[\frac{\partial}{\partial r}\left(\frac{1}{r}\frac{\partial}{\partial r}(r v_r)\right)\right.$$
$$\left. + \frac{1}{r^2}\frac{\partial^2 v_r}{\partial \theta^2} + \frac{\partial^2 u}{\partial x^2} - \frac{2}{r^2}\frac{\partial v_\theta}{\partial \theta}\right]$$

$$\rho\left(\frac{\partial v_\theta}{\partial t} + v_r\frac{\partial v_\theta}{\partial r} + \frac{v_\theta}{r}\frac{\partial v_\theta}{\partial \theta} + u\frac{\partial v_\theta}{\partial x}\right) = \rho g_\theta - \frac{1}{r}\frac{\partial P}{\partial \theta}$$
$$+ \mu\left[\frac{\partial}{\partial r}\left(\frac{1}{r}\frac{\partial}{\partial r}(r v_\theta)\right) + \frac{1}{r^2}\frac{\partial^2 v_\theta}{\partial \theta^2} + \frac{\partial^2 v_\theta}{\partial x^2} - \frac{2}{r^2}\frac{\partial u}{\partial \theta}\right]$$

$$\rho\left(\frac{\partial u}{\partial t} + v_r\frac{\partial u}{\partial r} + \frac{v_\theta}{r}\frac{\partial u}{\partial \theta} + u\frac{\partial u}{\partial x}\right)$$
$$= \rho g_x - \frac{\partial P}{\partial x} + \mu\left[\frac{1}{r}\frac{\partial}{\partial r}\left(r\frac{\partial u}{\partial r}\right) + \frac{1}{r^2}\frac{\partial^2 u}{\partial \theta^2} + \frac{\partial^2 u}{\partial x^2}\right]$$

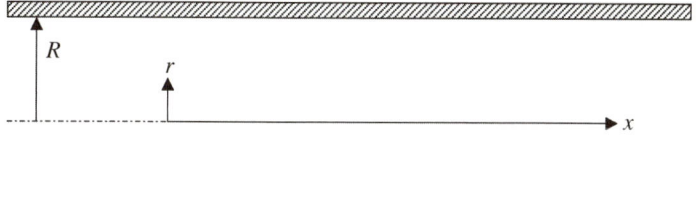

FIGURE 5.2 Sketch showing the coordinate system for finding the velocity profile in pipe flow.

Consider a pipe of outer radius R, with a steady mass flow rate $\dot{m}$. Also assume that there is no flow in the θ, or azimuthal, direction (that is, that there is no angular swirling flow), so that the problem is two-dimensional in x and r. Under this assumption the Navier-Stokes equations reduce to

$$0 = \left(\frac{1}{r}\frac{\partial}{\partial r}(rv_r) + \frac{\partial u}{\partial x}\right)$$

$$\rho\left(v_r\frac{\partial v_r}{\partial r} + u\frac{\partial v_r}{\partial x}\right) = -\frac{\partial P}{\partial r} + \mu\left[\frac{\partial}{\partial r}\left(\frac{1}{r}\frac{\partial}{\partial r}(rv_r)\right) + \frac{\partial^2 v_r}{\partial x^2}\right]$$

$$\rho\left(v_r\frac{\partial u}{\partial r} + u\frac{\partial u}{\partial x}\right) = -\frac{\partial P}{\partial x} + \mu\left[\frac{1}{r}\frac{\partial}{\partial r}\left(r\frac{\partial u}{\partial r}\right) + \frac{\partial^2 u}{\partial x^2}\right]$$

These equations can be simplified even further if we assume that the velocity profile does not change significantly in the x-direction for steady-state flow through a pipe of constant diameter:

$$0 = \left(\frac{1}{r}\frac{\partial}{\partial r}(rv_r)\right)$$

$$\rho\left(v_r\frac{\partial v_r}{\partial r}\right) = -\frac{\partial P}{\partial r} + \mu\left[\frac{\partial}{\partial r}\left(\frac{1}{r}\frac{\partial}{\partial r}(rv_r)\right)\right]$$

$$\rho\left(v_r\frac{\partial u}{\partial r}\right) = -\frac{\partial P}{\partial x} + \mu\left[\frac{1}{r}\frac{\partial}{\partial r}\left(r\frac{\partial u}{\partial r}\right)\right]$$

The boundary conditions are the no-slip velocity at the wall ($u = 0$), and that the wall is impermeable ($v_r = 0$). The velocity profile should also be symmetric since the geometry is symmetric ($du/dr = 0$ at $r = 0$). These conditions are sketched in Figure 5.3.

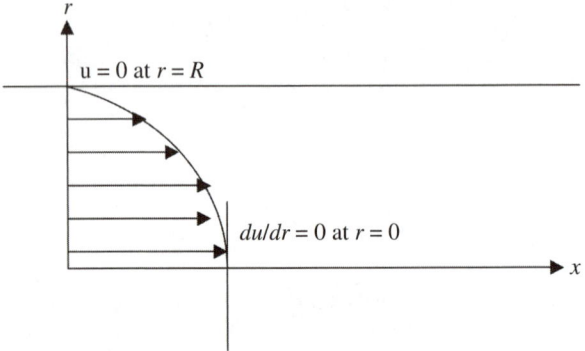

FIGURE 5.3 Sketch of boundary conditions to an unknown function for the velocity profile in pipe flow.

It seems logical that $v_r = 0$ since, in the steady state, flow cannot go through the wall or accumulate in the middle. So the conservation of mass and radial momentum equations will simplify even more:

$$0 = \frac{\partial v_x}{\partial x} \quad \text{and} \quad 0 = -\frac{\partial P}{\partial r}$$

All that is left is the momentum equation along the pipe axis, z:

$$\rho\left(v_x \frac{\partial v_x}{\partial x}\right) = -\frac{\partial P}{\partial x} + \mu\left[\frac{1}{r}\frac{\partial}{\partial r}\left(r\frac{\partial v_x}{\partial r}\right) + \frac{\partial^2 v_x}{\partial x^2}\right]$$

As before, we can substitute from continuity into the z-momentum:

$$\frac{\partial P}{\partial x} = \mu\left[\frac{1}{r}\frac{\partial}{\partial r}\left(r\frac{\partial v_x}{\partial r}\right)\right] = C_1$$

We now integrate to find the velocity profile, v_x, as a function of r,

$$\left(r\frac{\partial v_x}{\partial r}\right) = \frac{C_1}{\mu}\frac{r^2}{2} + C_2$$

and integrate once again with respect to r to obtain

$$v_z = \frac{C_1}{\mu}\frac{r^2}{4} + C_2\ln(r) + C_3$$

Once again, we have to apply boundary conditions to find the three constants. The range of possible values for r is $0 < r < R$, but $\ln(0)$ is undefined. So the only way the formula can work is if $C_2 = 0$. Next, $v(R) = 0$ by the no-slip boundary condition, so

$$0 = \frac{C_1}{\mu}\frac{R^2}{4} + C_3 \quad \text{or} \quad C_3 = -\frac{C_1}{\mu}\frac{R^2}{4} \quad \text{so} \quad v_z(r) = \frac{C_1}{4\mu}(r^2 - R^2)$$

Once again we can assume that the mass flow rate will be known, so that

$$\dot{m} = \int_A \rho V \, dA = \int_0^R \rho \left[\frac{C}{4\mu}(r^2 - R^2)\right](2\pi r \, dr) = \frac{\pi C \rho}{2\mu}\int_0^R (r^2 - R^2) r \, dr$$

and upon integrating we have

$$\dot{m} = \frac{\pi C \rho}{2\mu}\int_0^R (r^2 - R^2) r \, dr = \frac{\pi C \rho}{2\mu}\left(\frac{r^4}{4} - \frac{R^2 r^2}{2}\right)\bigg|_0^R = -\frac{\pi C \rho R^4}{8\mu}$$

so

$$C = -\frac{8\mu \dot{m}}{\pi \rho R^4}$$

and the velocity profile can finally be written as

$$v_x(r) = \frac{2\dot{m}}{\pi \rho R^4}(R^2 - r^2) = \frac{2\dot{m}}{\pi \rho R^2}\left[1 - \left(\frac{r}{R}\right)^2\right] \quad (5.1)$$

Equation 5.1 shows that the axial velocity profile in a laminar, steady, fully developed pipe flow is parabolic. The maximum velocity will be at the pipe centerline, where $r = 0$:

$$V_{max} = \frac{2\dot{m}}{\pi \rho R^2} \quad (5.2)$$

Thus the velocity profile can be expressed relative to the maximum velocity as

$$V(r) = V_{max}\left[1 - \left(\frac{r}{R}\right)^2\right] \quad (5.3)$$

5.1.2 Wall Shear Stress

Now with the velocity profile known, we can calculate the shear stress at the wall:

$$\tau_{wall} = \mu \left.\frac{du}{dr}\right|_{r=R} \tag{5.4}$$

Note that while the velocity goes to zero at the wall, the gradient of the velocity profile does not. The derivative of the velocity profile is

$$\frac{dv}{dr} = V_{max}\left(-\frac{2r}{R^2}\right) = -4V_{ave}\frac{r}{R^2}$$

So at the wall where $r = R$, the shear stress is

$$\tau_{wall} = \mu\left(-4V\frac{1}{R}\right) = -8\mu\frac{V}{D}$$

where V represents the average velocity of the flow in the pipe. The total frictional force along the pipe wall is obtained by integrating the shear stress over the area.

$$F = \int_A \tau \, dA \tag{5.5}$$

For steady-state, fully developed pipe flow, the velocity profile does not change in the axial direction. Thus the shear stress is constant along the length of the pipe, and Equation 5.5 will simplify to

$$F = \tau A = \tau(\pi D L) \tag{5.6}$$

Thus the total shear force on a section of pipe of length L in laminar flow is

$$F = -8\pi\mu V L$$

The sign is negative because the shear forces act opposite to the direction of flow.

The pressure drop can also be calculated for steady horizontal flow by drawing a control volume, as sketched in Figure 5.4. Note that the control volume is sketched so that its outer surface is just inside the pipe wall. We start by applying conservation of mass to a section of pipe of length L. Since the flow is steady state with a single inlet and a single outlet, the inlet and outlet mass flow rates must equal. Thus,

$$\dot{m}_1 = \dot{m}_2$$
$$\rho_1 A_1 V_1 = \rho_2 A_2 V_2$$

FIGURE 5.4 Schematic of steady horizontal pipe flow, with a control volume to be used to calculate the pressure drop.

Since the diameter of the pipe is constant, $A_1 = A_2$. We assume that the flow is incompressible, so $\rho_1 = \rho_2$. Therefore the velocities must also be equal:

$$V_1 = V_2$$

With this knowledge about the velocities, conservation of momentum can now be applied.

Since the inlet and outlet flow rates and velocities are equal, the thrust terms cancel out. The transient term is also obviously zero, so the only terms left are the external forces applied to the section of pipe. There are no solid objects crossing the control volume boundary; the only forces are those exerted by the fluid. The shear stress acts along the radial periphery of the control volume (for the center cut shown in Figure 5.4, these are the top and bottom surfaces), producing a net force to the left. This force was calculated previously in Equation 5.6 as $F_{shear} = \tau(\pi DL)$.

The pressure forces can only act normal to the control volume surfaces, which is at the left and right ends. The net pressure force on the control volume is

$$F_{pressure} = P_1 A_1 - P_2 A_2 = (P_1 - P_2)A = \frac{\pi}{4} D^2 (P_1 - P_2)$$

Since there are only two forces acting on the control volume and the right-hand side of the momentum equation is equal to zero, these two forces must be equal and opposite. Solving for the net pressure force, we have

$$F_{pressure} = -F_{shear}$$

$$\frac{\pi}{4} D^2 (P_1 - P_2) = \tau(\pi DL)$$

Now solving for the pressure drop gives

$$(P_1 - P_2) = \frac{\tau(\pi DL)}{\frac{\pi}{4} D^2}$$

which upon simplification gives a relation between pressure drop and shear stress:

$$(P_1 - P_2) = \tau 4 \frac{L}{D} \tag{5.7}$$

So if the velocity profile within the pipe is known, the shear stress can be calculated, and from that the pressure drop can be calculated. Equation 5.7 shows that the pressure drop varies linearly with the shear stress at the wall and with the length of the pipe, and inversely with the diameter of the pipe. This implies that for a given flow rate, the pressure drop due to friction can be reduced by using larger pipes. (Of course there is a limit as to how large you would want to make the pipes, as larger pipes contain more material and will cost more money to purchase and install. In some applications the weight of the piping system may also be a concern.) Specifically for the case of laminar flow, the pressure drop over a length of pipe L can be written as

$$\Delta P = \left(8\mu \frac{V}{D}\right) 4 \frac{L}{D} = 32\mu \frac{VL}{D^2} \tag{5.8}$$

5.1.3 Energy Loss Due to Friction

Note that the Bernoulli equation cannot be applied to this problem because there is a significant friction loss, which violates one of the key assumptions of the Bernoulli equation. However, the Bernoulli equation can be modified to account for friction losses, as will be discussed later in this chapter.

The calculated pressure drop for laminar flow is useful for determining the pumping power requirements for a system. With the pressure drop known, the pumping power can be calculated. From thermodynamics, and as discussed in Chapter 3, the general form of the energy equation is

$$\frac{dE}{dt} = \pm \dot{Q} \pm \dot{W} + \sum_{in} \dot{m}\left(h + \frac{V^2}{2} + gz\right) - \sum_{out} \dot{m}\left(h + \frac{V^2}{2} + gz\right)$$

The power needed solely to overcome the pressure drop can be found by eliminating all the other terms except external work and pressure drop:

$$\dot{W} = \dot{m}\left(\frac{P_1 - P_2}{\rho}\right)$$

The shear stress can be nondimensionalized into a friction factor by dividing by the dynamic pressure. The nondimensional friction factor for *internal* flow (also called the Darcy friction factor) is

$$f = \frac{8\tau}{\rho U_o^2} \tag{5.9}$$

Substituting in the value for shear stress we get the friction factor for laminar flow:

$$f = 8\left(\frac{8\mu\frac{V}{D}}{\rho V^2}\right) = 64\frac{\mu}{\rho VD} = \frac{64}{Re} \qquad (5.10)$$

For laminar flow this friction factor is a function solely of the Reynolds number. We can also express this friction loss in units of length, denoted by h_f, instead of units of shear stress:

$$h_f = f\frac{L}{D}\frac{V^2}{2g} \qquad (5.11)$$

Now that we have a formula for the friction loss, we can modify the Bernoulli equation to include it. The head loss, h_f, for internal flow is

$$h_f = \left(\frac{P_1}{\rho g} + \frac{V_1^2}{2g} + z_1\right) - \left(\frac{P_2}{\rho g} + \frac{V_2^2}{2g} + z_2\right) \qquad (5.12)$$

Note that Equation 5.12 is only valid for steady-state flow of a system with a single inlet and a single outlet.

EXAMPLE 5.1

Compute the friction loss in a section of horizontal capillary pipe under the following flow conditions: Water flows through a small glass pipe of inner diameter 2 mm, at a flow rate of 0.15 L/min. Compute the friction factor f and the pressure drop ΔP per unit length of pipe.

SOLUTION: First we convert the flow rate to standard units:

$$Q = 0.15\frac{L}{min}\frac{1\ m^3}{1000\ L}\frac{1\ min}{60\ s} = 2.5 \times 10^{-6}\frac{m^3}{s}$$

Now we can calculate the velocity:

$$V = \frac{Q}{A} = \frac{2.5 \times 10^{-6}\frac{m^3}{s}}{\frac{\pi}{4}(0.002\ m)^2} = 0.796\frac{m}{s}$$

We must check the Reynolds number to verify that the flow is indeed laminar:

$$\text{Re} = \frac{VD}{\nu} = \frac{(0.796 \text{ m/s})(0.002 \text{ m})}{1.0 \times 10^{-6} \text{ m}^2/\text{s}} = 1590$$

Now we calculate the friction factor, f, as

$$f = \frac{64}{\text{Re}} = \frac{64}{1590} = 0.040$$

The friction loss in units of length, or head, based on a 1-m length of pipe is

$$h_f = f \frac{L}{D} \frac{V^2}{2g} = 0.040 \frac{1 \text{ m}}{0.002 \text{ m}} \frac{(0.796 \text{ m/s})^2}{2(9.8 \text{ m/s}^2)} = 0.647 \text{ m}$$

The corresponding pressure drop due to friction is

$$\Delta P = \rho g h_f = \left(1000 \frac{\text{kg}}{\text{m}^3}\right)\left(9.8 \frac{\text{m}}{\text{s}^2}\right)(0.647 \text{ m}) = 6340 \text{ Pa}$$

Thus the flow loses 6340 Pa for every 1 m of pipe length.

If the same pipe is directed downward, what is the pressure change? The energy equation, in the form of the modified Bernoulli equation, can be used to calculate the pressure drop in this case. For flow going downward, the force of gravity acts to increase the pressure, while friction acts to reduce the pressure. The net effect is calculated as

$$\Delta P = \rho g (h_f - \Delta z) = \left(1000 \frac{\text{kg}}{\text{m}^3}\right)\left(9.8 \frac{\text{m}}{\text{s}^2}\right)(0.647 \text{ m} - 1 \text{ m}) = -3460 \text{ Pa}$$

For the case of downward vertical flow, the pressure would actually rise.

EXAMPLE 5.2

What is the power required to pump the water in the horizontal pipe of Example 5.1?

SOLUTION: Since we have already calculated the flow rate and pressure drop, all that remains is to multiply them together to get the power. From the first law of thermodynamics (conservation of energy) it can be shown that the power associated with flow through a pressure difference is

$$\dot{W} = \dot{m}\left(\frac{\Delta P}{\rho}\right) = Q \, \Delta P$$

Substituting the values calculated previously in standard metric units will provide the power in units of watts:

$$\dot{W} = \left(2.5 \times 10^{-6} \frac{m^3}{s}\right)(6340 \text{ Pa}) = 0.016 \text{ W}$$

Thus, very little power would be required to pump this capillary flow.

EXAMPLE 5.3

Consider a vertical pipe of diameter 1.5 mm and length 10 cm, with a steady flow of water going downward, and with the pressures at the inlet and outlet sections equal. What is the velocity?

SOLUTION: The energy equation can be written as

$$h_f = \left(\frac{P_1}{\rho g} + \frac{V_1^2}{2g} + z_1\right) - \left(\frac{P_2}{\rho g} + \frac{V_2^2}{2g} + z_2\right)$$

Note that the problem is steady-state flow with a single inlet and outlet. Let point 1 be the inlet at the top and point 2 be the outlet at the bottom. Then $P_1 = P_2$ and, by conservation of mass, $V_1 = V_2$. So the only terms remaining are the change in potential energy and the loss due to friction energy:

$$h_f = z_1 - z_2$$

We use the formula for the head loss due to friction, Equation 5.11:

$$h_f = f \frac{L}{D} \frac{V^2}{2g}$$

Everything on the right-hand side is known, except the velocity V and friction factor f, which is a function of the Reynolds number. Assuming laminar flow, $f = 64/\text{Re}$, so the head loss is

$$h_f = \frac{64}{\text{Re}} \frac{L}{D} \frac{V^2}{2g} = \frac{64\nu}{VD} \frac{L}{D} \frac{V^2}{2g} = \frac{32 L V}{D^2} \frac{\nu}{g}$$

Now the only unknown is the velocity, V, since $h_f = \Delta z = 0.10$ m. Solving for V, we have

$$V = \frac{\Delta z D^2 g}{32 L \nu} = \frac{(0.01\text{ m})(0.0015\text{ m})^2(9.8\text{ m/s}^2)}{32(0.01\text{ m})(1 \times 10^{-6}\text{ m}^2/\text{s})} = 0.69\,\frac{\text{m}}{\text{s}}$$

Finally, we check the Reynolds number to verify that the flow is indeed laminar.

$$\text{Re} = \frac{VD}{\nu} = \frac{(0.69\,\tfrac{\text{m}}{\text{s}})(0.0015\text{ m})}{1.0 \times 10^{-6}\text{ m}^2/\text{s}} = 1035$$

Since Re $<$ 2000 the flow is indeed laminar, and the friction factor for laminar flows can be used.

EXAMPLE 5.4

What is the largest velocity that can be used with water flow in a 1-cm pipe that will be laminar? What is the largest velocity for laminar air flow?

SOLUTION: We set the Reynolds number equal to the maximum value for laminar flow, 2000. The kinematic viscosity for water is 1×10^{-6} m^2/s, and the kinematic viscosity for air is 1.5×10^{-5} m^2/s at standard conditions. So the maximum possible velocity for water is

$$V = \text{Re}\,\frac{\nu}{D} = 2000\,\frac{1 \times 10^{-6}\text{ m}^2/\text{s}}{0.01\text{ m}} = 0.2\,\frac{\text{m}}{\text{s}}$$

and the maximum velocity for air is

$$V = \text{Re}\,\frac{\nu}{D} = 2000\,\frac{1.5 \times 10^{-5}\text{ m}^2/\text{s}}{0.01\text{ m}} = 3.0\,\frac{\text{m}}{\text{s}}$$

5.2 Turbulent Flow

As previously mentioned, when the Reynolds number of an internal pipe flow increases above a value of about 2000, the flow will begin to transition to turbulence. Turbulent flow differs from laminar flow in that the streamlines do not remain parallel, and even in a steady state there is a fluctuating nature to the velocity. So when we speak of the turbulent velocity profile, it is important to remember that we are referring to the *average velocity* of the flow at a point in space.

5.2 Turbulent Flow

5.2.1 Critical Reynolds Number

Typically for internal flow in pipes, the flow will cease to be entirely laminar at a Reynolds number *around* 2,000 and will become fully turbulent at a Reynolds number of *about* 10,000. However, under carefully controlled conditions laminar flow has been observed at Reynolds numbers as high as 100,000. Thus in reality the transition from laminar to turbulent flow depends not only on the Reynolds number, but also on flow conditions such as the uniformity of the velocity profile, the magnitude of any disturbances in the flow, and the conditions of the pipe surface. For most engineering applications it is safe to assume that significant disturbances (such as perturbations in velocity, pressure, or composition) exist in the flow, and that the transition takes place at a relatively low Reynolds number. It is also convenient to assume that there is a sharp transition between laminar and turbulent flow, and so a *critical Reynolds number* is employed. For internal pipe flows $Re_{cr} = 2,300$.

In the 1880s Osborne Reynolds constructed an apparatus, shown in Figure 5.5, to analyze the differences between laminar and turbulent flow in pipes. His sketches

FIGURE 5.5 Drawing of Osborne Reynolds with his device to create laminar and turbulent flows in pipes. From [Reynolds83].

FIGURE 5.6 Reynolds' sketches of his observations of pipe flow from the apparatus shown in Figure 5.5. From [Reynolds83].

of his visual observations are shown in Figure 5.6. The author has an apparatus similar to that of Reynolds, in which the flow rate is controlled through a valve. At low speeds the flow is laminar, as seen by the straight dye streak in Figure 5.7. As the velocity increases, the flow starts to become unstable, as shown in Figure 5.8, and at still higher speeds, the flow becomes completely turbulent, as seen in Figures 5.9 and 5.10. You can perform a similar experiment at home to check whether the flow in a water hose is laminar or turbulent. For the hose in Figure 5.11 the flow is clearly turbulent.

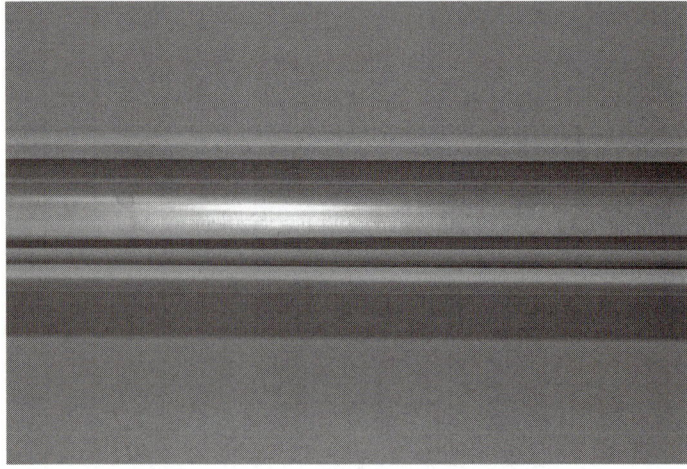

FIGURE 5.7 Picture of laminar flow, illustrated by dye, in a circular pipe.

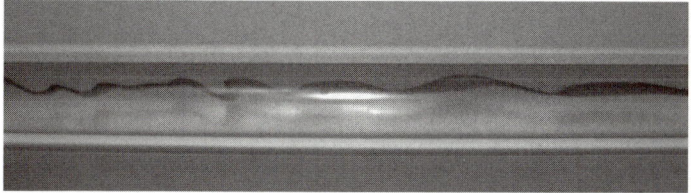

FIGURE 5.8 Picture of wavy flow showing the beginning of instability, illustrated by dye, in a circular pipe.

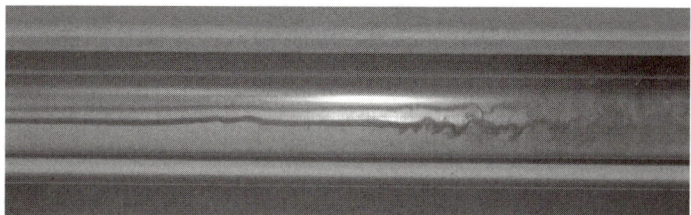

FIGURE 5.9 Picture of laminar flow suddenly bursting to turbulence, illustrated by dye, in a circular pipe.

In contrast to laminar flow, there is no simple way to derive the friction losses in turbulent pipe flow. This problem confounded hydraulic engineers for a long time: Most measurements in practical systems were of turbulent flows, but the only theory available was for laminar flows. It was only for the special case of capillary tubes, in which the flow was laminar, that agreement could be found between theory and experiments.

Figure 5.12 shows a typical curve of pressure loss versus Reynolds number for the friction loss in internal pipe flow for round pipes. Note that the pressure drop is much higher for turbulent flow than it would be for laminar flow at the same Reynolds number, and that it increases more steeply with increasing Reynolds number. Unfortunately, turbulence results in much higher friction losses than in laminar flow.

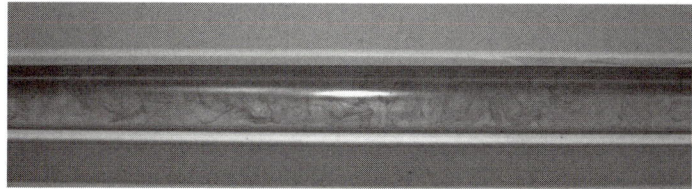

FIGURE 5.10 Picture of turbulent flow, illustrated by dye, in a circular pipe.

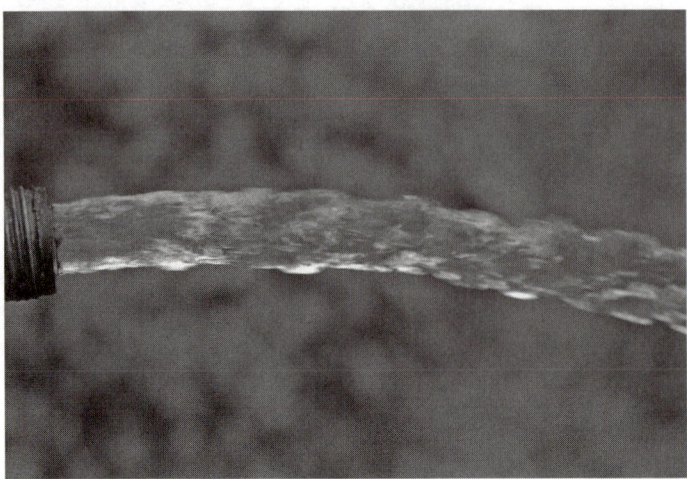

FIGURE 5.11 Picture of water flow coming out of a hose. The wrinkles in the water surface indicate that the flow within the hose is turbulent.

This is unfortunate since we often seek to maximize flow rates, which results in the high Reynolds numbers that lead to turbulent flow.

The average velocity profile of turbulent pipe flow differs quite a bit from that of laminar flow. A comparison of the two velocity profiles is shown in Figure 5.13.

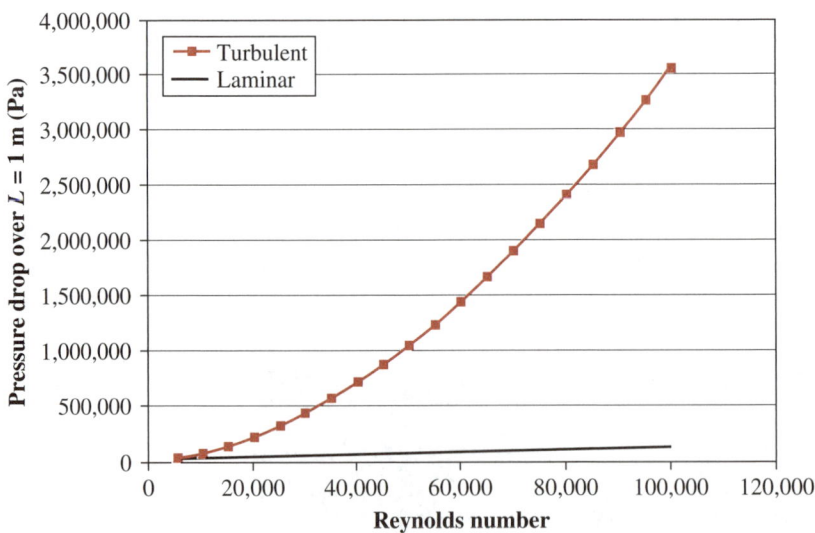

FIGURE 5.12 Pressure loss per unit length vs Reynolds number (nondimensional velocity) for turbulent water flow through a pipe of fixed diameter.

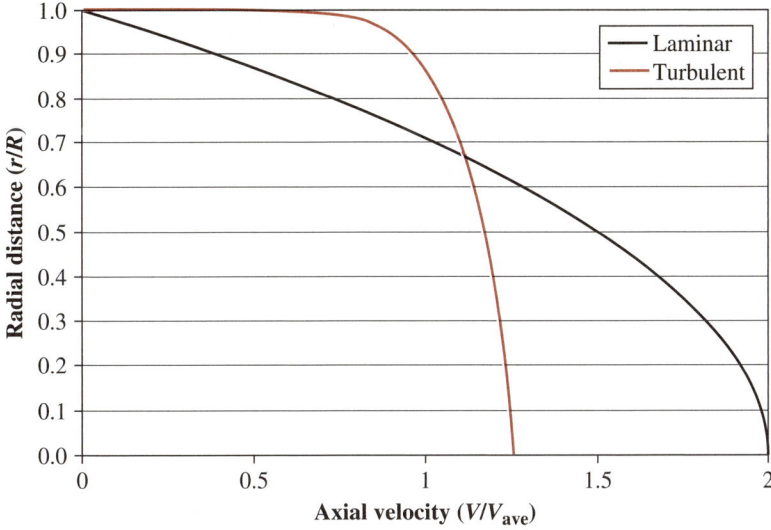

FIGURE 5.13 Comparison of laminar and turbulent velocity profiles on the same scale for the same flow rate.

5.2.2 Kinetic Energy Correction Factor

In most of the problems in Chapter 3, we assumed that the mass-averaged velocity could be used in the momentum equation and in the Bernoulli equation for solving flow problems. That is, we assumed that the kinetic energy per unit mass of flowing fluid could be expressed as $\frac{1}{2}V^2$, where V is the mass-averaged velocity defined as

$$V_{ave} = \frac{\dot{m}}{\rho A} = \frac{\int_A \rho V \, dA}{\rho A}$$

In real pipe flows the velocity is not constant. Rather, it varies from a value of zero at the walls (no-slip condition) to a maximum value at the flow centerline. This variation becomes important when considering the kinetic energy flow rate, because the kinetic energy is nonlinear with velocity. The flow rate of kinetic energy, in units of energy per time, is defined as

$$\dot{KE} = \int_A \rho \left(\frac{V^2}{2}\right)(V \, dA)$$

The kinetic energy flow rate that is calculated approximately using the mass-averaged velocity is

$$\dot{KE}_{app} = \dot{m}\left(\frac{1}{2}\overline{V}^2\right) = \frac{1}{2}\rho A \overline{V}^3$$

The *kinetic energy correction factor*, α, is defined as the ratio of the actual kinetic energy of the flow divided by the kinetic energy calculated using the mass-averaged velocity:

$$\alpha = \frac{\dot{KE}}{\dot{KE}_{app}} \quad (5.13)$$

If the velocity profile is known, the value of α can be calculated from

$$\alpha = \frac{\int_A \rho(\frac{1}{2}V^2)(V\,dA)}{\frac{1}{2}\rho A \overline{V}^3} = \frac{\int_A V^3\,dA}{A\overline{V}^3} \quad (5.14)$$

Here it has been assumed that the pipe walls and the flow streamlines are parallel, so the dot product of $\vec{V} \cdot d\vec{A}$ can be replaced by the scalar quantity $V\,dA$.

As an example, consider the velocity profile for laminar flow through a round pipe:

$$V(r) = V_{max}\left[1 - \left(\frac{r}{R}\right)^2\right] \quad \text{for } 0 < r < R$$

Substituting into the definition for the kinetic energy correction factor, we have

$$\alpha = \frac{\int_0^R \left\{V_{max}\left[1 - \left(\frac{r}{R}\right)^2\right]\right\}^3 2\pi r\,dr}{\pi R^2 \overline{V}^3}$$

Expanding out the polynomial yields

$$\alpha = \frac{2V_{max}^3}{R^2 \overline{V}^3}\int_0^R \left\{\left[1 - 3\left(\frac{r}{R}\right)^2 + 3\left(\frac{r}{R}\right)^4 - \left(\frac{r}{R}\right)^6\right]\right\}r\,dr$$

Recall that for this parabolic profile, the average velocity is half of V_{max}. Thus,

$$\alpha = \frac{16}{R^2} \left[\frac{r^2}{2} - \frac{3}{4}\left(\frac{r^4}{R^2}\right) + \frac{3}{6}\left(\frac{r^6}{R^4}\right) - \frac{1}{8}\left(\frac{r^8}{R^6}\right) \right]_{r=0}^{r=R}$$

and finally

$$\alpha = \frac{16}{R^2}(0.5 - 0.75 + 0.5 - 0.125)R^2 = 2.0$$

Thus for laminar pipe flow, the actual kinetic energy of the flow is twice that which is calculated using the mass-averaged velocity. Does this mean that all the calculations done thus far with the Bernoulli equation are incorrect? Fortunately not. The majority of flows of engineering interest are turbulent, and turbulent velocity profiles are much flatter than laminar ones. The kinetic energy correction factors for turbulent pipe flow are usually in the range of 1.05 to 1.10, which is often comparable to the uncertainty in experimental measurements and can usually be safely neglected. For laminar flows the velocity will usually be small, and in most laminar flows the kinetic energy term will be small compared to the other energy terms. Thus the error in using the mass-averaged velocity is again not noticeable. However, an engineer should always keep in mind that this and other approximations are usually being made in flow analysis. This is one reason why validation testing is important.

There is also a *momentum flow rate correction factor*, β, used for problems using the momentum equation. This factor is defined as

$$\beta = \frac{\int_A V \rho (V \, dA)}{\dot{m} \overline{V}} = \frac{\int_A V^2 \, dA}{A \overline{V}^2} \tag{5.15}$$

For laminar pipe flow through a circular pipe, $\beta = 4/3 = 1.33$. For turbulent pipe flows the value of the momentum flow rate correction factor is close enough to 1.0 that it can be neglected.

5.2.3 Effects of Surface Roughness

Another way in which turbulent flow differs from laminar flow is that turbulent flows are affected by the surface roughness, ε, of the pipe walls. Since the streamlines in a laminar flow are parallel, the core of the flow is unaffected by small irregularities in the wall, and the velocity profile and friction losses are independent of the surface roughness for reasonably small values of surface roughness. Turbulent flow, however, is composed of many small eddies, which can interact with the wall roughness.

Table 5.1 Typical Surface Roughness (ε) Values for Pipes

Type	Roughness (mm)	Roughness (in.)
Concrete	1.0	0.04
Cast iron	0.26	0.01
Galvanized iron	0.15	0.006
Wrought iron	0.045	0.002
Commercial steel	0.045	0.002
Rubber	0.01	0.0004
Fiberglass	0.005	0.0002
Stainless steel	0.002	0.0001
Copper, brass, or PVC	0.0015	0.00006
Glass	0.0	0.0

Table 5.1 lists typical surface roughness values for a variety of common pipe materials. These roughness values are based on new pipes, but should only be taken as approximate, since there is a large variation in surface roughness for a particular material as a result of different manufacturing processes and the final surface finish. Also, over time, corrosion of pipe surfaces or the buildup of deposits such as lime and scale can significantly increase the surface roughness.

Generally speaking, the higher the Reynolds number, the more likely the flow is to be turbulent, though the exact value of the Reynolds number for transition depends on the type of flow (whether external or internal) and if there are any disturbances in the flow. Traditionally the Moody chart (see Figure 5.14) has been used to find the friction factor for pipes in turbulent flow. The Moody chart is a created from a compilation of experimental data. However, with the common availability of powerful calculators, it is more efficient to use the rather complex curve fit equation that has been developed to model the data in the Moody chart. This curve fit is given by

$$f = \left\{ -1.8 \log \left[\frac{6.9}{\text{Re}} + \left(\frac{\varepsilon/D}{3.7} \right)^{1.11} \right] \right\}^{-2} \tag{5.16}$$

This equation is valid for Re > 2300. Note: Whenever an equation includes a decimal exponent, that is a sure sign that it represents a curve fit to data and is not derived from theory.

Why is the surface roughness included in the equation for the turbulent friction factor, when it does not appear in laminar flow? For reasonably small values of the surface roughness, the laminar velocity profile depends only on the average pipe

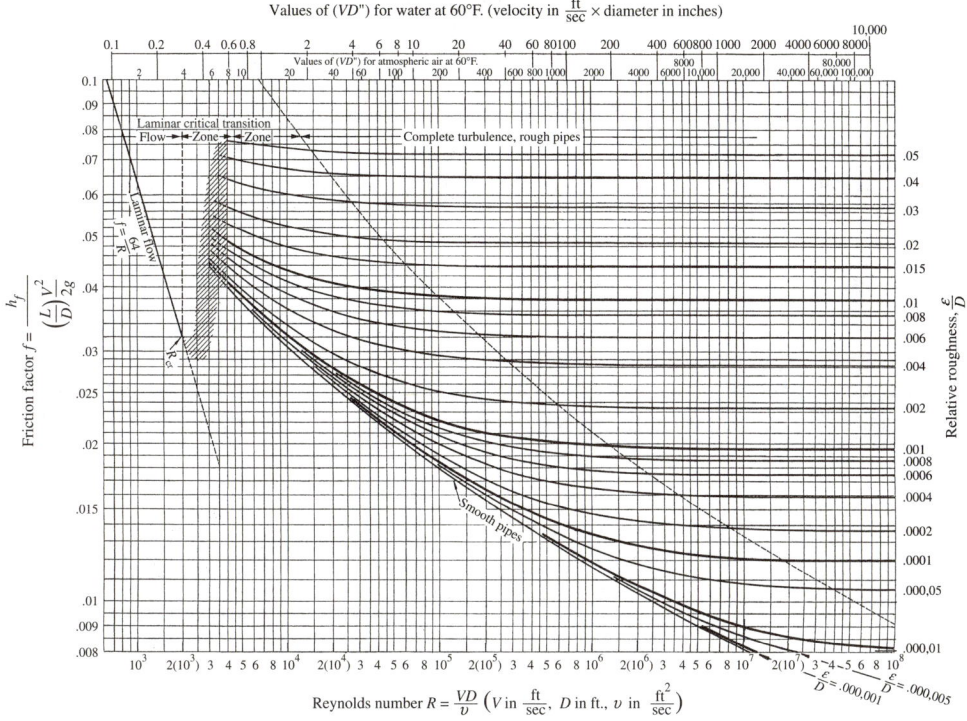

FIGURE 5.14 Moody chart, showing the nondimensional friction factor as a function of the Reynolds number and nondimensional surface roughness, from [Moody44]. Used with permission of ASME.

diameter, so the velocity profile at the wall does not change as the surface roughness is increased. As stated earlier, the flow stress is related directly to the shear stress at the wall:

$$\tau_{wall} = \mu \left.\frac{\partial u}{\partial y}\right|_{wall}$$

and

$$\Delta P = 4\frac{L}{D}\tau_{wall}$$

In turbulent flows, there are small-scale eddies that can interact with the bumps along the walls of a rough pipe. The effect of surface roughness is to enhance the turbulence by generating more eddies, which enhances mixing and causes the velocity profile to be steeper and the shear stress to increase.

EXAMPLE 5.5

Consider a circular stainless steel pipe of diameter of 60 cm. Calculate the friction factor for air flowing through the pipe at standard conditions and a velocity of 40 m/s.

SOLUTION: First we calculate the Reynolds number:

$$\text{Re} = \frac{VD}{\nu} = \frac{(40 \text{ m/s})(0.6 \text{ m})}{1.5 \times 10^{-5} \text{ m}^2/\text{s}} = 1{,}600{,}000$$

Since Re > 2,000, the flow is turbulent. Thus we also need to calculate the relative roughness of the pipe. The relative roughness is $\varepsilon/D = 0.002 \text{ mm}/600 \text{ mm} = 3.33 \times 10^{-6}$. We can now calculate the friction factor using Equation 5.16:

$$f = \left\{-1.8 \log\left[\frac{6.9}{\text{Re}} + \left(\frac{\varepsilon/D}{3.7}\right)^{1.11}\right]\right\}^{-2}$$

$$= \left\{-1.8 \log\left[\frac{6.9}{1{,}600{,}000} + \left(\frac{3.33 \times 10^{-6}}{3.7}\right)^{1.11}\right]\right\}^{-2} = 0.0108$$

EXAMPLE 5.6

Consider a horizontal commercial steel pipe of length 1 m and diameter 5 cm carrying water flowing at an average velocity of 3 m/s. Calculate the friction factor, the head loss, and the pressure drop due to friction.

SOLUTION: The Reynolds number is

$$\text{Re} = \frac{VD}{\nu} = \frac{(3 \text{ m/s})(0.05 \text{ m})}{1 \times 10^{-6} \text{ m}^2/\text{s}} = 150{,}000$$

So the flow is clearly turbulent. The relative roughness is $\varepsilon/D = 0.045 \text{ mm}/50 \text{ mm} = 0.0009$. The turbulent friction factor is

$$f = \left\{-1.8 \log\left[\frac{6.9}{\text{Re}} + \left(\frac{\varepsilon/D}{3.7}\right)^{1.11}\right]\right\}^{-2}$$

$$= \left\{-1.8 \log\left[\frac{6.9}{150{,}000} + \left(\frac{0.0009}{3.7}\right)^{1.11}\right]\right\}^{-2} = 0.0209$$

The frictional head loss is

$$h_f = f\frac{L}{D}\frac{V^2}{2g} = 0.0209\frac{1\text{ m}}{0.05\text{ m}}\frac{(3\text{ m/s})^2}{2(9.8\text{ m/s}^2)} = 0.192\text{ m}$$

The pressure loss over the 1-m length of the pipe is

$$\Delta P = \rho g h_f = \left(1{,}000\,\frac{\text{kg}}{\text{m}^3}\right)\left(9.8\,\frac{\text{m}}{\text{s}^2}\right)(0.192\text{ m}) = 1{,}880\text{ Pa}$$

5.2.4 Entrance Length

The velocity profile expressed in Equation 5.3 is valid only for fully developed flow—that is, flow in a long section of straight pipe, sufficiently far away from any disturbances such as valves or fittings. Now consider a section of pipe whose entrance connects to a large reservoir of liquid. In the vicinity of the entrance, the velocity profile will not be the fully developed profile since it is affected by the incoming flow. The distance over which the incoming flow effects are felt is termed the *entrance length*. The correlation for entrance length is

$$L_e = 0.06 D\,\text{Re} \tag{5.17}$$

for laminar flow. Assuming a transition Reynolds number of 2000, the entrance length can be as long as 120 pipe diameters for laminar flow. (See Figure 5.15.) The enhanced mixing in turbulent flows causes more rapid growth of the boundary layers from the entrance, and the entrance length is shorter, usually around 25 to 40 diameters. A common correlation for entrance length for turbulent flow is

$$L_e = 4.4 D\,\text{Re}^{1/6} \tag{5.18}$$

In the initial development region, the velocity gradient at the walls will be sharper than in the fully developed region. Thus the pressure drop per unit length of pipe is larger in the entrance length than it is in the rest of the pipe. As long as the section of pipe is long compared to the entrance length, the effects of the entrance region can be safely neglected for engineering purposes. For cases where the pipe exits into another large reservoir, the effects of the exit are not felt back into the pipe for any significant length, so there is no corresponding "exit length" in engineering practice. In the entrance region the velocity profile is a function of both axial and radial distance, so $V = V(x, r)$, whereas in the fully developed region for steady-state flow through a constant-diameter pipe, $V = V(r)$ only.

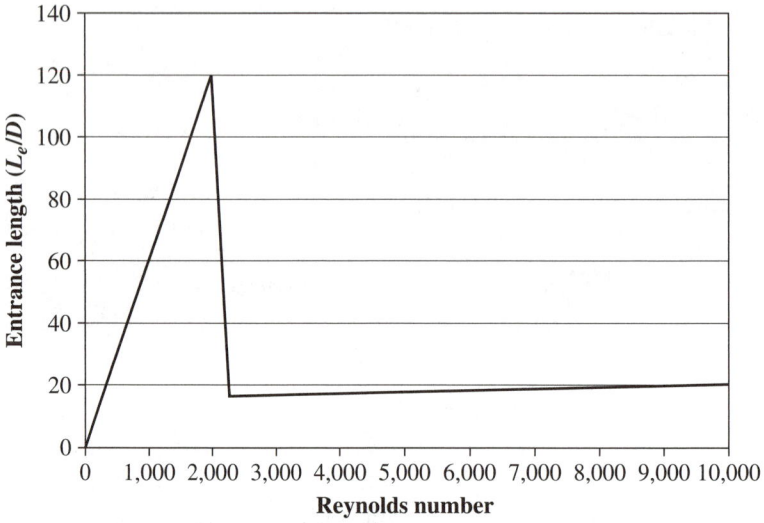

FIGURE 5.15 Nondimensionalized entry length vs Reynolds number for a fixed inlet geometry, assuming a sharp transition from laminar to turbulent flow at Re = 2,000.

EXAMPLE 5.7

How long is the entrance length for a pipe of diameter 4 cm carrying water at a flow rate of 35 L/min? If the pipe is a meter long, does the entrance length significantly affect the pressure drop in the pipe?

SOLUTION: As before, we need to calculate the Reynolds number to determine whether the flow is laminar or turbulent. To do so, we first must calculate the velocity from the flow rate and the diameter:

$$V = \frac{Q}{A} = \frac{150 \frac{L}{min}}{\frac{\pi}{4}(0.04 \text{ m})^2} \frac{1 \text{ min}}{60 \text{ s}} \frac{1 \text{ m}^3}{1,000 \text{ L}} = 1.99 \frac{\text{m}}{\text{s}}$$

The Reynolds number is then

$$\text{Re} = \frac{VD}{\nu} = \frac{(1.99 \text{ m/s})(0.04 \text{ m})}{1 \times 10^{-6} \text{ m}^2/\text{s}} = 79,600$$

So the flow is turbulent, and the turbulent correlation for entrance length (Equation 5.18) must be used:

$$L = 4.4D\text{Re}^{1/6} = 4.4(0.04 \text{ m})(79,600)^{1/6} = 28.9(0.04 \text{ m}) = 1.15 \text{ m}$$

If the pipe is only 1 m long, then the entire pipe is in the entrance region, and the standard correlations for pipe friction factor will underpredict the friction loss.

5.2.5 Noncircular Pipes

Thus far we have considered flow through circular pipes only. However, in practice pipes and ducts of other shapes are used. For example, rectangular ducts are used for ventilation shafts. In round pipes lines of constant velocity form concentric circles, but in noncircular pipes the lines of constant velocity are distorted due to corner effects. Figure 5.16 shows the constant-velocity contours in a square duct, and Figure 5.17 shows the velocity contours in a triangular duct. For noncircular pipes we can define the equivalent hydraulic diameter of the pipe as

$$D_h = 4A/P \tag{5.19}$$

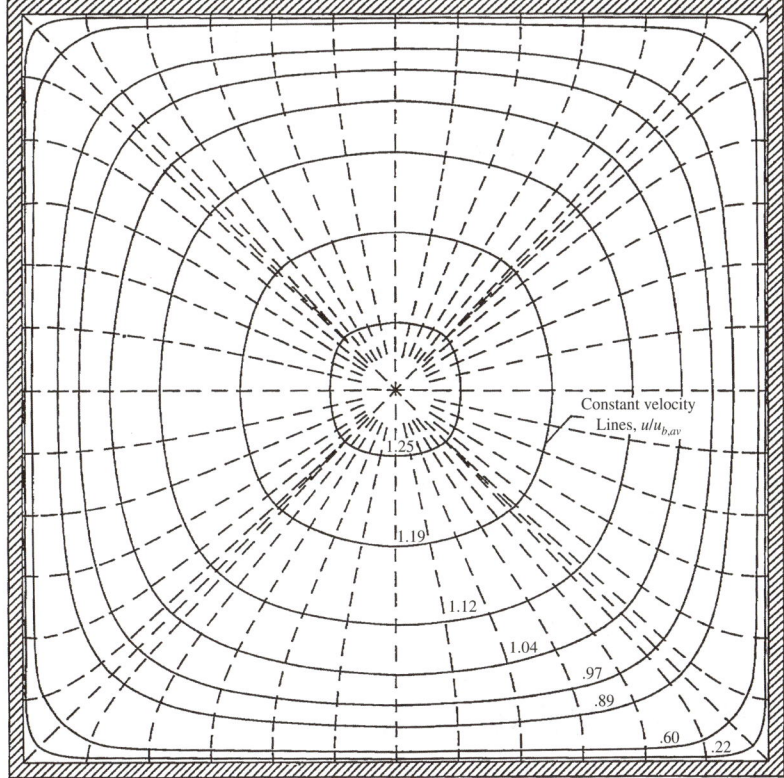

Square passage; Reynolds number, 24,000 or 900,000.

FIGURE 5.16 Contours of constant velocity in a square duct for turbulent flow. From [NACA58].

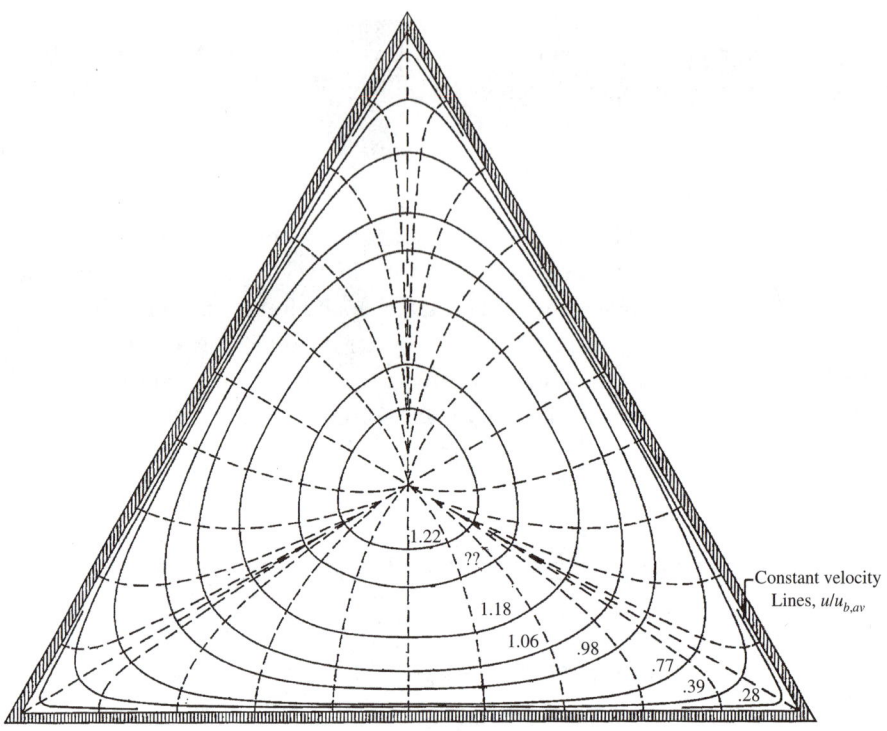

Triangular passage; Reynolds number, 24,000 or 900,000.

FIGURE 5.17 Contours of constant velocity in a triangular duct for turbulent flow. From [NACA58].

where P is the perimeter of the channel. For a circular pipe, the area is $(\pi/4)D^2$, and the perimeter is πD, so the hydraulic diameter is the same as the actual diameter. Note that the perimeter, P, is the wetted perimeter. This distinction is important in open channel flows. A is the cross-sectional area. For computing the relative roughness and the Reynolds number, use the hydraulic diameter for noncircular pipes shown in Table 5.2.

Table 5.2 Hydraulic Diameter for Different Common Shapes

Object	Hydraulic Diameter
Square of side length a	a
Rectangle of size a by b	$2ab/(a + b)$
Equilateral triangle of length a	$0.577a$
Semicircle of radius R	$2\pi R/(\pi + 2)$

EXAMPLE 5.8

Oil (SG = 0.8, kinematic viscosity = 1×10^{-5} m²/s) flows in the outer shell of a concentric tube counterflow heat exchanger. The oil flow rate is 0.25 kg/s, the inner diameter is 27 mm, and the outer diameter is 49 mm. Determine the pressure drop and pumping power required if the heat exchanger is 2 m long.

SOLUTION: To calculate the hydraulic diameter, we must first find the area and wetted perimeter. The area is

$$A = \frac{\pi}{4}(D_o^2 - D_i^2) = \frac{\pi}{4}[(0.049 \text{ m})^2 - (0.027 \text{ m})^2] = 0.00114 \text{ m}^2$$

The wetted perimeter is

$$P = \pi D_o + \pi D_i = \pi(0.049 \text{ m} + 0.027 \text{ m}) = 0.239 \text{ m}$$

Now we can calculate the hydraulic diameter:

$$D_h = \frac{4A}{P} = \frac{4(0.00114 \text{ m}^2)}{0.239 \text{ m}} = 0.019 \text{ m} = 19 \text{ mm}$$

The velocity is

$$V = \frac{\dot{m}}{\rho A} = \frac{0.25 \text{ kg/s}}{(800 \frac{\text{kg}}{\text{m}^3})(0.00114 \text{ m}^2)} = 0.274 \frac{\text{m}}{\text{s}}$$

The Reynolds number is

$$\text{Re} = \frac{VD_h}{\nu} = \frac{(0.274 \text{ m/s})(0.019 \text{ m})}{1 \times 10^{-5} \text{ m}^2/\text{s}} = 521$$

So the flow is laminar. The friction factor is $f = 64/\text{Re} = 64/521 = 0.123$. The head loss is

$$h_f = f\frac{L}{D_h}\frac{V^2}{2g} = 0.123\frac{2 \text{ m}}{0.019 \text{ m}}\frac{(0.274 \text{ m/s})^2}{2(9.8 \text{ m}^2/\text{s})} = 0.050 \text{ m}$$

The pressure drop is

$$\Delta P = \rho g h_f = \left(800 \frac{\text{kg}}{\text{m}^3}\right)\left(9.8 \frac{\text{m}}{\text{s}^2}\right)(0.05 \text{ m}) = 392 \text{ Pa}$$

The required pumping power is

$$\dot{W} = \dot{m}\frac{\Delta P}{\rho} = (0.25 \text{ kg/s})\frac{392 \text{ Pa}}{800 \text{ kg/m}^3} = 0.12 \text{ W}$$

EXAMPLE 5.9

A counterflow heat exchanger (see Figure 5.18) has an outer shell divided by fins to enhance the heat transfer. The outer diameter of the outer shell is 6 cm, and the inner diameter is 3 cm. There are six equally spaced fins dividing the annular passage. What is the hydraulic diameter of the passageways? What would be the hydraulic diameter if there were no fins?

SOLUTION: Assume the fins are of negligible thickness, so the sector area of each passageway is 60°. The cross-sectional area of the passageway is

$$A = \pi \frac{\theta}{360°}(R_{outer}^2 - R_{inner}^2)$$

$$= \frac{\pi}{6}\left[(0.03 \text{ m})^2 - (0.015 \text{ m})^2\right] = 3.53 \times 10^{-4} \text{ m}^2$$

The perimeter of the section is the sum of the two straight side sections, plus the inner arc plus the outer arc. The straight sections each have length (6 cm − 3 cm)/2 = 1.5 cm = 0.015 m. The length of the inner arc is $2\pi R_{inner}/6 = 2\pi(0.015 \text{ m})/6 = 0.0078$ m. The length of the outer arc is $2\pi R_{outer}/6 = 2\pi(0.03 \text{ m})/6 = 0.0157$ m. And so the total wetted perimeter is

$$P = 2(0.015 \text{ m}) + 0.0078 \text{ m} + 0.0157 \text{ m} = 0.0535 \text{ m}$$

and the hydraulic diameter is

$$D_h = \frac{4A}{P} = \frac{4(3.53 \times 10^{-4} \text{ m}^2)}{0.0535 \text{ m}} = 0.0264 \text{ m} = 2.64 \text{ cm}$$

The perimeter for a sector with no side walls is $P = 0.0078 \text{ m} + 0.0157 \text{ m} = 0.0235$ m. Then the hydraulic radius would be

$$D_h = \frac{4A}{P} = \frac{4(3.53 \times 10^{-4} \text{ m}^2)}{0.0235 \text{ m}} = 0.0601 \text{ m} = 6.01 \text{ cm}$$

If there were no fins the hydraulic diameter would be that for an annulus, similar to the previous problem.

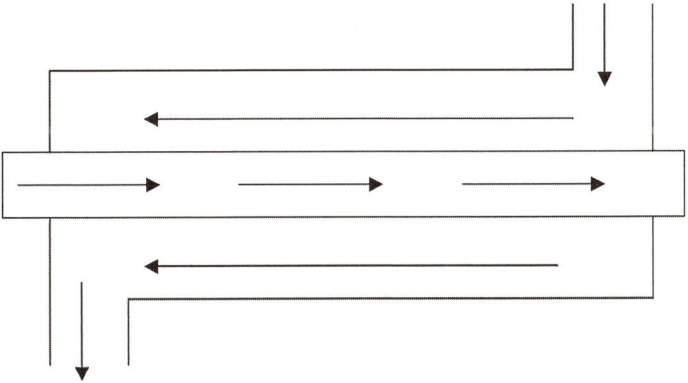

I FIGURE 5.18 Sketch of a counterflow heat exchanger divided by fins.

5.2.6 Solving Pipe Flow Problems

There are three common classes of problems encountered in pipe flow. The first is to compute the flow loss in a given system, in order to determine the pumping power needed to overcome the loss and to decide what size pump must be used. The second is to compute the flow rate delivered by a system, given measured pressure losses. The third is to compute the diameter of pipe to use in a given system.

The flow loss decreases as the diameter of a pipe increases if all other factors are kept constant. So the pumping power can be reduced as the diameter increases. However, larger pipes also cost more to purchase, so there is some economically optimum pipe size.

For a fixed flow rate, Q, how does the flow loss vary with changing diameter, while the length of the pipe and fluid properties are kept constant? The flow loss, in units of length, is expressed as

$$h_f = f \frac{L}{D} \frac{V^2}{2g} \tag{5.20}$$

The flow rate is $Q = \frac{1}{4}\pi D^2 V$, so the velocity can be expressed as

$$V = \frac{4Q}{\pi D^2}$$

If we also assume that the flow is laminar, then $f = 64/\text{Re}$. Substituting for f and V into the expression for the flow loss gives

$$h_f = f \frac{L}{D} \frac{V^2}{2g} = \frac{64\mu}{\rho V D} \frac{L}{D} \frac{V^2}{2g} = \frac{32\mu L V}{\rho D^2 g} = \frac{128 \mu L Q}{\pi \rho g D^4} \tag{5.21}$$

Thus the flow loss is inversely proportional to the diameter, all other factors held constant. While the exact dependency will change for turbulent flow, the trend remains the same. The smaller the pipe, the higher the friction loss. So the flow loss can be reduced by increasing the diameter of the pipe. Equation 5.21 is also known as *Poiseuille's law*.

EXAMPLE 5.10

Suppose water is to be transferred from one reservoir to another that is 3 m higher, with a total length of 5 m between them. No pumps are to be used. What is the minimum pipe size that will provide a flow rate of 0.2 L/min (200 L/min)?

SOLUTION: First we convert the flow rate to standard metric units:

$$Q = 0.2 \frac{\text{L}}{\text{min}} \left(\frac{1 \text{ min}}{60 \text{ s}}\right)\left(\frac{0.001 \text{ m}^3}{1 \text{ L}}\right) = 3.33 \times 10^{-6} \frac{\text{m}^3}{\text{s}}$$

Assume that the surface of each reservoir is open to the atmosphere. Since the pipe is of constant diameter and the fluid is incompressible, the velocity will be constant throughout the pipe. Therefore the only terms left in the energy equation are the change in height and the friction loss. Assuming the flow is laminar, we can use Equation 5.21 for flow loss:

$$\Delta z = h_f = \frac{128 \mu L Q}{\pi \rho g D^4}$$

The only unknown in this equation is the diameter, D. Solving for D yields

$$D = \sqrt[4]{\frac{128 \mu L Q}{\pi \rho g \Delta z}} = \sqrt[4]{\frac{128(0.001 \frac{\text{kg}}{\text{m} \cdot \text{s}})(5 \text{ m})(3.33 \times 10^{-6} \frac{\text{m}^3}{\text{s}})}{\pi (1000 \frac{\text{kg}}{\text{m}^3})(9.8 \frac{\text{m}}{\text{s}^2})(3 \text{ m})}}$$

$$= 2.22 \times 10^{-3} \text{ m} = 2.22 \text{ mm}$$

Now we check the Reynolds number to verify that the flow is indeed laminar. Knowing the diameter and the flow rate, we can calculate the velocity:

$$V = \frac{Q}{A} = \frac{Q}{\frac{\pi}{4} D^2} = \frac{3.33 \times 10^{-6} \frac{\text{m}^3}{\text{s}}}{\frac{\pi}{4}(2.22 \times 10^{-3} \text{ m})^2} = 0.863 \frac{\text{m}}{\text{s}}$$

Thus the Reynolds number is

$$\text{Re} = \frac{\rho VD}{\mu} = \frac{(1000\,\tfrac{\text{kg}}{\text{m}^3})(0.863\,\tfrac{\text{m}}{\text{s}})(2.22\times 10^{-3}\,\text{m})}{0.001\,\tfrac{\text{kg}}{\text{m}\cdot\text{s}}} = 1920$$

Since Re < 2000, the flow is indeed laminar as assumed.

Note that if the flow had been turbulent, an *explicit* solution would not have been possible, and an *iterative* method would have to be used to find the desired diameter.

Iterative Method for Solving Pipe Flow Problems

Step 1: Write the energy equation in units of length.

Step 2: Identify the unknown quantity you need to solve for (P, V, D, Q, etc.).

Step 3: Use the relation $h_f = f(\text{Re}, \varepsilon/D)$ for major flow loss.

Step 4: Determine if you can solve for the unknown variable explicitly (directly):

If the unknown is P, z, or h_f, then you can.

If the unknown is D or V, then you cannot.

If you can, then plug all known values into the energy equation and solve.

If you cannot, then set up an iterative solution and solve either by hand or with software:

Step 4.1: Write all equations in terms of the unknown variable.

Step 4.2: Determine which variable or parameter is the best to iterate on. It is usually f, since this varies the least.

Step 4.3: Guess an initial value for f.

Step 4.4: Calculate the unknown (D or V) based on the assumed value of f. Compute the Reynolds number and the relative roughness.

Step 4.5: Calculate the new value of f, and see how it compares to the guessed value. If within 1%, then stop. If not, return to step 4.4.

Why does the iterative method work? If f is guessed too high, V is adjusted lower to meet the value of h_f. This lower value of V leads to a new value of f that is closer to the answer. This type of iterative technique is useful in other fields of engineering as well. Other examples of iterative techniques include Newton's method for solving nonlinear equations. In fact, iterative techniques are at the heart of CFD codes.

EXAMPLE 5.11

For a concrete pipe of diameter 23 cm and length 500 m carrying water down an elevation change of 10 m with negligible change in water pressure, what is the flow rate in liters per minute?

SOLUTION: In this problem the Reynolds number and friction factor cannot yet be calculated because the velocity is unknown. Since the pipe is of constant diameter and the fluid is incompressible, the velocity must be constant through the pipe and hence the kinetic energy does not change. The pressure is also constant, so the only terms left in the energy equation are the elevation change and the friction loss:

$$\Delta z = h_f$$

The flow loss is written as

$$h_f = f \frac{L}{D} \frac{V^2}{2g}$$

Using the iterative method, we guess a value for f, then calculate a value for V, then calculate a better value for f, and so on until convergence is obtained.

Combining the previous two equations to solve for V yields

$$V = \sqrt{\frac{2gD\Delta z}{fL}}$$

Let us start with an initial guess of $f = 0.01$. Then the corresponding value of V would be

$$V = \sqrt{\frac{2(9.8 \frac{m}{s^2})(0.23 \text{ m})(10 \text{ m})}{(0.01)(500 \text{ m})}} = 3.00 \frac{m}{s}$$

Now we calculate the Reynolds number as

$$\text{Re} = \frac{VD}{\nu} = \frac{(3 \text{ m/s})(0.23 \text{ m})}{1 \times 10^{-6} \text{ m}^2/\text{s}} = 691{,}000$$

The relative roughness is $\varepsilon/D = 1.0 \text{ mm}/230 \text{ mm} = 0.00435$. Plugging these numbers in gives us a new and improved value for the friction factor:

$$f = \left\{ -1.8 \log \left[\frac{6.9}{\text{Re}} + \left(\frac{\varepsilon/D}{3.7} \right)^{1.11} \right] \right\}^{-2}$$

$$= \left\{ -1.8 \log \left[\frac{6.9}{691{,}000} + \left(\frac{0.00435}{3.7} \right)^{1.11} \right] \right\}^{-2} = 0.0293$$

Then the new value for the velocity is

$$V = \sqrt{\frac{2(9.8 \frac{m}{s^2})(0.23 \text{ m})(10 \text{ m})}{(0.0293)(500 \text{ m})}} = 1.75 \frac{m}{s}$$

The updated Reynolds number is

$$\text{Re} = \frac{VD}{\nu} = \frac{(1.75 \text{ m/s})(0.23 \text{ m})}{1 \times 10^{-6} \text{ m}^2/\text{s}} = 403{,}000$$

and the next value for the friction factor is

$$f = \left\{-1.8 \log\left[\frac{6.9}{403{,}000} + \left(\frac{0.00435}{3.7}\right)^{1.11}\right]\right\}^{-2} = 0.0294$$

and the recalculated velocity is

$$V = \sqrt{\frac{2(9.8 \frac{m}{s^2})(0.23 \text{ m})(10 \text{ m})}{(0.0294)(500 \text{ m})}} = 1.75 \frac{m}{s}$$

To three significant digits, the velocity has not changed since the last iteration, so the solution has converged to sufficient accuracy. We can now calculate the flow rate:

$$Q = AV = \frac{\pi}{4}(0.23 \text{ m})^2 \left(1.75 \frac{m}{s}\right)$$

$$= 0.0727 \frac{m^3}{s} \times \frac{1000 \text{ L}}{1 \text{ m}^3} \times \frac{60 \text{ s}}{1 \text{ min}} = 4360 \text{ L/min}$$

Almost all problems of this sort converge within two or at most three iterations. For this particular example the initial guess for f was pretty low considering the high roughness of concrete pipes, yet the iteration still quickly converged to a solution. Also note that asking for the flow rate with diameter given is equivalent to asking for velocity.

EXAMPLE 5.12

Used motor oil (SG = 0.888, $\nu = 4.0 \times 10^{-5}$ m²/s) flows through a cast iron pipe of length 24 m. The flow is level with an available supply pressure of 340 kPa, gage, and it is necessary to have a flow rate of at least 15.0 kg/s. Find the smallest diameter that will provide the necessary flow rate for given pressure head.

SOLUTION: In this problem the diameter is not known, so the Reynolds number and friction factor cannot be calculated. Thus we must use the iterative method. Applying the energy equation shows that the friction head must be balanced by the pressure head. The pressure head is

$$\frac{\Delta P}{\rho g} = \frac{340{,}000 \text{ Pa}}{(888 \frac{\text{kg}}{\text{m}^3})(9.8 \frac{\text{m}}{\text{s}^2})} = 39.1 \text{ m}$$

$$\frac{\Delta P}{\rho g} = h_f = f \frac{L}{D} \frac{V^2}{2g}$$

Unlike the previous problem, not only are f and V unknown on the right-hand side, but D is also unknown. Fortunately the flow rate is specified and we can use that to relate D and V. The volume flow rate is $Q = (15 \text{ kg/s})/(888 \text{ kg/m}^3) = 0.0169$ m³/s, and $Q = AV$. Since the diameter is the quantity we are after, let us express the velocity in terms of the diameter:

$$V = \frac{Q}{A} = \frac{0.0169 \text{ m}^3/\text{s}}{\frac{\pi}{4} D^2} = \frac{0.0215 \text{ m}^3/\text{s}}{D^2}$$

Now we substitute into the head loss equation:

$$h_f = 39.1 \text{ m} = f \frac{L}{D} \frac{V^2}{2g} = f \frac{L}{D} \frac{(\frac{0.0215}{D^2})^2}{2g} = 0.000463 f \frac{L}{2gD^5}$$

Rearranging to solve for D in terms of f gives

$$D^5 = \frac{0.000463 \frac{\text{m}^6}{\text{s}^2} (24 \text{ m})}{2(9.8 \frac{\text{m}}{\text{s}^2})(39.1 \text{ m})} f = (1.45 \times 10^{-5} \text{ m}^5) f$$

so we take the fifth root of each side:

$$D = (0.108 \text{ m})\sqrt[5]{f}$$

Now we are in a position to start the iterative solution. Let is start with an initial guess of $f = 0.02$. Then the corresponding value of D is

$$D = (0.108 \text{ m})\sqrt[5]{0.02} = 0.0494 \text{ m}$$

The velocity at that diameter is

$$V = \frac{0.0215 \text{ m}^3/\text{s}}{D^2} = \frac{0.0215 \text{ m}^3/\text{s}}{(0.0494 \text{ m})^2} = 8.81 \frac{\text{m}}{\text{s}}$$

The Reynolds number is

$$\text{Re} = \frac{VD}{\nu} = \frac{(8.81 \text{ m/s})(0.0494 \text{ m})}{4.0 \times 10^{-5} \text{ m}^2/\text{s}} = 10{,}880$$

and the relative roughness is $\varepsilon/D = 0.26$ mm/49.4 mm $= 0.00526$. Now we calculate the friction factor:

$$f = \left\{-1.8 \log\left[\frac{6.9}{10{,}880} + \left(\frac{0.00526}{3.7}\right)^{1.11}\right]\right\}^{-2} = 0.0373$$

With this new value of f, the new diameter is

$$D = (0.108 \text{ m})\sqrt[5]{0.0373} = 0.0559 \text{ m}$$

the new velocity is

$$V = \frac{0.0215 \text{ m}^3/\text{s}}{(0.0559 \text{ m})^2} = 6.88 \frac{\text{m}}{\text{s}}$$

and the updated Reynolds number is

$$\text{Re} = \frac{VD}{\nu} = \frac{(6.88 \text{ m/s})(0.0559 \text{ m})}{4.0 \times 10^{-5} \text{ m}^2/\text{s}} = 9{,}615$$

Now the relative roughness is $\varepsilon/D = 0.26$ mm/55.9 mm $= 0.00465$, and the new friction factor is

$$f = \left\{-1.8 \log\left[\frac{6.9}{9{,}615} + \left(\frac{0.00465}{3.7}\right)^{1.11}\right]\right\}^{-2} = 0.0372$$

With this value of f, the updated value for diameter is

$$D = (0.108 \text{ m})\sqrt[5]{0.0372} = 0.0559 \text{ m}$$

which is the same as the last value, to three digits, and so the solution has converged.

5.3 Open Channel Flow

Particularly important in civil engineering are *open channel flows*, where the top side of the flow is not enclosed but rather is open to the atmosphere. The liquid surface that is open to the atmosphere is called the *free surface*. Examples of these flows include the flow of water in rivers, canals, storm drains, sewers, and aqueducts. Obviously open channel flow refers only to liquids, since an unconstrained gas would diffuse out of the open channel.

Open channel flow was called *eaux courantes* by French engineers in the 19th century, when many new canals were being built. Much of the advance in both theoretical and empirical hydrodynamics at that time was driven by the need to improve the navigability of the canal systems. Darrigol suggests that the study of turbulence effects on fluid flows was undertaken first by engineers concerned with open channel flows because the turbulence was visible and thus more readily studied than in pipe flows [Darrigol05].

One complication of open channel flow is that, depending on the flow conditions, the liquid depth may not be constant. The pressure at the liquid surface in the open channel flow must be equal to the surrounding atmospheric pressure. It is also usually assumed that the shear stress at the surface is zero (i.e., friction between the air and the liquid is negligible). While flows in a closed pipe may be driven either by pressure or gravity, flows in open channels are driven by gravity alone, and are opposed by friction along the wetted solid surfaces.

Figure 5.19 shows several types of open channel flow. Most open channel flows occur on relatively large scales (1 m or more), so they are usually turbulent. As with most turbulent flows, this means that calculations rely on empirical correlations of shear stress and other flow characteristics. However, there is much qualitative insight to be gained through the use of simplified analytical methods.

We will start with an outline of the basic terminology of open channel flow and some of the laminar flow solutions. Then we will present the analysis techniques for the simplest type of open channel flow—uniform flow. In uniform flow the water depth in the channel is constant over the length of the channel. Then we will analyze gradually varying flows, and finally flows with rapid changes, such as hydraulic jumps.

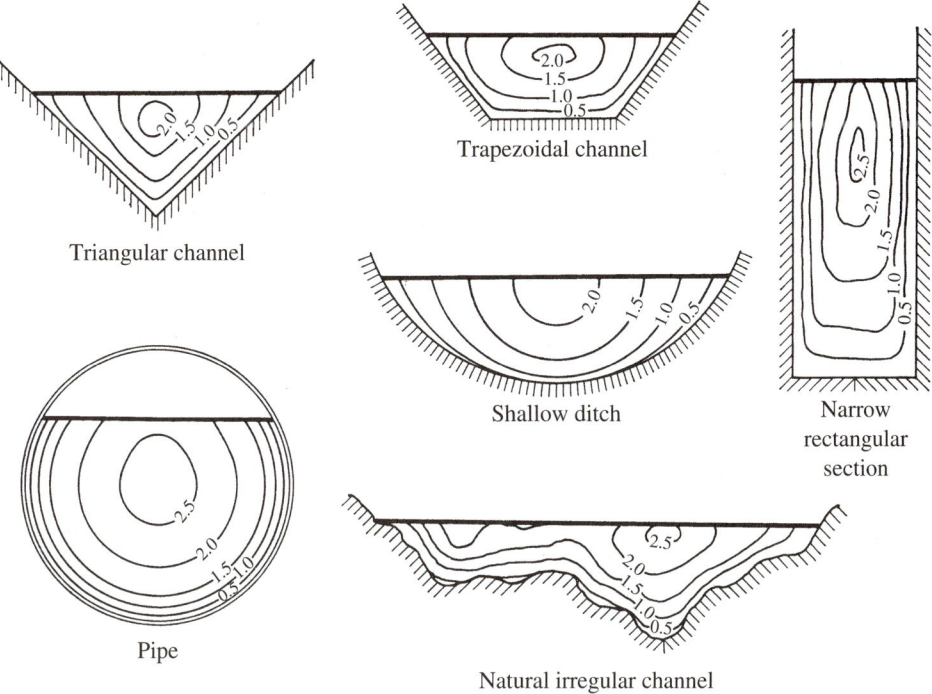

FIGURE 5.19 Diagrams of flow contours in cross sections of open channel flow. Note that the maximum velocity is slightly below the surface. From [Chow59]. Used with permission.

Terminology

In analyzing open channel flow it is useful to define the *x*-axis as being parallel and in line with the bottom floor of the channel, and the *y*-axis as being perpendicular to *x*. The depth of the channel at any location *x* can be defined as h, so that $h = h(x)$. The slope of the bottom of the channel can be denoted by the angle α. (See Figure 5.20.) In all cases it is assumed that the flow is in steady state. In most cases the flow is modeled as being one-dimensional, so only the bulk average velocity, V, is used.

It is common practice when dealing with open channel flows to define a *hydraulic radius*, R_h, which is the ratio of the cross-sectional area to the wetted perimeter where solid surfaces come in contact with the fluid. Thus,

$$R_h = \frac{A}{P} \tag{5.22}$$

This is slightly different from the hydraulic diameter for pipe flow defined in Equation 5.19. In open channel flows the top surface of the liquid is open to the atmosphere and is not included in the calculation of the wetted perimeter.

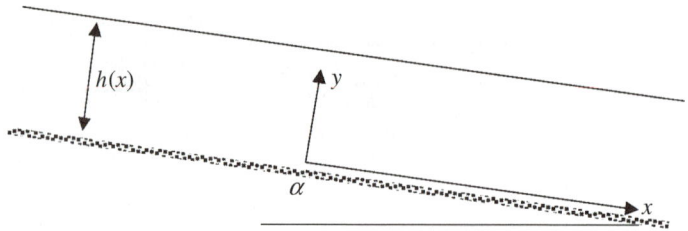

FIGURE 5.20 Schematic illustrating the terminology for open channel flows.

As stated above, it is also common practice to define the *x*-axis as parallel to the floor of the channel and the *y*-axis as perpendicular upward from it. So the depth of flow, *y*, is defined perpendicular from the sloping base of the channel, not vertically downward. Note that the slope of the water height need not be the same as the slope of the base of the channel. The slope of the channel bed is usually denoted by the angle θ. The width of the channel is *b*. For nonrectangular channels, an effective *hydraulic depth*, y_h, can be defined as

$$y_h = \frac{A}{b_s} \tag{5.23}$$

where b_s is the width of the channel at the water surface.

Another parameter that is useful in the analysis is the nondimensional *Froude number*:

$$\mathrm{Fr} = \frac{V}{\sqrt{gy}} \tag{5.24}$$

where *V* is a characteristic velocity, either of the bulk flow itself or of some artifact such as the propagation velocity of a wave; *g* is the acceleration due to gravity; and *y* is the channel depth. It has been found that open channel flows can be classified according to the value of the Froude number, where Fr = 1 represents critical flow. For values of Fr > 1 the flow is supercritical, convection dominates, no downstream disturbances can be felt upstream, and no waves can travel upstream. For Fr < 1, the flow is subcritical and disturbances such as waves can travel upstream.

5.3.1 Laminar Flow Solutions in Uniform Flow

In open channel flow, the pressure on the bottom of the channel is $P = \rho g h \cos(\theta)$. For small values of the slope angle θ, the pressure is approximately $P = \rho g h$. By a force balance, it can be shown that for uniform flow (flow of constant velocity and fluid depth), the shear stress acting along the surface is

$$\tau = \frac{\rho A}{P} g \sin(\theta) = \rho g R_h \sin(\theta)$$

where P is the wetted perimeter of the channel, and R_h is the hydraulic radius. For a wide, rectangular, flat-bottom channel the velocity profile is

$$\frac{u}{U_{max}} = 1 - \left(\frac{y}{h}\right)^2 \qquad (5.25)$$

where h is the channel depth. In general $h = h(x)$, but for uniform flow h is a constant. For this velocity profile the average velocity is 2/3 of the maximum velocity. In reality it has been observed that the maximum velocity occurs somewhat below the surface and not right at the surface.

If the flow is approximately two-dimensional (that is, the channel is wide compared to its depth), then the governing equations for laminar flow can be written as

$$\frac{dP}{dx} = \mu \frac{d^2 u}{dy^2} \qquad (5.26)$$

where the velocity profile is

$$u(y) = \frac{g \sin \alpha}{2\nu} y(2h - y) \qquad (5.27)$$

The assumption of laminar flow for such a geometry is only valid to a Reynolds number of about 330, where the Reynolds number is defined based on the mean flow velocity and the channel height:

$$\mathrm{Re} = \frac{U_{mean} h}{\nu} \qquad (5.28)$$

General Analysis for Uniform Flow

Flow at constant fluid depth in the channel is called *uniform flow* or normal depth. At normal depth in uniform flow, the velocity is constant ($V_1 = V_2$) and the pressure is also constant, so the energy equation becomes a balance between potential energy and energy loss due to friction:

$$h_f = z_1 - z_2 = L S_b$$

where L is the length of the channel between points 1 and 2, and S_b is the slope of the bottom of the channel and $S_b = \sin(\theta)$. Even though uniform flow is not encountered often in practice, it makes a useful approximation to real flows. However, most real flows are turbulent, and even for the case of uniform turbulent flows, all of the analysis is highly empirical.

The critical Reynolds number for transition to turbulence for open channel flow is

$$\text{Re}_{\text{crit}} = \frac{VD_h}{\nu} = 6000 \tag{5.29}$$

The friction force is the integral of the wall shear stress over the wetted area. Since open channel flows are typically turbulent, there is not a simple relationship between the shear stress and the flow rate and channel geometry, and empirical correlations are needed. The head loss for open channel flow is defined as

$$h_f = f \frac{L}{4R_h} \frac{V^2}{2g} \tag{5.30}$$

The head loss has units of length, but it can also be thought of as the energy loss due to friction per unit weight of the fluid. Formulas for friction loss include Chezy's equation:

$$V = C\sqrt{R_h S_b} \tag{5.31}$$

where C is an empirical constant, R is the hydraulic radius, and S is the slope of the bottom of the channel. Values of C were curve fit by Manning to the formula:

$$C = \sqrt{\frac{8g}{f}} = \frac{R_h^{1/6}}{n} \tag{5.32}$$

where values of n depend on the surface roughness of the material in the channel. Be very careful with the units for n!

The flow rate is $Q = AV$. Note that to maximize the flow rate, we want to minimize the wetted perimeter, P. Another way to write Chezy's equation is as

$$V = \sqrt{\frac{2g}{f} D_h S} = \sqrt{\frac{8g}{f} R_h S} \tag{5.33}$$

Alternatively, Manning's formula can be written as

$$V = \frac{1}{n}\left(\frac{D_h}{4}\right)^{2/3} S^{1/2} \tag{5.34}$$

If all terms are kept in standard kg-m-s units, then the flow velocity will be in m/s. The parameter n represents the roughness of the channel surface, and depends on the material of which the channel is made.

The friction factor, f, in Chezy's equation is a function of the Reynolds number and the relative roughness of the channel surface, much the same as for enclosed pipe flow, while Manning's n is a function solely of the channel material and condition. For fully turbulent flows (Re > 8000), the correlation for the friction factor in Chezy's equation is

$$f = \frac{0.25}{[0.57 - \log(\frac{\varepsilon}{D})]^2} \tag{5.35}$$

For transition flow in the range $2000 < \text{Re} < 8000$, the correlation for the friction factor is

$$f = \frac{0.25}{[\log(\frac{\varepsilon}{3.7D} + \frac{5.74}{\text{Re}^{0.9}})]^2} \tag{5.36}$$

Extensive values of Manning's n value can be found in [Chow59], and are reproduced here in Table 5.3.

The concept of the most efficient cross section is the shape that maximizes the volume flow rate for a given slope. The most efficient shape is a semicircle. However, in practice it is more expensive to build a semicircular channel than a rectangular or trapezoidal one.

5.3.2 Gradually Varying Flows

In gradually varying flows, the depth of the flow has not reached equilibrium with the channel floor, and so the channel depth, h, changes along x. This could be caused by a change in the slope or shape of the channel or by the presence of an obstruction such as a dam or weir. The term *gradually varying* implies there are no sudden changes in the flow, and so it is also implied that the slopes are relatively small, the streamlines nearly parallel, and the correlations for friction loss in uniform flow can also be applied here. Examples of gradually varying flows are the drawdown of the fluid surface as the channel approaches a sudden drop, the backwater behind an obstruction or bump, or the flow approaching a sluice gate. In each of these the flow may be either subcritical, critical, or supercritical.

In some cases the approximation of frictionless flow (like the Bernoulli equation) can provide useful insight. If friction can be neglected, then the energy of the fluid per unit volume of the fluid is

$$e = \frac{1}{2}\rho V^2 + \rho g h$$

Table 5.3 Manning's *n* Value for Several Surfaces for Use with Equation 5.34

Surface material	Minimum	Normal	Maximum
Closed conduits flowing partly full			
Brass, smooth	0.009	0.010	0.013
Steel, lockbar and welded	0.010	0.012	0.014
Steel, riveted and spiral	0.013	0.016	0.017
Cast iron, coated	0.010	0.013	0.014
Cast iron, uncoated	0.011	0.014	0.016
Wrought iron, black	0.012	0.014	0.015
Wrought iron, galvanized	0.013	0.016	0.017
Corrugated metal sub drain	0.017	0.019	0.021
Corrugated metal storm drain	0.021	0.024	0.030
Lucite	0.008	0.009	0.010
Glass	0.009	0.010	0.013
Cement, neat	0.010	0.011	0.013
Cement, mortar	0.011	0.013	0.015
Concrete, culvert straight	0.010	0.011	0.013
Concrete, culvert with bends	0.011	0.013	0.014
Concrete, finished	0.011	0.012	0.14
Concrete, sewer with manholes	0.013	0.015	0.017
Concrete, unfinished steel form	0.012	0.013	0.014
Concrete, unfinished, smooth wood form	0.012	0.014	0.016
Concrete, unfinished, rough wood form	0.015	0.017	0.020
Wood, stave	0.010	0.012	0.015
Wood, laminated	0.015	0.017	0.020
Clay, drainage tile	0.011	0.013	0.017
Clay, vitrified sewer	0.011	0.014	0.017
Clay, vitrified sewer, with manholes, inlet, etc.	0.013	0.015	0.017
Clay, vitrified sewer with open joint	0.014	0.016	0.018
Brickwork, glazed	0.011	0.013	0.015
Brickwork, lined with cement mortar	0.012	0.015	0.017
Sanitary sewers	0.012	0.013	0.016
Paved invert, smooth bottom	0.016	0.019	0.020
Rubble masonry, cemented	0.018	0.025	0.030
Lined or built-up channels			
Steel, unpainted	0.011	0.012	0.014
Steel, painted	0.012	0.013	0.017
Steel, corrugated	0.021	0.025	0.030
Cement, neat	0.010	0.011	0.013

(Continued)

Table 5.3 Continued

Surface material	Minimum	Normal	Maximum
Cement, mortar	0.011	0.013	0.015
Wood, planed, untreated	0.010	0.012	0.014
Wood, planed, creosoted	0.011	0.012	0.015
Wood, unplaned	0.011	0.013	0.015
Wood, plank with battens	0.012	0.015	0.018
Wood, lined with roofing paper	0.010	0.014	0.017
Concrete, trowel finish	0.011	0.013	0.015
Concrete, float finish	0.013	0.015	0.016
Concrete, finished with gravel on bottom	0.015	0.017	0.020
Concrete, unfinished	0.014	0.017	0.020
Concrete, gunite	0.016	0.019	0.023
Concrete, gunite, wavy	0.018	0.022	0.025
Concrete on good excavated rock	0.017	0.020	—
Concrete on irregular excavated rock	0.022	0.027	—
Concrete bottom float finished with sides of dressed stone in mortar	0.015	0.017	0.020
Concrete bottom float finished with sides of random stone in mortar	0.017	0.020	0.024
Concrete bottom float finished with sides of cement rubble masonry	0.020	0.025	0.030
Concrete bottom float finished with sides of dry rubble	0.020	0.030	0.035
Gravel bottom with sides of formed concrete	0.017	0.020	0.025
Gravel bottom with sides of random stone	0.020	0.023	0.026
Gravel bottom with sides of dry rubble	0.023	0.033	0.036
Brick, glazed	0.011	0.013	0.015
Brick, in cement mortar	0.012	0.015	0.018
Masonry, cemented rubble	0.017	0.025	0.030
Masonry, dry rubble	0.023	0.032	0.035
Dressed ashlar	0.013	0.015	0.017
Asphalt, smooth	0.013	0.013	—
Asphalt, rough	0.016	0.016	—
Vegetal lining	0.030	—	0.500
Excavated or dredged			
Earth, straight, uniform, clean, recently completed	0.016	0.018	0.020
Earth, straight, uniform, clean, after weathering	0.018	0.022	0.025
Earth, straight, uniform, gravel, clean	0.022	0.025	0.030

(*Continued*)

Table 5.3 Continued

Surface material	Minimum	Normal	Maximum
Earth, straight, uniform, short grass	0.022	0.027	0.033
Earth, winding, no vegetation	0.023	0.025	0.030
Earth, winding, grass	0.025	0.030	0.033
Earth, winding, dense weeds	0.030	0.035	0.040
Earth, winding, with rubble sides	0.028	0.030	0.035
Earth, winding, stony bottom and weedy banks	0.025	0.035	0.040
Earth, winding, cobble bottom and clean sides	0.030	0.040	0.050
Dragline excavated, no vegetation	0.025	0.028	0.033
Dragline excavated, light brush on banks	0.035	0.050	0.060
Rock cuts, smooth and uniform	0.025	0.035	0.040
Rock cuts, jagged and irregular	0.035	0.040	0.050
Channels not maintained, dense weeds as high as flow depth	0.050	0.080	0.120
Channels not maintained, clean bottom, brush on sides	0.040	0.050	0.080
Natural streams			
Minor stream			
Stream on plain; clean, straight	0.025	0.030	0.033
Stream on plain; clean, straight with stones and weeds	0.030	0.035	0.040
Stream on plain; clean, winding	0.033	0.040	0.045
Stream on plain; clean, winding with stones and weeds	0.035	0.045	0.050
Stream on plain; very weedy, deep pools	0.075	0.100	0.150
Mountain stream, bottom with gravels and cobble	0.030	0.040	0.050
Mountain stream, bottom with large boulders	0.040	0.050	0.070
Flood plains			
Pasture, no brush, short grass	0.025	0.030	0.035
Pasture, no brush, high grass	0.030	0.035	0.050
Cultivated area, no crop	0.020	0.030	0.040
Cultivated area, mature row crops	0.025	0.035	0.045
Cultivated area, mature field crops	0.030	0.040	0.050
Scattered brush, heavy weeds	0.035	0.050	0.070
Light brush and trees, in winter	0.035	0.050	0.060
Light brush and trees, in summer	0.040	0.060	0.080
Medium to dense brush and trees, in winter	0.045	0.070	0.110
Medium to dense brush and trees, in summer	0.070	0.100	0.160
Dense willows, summer, straight	0.110	0.150	0.200

(Continued)

Table 5.3 Continued

Surface material	Minimum	Normal	Maximum
Cleared land with tree stumps, no sprouts	0.030	0.040	0.050
Cleared land with tree stumps, heavy sprouts	0.050	0.060	0.080
Heavy stand of timber, little undergrowth, flood below branches	0.080	0.100	0.120
Heavy stand of timber, little undergrowth, flood reaches branches	0.100	0.120	0.160
Major streams (top width at flood stage > 100 ft)			
Regular section with no boulders or brush	0.025	—	0.060
Irregular and rough section	0.035	—	0.100

From [Chow59]. Used with permission.

Critical flow (Fr = 1) is the flow that provides the minimum possible energy, e, of the fluid for a given flow rate or, conversely, that provides the maximum flow rate for a given flow energy.

For frictionless flow over a bump of height $h(x)$, the slope of the water in the channel changes according to

$$\frac{dy}{dx} = \frac{1}{\text{Fr}^2 - 1} \frac{dh}{dx} \tag{5.37}$$

5.3.3 Hydraulic Jumps

Normally flows go from higher elevations to lower elevations, but there are a few cases in which flows can briefly go uphill. This can occur either gradually, in what is termed *adverse flow*, or suddenly, as in a *hydraulic jump*. Mathematically, a hydraulic jump (Figure 5.21) is the transition from a supercritical flow upstream to a subcritical flow downstream. The upstream condition must always be supercritical for a hydraulic jump to occur. A hydraulic jump also dissipates the fluid energy, e. Momentum is conserved, but some energy is lost to dissipation and frictional effects. The physical structure of the hydraulic jump can be either undulating, at low Froude numbers, or more direct with a strong turbulent rolling motion, at higher Froude numbers.

For flow in a uniform rectangular channel, the relation between the depth of the water going into the hydraulic jump, h_1, and the water at the higher level, h_2, is

$$\frac{h_2}{h_1} = \frac{1}{2}\left(\sqrt{1 + 8\,\text{Fr}_1^2} - 1\right) \tag{5.38}$$

FIGURE 5.21 Picture of a hydraulic jump in a converging–diverging section of flow in a water table.

Here $h_2 > h_1$ and the Froude number is based on the incoming flow conditions. Equation 5.38 is derived by combining the conservation of mass, momentum, and energy equations for frictionless flows.

Hydraulic jumps can also occur after sudden expansions in a channel. The head loss in such a hydraulic jump is

$$\frac{h_f}{(h_1 + \frac{V_1^2}{2g})} = \frac{(\sqrt{1 + 8\,\text{Fr}_1^2} - 3)^3}{8(\sqrt{1 + 8\,\text{Fr}_1^2} - 3)(\text{Fr}_1^2 + 2)} \tag{5.39}$$

5.3.4 Weirs

Weirs are usually built for the purpose of raising the fluid level in an open channel, or sometimes to measure the fluid level or flow rate in streams and rivers where water management is a concern. Weirs, which are usually completely submerged by the flow, create a backwater in the approaching flow. The flow over a weir is called a *nappe*. (In contrast, dams usually allow the flow to go underneath them via a sluice gate or around them via a spillway, rather than over them.) Under normal circumstances the flow around a weir is influenced only by the upstream, not the downstream, conditions.

The flow rate over a weir can be calculated as

$$Q = C_d b \frac{L}{b} \sqrt{g} (h_1 - h_{\text{weir}})^{3/2} \tag{5.40}$$

Here b is the channel width, L is length of the cutout in the crest of the weir, C_d is an empirical coefficient to account for flow losses over the weir, and $h_1 - h_{\text{weir}}$ is the depth of the water immediately over the weir.

For small discharges of water, V-notch weirs are used to measure flow rates. For medium discharges, full-width thin-plate weirs are used, and for large discharges, Crump triangular profile weirs and other long-base structures are used [Ackers78]. For flow over a V-notch weir of notch angle θ, it can be shown from dimensional analysis that the flow rate, Q, is

$$Q = C_d \tan\left(\frac{\theta}{2}\right)\sqrt{g}\, h_1^{5/2}$$

where h_1 is the upstream height of the water above the vertex of the notch.

5.4 Complex Pipe Systems

Many engineering systems are more complex than just a straight piece of constant-diameter pipe. The more complex systems include intake and exhaust manifolds for car engines, piping systems in houses or buildings, fire protection systems, water supply systems for cities, and fuel supply systems on planes with multiple engines, just to name a few. We often classify these complex systems as being of one of two types: pipes in series or pipes in parallel. Figures 5.22 through 5.24 show some different possible configurations of multipipe systems.

If the pipes are in series (as in Figure 5.22), then the problem is simple—we just add the resistance for each section of pipe to get the total resistance, much like adding the resistances for electrical resistors in series in a circuit. Thus,

$$h_f = \sum_i h_{f,i} \tag{5.41}$$

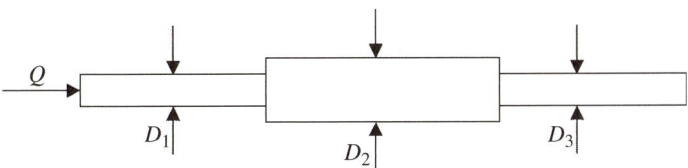

FIGURE 5.22 Schematic of pipes in series connection.

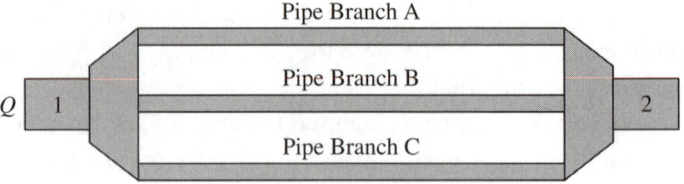

FIGURE 5.23 Schematic of pipes in parallel, where the pipes connect at the same end node.

In other words, the total friction head is equal to the sum of the individual friction heads in each section of connected pipe. Likewise, the total pressure loss is the sum of the pressure losses in each section:

$$h_{f,\text{total}} = f_1 \frac{L_1}{D_1} \frac{V_1^2}{2g} + f_2 \frac{L_2}{D_2} \frac{V_2^2}{2g} + f_3 \frac{L_3}{D_3} \frac{V_3^2}{2g} + \cdots \qquad (5.42)$$

For pipes in parallel, the math is a little more complex. See Figure 5.23 for a schematic of a simple system with three pipes in parallel. The problem here is to calculate the total pressure drop between points 1 and 2. The parameters L, D, V, and f may be different for each section of parallel pipe. When the three pipes converge at point 2 the pressure will be the same, and hence the pressure drop in each section of pipe must be the same, though again the flow rates may be different. Since the pressure drop is not linear with flow rate (it is roughly quadratic for fully turbulent flow), there is no direct analytical solution. Thus an iterative method, which can take some time when solving problems by hand, must be employed. Fortunately modern computer software tools are available that speed up the process. (It is important to remember that a computer cannot do anything that an engineer cannot do, it just does the calculations faster.) In general, for pipes in parallel you will get the most flow through the pipe of least resistance.

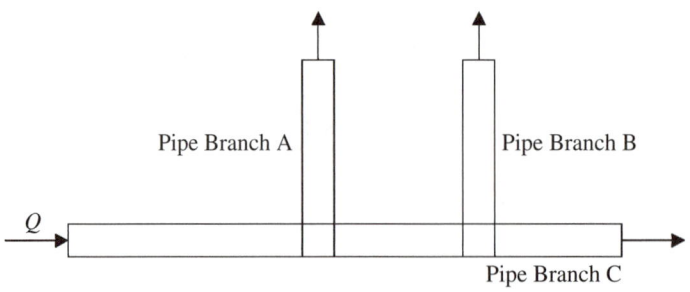

FIGURE 5.24 Schematic of pipes in parallel, where the pipes have different ending points.

5.4 Complex Pipe Systems

EXAMPLE 5.13

Consider two pipes in parallel, labeled A and B, between pipe junctions 1 (the inlet) and 2 (the outlet). The diameter of pipe A is 5 cm and the diameter of pipe B is 3 cm. The length of pipe A is 1 m, and the length of pipe B is 5 m. Both pipes are made out of copper. The total flow into junction A is $Q = 0.04$ m^3/s. How much flow goes through each of the pipes?

SOLUTION: Whichever path the fluid takes, it must go in at point 1 and end up at point 2, so it will go through a pressure drop of $\Delta P = P_1 - P_2$. Hence the friction loss must also be the same in each pipe, so $h_{f,A} = h_{f,B}$. Let us first determine how many equations and how many unknowns there are. There are two unknowns (V_A and V_B), and we apply conservation of mass and conservation of energy at each junction. As previously mentioned, we must use an iterative method for parallel pipe flow problems.

To start the problem, we assume equal flow rates in each of the two pipes: $Q_A = Q_B = \frac{1}{2}Q = 0.02$ m^3/s. Then we calculate the velocity, Reynolds number, and friction factor for each pipe. The area of pipe A is 0.00196 m^2, and the area of pipe B is 0.000707 m^2. So $V_A = 0.02/0.00196 = 10.2$ m/s, and $V_B = 0.02/0.000707 = 28.3$ m/s. $\text{Re}_A = (10.2)(0.05)/1 \times 10^{-6} = 510{,}000$, and $\text{Re}_B = (28.3)(0.03)/1 \times 10^{-6} = 849{,}000$. The relative roughnesses are $\varepsilon/D_A = 0.0015/50 = 0.00003$ and $\varepsilon/D_B = 0.0015/30 = 0.00005$. The corresponding friction factors are $f_A = 0.0134$ and $f_B = 0.0128$.

Now it is time to iterate, using a relation between pipes A and B. The head losses in A and B must be equal, so

$$\left(f\frac{L}{D}\frac{V^2}{2g}\right)_A = \left(f\frac{L}{D}\frac{V^2}{2g}\right)_B$$

Simplifying gives

$$V_A = V_B\sqrt{\frac{f_B}{f_A}\frac{(L/D)_B}{(L/D)_A}} = 2.89 V_B\sqrt{\frac{f_B}{f_A}}$$

Plugging in the values for the friction factors yields $V_A = 2.82 V_B$. Applying conservation of mass to solve gives $V_A = 18.1$ m/s and $V_B = 6.4$ m/s. Now we repeat the process and calculate Reynolds numbers and friction factors again: $\text{Re}_A = (18.1)(0.05)/1 \times 10^{-6} = 905{,}000$, and $\text{Re}_B = (6.4)(0.03)/1 \times 10^{-6} = 192{,}000$. The corresponding friction factors are $f_A = 0.0123$ and $f_B = 0.0159$. The new velocity relationship is $V_A = 3.28 V_B$. Applying conservation of mass to solve gives $V_A = 18.4$ m/s and $V_B = 5.6$ m/s. The process could be repeated again, but this is probably close to the final answer. The corresponding flow rates are $Q_A = 0.036$ m³/s and $Q_B = 0.004$ m³/s.

5.5 Secondary Losses

The loss of pressure due to friction with the pipe walls is termed the *primary* or *major loss*. In addition to this loss, there are also pressure losses present when the flow goes around bends or through open valves and expansions and contractions. Such losses are called *secondary* or *minor losses*.

The loss coefficient, K, for secondary loss across a device is defined as

$$K = \frac{\Delta P}{\frac{1}{2}\rho V^2} \tag{5.43}$$

Such losses tend to correlate well with the dynamic pressure, $\frac{1}{2}\rho V^2$. Equivalently, the loss coefficient can be defined in terms of the head loss as

$$K = \frac{h_L}{(V^2/2g)} \tag{5.44}$$

So then in a piping system the total flow loss between two points can be expressed as

$$h_{\text{total}} = \sum h = \left(f\frac{L}{D} + \sum K\right)\frac{V^2}{2g} \tag{5.45}$$

The loss coefficient K usually cannot be calculated from theory, so empirical methods are used to find it. Data from experiments have been collated and presented in graphical form for many commonly encountered flow devices that result in a pressure loss. Such devices include orifices, entrances and exits, valves, gates, diffusers and nozzles, Venturis, and so on. The data in Table 5.4 and Figure 5.25, and the correlations in Equations 5.46 through 5.49 give loss coefficients for a great many of these. Fluids handbooks, such as those listed in Appendix E, give data for specific cases.

One of the few loss coefficients that can be determined theoretically is that for a *sudden expansion*. For this case the loss coefficient is

$$K = \left(1 - \frac{A_{\text{small}}}{A_{\text{large}}}\right)^2 \tag{5.46}$$

For the discharge of a fluid from a relatively small pipe into a large reservoir or chamber, the loss coefficient is $K = 1.0$, since essentially all of the kinetic energy of the fluid is lost and cannot do useful work.

Table 5.4 Typical Range of Loss Coefficients for Various Pipe Fittings

Fitting description	K-value
Globe valve, fully open	3–8
Ball valve, fully open	0.04–0.10
Check valve, fully open	2
Gate valve, fully open	0.03–0.20
Butterfly valve, fully open	0.5–2.0
Square-edged inlet from tank	0.5
Bell-mouth inlet from tank	0.05
Discharge into tank	1.0
Standard elbow	0.2–0.3
Long radius elbow	<0.1–0.3

From [DOE07].

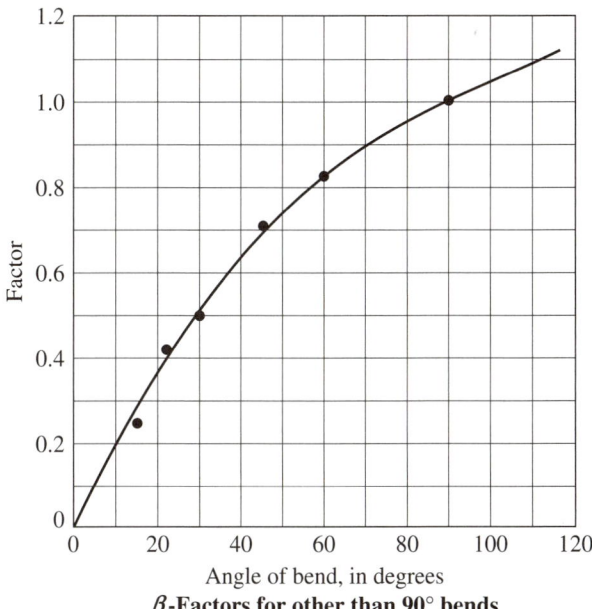

β-Factors for other than 90° bends

FIGURE 5.25 Loss coefficients in bends of different angles, normalized to the loss coefficient of a 90° bend. From [USBR92].

For loss through a *sudden contraction*, a correlation for the loss coefficient is [White08]

$$K = 0.42\left(1 - \frac{A_{small}}{A_{large}}\right) \tag{5.47}$$

For the more *gradual expansion* through a diffuser, a correlation for the loss coefficient is [Gibson10]

$$K = 2.61 \sin(\theta)\left(1 - \frac{A_{small}}{A_{large}}\right)^2 + f_{ave}\frac{L}{D_{ave}} \tag{5.48}$$

where θ is the half angle of the diffuser, L is the length of the diffuser, $D_{ave} = (D_1 + D_2)/2$, and f_{ave} is the average of the friction factors of the inlet and outlet pipes.

The loss encountered in a sudden expansion or contraction can be reduced by smoothing the flow transition with a nozzle or diffuser section between the different diameter pipes. Note that diffuser performance can be diminished if the spread angle is too large and flow separation occurs at the walls.

When a pipe goes through a bend, the head or pressure loss is larger than it would be at a corresponding section of straight pipe of equivalent length. This is because of a secondary three-dimensional flow that is established in the bend. A vortex forms in the bend, increasing the effective distance each particle must travel and the relative velocity between the fluid and the pipe wall. The loss coefficient for a bend depends on the angle and the radius of curvature of the bend. A correlation for the loss coefficient through a 90° bend is [Ito60]

$$K = \frac{0.388\alpha}{\text{Re}^{0.17}}\left(\frac{r}{D}\right)^{0.84} \tag{5.49}$$

where r is the radius of the bend, and the value of α is given by

$$\alpha = 0.95 + 4.42\left(\frac{r}{D}\right)^{-1.96}$$

This formula is only valid for turbulent flow. For miter bends, which are popular because of their low cost, the loss coefficient is higher than it would be for a smooth bend of the same angle. For elbows and bends of angles other than 90°, Figure 5.25 shows a correlation factor that is used by multiplying it by the loss coefficient calculated using Equation 5.49.

Even fully open valves can contribute significantly to the overall flow losses in a system. Partially closed valves can have extremely high loss coefficients. Table 5.4 shows the loss coefficients for some types of valves. A fluids reference book (see Appendix E) should be consulted if more extensive data is needed.

EXAMPLE 5.14

What is the pressure loss associated with a sudden enlargement in an air pipe from a diameter of 10 cm to a diameter of 30 cm? The average velocity of air in the smaller pipe is 25 m/s. How much would the pressure loss be reduced by using a gradual conical enlargement section (a diffuser) of angle 20°?

SOLUTION: For the sudden enlargement, Equation 5.46 applies, and the loss coefficient is

$$K = \left(1 - \frac{A_1}{A_2}\right)^2 = \left[1 - \left(\frac{10 \text{ cm}}{30 \text{ cm}}\right)^2\right]^2 = 0.79$$

The pressure loss associated with the sudden expansion is

$$\Delta P = K \frac{1}{2}\rho V_1^2 = (0.79)\frac{1}{2}\left(1.2 \frac{\text{kg}}{\text{m}^3}\right)\left(25 \frac{\text{m}}{\text{s}}\right)^2 = 296 \text{ Pa}$$

If a conical expansion joint had been used, the loss coefficient would be $K = 0.38$ and the associated pressure drop would be

$$\Delta P = K \frac{1}{2}\rho V_1^2 = (0.38)\frac{1}{2}\left(1.2 \frac{\text{kg}}{\text{m}^3}\right)\left(25 \frac{\text{m}}{\text{s}}\right)^2 = 143 \text{ Pa}$$

So using a diffuser reduces the pressure drop by half. Note that in either case, the overall static pressure will actually rise in the expansion because of the conversion of kinetic energy to pressure energy in Bernoulli's equation.

■ Summary

This chapter presented the tools used by modern engineers to calculate the pressure loss due to friction in internal flows. Correlations for both laminar and turbulent flow were presented. The tools for dealing with noncircular ducts and open channel flows were also presented. After reading this chapter, you should be able to calculate the friction loss in internal flows and solve basic pipe flow problems.

■ References

[Reynolds83] Reynolds, Osborne. 1883. "An Experimental Investigation of the Circumstances Which Determine Whether the Motion of Water Shall be Direct or Sinuous, and of the Law of Resistance in Parallel Channels." *Philosophical Transactions of the Royal Society of London*, Vol. 174, pp. 935–982.

[Gibson10] Gibson, A. 1910. "On the Flow of Water through Pipes and Passages." *Proceedings of the Royal Society of London A*, Vol. 83, pp. 366–378.

[Moody44] Moody. 1944. "Friction Factors for Pipe Flow." *Transactions of the ASME*, Vol. 66.

[NACA58] Deissler, R. and M. Taylor. 1958. Analysis of Turbulent Flow and Heat Transfer in Noncircular Passages. NACA-TN-4384.

[Chow59] Chow, V. 1959. *Open Channel Hydraulics*. McGraw-Hill.

[Ito60] Ito, H. 1960. "Pressure Losses in Smooth Pipe Bends." *Journal of Basic Engineering*, pp. 131–143.

[Ackers78] Ackers, P. 1978. *Weirs and Flumes for Flow Measurement*. Wiley.

[USBR92] U.S. Bureau of Reclamation. 1992. *Friction Factors for Large Conduits Flowing Full*. Engineering Monograph No. 7.

[Darrigol05] Darrigol, O. 2005 *Worlds of Flow*. Oxford University Press.

[DOE07] U.S. Department of Energy. 2007. "Energy Tips—Pumping Systems." DOE/GO-102007-2228.

[White08] White, F. 2008. *Fluid Mechanics*, 6th edition. McGraw-Hill.

■ Problems

1. A large container filled with water 1 m deep is open to the atmosphere at the top. What is the largest diameter pipe that can be attached to the bottom and still have laminar flow? Assume that the length of the pipe is small compared to the height of the basin and that the bottom end opens to the atmosphere. Neglect fluid friction.

2. Repeat Problem 1 including frictional effects. (The length of the pipe is 10 cm.)

3. Repeat Problem 1 including wall friction *and* entrance loss effects. (The length of the pipe is 10 cm, and for the entrance $K = 0.5$.)

4. A horizontal pipe of diameter D is partially clogged with debris. If the bottom half of the pipe is blocked off, what is the effective hydraulic diameter for the flow through the top half (a semicircular cross section)?

5. Why is the drain pipe in your kitchen sink larger in diameter than the drain pipe in your bathroom sink?

6. Compare the Reynolds numbers for flow through a 2-mm ID fuel line at 0.03 gal/min for the following fuels: gasoline, diesel, ethanol, DME.

7. Air flows through a converging nozzle at 1 m/s at the inlet, which is 25 cm in diameter. The outlet is 10 cm in diameter. What are the Reynolds numbers at the inlet and the exit? When calculating the K-value for the loss, which velocity do you use, the inlet or the outlet?

8. Air flows through an air-conditioning system with rectangular ducts of size 12 in. by 20 in. If the overall length of the duct system is 100 ft and the total flow rate is 50 ft^3/s (SCFM), what is the required fan power?

9. What is the definition of SCFM? Of SCF? If your flow meter is calibrated in SCFM, what does this mean for your use of it? A SCFM corresponds to what flow rate of air, in units of kg/s?

10. Engineers designing an air flow system are trying to decide whether to use a pushing fan at the entrance or a suction fan at the exit. If the desired flow rate is 1 kg/s and the duct has size 1 ft by 2 ft, estimate the longest length of ducting for which a suction fan could possibly work.

11. Air flows through a pipe of diameter 1 in. at an average velocity of 20 ft/s. If the air is replaced by water flowing through the same pipe at the same Reynolds number, will the friction losses increase or decrease?

12. Calculate the pressure loss per mile of the Trans-Alaskan Pipeline.

13. Compare the difference in friction losses for water flow in glass, PVC, rubber, and stainless steel pipes for fixed flow conditions of $D = 2$ cm, $L = 1$ m, and $V = 20$ m/s.

14. Repeat Problem 13, but with the mass flow rate of water increased by a factor of 10, and increased by a factor of 100.

15. A section of a canal to bring fresh water to a desert area is 20 km long and goes down an elevation change of 75 m. If the canal is 12 m wide and 9 m deep, and the canal walls are made of concrete, what is the maximum flow rate that the canal can handle?

16. Consider coannular flow in a counterflow heat exchanger. The outer shell has a diameter of 10 in. and the inner shell has a diameter of 7.1 in., and both are 3 ft long. Water flows through both sides at a flow rate of 85 gal/min. Which side has more friction loss?

17. Which would be more efficient for delivering a flow of water of 1 L/s: one large pipe with turbulent flow or a bunch of little pipes with laminar flow?

18. An overzealous safety engineer added 5 shutoff valves to 25 ft of steel piping of diameter 3 inches. If the typical flow rate through the system is 10 gal/min, what is the friction loss associated with the open valves? If the cost of electrical power is $0.10/kW-hr, how much money will this cost over the course of 1 year? (Neglect any inefficiencies in the pump.)

19. In terms of minimizing flow losses, is it better to have a short abrupt expansion or a long gradual expansion from a narrow pipe to a wide one?

20. A common-rail fuel system for a diesel engine delivers fuel at 200 MPa and a flow rate of 0.016 kg/s of diesel fuel through a stainless steel rail with internal diameter of 1 mm. What is the friction loss through 20 cm of common-rail pipe?

21. In many cities, water mains were originally made out of wood. Find a value of surface roughness for wood and calculate the flow loss through an 8 in. ID section of pipe 50 ft long for a velocity of 4.0 ft/s.

22. The rubber tube of a Super-soaker water gun has internal diameter of 2 mm and delivers water with an average velocity of 10 m/s. What is the pressure loss over 28 cm of tubing?

23. What is the viscosity of molten milk chocolate? When molten milk chocolate is being poured down an open channel flow into molds, do you think the flow is laminar or turbulent?

24. Explain how weirs improve the navigability of rivers.

25. The water supply in a house has a pressure of 50 psig. With a 100-ft-long outdoor hose of inner diameter 5/8 inch, the homeowner can spray water another 20 ft horizontally when the nozzle is aimed level with the ground at a height 4 ft above the ground. The diameter of the nozzle is 5/16 in. Estimate the magnitude of the major head loss at the spigot.

26. For a 100-ft-long vinyl water hose, compare the flow rate if it is straight to that if it is coiled with 2 ft average coil diameter. The inner diameter of the pipe is $\frac{3}{4}$ in. The supply pressure is 30 psig.

27. A large cylindrical reservoir has a small horizontal cast iron pipe coming out of the bottom. The reservoir is 10 m high and 10 m in diameter, and the pipe is 50 m long and 50 mm in diameter. The reservoir is filled with SAE 30 oil at STP. Calculate the initial velocity of oil moving through the pipe if (a) friction losses are ignored, (b) friction losses in the pipe are included.

28. Using the information in the previous problem, calculate the velocity in the pipe as a function of time until the tank is emptied, neglecting the friction losses. Repeat the calculations when including the friction losses. (Hint: You will want to write a computer program to do the second part of this problem.)

29. Water flows through a cast iron pipe of 10 cm diameter at a velocity of 50 m/s. Calculate the pressure loss through 100 m of pipe.

30. A laminar flow element (LFE) is used to measure the flow rate through a pipe. For air at standard conditions, calculate the total flow rate (Q) of air knowing that the measured pressure loss $\Delta P = 20$ Pa over the laminar flow element with a length of $L = 5$ cm. The diameter of the main pipe, D, is 10 cm, the diameter of the laminar flow elements, D_L, is 3 mm, and the pressure drop is measured. You may assume that the total cross-sectional area of the laminar flow element is essentially the same as the cross-sectional area in the main flow pipe.

31. Water flows through a commercial steel pipe at an average speed of 2 m/s. The pipe has a diameter of 23 cm and is slanted at an angle of 25° with the flow

going downward. Will the pressure increase or decrease along the length of the pipe?

32. Crude oil (density 928 kg/m³, viscosity 0.017 kg/m-s) flows downward through an inclined (15°) galvanized iron pipeline with a diameter of 8 cm. The flow rate is 250 L/min. Determine (a) the pressure drop between two points 5 m apart, (b) the shear stress between the oil and the pipe.

33. Crude oil (density 928 kg/m³, viscosity 0.017 kg/m-s) is extracted from an underground reservoir by pumping pressurized steam on top of the oil. If the diameter of the extraction pipe is 10 cm, its length is 1200 m, the depth from the top of the pipe to the oil surface is 1000 m, and the surface roughness of the pipe is 0.1 mm, find the required pressure to extract the oil.

34. A cell-sorting device sorts cells based on the fluorescence properties of substances in the cells. The cells are put into solution with water and then pass through a vertical lead pipe before going through the probe volume. It is critical that each cell pass through the probe volume only once, therefore turbulence must be avoided at all costs. If the maximum allowable Reynolds number in the lead pipe is 2000, what diameter of tube should be used? The pipe length is 10 cm and you may assume that the pressure at the entrance and exit of the lead pipe are both nearly atmospheric. You may assume that the density and viscosity of the solution are that of water.

35. SAE 50 oil flows through a 12-cm-diameter horizontal pipe at a mean velocity of 10 m/s. Two ends of a U-tube manometer filled with mercury are connected to the pipe 22 cm apart. Predict the difference in the height of the mercury in the two legs.

36. Crude oil flows through a horizontal pipeline from Alaska to the continental United States. The flow rate is $Q = 2.5$ m³/s, and the diameter is $D = 1$ m. The pipe is galvanized iron. If the maximum allowable pressure drop between pumps is 7.5 MPa, determine the maximum allowable distance between pumps, in units of km.

37. Water flows through a 11-cm-diameter cast iron pipe at a velocity of 2.2 m/s. The pipe is also angled upward so that the exit is 10 m higher than the entrance of the pipe where the pump is located. What is the power required in kW to pump the water through 100 m of pipe?

38. If you can drink a 12-oz soda through an 8-mm straw of length 20 cm in 2 min, estimate the pressure loss in the straw. How does this compare to the vacuum pressure that must be generated to move the soda up a height of 19 cm (when the can is nearly empty)?

39. A diesel engine of displacement 1.6 L runs at a steady speed of 2800 rpm. If the exhaust gas conditions are approximately 120 kPa (absolute) and 200°C, and the tailpipe is approximated as a section of straight galvanized steel pipe 1.4 m long with a diameter of 4.5 cm, calculate the pressure loss through the exhaust pipe.

40. Verify that the momentum flow rate correction factor is in fact equal to $\frac{4}{3}$ for laminar flow through a round pipe.

41. For a pipe of 1 cm diameter, what is the velocity that corresponds to a Reynolds number of 2000 for water and for air?

42. At its maximum flow rate, is the flow in your kitchen faucet laminar or turbulent? (Hint: Use an empty soda bottle or milk jug and a stopwatch to measure the flow rate.)

43. Is the flow of gasoline in the fuel lines of your car laminar or turbulent?

44. Is the flow of air in the ducts of the air conditioning system of your home laminar or turbulent?

45. Is the flow of liquid oxygen in the fuel lines of the space shuttle's main engines laminar or turbulent?

46. Two pressure taps are placed 10 m apart on a pipe of diameter 15 cm. The measured pressure drop is 20 kPa. Calculate the shear stress that acts on the walls of the pipe. Is the flow laminar or turbulent?

47. Find the kinematic viscosity at STP for the following common fuels: gasoline, diesel fuel, jet fuel A, propane, natural gas, hydrogen, and ethanol.

48. One proposed alternative fuel is dimethyl ether (DME). At room temperature and pressure it is a gas, but when slightly compressed it liquefies. Find the viscosity of liquid DME. How does its viscosity compare to that of traditional fuels, and what does this imply for the modifications needed in an existing fuel system to handle DME?

49. Determine the kinetic energy correction factor and the momentum flux correction factor for laminar flow through an annulus where the outer diameter is twice the inner diameter.

50. Do some research to find the definition of the Fanning friction factor. How is the Fanning friction factor related to the Darcy friction factor?

51. Air at standard conditions flows through a sudden enlargement from a 5-cm pipe to a 15-cm pipe. Determine the pressure rise (a) neglecting any losses, and (b) including the secondary loss associated with the expansion.

52. Water at 22°C flows through a concrete pipe of diameter 12 cm at a rate of 10 kg/s. Determine the pressure drop for 1 km of pipe.

53. SAE 30 oil flows through a steel pipe of diameter 3 cm. If the desired flow rate is 100 L/min, at what slope should the pipe be set so that the pressure remains constant?

54. It is desired to have an air flow rate of 0.5 m³/s in an office building's air conditioning system. The cross-sectional area is to be 400 cm². The following pipe ducts are to be considered: circular, square, and rectangular with a 2:1 ratio. Find the pressure loss per meter of length for each. All the ducts are made out of aluminum.

55. A racing league mandates the use of a restrictor on the intake system of its cars. What is the maximum expansion angle that could be used between the restrictor and the airbox to ensure that flow separation does not occur? (The angle should be maximized to reduce the length and weight of the intake system.)

56. The intake system of race car features 0.5 m of straight pipe of 30 mm diameter, two 90° bends, a sudden expansion into an airbox of diameter 100 mm, and a sudden contraction into the intake manifold with diameter of 20 mm. If the car is running at wide open throttle at 10,000 rpm, with an engine displacement of 600 cc, estimate the pressure loss in the intake system.

57. What are typical values of loss coefficient for intake air filters for automobiles?

58. Water is supplied at a constant elevation head of 50 cm. Three different tubes of length 1 m, having diameters of 1 cm, 1.5 cm, and 2 cm are connected to the water supply. Find the flow rate in L/min for each, and calculate the friction factor in each of the three pipes. What is the Reynolds number for each tube? Put your results in a table.

59. Calculate the Reynolds number for the flow of air in the inlet duct to a boiler in a 300 MW power plant. The duct cross section is a rectangle of size 9 ft by 20.5 ft. The air mass flow rate is 1845 lbm/hr, with measured conditions of 600°F and a vacuum pressure of 2.7 in. of water.

60. Calculate the Reynolds number for 5 mm ID artery in the human body, with a velocity of the blood of 2 cm/s. Is the flow laminar or turbulent?

61. Calculate the Reynolds number of the flow through an expressway drainage pipe of diameter 16 in., operating at a flow rate of 9900 gal/min. Assume the water is at a temperature of 70°F. Is the flow laminar or turbulent?

62. An open flow wind tunnel has a rectangular inlet section of size 35 in. by 39 in, and a rectangular test section of size 10 in. by 14 in. If the maximum velocity in the test section is 130 mph, calculate the Reynolds number and the friction factor at both the inlet and the test section. If the test section is 20 in. long, what is the pressure loss due to wall friction in that section? (Keep in mind that the pressure in the test section is not atmospheric, and that will affect the density of the air. You can assume isothermal flow.)

63. Consider the engine of a scramjet. Assume the engine area is roughly cylindrical with a diameter of 9.5 in. and a length of 5.2 ft. The inlet air is coming in at Mach 7.0 at a flight altitude of 60,000 ft. Neglect any shock waves or other compressible flow effects. Calculate the Reynolds number for the flow through the engine.

64. What pressure is required for a flow of 10 gal/min of water through a $\frac{3}{4}$-in. pipe that is 500 ft long, if the outlet is at atmospheric pressure?

6 External Flow

In This Chapter
- Introduction to Aerodynamics
- Viscous Drag
- Form (Pressure) Drag
- Lift
- Vehicle Aerodynamics
- Transient Drag

The objective of this chapter is to develop an understanding of aerodynamics and the ability to solve basic aerodynamics problems.

■ 6.1 Introduction to Aerodynamics

Drag is the name for the resistance an object experiences as it moves through a fluid. When the object moves through air, the force is called *air resistance*, and when it moves through water, it is called *hydrodynamic drag*. In the steady state, the drag force can be broken down into two components, as shown in Figure 6.1.

The *viscous drag* is the result of shear stress acting along the surface, while the *form drag* is the net of the normal pressure force across the surface area. In a well-streamlined object, most of the drag will be viscous drag.

$$F_{drag} = F_{shear} + F_{normal} = \int_A \tau \, dA + \int_A -P \, dA \tag{6.1}$$

A dimensional analysis (see Chapter 9) is often helpful in aerodynamics. What properties or variables affect the magnitude of the drag force on an object? The speed and size of the object obviously come into play, and the viscosity of the fluid and its density should also be important. The denser a fluid is, the greater the amount of mass that has to be moved around the object. So then we could say

$$F = f(V, L, \mu, \rho) \tag{6.2}$$

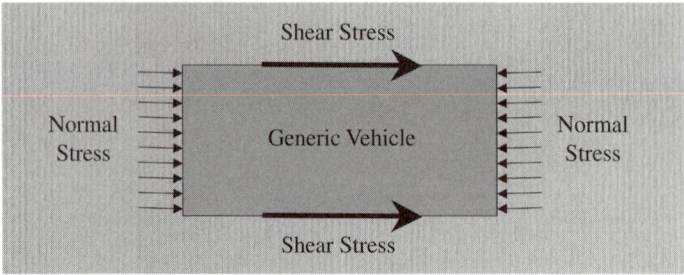

FIGURE 6.1 Drag on a generic rectangular object.

Dimensional analysis allows us to simplify the variable relationships. The nondimensional *drag coefficient* is defined as

$$C_D = \frac{F_D}{\frac{1}{2}\rho A V^2} \tag{6.3}$$

and the *lift coefficient* is

$$C_L = \frac{F_L}{\frac{1}{2}\rho A V^2} \tag{6.4}$$

where C_D and C_L are both functions of the Reynolds number. That is, $C_D = C_D(\text{Re})$. (See Example 9.5 for verification that $C_D = C_D(\text{Re})$.) In high-speed flows the Mach number is also important:

$$\text{Ma} = \frac{V}{a} \tag{6.5}$$

so that $C_D = C_D(\text{Re}, \text{Ma})$ for $\text{Ma} > 0.3$. (The Mach number is named after Ernst Mach, who was the first to photograph a shock wave when, in 1887, he captured an image of one coming off a bullet using the shadowgraph technique [see Chapter 9 for more information about experimental techniques].) It has been found through experience that there are several flight regimes that aircraft experience, depending on their speed. These regimes, which are listed as a function of Mach number in Table 6.1, affect the fluid mechanics of the flow of air around the aircraft. Figure 6.2 shows a picture of the shock wave coming off a plane in a supersonic regime. Figure 6.3 shows a diagram of a plane in the transonic flight regime.

Table 6.1 Flight Regimes as a Function of Mach Number

Ma	Regime	Notes
Ma < 0.3	Incompressible	Density does not change with velocity.
0.3 < Ma < 0.8	Compressible	Density changes are significant.
0.8 < Ma < 1.0	Transonic	Local shock waves may form over wings.
1.0 < Ma < 5.0	Supersonic	Shock waves occur on front of vehicle.
5.0 < Ma < 1.0	Hypersonic	Air coming into engines may ionize.

Another form of drag is present for surface-going boats and other vessels: *wave drag*. For flows of surface vessels on water, the Froude number is also important. The Froude number is the ratio of inertial forces to gravitational (wave) forces:

$$\text{Fr} = \frac{V^2}{gL} \tag{6.6}$$

Wave drag and the hydrodynamics of boats will be discussed in further detail in Section 6.5.

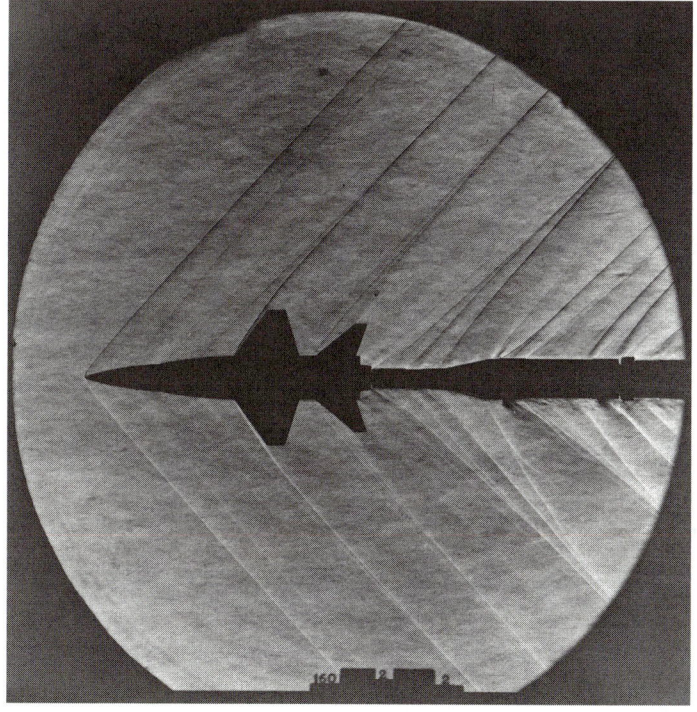

FIGURE 6.2 Shock wave coming off a plane traveling supersonically. Courtesy of NASA.

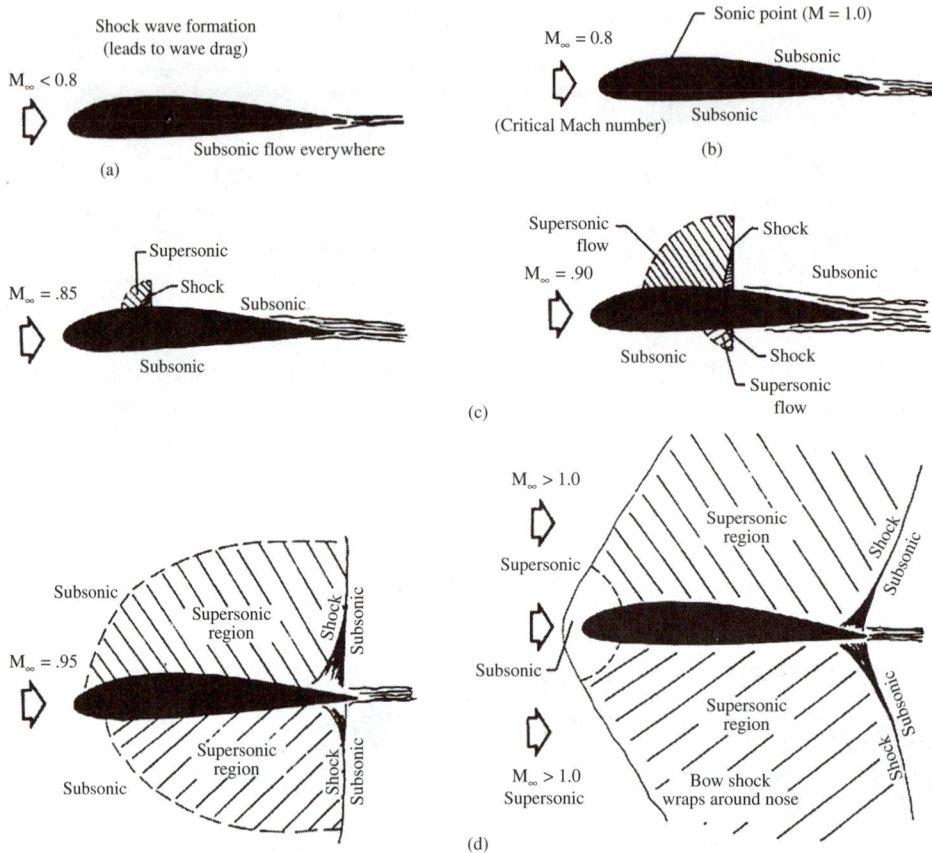

FIGURE 6.3 Schematic illustrating a small local shock wave visible over the wings of a plane in the transonic flight regime.

In total, there are actually six different phenomena that give rise to drag resistance: viscosity, pressure, surface waves on a liquid, shock waves, lift-induced drag, and drag due to acceleration or changes in velocity.

For ground vehicles, only viscous and pressure drag need be considered. Remember, pressure always acts *perpendicular* to a surface, while shear stress acts *parallel* to it. These two forces will be discussed in the next two sections. The general procedure for calculating the drag force on an object is as follows:

1. Calculate the Reynolds number of the object.
2. Find a correlation equation, chart, or tabular data that gives the drag coefficient for the object in the appropriate range of Reynolds numbers.

3. Calculate the drag force using the definition of the drag coefficient: $F_D = \frac{1}{2} C_D \rho A V^2$.

4. If the power to overcome drag is needed, that can be calculated by the equation Power $= FV = \frac{1}{2} C_D \rho A V^3$.

Drag coefficients for a variety of objects will be presented in Section 6.3.

■ 6.2 Viscous Drag

The skin friction, or viscous drag, over a flat plate was discussed in Chapter 4. The *viscous drag* is due to motion *parallel* to the object's surface. Within the *boundary layer* over an object, the fluid velocity changes rapidly from zero in the free stream to the velocity of the object at the surface due to the no-slip boundary condition (or in the case of a stationary object in a moving fluid, the velocity changes across the boundary layer from the free-stream velocity to zero at the surface). To get the value of the shear stress at the object's surface, the shape of the velocity profile within the boundary layer must be known so that Newton's law of viscosity may be applied at the object's surface:

$$\tau = \mu \frac{\partial u}{\partial y}$$

Unfortunately, analytical solutions to the Navier-Stokes equations can be found only for the simplest of geometries, such as a flat plate or a perfect sphere, and even then the analytical solution is valid only for laminar flow.

Recall from Chapter 4 that for the simple case of laminar flow parallel to a flat plate, the local shear stress can be found from the velocity profile as

$$\tau = \mu \frac{\partial u}{\partial y}\bigg|_{y=0} = 0.332 \frac{\mu U_o}{x} \sqrt{\frac{\rho U_o x}{\mu}} = 0.332 \sqrt{\frac{\rho \mu U_o^3}{x}} \qquad (4.15)$$

The frictional drag force is equal to the integral of the shear stress over the area, but the shear stress does not stay constant in external flow.

$$F_D = \int_A \tau \, dA = \int_0^L \tau(x) W \, dx \qquad (6.7)$$

So substituting in for $\tau(x)$, the integration can be performed, giving the equation for drag force on a flat plate exposed to a steady flow as:

$$F_D = \int_0^L 0.332\sqrt{\frac{\rho\mu U^3}{x}}\, W\, dx = 0.332 W\sqrt{\rho\mu U^3}\int_0^L \sqrt{\frac{1}{x}}\, dx$$

$$= 0.332 W\sqrt{\rho\mu U^3}\, (2\sqrt{x})\bigg|_{x=0}^{x=L}$$

$$F_D = 0.664 W\sqrt{\rho\mu L U^3}$$

This is valid when the flow is laminar. For internal flow through a pipe, the flow is laminar up to a Reynolds number of about 2,300, but for *external* flow, the flow can stay laminar up to a critical Reynolds number of

$$\text{Re}_{\text{crit}} = 500{,}000 \tag{4.16}$$

Below this critical Reynolds number the viscous forces are strong enough to keep the flow in the boundary layer laminar, but above it the flow will transition to turbulence.

Also note that the nondimensional drag coefficient is defined as

$$C_D = \frac{F_D}{0.5\rho A U_o^2} \tag{4.20}$$

So the drag coefficient for the laminar flow around a flat plate is

$$C_D = \frac{0.664 W\sqrt{\rho\mu L U^3}}{0.5\rho(WL)U^2} = 1.328\sqrt{\frac{\mu}{\rho L U}} = \frac{1.328}{\sqrt{\text{Re}_L}} \tag{6.8}$$

where the Reynolds number

$$\text{Re}_L = \frac{\rho V L}{\mu}$$

is based on the total length, L, of the plate. The drag force is then defined as $F_D = C_D 0.5\rho A U_o^2$, where $A = W \times L$ for a rectangular plate. Note that this is all based on the assumption of fluid flow over only one side of the plate. If both sides are exposed to fluid flow then the drag force is doubled.

For turbulent boundary layers no satisfactory theoretical solution exists for the velocity profile, so empirical equations must be used. A popular correlation for the drag coefficient in turbulent boundary layers is

$$C_D = \frac{0.031}{\text{Re}_L^{1/7}} \tag{4.21}$$

This correlation is accurate for a range of Reynolds numbers from $500{,}000 < \text{Re}_x < 10^7$. For cases of mixed flow, where the front part of the boundary layer is laminar and the rear part is turbulent, the overall drag coefficient is

$$C_D = \frac{0.031}{\text{Re}_L^{1/7}} - \frac{1440}{\text{Re}_L} \qquad (4.22)$$

The total drag on any object is equal to the sum of the viscous drag and the pressure drag. If the sides of the vehicle (car, boat, plane, truck, submarine, space ship, etc.) are parallel or nearly parallel to the flow direction, then we can approximate the friction drag with that for flow over a perfectly flat plate. For shapes where the pressure drag is not negligible—such as bluff bodies like tractor-trailers and even spheres—we usually need to look up an empirical correlation for C_D. Extensive data is recorded in Blevins' *Handbook of Fluid Dynamics* [Blevins92].

EXAMPLE 6.1

A pizza delivery car travels at a speed of 15 m/s. The pizza company sign on the top of the car can be modeled as a flat plate mounted vertically, with a height of 0.6 m and length of 1.0 m. Calculate the drag force on the plate for air at sea-level conditions.

SOLUTION: The Reynolds number based on the length of the plate is

$$\text{Re}_L = \frac{VL}{\nu} = \frac{(15 \text{ m/s})(1 \text{ m})}{(0.000015 \text{ m}^2/\text{s})} = 1{,}000{,}000$$

This is a mixed laminar and turbulent flow regime—the front half of the sign will be laminar and the back half will be turbulent—so we use Equation 4.22:

$$C_D = \frac{0.031}{(1{,}000{,}000)^{1/7}} - \frac{1{,}440}{1{,}000{,}000} = 0.0043 - 0.0014 = 0.0029$$

Then we calculate the force on one side of the sign as

$$F_D = \frac{1}{2} C_D \rho A V^2 = \frac{1}{2}(0.0029)\left(1.2 \frac{\text{kg}}{\text{m}^3}\right)(0.6 \text{ m}^2)\left(15 \frac{\text{m}}{\text{s}}\right)^2 = 0.235 \text{ N}$$

The total force on both sides of the sign would be $2(0.235 \text{ N}) = 0.47 \text{ N}$.

> ### EXAMPLE 6.2
>
> The wing on a small high-wing personal airplane has a wingspan of 35 ft and a wing area of 175 ft². The wing is rectangular in shape. If the wing is approximated as flat plate, find the drag over the wing at a speed of 120 mph at sea level.
>
> SOLUTION: The chord of the wing can be calculated as L = 175 ft²/35 ft = 5 ft. The velocity in units of ft/s is (120 mph)(5,280 ft/mi)/(3,600 s/hr) = 176 ft/s. The Reynolds number based on length is
>
> $$\text{Re}_L = \frac{VL}{\nu} = \frac{(176 \text{ ft/s})(5 \text{ ft})}{(0.00016 \text{ ft}^2/\text{s})} = 5{,}500{,}000$$
>
> More than 90% of the flow over the wing is turbulent, so we approximate the flow as being completely turbulent to simplify the calculation. The drag coefficient is then calculated as
>
> $$C_D = \frac{0.031}{\text{Re}^{1/7}} = \frac{0.031}{(5{,}500{,}000)^{1/7}} = 0.0034$$
>
> and the drag force is
>
> $$F_D = \frac{1}{2} C_D \rho A V^2 = \frac{1}{2}(0.0034)\left(0.0075 \frac{\text{lbm}}{\text{ft}^3}\right)(175 \text{ ft}^2)$$
> $$\times \left(176 \frac{\text{ft}}{\text{s}}\right)^2 \frac{1 \text{ lbf}}{32.2 \text{ lbm} \cdot \text{ft/s}^2} = 21.5 \text{ lbf}$$
>
> The density of air at sea level is about 1.2 kg/m³ = 0.075 lbm/ft³ = 0.0023 slug/ft³. In this example, the drag coefficient for a flat plate was used to approximate the drag on the airplane wing. In reality, the actual drag would be higher due to the curvature and thickness of the wing.

If the surface of the object is slightly curved—that is, not perfectly parallel to the flow—the drag will be higher than that predicted for a flat plate of the same surface area. But Equations 4.21 and 4.22 can still be used to give approximate drag coefficients. The drag on a curved surface will be higher due to form drag, variation of the boundary layer profile, three-dimensional shape effects, angle of attack not zero, and the curvature of the wing. If the surface has significant curvature, so that it has a significant amount of projected area perpendicular to the oncoming fluid flow, then the effects of pressure on the normal surface become important. These effects are discussed in the next section.

What if there are imperfections on the surface of a flat, smooth plate? How small do bumps or indentations have to be before they can be considered negligible? Goldstein (1965) states that the effect of a bump on a flat surface's drag is negligible if the Reynolds number of the bump is less than 30 for a pointed bump and less than 50 for a rounded bump. The length scale is the height of the bump, ε, above the surrounding surface, and the velocity scale remains the free-stream velocity [Goldstein38].

6.3 Form (Pressure) Drag

The *form drag* is also called the *pressure drag*. The pressure drag is most significant with objects that have a flat surface perpendicular to the flow, either at the front or the back of the object. When the flat region is at the front of the object, a *stagnation flow* results. When the flat region is at the rear of the object, a *separated wake flow* results. The low pressure in the wake region adds to the drag of the object. The total drag on an object moving at steady speed is the sum of the viscous and pressure drags. Table 6.2 lists drag coefficients for a variety of basic shapes. Notice that the objects with blunt shapes tend to have higher drag coefficients than the streamlined ones, due to the high form drag.

A general note on drag coefficients: Be careful to note which length scale is used in the Reynolds number and which area is used in the calculation of the drag force from the drag coefficient. For different objects, either the height, width, or length is used for the length scale, and either the forward projected area or the planform (wing) area is used. In some cases the wetted area or volume raised to the $\frac{2}{3}$ power is used for the reference area.

Generally speaking, the higher the Reynolds number, the more likely the flow is to be turbulent, although the exact value of the Reynolds number for transition depends on the type of flow (external or internal) and whether there are any disturbances in the flow.

6.3.1 Drag on a Sphere

For a sphere moving through a fluid, or for a fluid flowing over a sphere, the drag depends on the density and viscosity of the fluid; the *relative velocity* between the sphere and the fluid, V; and the size of the sphere, as quantified by the diameter. Thus, like for most objects, the drag coefficient is a function of the Reynolds number. The critical Reynolds number for transition from laminar to turbulent flow is around 350,000 for a smooth sphere, which is somewhat less than the critical Reynolds number for a flat plate.

For flows at Re $<$ 1, the flow is called *Stokes flow*. Stokes solved the Navier–Stokes equations for this case to get the flow field around the sphere. For Re $<$ 1

Table 6.2 Drag Coefficients for Various Bluff Shapes in the Reynolds Number Range Where the Drag Coefficient Is Constant

Type	C_D	Shape
Sphere	0.45	○
Cylinder	1.2	
Hemisphere—facing toward flow	0.4	
Hemisphere—facing away from flow	1.4	
Cube	1.05	
Flat plate perpendicular, square and rectangle		
Square	1.05	
2 by 1 rectangle	1.1	
Long rectangle	2.0	
Circular disk flat plate	1.1	○
Cone—120° angle	0.92	
Streamlined body	0.025	
Cylinders of various cross sections		
Square	2.1	
Diamond	1.6	◇
Ellipse (2:1 ratio)	0.6	
Ellipse (8:1 ratio)	0.2	
Triangle (equilateral)	1.3	△
Triangle (30° apex)	1.0	
Open arc (C-section) facing away from flow	1.1	∩
Open arc (C-section) facing toward flow	2.3	U

Data from [NACA34], [NACA36], [NACA38], [NACA53], [Goldstein65], [Hoerner65], [Schlicting68], and [NASA71].

viscosity dominates the flow and the streamlines merge smoothly around the back, with the effect that the pressure profile is symmetric and there is no net pressure drag. All the drag is viscous drag (no form drag), and there is no significant wake behind the sphere. Stokes integrated the surface shear stress to get

$$F = 3\pi\mu DV \tag{6.9}$$

From this, the drag coefficient can be obtained for very slow flows around a sphere:

$$C_D = \frac{24}{\text{Re}} \tag{6.10}$$

Stokes' formula (see Chapter 4) works well up to Re = 1. At higher Reynolds numbers, a wake starts to form in the back and there is pressure drag due to the low pressure in the wake, so the drag will be higher than that predicted by Stokes' formula. For particles of size less than 1 μm in air at standard conditions, the size of the particle becomes comparable to the mean free path of the gas as the sphere size is decreased, and the no-slip condition no longer holds. The drag force can be corrected with a slip correction factor, C_C,

$$F = \frac{3\pi\mu DV}{C_C} \tag{6.11}$$

A correlation for the slip correction factor is

$$C_C = 1 + \frac{\lambda}{D}\left[2.514 + 0.800 \exp\left(-0.55\frac{D}{\lambda}\right)\right] \tag{6.12}$$

where λ is the mean free path of the gas. This correlation is valid for spheres down to sizes of 0.001 μm [Hinds82].

At higher Reynolds numbers, even before the flow becomes turbulent, Stokes' theory does not adequately represent the flow around the sphere, and we must resort to empirical correlations. The drag is higher than what is predicted by Stokes' formula because the flow separates on the back side of the sphere and a low-pressure wake region is formed, which increases the form drag. For 3 < Re < 1000, the following correlation works well [Hinds82]:

$$C_D = \frac{24}{\text{Re}}\left(1 + \frac{\text{Re}^{2/3}}{6}\right) \tag{6.13}$$

Other correlations are also available that give similar accuracy. The standard drag curve should only be considered accurate to ±10%. Figure 6.4 shows the pressure

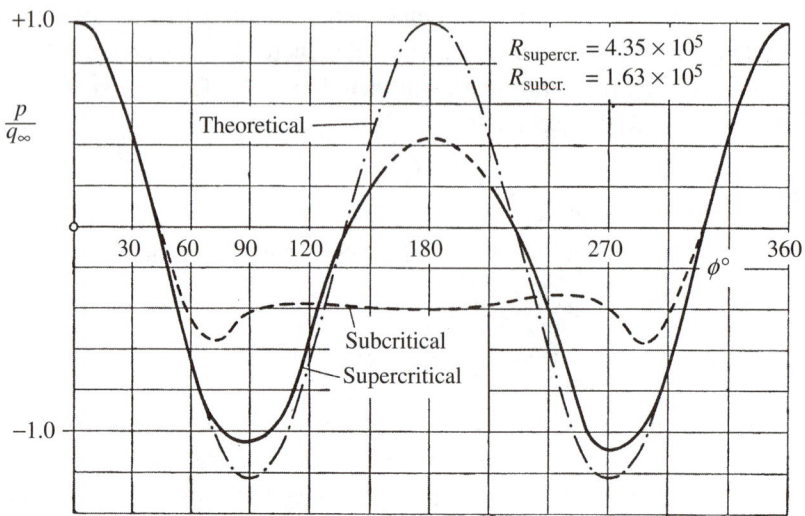

FIGURE 6.4 Pressure profile over a sphere. From [Schlicting68].

profile over a sphere, with the theoretical profile of Stokes for Re < 1 and measurements at Reynolds numbers both above and below the critical Reynolds number for transition to turbulence.

As the Reynolds number increases above 1.0, the flow starts to separate from the back end of the sphere and the pressure profile is no longer symmetric. The local nondimensional pressure coefficient is defined as

$$C_P = \frac{P - P_\infty}{\frac{1}{2}\rho V^2} \tag{6.14}$$

Thus at the stagnation point on the very front of the sphere, $C_P = 1.0$, and at a location where the pressure equals the free-stream pressure, $C_P = 0.0$. Figure 6.5 shows the drag coefficient of a sphere as a function of the Reynolds number for a wide range of Reynolds numbers. This curve is also called the *standard drag curve*.

As noted by Crowe, there is actually quite a bit of variation in measured drag coefficients over a sphere, depending on the Mach number, the transition to turbulence, and the exact test conditions [Crowe98]. The plot in Figure 6.5 is also only for steady-state drag. Drag coefficients should be viewed as only accurate to ±10%. Table 6.3 delineates the different flow regimes that affect the drag coefficient.

6.3.2 Strouhal Number

Bluff bodies shed vortices in a Von Karman *vortex trail* or *vortex street*. The vortices alternate from side to side. By Newton's third law, the object that sheds these vortices

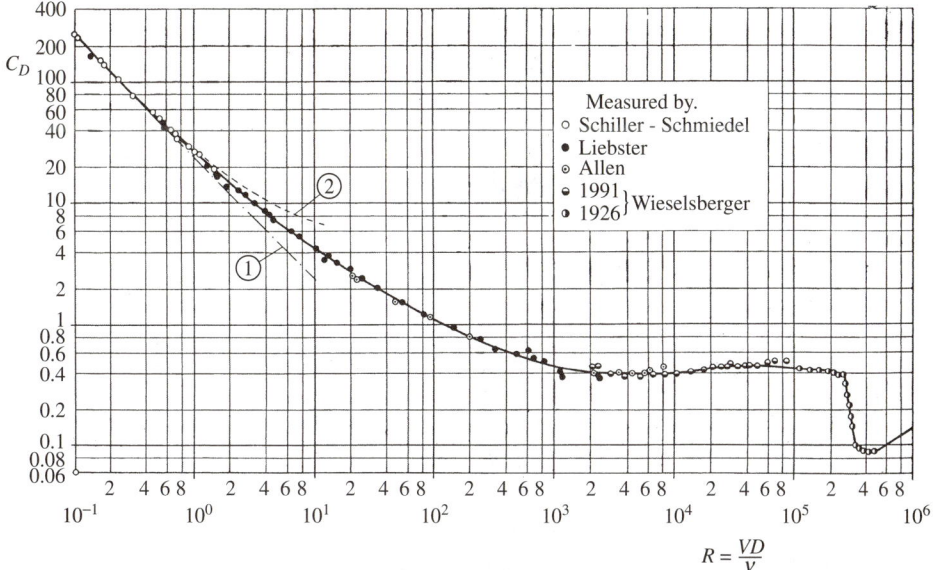

FIGURE 6.5 Drag coefficient of a sphere as a function of Reynolds number. From [Schlicting68].

feels a periodic buffeting force from them that increases the drag. The characteristic frequency of shedding of the vortices is characterized by a nondimensional *Strouhal number*:

$$\text{St} = \frac{fL}{V} \tag{6.15}$$

where f is the frequency of shedding in units of Hertz, and L is a characteristic length scale, which would be equal to the diameter D for a sphere or cylinder. Examples of bluff bodies that experience vortex shedding are trucks, vans, buses, tractor-trailers, and lifting bodies (which will be discussed later). From a Reynolds number of about 300 up to 30,000, the Strouhal number is constant at St = 0.2. Vortex shedding has been

Table 6.3 Flow Regimes Around a Sphere as a Function of Reynolds Number

Re	C_D	Regime/Notes
Re < 1	24/Re	Stokes flow; drag is solely due to viscous forces
1 < Re < 40	Eq. 6.13	Separated flow; attached wake
40 < Re < 300	Eq. 6.13	Vortex shedding
1,000 < Re < 350,000	0.45	Vortex shedding; constant Strouhal number
Re > 350,000	0.2	Turbulent

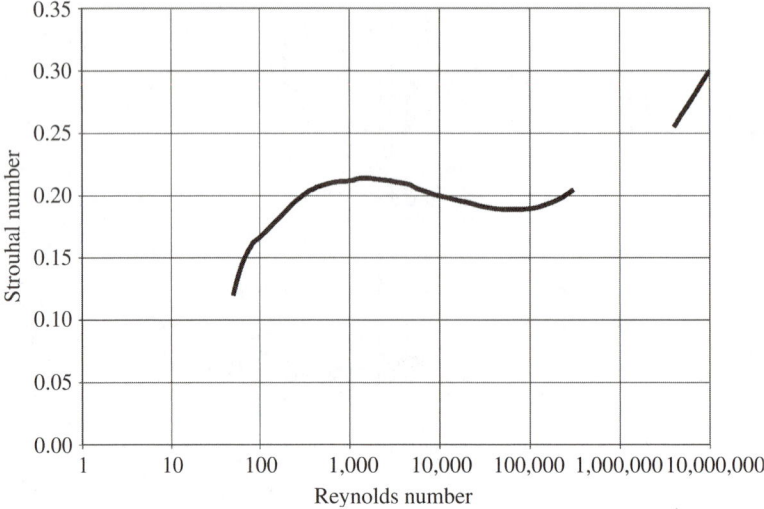

FIGURE 6.6 Plot of the Strouhal number as a function of the Reynolds number for vortex shedding behind blunt objects.

observed at Reynolds numbers as high as 10^{10} in a satellite photo of a mountain on Guadalupe Island. Around Re = 100, vortex shedding starts to occur. Figure 6.6 shows the trend of Strouhal number versus Reynolds number, and Figure 6.7 shows a visual observation of vortex shedding.

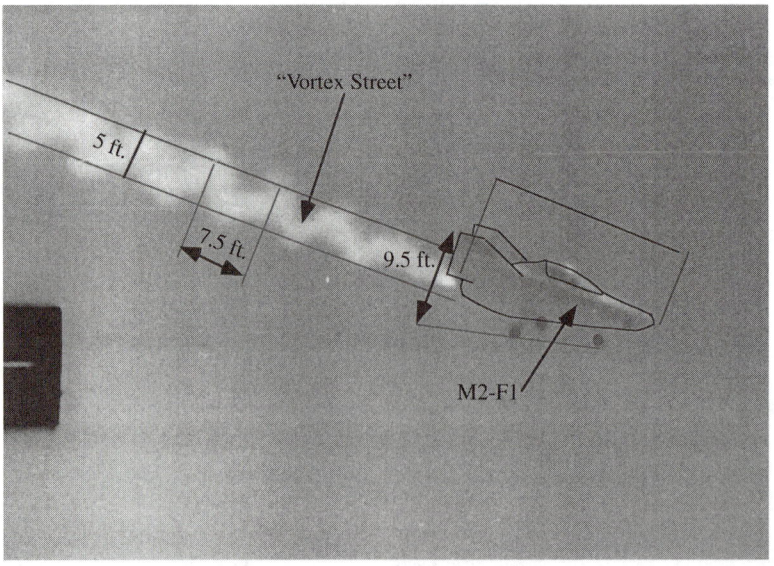

FIGURE 6.7 Vortex shedding behind an M2-F1. From [NASA01].

EXAMPLE 6.3

What size (diameter) of steel ball will fall through the atmosphere at a terminal velocity of 25 m/s? The density of steel is about 7800 kg/m³.

SOLUTION: We calculate the terminal velocity by letting the acceleration go to zero so that the maximum free-fall velocity is obtained. Thus,

$$\sum F = W - F_D = ma = 0$$

$$\rho_{sphere} \frac{\pi}{6} D^3 g = \frac{1}{2} C_D \rho_{air} \frac{\pi}{4} D^2 V^2$$

In this problem V is known, so we will rearrange the equation to solve for D:

$$D = \frac{3}{4} \frac{\rho_{air}}{\rho_{sphere}} C_D \frac{V^2}{g}$$

Note that C_D depends on the Reynolds number, which depends on D. The solution is likely to require an *iterative* technique. That is, we must guess a value for C_D, and then determine if the guess is correct. Let us try an initial guess of $C_D = 0.44$, which is in the upper end of the laminar regime in the flat portion of the drag curve. This results in a value of D of

$$D = \frac{3}{4} \frac{1.2 \text{ kg/m}^3}{7{,}800 \text{ kg/m}^3}(0.44)\frac{(25 \text{ m/s})^2}{9.8 \text{ m/s}^2} = 0.00324 \text{ m} = 3.24 \text{ mm}$$

Now we have to check if the guess was correct:

$$\text{Re} = \frac{VD}{\nu} = \frac{(25 \text{ m/s})(0.00324 \text{ m})}{(1.5 \times 10^{-5} \text{ m}^2/\text{s})} = 5{,}400$$

From the drag curve (Figure 6.5), this is indeed in the flat part of the drag curve where $C_D \approx 0.44$ ($\pm 10\%$). Note that if the initial guess had been $C_D = 1.0$, the calculated values would have been $D = 7.36$ mm and Re $= 12{,}300$. The drag coefficient at that Reynolds number is 0.44, and repeating the calculation with this new guess for C_D would result in the correct answer.

Note: if the buoyant force is included, the terminal velocity is

$$V = \frac{2D^2 g(\rho - \rho_0)}{9\mu}$$

for Stokes flow (Re < 1). For the more general case of arbitrary Reynolds number, the terminal velocity calculated from a force balance including buoyancy would be

$$V = \sqrt{\frac{4(\rho - \rho_0)gD}{3\rho_0}}$$

For most objects falling through air, the buoyant force is small compared to the drag force and can be neglected. For objects falling in water, however, the buoyant force is usually significant and should be accounted for.

Also note that for an object falling through a confined liquid in a container, unless the object is small compared to the internal dimensions of the container, the boundary layer around the object will interact with the walls, retarding the motion and increasing the resistance relative to true free fall. For a sphere falling inside a cylinder, Sprakling gives a correction factor of

$$\mu_{apparent} = \mu\left(1 + 2.4\frac{r_{sphere}}{R_{cylinder}}\right) \tag{6.16}$$

In the limit of $R_{cylinder}$ approaching infinity, the correction factor goes to 1.0 [Sprakling85].

6.3.3 Parachute Aerodynamics

For hemispherically shaped parachutes, engineers must note whether a reference defines the drag coefficient based on the cross-sectional area of the parachute or on the surface area of the parachute. The surface area of sphere is $4\pi r^2$, so the surface area of a complete hemisphere would be $2\pi r^2$. The area of the circular cross section is πr^2, so there is a factor of 2 difference between the two definitions of the drag coefficient. While the drag coefficient for a given parachute depends on its exact shape; the porosity of the fabric; and the number, shape, size, and location of holes, general numbers for C_D (based on cross-sectional area) are about 1.5 for a dome-shaped parachute and close to 0.75 for a flatter rectangular parachute [NACA49]. Figure 6.8 shows an example of a domed parachute.

6.3.4 Sports Balls Aerodynamics

Most balls used in sports—such as the balls used in cricket, tennis, baseball, basketball, and soccer—have basically spherical shapes. The nonuniformities on the surfaces of these balls usually make it easier for the athletes to generate a lift force (called the *Magnus force*) by putting spin on the ball. In soccer the players try to bend the ball to score a goal, while in baseball a pitcher may throw a curveball to fool the batter. Table 6.4 gives the parameters for the aerodynamics of sports balls. Figure 6.9 shows the wake behind a non-spinning baseball. Flow is from left to right.

FIGURE 6.8 Parachute being tested behind an SR-71. Courtesy of NASA-Dryden Flight Research Center.

The critical Reynolds number for transition to turbulence for a golf ball is 40,000, for a baseball it is 160,000, for a soccer ball it is 210,000, and for a smooth sphere it is around 350,000 to 400,000. A golf ball's flight is always above the critical Reynolds number, but baseballs and soccer balls may cross that transition Reynolds number in flight, which will affect the trajectory of the ball. For a soccer ball, the drag coefficient drops from 0.5 to 0.25; for a baseball, it drops from 0.5 to

Table 6.4 Aerodynamics of Sports Balls

Type	Diameter (cm)	Mass (g)	V (m/s)	C_D	ω (rev/s)	Source
Baseball	7.3	150	45	0.31	30	[Adair02]
Cricket	7.2	156	45	0.45	15	[Mehta85]
Tennis	6.6	57	67	0.62	70	[Mehta01]
Soccer	21.9	427	30	0.25	10	[Bray03]
American football	17.3	410	20	0.05	10	[Watts03]
Basketball	24.0	600	10	0.25	2	[Fontanella06]
Golf	4.2	45	75	0.25	60	[Bearman76]

Typical maximum velocities and rotational rates encountered are shown. Data from various sources—see above.

FIGURE 6.9 Picture of flow past a baseball with no spin.

about 0.3. The C_D values given in Table 6.4 are for post-transition Reynolds numbers. No value of the critical Reynolds number has yet been found for a tennis ball, but given the extreme roughness of the surface of a new tennis ball, it is probably safe to assume that the tennis ball is in turbulent flow for its entire flight. Note there is a wide range of reported values for C_D for sports balls, just as Crowe et al. [Crowe98] report a wide range of values for smooth spheres. The differences in values could be due to the condition of the surface of the ball, the spin on the ball, the orientation of the ball, and/or whether the test was conducted in a wind tunnel or in free-flight.

Note a cricket ball has six rows of primary stitching going across the equator, so it is not a smooth sphere. A golf ball will see a maximum Re of 200,000 when launched, and a maximum spin parameter, S, of over 1.0 when launched with an iron. Because a basketball is large with relatively low weight for its size, the buoyant force cannot be neglected, as it can for baseballs. The buoyant force is about 1.5% of the weight of the basketball. The typical velocity of a basketball in a game is 10 m/s, or 22 mph. For typical basketball shots, the rotational speed at release is around 2.0 rev/s, regardless of the type of shot (short jump shot, free throw, three-pointer). Typical launch speed for a free throw is 7 m/s (15 mph), with a launch angle of around 50° to horizontal.

The lift force on a golf ball is generated by backspin. Bearman and Harvery have shown that for spin rates above 1000 rpm, the lift force is large enough that a golf ball will actually travel farther in air than it would in a vacuum [Bearman76]. Typical golf ball velocities are above that needed for transition to turbulence above the critical Reynolds number. For a tennis ball, the initial transient stage where the ball is deformed is short compared to its overall transit time and can be neglected. C_D for a tennis ball is largely independent of the Reynolds number.

Effects of Spin

For the range of velocities where the drag coefficient is constant, one correlation for the Magnus force is

$$F_M = \frac{1}{2} C_L \rho A V^2 \tag{6.17}$$

where C_L is a function of the nondimensional spin parameter, S (and not the Reynolds number):

$$S = R\omega/V \tag{6.18}$$

where R is the radius of the sphere. A popular correlation for the lift on a spherical baseball as a function of spin is given by Hubbard [Hubbard03]:

$$\begin{aligned} C_L &= 1.5S & (S < 0.1) \\ C_L &= 0.09 + 0.6S & (S > 0.1) \end{aligned} \tag{6.19}$$

Data for soccer balls show a similar trend for C_L versus S. Goldstein's data for the lift coefficient of spinning smooth spheres show a value of around 0.3 at a spin number of $S = 1$ [Goldstein65], while for a baseball at $S = 1$ the lift coefficient would be $C_L = 0.09 + 0.6(1.0) = 0.69$, or about twice as high. This illustrates that the effects of a baseball's seams on its aerodynamics are quite significant.

EXAMPLE 6.4

Calculate the terminal velocity of a basketball dropped from the top of a tall building into still air at standard atmospheric conditions.

SOLUTION: We calculate the terminal velocity by letting the acceleration go to zero so that the maximum free-fall velocity is obtained:

$$\sum F = W - F_D = ma = 0$$

$$m_{\text{ball}} g = \frac{1}{2} C_D \rho_{\text{air}} \frac{\pi}{4} d^2 V^2$$

Rearranging to solve for the terminal velocity, V, gives

$$V = \sqrt{\frac{8}{\pi} \frac{m_{\text{ball}}}{\rho_{\text{air}} d^2} \frac{g}{C_D}}$$

Plugging in the numerical values from Table 6.4 will yield

$$V = \sqrt{\frac{8}{\pi} \frac{(0.6 \text{ kg})}{(1.2 \text{ kg/m}^3)(0.24 \text{ m})^2} \frac{(9.8 \text{ m/s}^2)}{0.25}}$$

which results in a terminal velocity of 29.4 m/s.

EXAMPLE 6.5

Calculate and plot the height as a function of time of a spherical ball dropped from a height of 7 m into still air at standard atmospheric conditions. The ball has a diameter of 20 cm and a mass of 0.20 kg.

SOLUTION: We will work this problem by first assuming a constant C_D at the value of the terminal velocity of the ball. We calculate the terminal velocity by letting the acceleration go to zero so that the maximum free-fall velocity is obtained:

$$\sum F = W - F_D = ma = m\frac{dV}{dt}$$

$$\rho_{\text{sphere}}\frac{\pi}{6}d^3 g - \frac{1}{2}C_D \rho_{\text{air}}\frac{\pi}{4}d^2 V^2 = \rho_{\text{sphere}}\frac{\pi}{6}d^3\frac{dV}{dt}$$

Rearranging to solve for dV/dt gives

$$\frac{dV}{dt} = \frac{\rho_{\text{sphere}}(\frac{\pi}{6})d^3 g - \frac{1}{2}C_D \rho_{\text{air}}(\frac{\pi}{4})d^2 V^2}{\rho_{\text{sphere}}(\frac{\pi}{6})d^3}$$

which can be simplified to

$$\frac{dV}{dt} = g - \frac{4}{3}\frac{\rho_{\text{air}}}{\rho_{\text{sphere}}}C_D\frac{V^2}{d}$$

Even if we assume a constant value of $C_D = 0.45$, the above equation is a nonlinear differential equation because of the V^2 term and is difficult to solve analytically. Therefore, we should use a numerical method, such as described in Algorithm 3.1 in Chapter 3, to solve this first-order ordinary differential equation. For this example, we used MATLAB to find the approximate numerical solution, which is plotted in Figure 6.10. It takes 1.2550 seconds for the ball to hit the ground.

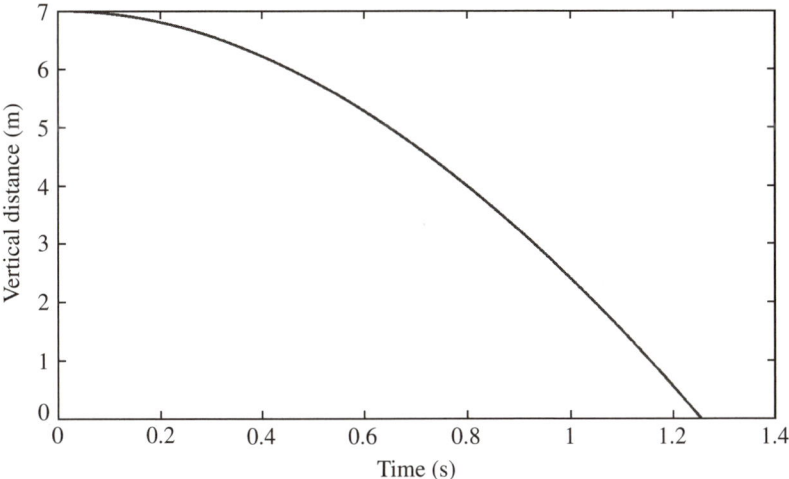

FIGURE 6.10 MATLAB solution to Example 6.5.

EXAMPLE 6.6

Calculate the change in velocity of a fastball from when the pitcher releases it to when it crosses home plate. The velocity when the pitcher releases the ball is 100 mph (44.7 m/s), and it has to travel 60 ft (18.3 m) to home plate. The dimensions of a baseball are given in Table 6.4.

SOLUTION: We will neglect the effects of spin on the drag coefficient, assume the ball travels in a straight line, and neglect the effects of gravity. Thus the problem reduces to a one-dimensional problem that can be represented by the force balance,

$$F_D = ma$$

which results in the differential equation

$$\frac{dV}{dt} = -\frac{C_D \rho A V^2}{2m}$$

The minus sign is present because the drag force acts opposite to the direction of motion. This differential equation can be solved analytically by separation of variables.

$$\frac{dV}{V^2} = -\frac{C_D \rho A}{2m} dt$$

$$\int \frac{dV}{V^2} = -\frac{C_D \rho A}{2m} \int dt$$

$$-\frac{1}{V} = -\left(\frac{C_D \rho A}{2m}\right)t + C$$

To eliminate the constant of integration, C, the initial condition $V(t = 0) = V_0$ is used. This results in $C = -1/V_0$. Thus the equation for $V(t)$ is

$$V = \frac{1}{\frac{C_D \rho A}{2m}t + \frac{1}{V_0}}$$

The Reynolds number for the ball is

$$\text{Re} = \frac{VD}{\nu} = \frac{(44.7 \text{ m/s})(0.073 \text{ m})}{(1.5 \times 10^{-5} \text{ m}^2/\text{s})} = 218{,}000$$

This is probably above the transition to a turbulent boundary layer, so we use the lower value of $C_D = 0.31$. Now we substitute in the numerical values of $m = 150$ g $= 0.15$ kg, $A = (\pi/4)(7.3 \text{ cm})^2 = 41.9 \text{ cm}^2 = 0.00419 \text{ m}^2$, so that $V(t)$ is

$$V = \frac{1}{0.0224 + 0.00520t}$$

For t in seconds, V will be in m/s. The distance traveled is x, so

$$x(t) = \int V \, dt$$

We use a handbook of mathematical tables of integrals to find an analytical expression for $x(t)$:

$$x = \frac{1}{0.0052} \ln(0.0224 + 0.0052t) + C$$

Applying the initial condition $x(0) = 0$ gives a value of $C = 730.52$. The value of t that corresponds to $x = 18.3$ m is $t = 0.43$ s. At this time, the velocity is $V = 1/[0.0224 + 0.0052(0.43)] = 40.6$ m/s.

It is just as easy to find a numerical solution to this problem using Algorithm 3.1 as it is to go through the analytical solution. Solving with Excel, we find that it takes 0.43 seconds for the ball to cross home plate, and its velocity is 40.6 m/s (91 mph) when it crosses, so that the fastball loses almost 10% of its velocity from release to the time it crosses the plate.

6.3.5 Numerical Simulation of Two-Dimensional Trajectories

For two-dimensional problems there is no choice but to use numerical methods to find a solution. The technique for solving two-dimensional trajectories is presented in Algorithm 6.1 and applied in the next two examples.

ALGORITHM 6.1   Two-dimensional trajectory

To compute the two-dimensional trajectory of a projectile with drag included, the vector form of Newton's second law must be solved.

$$\vec{F} = m\vec{a} = m\frac{d\vec{V}}{dt}$$

The vector equation in two dimensions can be written as two scalar differential equations that must be solved simultaneously.

$$F_x = m\frac{dV_x}{dt}$$

$$F_y = m\frac{dV_y}{dt}$$

The position of the projectile at each point in time is found by solving the following two additional differential equations:

$$\frac{dx}{dt} = V_x$$

$$\frac{dy}{dt} = V_y$$

This gives us a total of four first-order ordinary differential equations, all functions of time, that can be integrated simultaneously to solve for the four unknown variables (x, y, V_x, V_y) as functions of time. The Euler method, which was introduced in Algorithm 3.1, will be used. For example, to solve for x, the Euler algorithm is

$$x(t + \Delta t) = x(t) + \Delta t \cdot V_x(t)$$

and starts at $t = 0$. The value of Δt must be chosen small enough to provide an accurate reconstruction of the profile. For example, if the projectile is in the air for about 1 s before hitting the ground, a value of $\Delta t = 0.001$ s might be a good choice. Also note that the drag force is proportional to the square of the magnitude of the vector velocity. That is,

$$F_D = \frac{1}{2} C_D \rho A V^2$$

where

$$V = \sqrt{V_x^2 + V_y^2}$$

Algorithm Steps

1. Set the initial conditions x_0, y_0, V_{x0}, V_{y0}. (If the initial velocity's magnitude V_0 and angle θ are specified instead, then $V_{x0} = V_0 \cos(\theta)$ and $V_{y0} = V_0 \sin(\theta)$.)
2. Set $x(0) = x_0$, $y(0) = y_0$, $V_x(0) = V_{x0}$, $V_y(0) = V_{y0}$.
3. Set the time, $t = 0$, and select the numerical time step, Δt.
4. Specify the mass m and cross-sectional area A of the projectile, and the density ρ and viscosity ν of air for local atmospheric conditions.
5. LOOP:

    Compute the magnitude of the velocity: `V(t)` $= \sqrt{V_x^2 + V_y^2}$

    Compute the Reynolds number: `Re = V * d/`ν

    Compute the drag coefficient from the appropriate correlation: `C`$_D$ `= C`$_D$`(Re)`

    Compute the drag force: `F` $= \frac{1}{2} C_D \rho A V^2$

    Compute the x and y components of the drag force:

    ```
    Fx = F * Vx(t)/V
    Fy = F * Vy(t)/V
    ```
    Perform the integrations of the four variables:

    ```
    Vx(t + Δt) = Vx(t) - (Fx/m) * Δt
    Vy(t + Δt) = Vy(t) - (g + Fy/m) * Δt
    x(t + Δt) = x(t) + Vx(t) * Δt
    y(t + Δt) = y(t) + Vy(t) * Δt
    ```
    Increment the time step: `t = t + Δt`
6. END LOOP when `y < 0`.

To check that your choice of the time step, Δt, was sufficiently small, it is a good idea to repeat the computation with the value of Δt reduced by a factor of 2 or more to see if it makes a significant change to the computed trajectory.

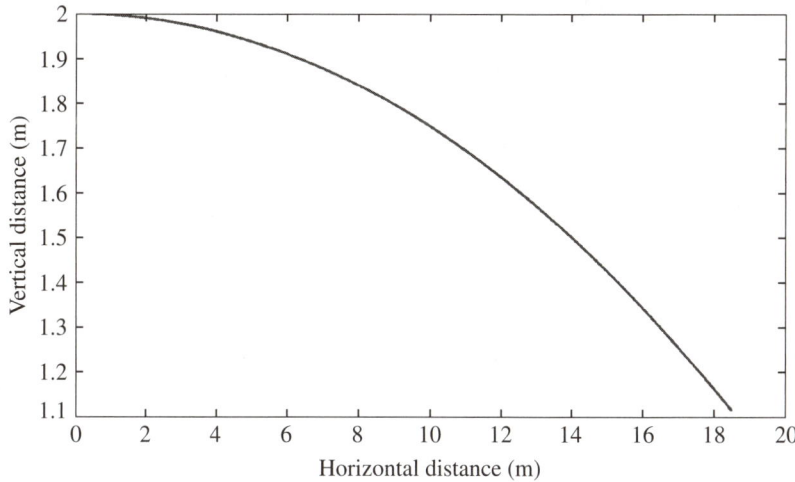

FIGURE 6.11 Two-dimensional trajectory of a baseball.

EXAMPLE 6.7

Consider a baseball released with an initial velocity of 100 mph. Calculate the drop in vertical height above the ground from the time the ball is released until it crosses the plate. Neglect effects of spin on the trajectory and assume there is no wind.

SOLUTION: The initial Reynolds number of the ball at 100 mph is 219,000, which puts the ball in the regime of constant C_D. Based on the results of Example 6.6, the velocity only changes 10%, so it is safe to assume a constant C_D. Algorithm 6.1 can be used for this two-dimensional problem. We can assume an initial release height of 6 ft (1.8 m) above the ground for the purposes of this problem, and that when the pitcher releases the ball it is 60 ft (18.3 m) in horizontal distance from home plate. The calculated trajectory is shown in Figure 6.11, which was generated in MATLAB (Excel, EES, or other software could also be used with Algorithm 6.1). The vertical scale has been exaggerated to show the motion.

EXAMPLE 6.8

During a professional game, a radar gun captured a baseball leaving the bat at 105 mph. If the initial launch angle was 30°, calculate the distance the ball travels. The game was played inside a dome at a city near sea level, so that the conditions inside the stadium can be taken as standard atmospheric conditions with no wind.

> SOLUTION: If you do not have a software package that can perform the integration automatically, you will have to write a short program using the Euler method of numerical integration, which can be done in Excel or MATLAB fairly easily. The Euler method is performed iteratively until $t > t_{final}$ or some other ending condition is met. Theoretically the error in using the Euler method disappears as $\Delta t \to 0$, but problems arise when the limited numerical precision of the computer starts to be a factor. More complex methods (such as Runge-Kutta methods) are used when high precision is needed. For this problem we used the MATLAB code in Appendix F, and calculated a value of the final landing distance of the ball as $x = 119.9$ m $= 393$ ft.

6.4 Lift

Any lift-generating device must force the moving fluid downward. All things that fly, whether birds or insects or airplanes or other vehicles, must produce lift in some way—most flying things use wings. For most man-made aircraft the wings are rigid; for birds the wings are flexible.

The lift coefficient typically depends strongly on the angle of attack and only weakly on the Reynolds number. The most significant effects of the Reynolds number are at high angles of attack near stall, since separation is influenced by the boundary layer characteristics. The lift force of a wing can be calculated either by integrating the pressure force over the total area of the wing and taking the vertical component, or by taking detailed measurements of the air flow in the wake and calculating the change in the vertical momentum of the air as it flows past the wing.

6.4.1 Airfoils

A *wing* is a three-dimensional lifting surface used to provide lift. An *airfoil* is the two-dimensional cross section of a wing. (See Figure 6.12.) The *chord* of a wing is the distance from the leading edge to the trailing edge, and the *chord line* is the line that connects those two points. The *camber* is the curvature of the wing above the chord line. The *angle of attack* is the angle between the chord line and the velocity vector from the relative wind. Figure 6.13 shows a generic aircraft, with relevant terminology labeled. One way to analyze lift is to use the momentum equation to calculate the upward force. Another is to integrate the pressure force over the surface of a wing.

NACA Airfoil Sections

The first attempt to systematically characterize airfoil shapes was the NACA four-digit series of airfoils. The National Advisory Committee on Aeronautics (NACA) was the predecessor organization to NASA. In the NACA four-digit series, a four-number

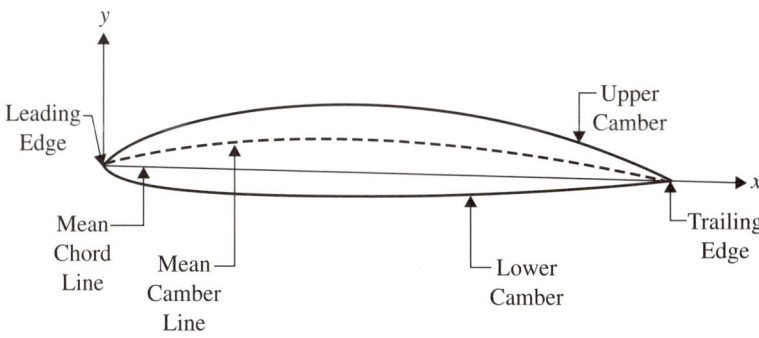

FIGURE 6.12 Diagram of an airfoil with the relevant terms labeled. From [NASA75].

designation is used to define each airfoil. The first number specifies the maximum camber, m, of the airfoil as a percentage of the chord length. The second number specifies the position, p, of the maximum camber from the leading edge, in tenths of the chord length (or percentage of chord length $\times$ 10). The last two numbers together specify the thickness, t, of the airfoil in percentage of chord. So as an example, a NACA 4515 airfoil has a maximum camber of 4%, located 50% of the chord back from the leading edge (halfway back), with a maximum thickness of 15% of the chord. Thus the four-digit number is sufficient to generate the shape of the airfoil (See Figure 6.14.)

In calculations, the four digits can be represented as NACA *mptt*. From the values of m and p, the equation for the *mean camber line* can be generated as

$$y_c = \frac{m}{p^2}(2px - x^2) \quad \text{for } 0 < x < p$$

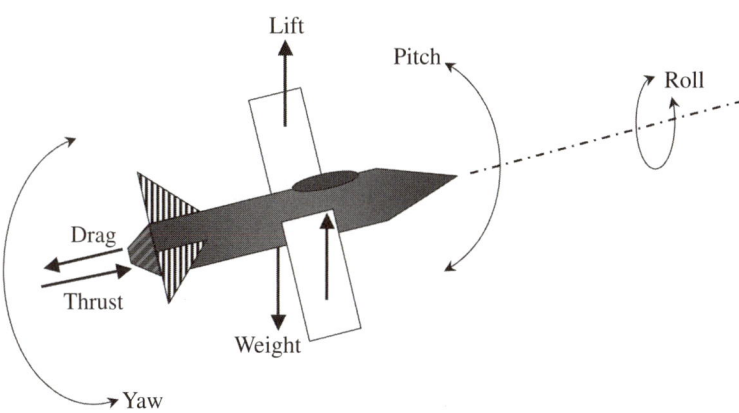

FIGURE 6.13 Schematic of a general flying vehicle with terminology labeled.

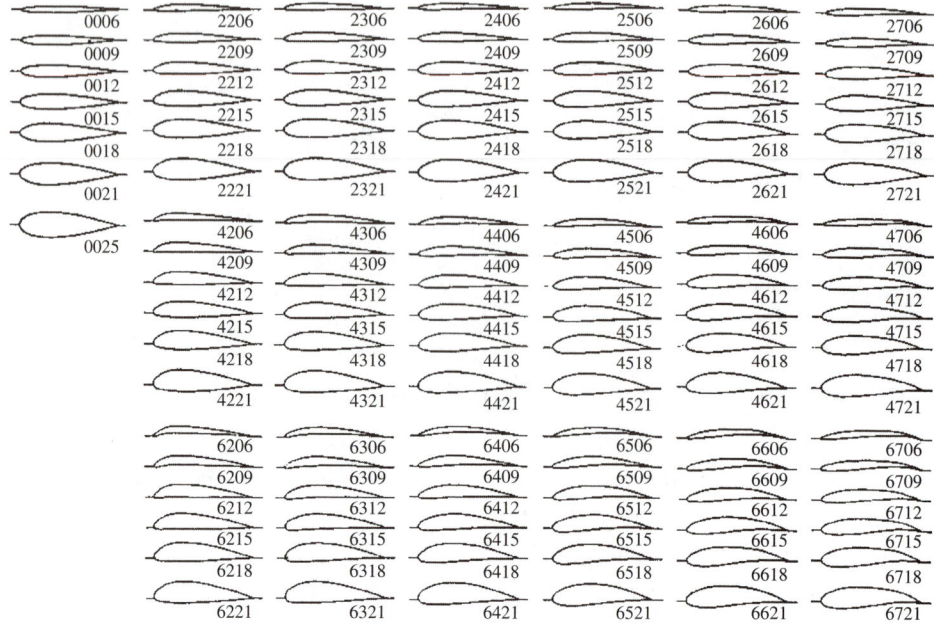

FIGURE 6.14 Representation of the NACA four-digit series of airfoils. From [NACA33].

and

$$y_c = \frac{m}{(1-p)^2}(1 - 2p + 2px - x^2) \quad \text{for } p < x < c$$

Here the coordinates are defined as illustrated in Figure 6.12, where x is the axis along the length of the airfoil running from the leading edge to the trailing edge, and y is the height above (or below) the x-axis. To generate the profile of the airfoil the thickness above and below the mean camber line must be known. By definition, the thickness above and below the mean camber line at each point x is the same. The equation for the thickness as a function of the x location is

$$\pm y_t = \frac{tt}{0.2}(0.2969\sqrt{x} - 0.1260x - 0.3516x^2 + 0.2843x^3 - 0.1015x^4) \quad (6.20)$$

The locations for the upper and lower surfaces of the airfoil at each axial location x is taken by adding or subtracting y_t and y_c, respectively. There were also more complex NACA five- and six-digit series airfoils, but these will not be discussed here.

As another example, a NACA 0012 airfoil is a symmetric airfoil, with a maximum thickness 12% of the chord. A symmetric airfoil generates no lift at zero angle of attack, and thus must be flown at positive angle of attack to generate lift. Symmetric airfoils are popular on aerobatic planes, which often fly upside down at airshows.

Further constraints on the NACA four-digit airfoil shapes are that $y_c = 0$ at $x/c = 0$ and $x/c = 1$. The maximum value of $y_c = mc/100$ occurs at $x/c = p/10$, and also at this point $dy_c/dx = 0$ [NACA33].

6.4.2 Lift-Induced Drag

It may seem straightforward enough to design an aircraft: Drag equals thrust, lift equals weight, and lift can be increased by increasing the angle of attack until the stall point is reached. However, there is an additional complication in that the amount of aerodynamic drag an airplane encounters is in fact related to the lift of the wings. In addition to the viscous drag (which is the largest component of drag for airplanes) and the pressure drag, there is an additional component of drag due to lift, called the *induced drag*. Because the pressure on the top surface of a wing is less than the pressure on the bottom surface, the air in the high-pressure region on the bottom tries to escape to the low-pressure region on top by moving around the tip of the wing. This results in the formation of swirling wingtip vortices at the end of wings. (See Figures 6.15 through 6.18.) These wingtip vortices create a downwash of air behind the wing, which increases the local angle of attack, which in turn results in an additional component of the aerodynamic force on the wing that points in the downstream direction [NASA75].

I FIGURE 6.15   Photo of wingtip vortices on an F-18 in flight. Courtesy of NASA-Dryden Flight Research Center.

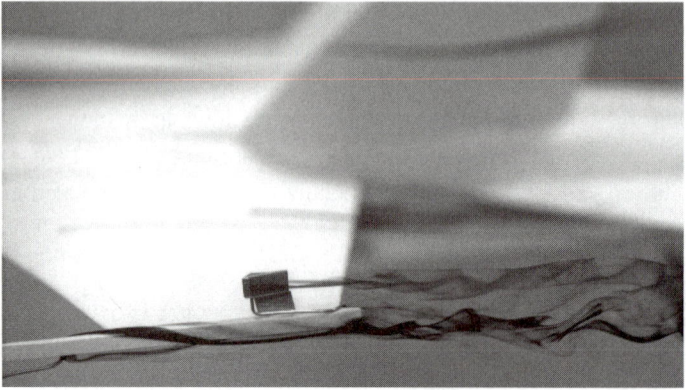

FIGURE 6.16 Photo of wingtip vortices on a model of an F-18. Courtesy of NASA-Dryden Flight Research Center.

The total drag coefficient, C_D, for an airplane is equal to sum of the drag coefficient at zero lift, $C_{D,0}$, plus the induced drag coefficient, $C_{D,i}$, which will increase with increasing angle of attack. Thus,

$$C_D = C_{D,0} + C_{D,i} \qquad (6.21)$$

As with all drag and lift coefficients for airplanes, the drag coefficients in this equation use the wing planform area as the reference area. The induced drag coefficient

FIGURE 6.17 Photo of wingtip vortices on a 727 in flight. Courtesy of NASA-Dryden Flight Research Center.

FIGURE 6.18 Photo of wingtip vortices on a crop duster. Courtesy of NASA.

$C_{D,i}$ is equal to the lift coefficient C_L squared, divided by π times the aspect ratio AR and a nondimensional efficiency factor ε [NASA75]:

$$C_{D,i} = \frac{C_L^2}{\varepsilon \pi (\text{AR})} \tag{6.22}$$

The aspect ratio AR of an airplane is defined as the square of the span S divided by the wing planform area A.

$$\text{AR} = \frac{S^2}{A} \tag{6.23}$$

For a rectangular wing the area is $A = S \times C$, where C is the chord length. Thus the aspect ratio is simply the ratio of the span to the chord.

$$\text{AR} = \frac{S^2}{A} = \frac{S^2}{S \times C} = \frac{S}{C} \tag{6.24}$$

From Equations 6.22–6.24, it can be seen that the induced drag can be reduced by increasing the wingspan (S). Induced drag is also greater for planes flown at relatively large angles of attack. Gliders and planes designed to fly at high altitudes where the air is not as dense (such as the U-2) tend to have large angles of attack. In the case of

FIGURE 6.19 Photo of NASA's ER-2 aircraft, which has the same body as the U-2 spyplane. Courtesy of NASA-Dryden Flight Research Center.

the glider the large angle of attack is needed because of the low forward flight speeds. Figure 6.19 shows the relatively high aspect ratio of the wings of such a plane.

As can be seen from Equations 6.22–6.24, an airplane with a long wingspan and short chord will have less induced drag than an airplane with the same wing area but a smaller aspect ratio. Of course, there are structural limitations to how long and thin a wing can be built [NASA75].

With the induced drag known, the total drag can be calculated as a function of the angle of attack. This explains the shape of the drag polar. Figures 6.20 and 6.21 show experimental data from a finite-span wing. Figure 6.20 is a traditional drag polar, in which the lift coefficient is plotted against the drag coefficient for various angles of attack. In Figure 6.21 the lift and drag coefficients are plotted directly against the angle of attack. Figures 6.22 through 6.24 show the air flow over wings at various angles of attack.

For a plane in level flight, increasing the velocity will increase the lift force so that the angle of attack may be reduced, which will reduce the induced drag. Unfortunately the skin friction drag force also increases since it is proportional to the velocity squared. In the linear region of small angles of attack, the lift coefficient is linearly proportional to the angle of attack (as illustrated by the data in Figure 6.21), so that

$$C_L \approx \alpha$$

Since the drag coefficient of an airplane is the sum of the zero lift drag plus the induced drag, and the induced drag is proportional to the lift coefficient squared, then

$$C_D = C_{D,0} + k\alpha^2$$

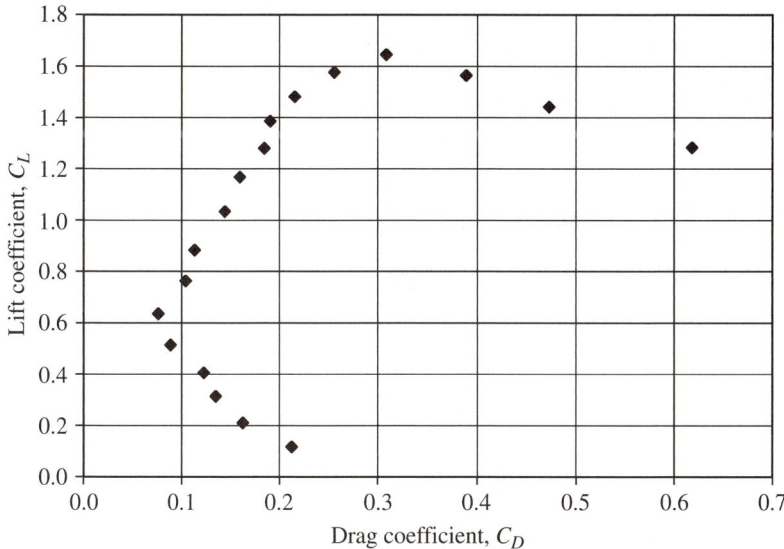

FIGURE 6.20 Drag polar plot. Data from the Bradley subsonic wind tunnel.

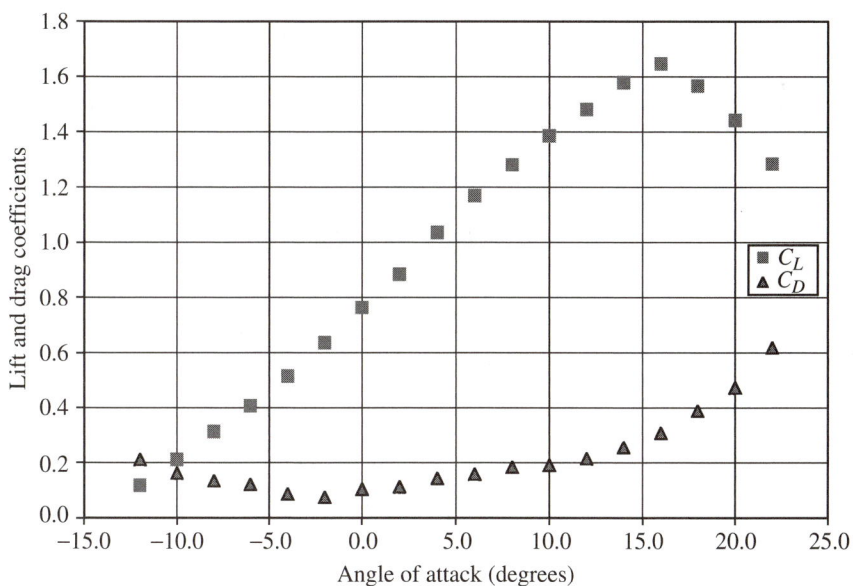

FIGURE 6.21 Drag and lift coefficients as a function of the angle of attack. Data from the Bradley subsonic wind tunnel.

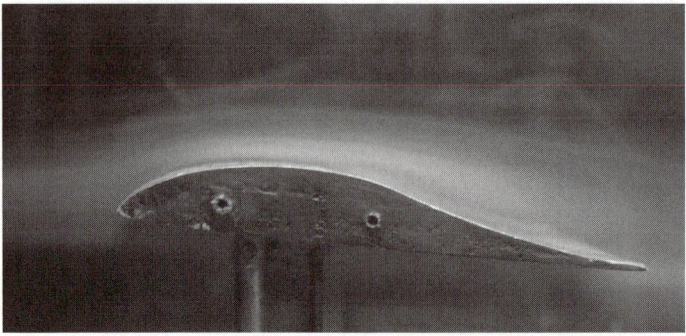

FIGURE 6.22 Photo of air flow over a wing at a small angle of attack.

where k is a proportionality constant. For a plane of fixed weight W and wing size A, the four forces must balance in level flight so that $F_D = F_T$ and $F_L = W$. Then, once in steady level flight, the only variables that the pilot can change are the angle of attack (by setting the trim) or the forward flight speed (by changing the engine thrust). So at low speeds a high angle of attack must be used to keep the plane in the air, which results in a high induced drag, while at high speeds the induced drag is low but the skin friction drag is high. There is an intermediate speed, known as the speed for minimum power, at which the total drag is minimized, as shown Figure 6.25.

6.4.3 Lift Distribution for Minimum Drag

How should the lifting force be arranged along the span of a wing so as to minimize the induced drag? It would seem logical to reduce the amount of lift per unit width as the wingtip is approached to reduce the strength of the wingtip vortex that causes the induced drag. But what shape of lift distribution would do this while still maintaining

FIGURE 6.23 Photo of air flow over a wing at a moderate angle of attack.

FIGURE 6.24 Photo of air flow over a wing at a large angle of attack, past the stall point.

high lift for the plane? Prandtl was the first to approach this problem, and for the constraints of a fixed span length and fixed total lift, he calculated that the lift distribution giving minimum induced drag is elliptical (cf. [Prandtl21] and [Munk24]), which is what is taught in most aerodynamics textbooks. In fact, the induced drag coefficient is

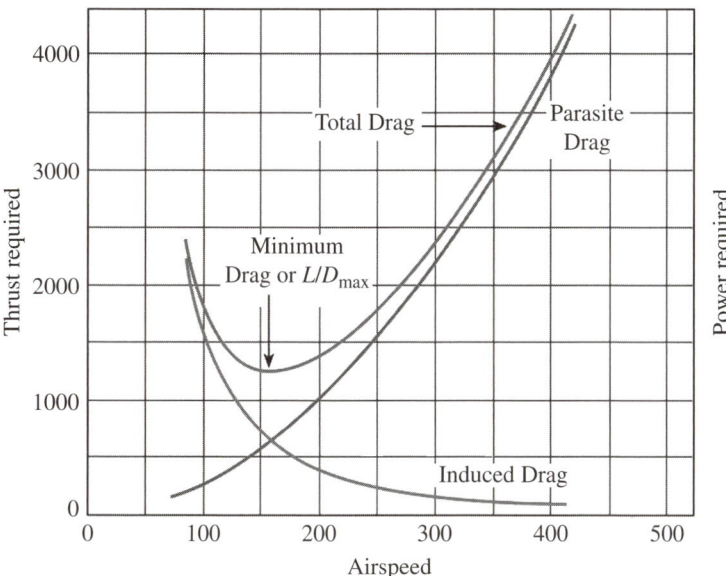

FIGURE 6.25 Plot of the total drag coefficient of an airplane as a function of the forward flight speed, for fixed plane weight and size. From [FAA04].

defined with the elliptical lift distribution having an efficiency factor of $\varepsilon = 1$. An elliptical lift distribution can be obtained by having a straight wing that has an elliptically shaped area (such as with the Supermarine Spitfire of WWII) or by changing the angle of attack along the span of the wing. A more typical value is $\varepsilon = 0.7$ for a straight rectangular wing.

The lift distribution along a wing can be changed by the designer in one of three ways [NASA75]:

1. Planform taper. The chord length may change along the span of the wing from the root to the tip, usually getting smaller as the wingtip is approached.
2. Change in airfoil section. Although it is easiest to manufacture a wing with a constant shape along the span, the shape of the airfoil section can change along the span.
3. Washin/washout. The angle of attack of the wing can change along the span. Washout is a decreasing angle of attack as the tip is approached; washin is an increase in angle of attack toward the tip.

Some designers use a combination of all three strategies to modify the lift distribution to minimize the induced drag.

However, if the problem of finding the lift distribution for minimum drag is formulated so that the span of the wing is allowed to vary but the root bending moment is kept constant, a different answer is obtained. It was discovered by Prandtl, and later independently by Jones [Jones50], that a bell-shaped curve for the lift distribution actually gives a lower induced drag than the elliptical distribution, with a value of $\varepsilon > 1.0$. Figure 6.26 shows the computed results of Jones.

6.4.4 Gliding Flight

The Wright Brothers started their research with glider flights at Kitty Hawk from 1900–1902. While gliders today are quite popular due to the low fuel cost, glide performance and the gliding range is also important in other aircraft in the case of a propulsion failure. In gliding flight, there are only three forces to be considered: lift, drag, and weight. (See Figure 6.27.) In steady state, these three forces must balance. The only way that can happen is if the nose of the plane points downward, so that part of the lift vector points forward, to counteract the drag. This means that the gliding plane will continually lose altitude until it comes to rest on the ground.

What angle of inclination will keep the plane at a steady speed? The force balance in the horizontal direction is

$$L \sin(\theta) - D \cos(\theta) = 0$$

and the force balance in the vertical direction is

$$L \cos(\theta) + D \sin(\theta) - W = 0$$

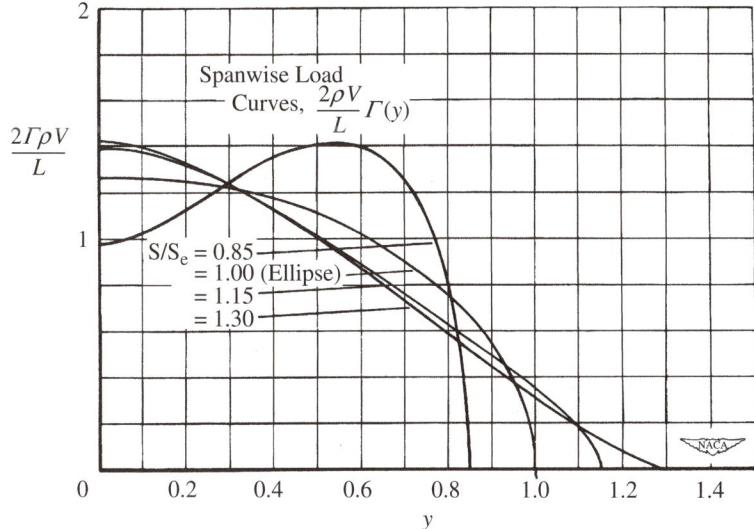

FIGURE 6.26 Efficiency factors for different spanloads. From [Jones50].

From the horizontal force balance, the following relationship between lift and drag can be obtained:

$$L \sin(\theta) = D \cos(\theta)$$
$$L/D = \cos(\theta)/\sin(\theta) = 1/\tan(\theta) \qquad (6.25)$$

Thus if the aerodynamic L/D ratio of the plane is known, the glide angle is also known. Although the L/D ratio is important in all planes, it is particularly important for gliders.

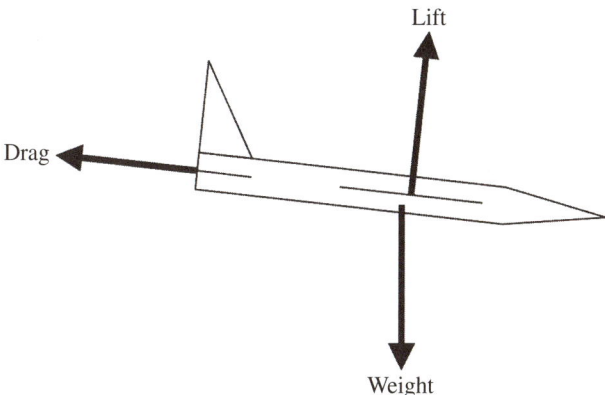

FIGURE 6.27 Schematic of a gliding plane.

With the glide angle known, the gliding range, Δx, of the aircraft at a given initial height, H, can be found:

$$\Delta x = H/\tan(\theta) \tag{6.26}$$

EXAMPLE 6.9

Calculate the glide angle of a plane with an $L/D = 16$. At an altitude of 20,000 ft, what is the gliding range of the plane?

SOLUTION: From Equation 6.25, the glide angle θ is

$$\theta = \arctan\left(\frac{D}{L}\right) = \arctan\left(\frac{1}{16}\right) = 3.58°$$

From this, we can calculate the range as

$$\Delta x = \frac{H}{\tan(\theta)} = \frac{20{,}000 \text{ ft}}{\tan(3.58°)} = 320{,}000 \text{ ft} = 60.6 \text{ miles}$$

6.5 Vehicle Aerodynamics

In this section we will focus on aerodynamics of road vehicles such as passenger cars and tractor-trailer trucks. Our discussion will include the wave drag on surface-going boats, supersonic effects on airplanes, transonic drag, the use of wind tunnels to study the aerodynamics of both planes and ground vehicles, the stability of flying vehicles, and the use of lifting bodies for aircraft and spaceships.

6.5.1 Aerodynamics of Passenger Cars

Most passenger cars, and certainly pickup trucks and SUVs, can be considered as blunt-base vehicles, as most of the drag comes from pressure or form drag. In addition to the main body shape of a car, significant sources of drag include the flow into the radiator, flow around the underbody, and flow around the side mirrors. For pickup trucks, the tailgate is also a source of drag that effects gas mileage. According to some studies, pickups get better mileage with the tailgate up, though it is not a very large difference [Cooper04].

> **EXAMPLE 6.10**
>
> How much engine power is spent on overcoming air resistance for steady driving at $V = 70$ mph on a level highway in a typical sedan with a cross-sectional area of 24 ft² and $C_D = 0.35$? How does this compare to a car with a low-drag design with cross-sectional area of 21 ft² and $C_D = 0.25$?
>
> SOLUTION: Converting the information to metric units gives $V = 31.3$ m/s and $A = 2.23$ m² for the sedan. The power to overcome drag resistance is
>
> $$\dot{W} = FV = \frac{1}{2}C_D \rho A V^3 = \frac{1}{2}(0.35)\left(1.2\,\frac{\text{kg}}{\text{m}^3}\right)(2.23\,\text{m}^2)\left(31.3\,\frac{\text{m}}{\text{s}}\right)^3 = 14{,}360\text{ W}$$
>
> This is equivalent to 19.2 hp.
>
> For the low-drag design, $A = 1.95$ m², and the required power is
>
> $$\dot{W} = \frac{1}{2}C_D \rho A V^3 = \frac{1}{2}(0.25)\left(1.2\,\frac{\text{kg}}{\text{m}^3}\right)(1.95\,\text{m}^2)\left(31.3\,\frac{\text{m}}{\text{s}}\right)^3 = 8{,}970\text{ W}$$
>
> which is equal to 12.0 hp.

6.5.2 Aerodynamics of Tractor-Trailers

In the United States there are around 2 million tractor-trailers in service that travel over 120 billion miles per year and consume over 20 billion gallons of fuel per year. A small change in C_D makes a big change in fuel costs because of the long distances driven. In recent years much progress has been made in improving the aerodynamics of the cab. (See Figure 6.28.) Improving the aerodynamics of the trailer is a bigger challenge because operators want to maximize the storage volume available and the rear doors need to provide for easy access. (See Figure 6.29.) For a typical trailer the cross section is around 2.6 m by 4.0 m, for a frontal area of 10.4 m². Mileage is typically in the range of 6 to 8 mpg for a fully loaded tractor-trailer.

> **EXAMPLE 6.11**
>
> Calculate the drag force and the power to overcome air resistance for a tractor-trailer traveling at 25 m/s with an overall C_D of 0.7 and cross-sectional dimensions of 2.6 m by 4.0 m.

304   Chapter 6   External Flow

FIGURE 6.28 Picture of a research truck with a cab-over-engine (COE) design with aerodynamic fairings. Courtesy of NASA-Dryden Flight Research Center.

FIGURE 6.29 Photo of an aerodynamic test van. With the front corners rounded and a rear boat tail added, the drag coefficient was greatly reduced. Courtesy of NASA-Dryden Flight Research Center.

SOLUTION: The drag force on the tractor-trailer is

$$F = \frac{1}{2}C_D \rho A V^2 = \frac{1}{2}(0.7)\left(1.2 \, \frac{\text{kg}}{\text{m}^3}\right)(10.4 \, \text{m}^2)\left(25 \, \frac{\text{m}}{\text{s}}\right)^2 = 2{,}730 \, \text{N}$$

which is equal to 613 lbf. The power to overcome drag is

$$\dot{W} = \frac{1}{2}C_D \rho A V^3 = \frac{1}{2}(0.7)\left(1.2 \, \frac{\text{kg}}{\text{m}^3}\right)(10.4 \, \text{m}^2)\left(25 \, \frac{\text{m}}{\text{s}}\right)^3 = 68{,}250 \, \text{W}$$

which is equal to 91.5 hp. Note that if the tractor-trailer were traveling at 70 mph (31.3 m/s) instead, the drag force would be 4,280 N (962 lbf) and the power would be 133,940 W (179.5 hp)—a considerable difference.

6.5.3 Wave Drag

As a boat moves through water, the water must move aside to make way for the boat. As the water moves to the side, the water level rises in the local vicinity of the boat. Gravity then acts on the water to pull it down, creating a wave. The energy lost to pushing the water up and to the side, creating the wave, is called *wave drag* and is above and beyond the friction and form drag that the underwater part of the boat experiences. As the boat speed increases, the length of the waves generated increase. When the wavelength is the same as the boat length, the wave is large and strong with peaks at both the front and rear (bow and stern) of the boat, and the wave drag increases rapidly with further increases in boat speed.

The drag of a surface-going vessel can be written as

$$F_D = \frac{1}{2}\rho A V^2 (C_F + C_W)$$

where A is the wetted area of the boat and C_W is the wave-making coefficient of drag. C_W varies with velocity to the fourth power, so that the force due to wave drag varies as V^6. The wave-making resistance of a boat is proportional to V^6 [Harvald83], compared to the frictional resistance, which rises only as V^2. In the equation, C_F is a function of the Reynolds number, and C_W is a function of the Froude number. The Froude number is defined as

$$\text{Fr} = \frac{V}{\sqrt{gL}} \tag{6.27}$$

where L is the waterline length of the boat. Note that some references use the square of this quantity as the Froude number, so it is sometimes defined as $\text{Fr} = V^2/(gL)$.

A quantity called the *celerity*, c, is defined for waves as

$$c = \frac{\lambda}{t} \tag{6.28}$$

where λ is the wavelength and t is the period of the wave. For gravity-driven waves the celerity is

$$c = \sqrt{\frac{g\lambda}{2\pi}} \tag{6.29}$$

This is equal to the wave propagation velocity for waves in deep water. For small waves where surface tension, σ, dominates, the celerity is

$$c = \sqrt{\frac{2\pi\sigma}{\rho\lambda}} \tag{6.30}$$

The *hull speed* of a boat is the speed at which the wave propagation speed equals the boat speed. This is obtained when

$$V_{\text{hull}} = \sqrt{\frac{gL}{2\pi}}$$

This is equivalent to a Froude number of

$$\text{Fr} = \frac{V}{\sqrt{gL}} = \frac{1}{\sqrt{2\pi}} \approx 0.4$$

Exceeding the hull speed takes a lot of power because of the rapid increase in wave drag beyond that speed. For boats at low speeds, the skin friction resistance dominates drag, whereas for boats at very high speeds, wave-making resistance is the largest component of the drag.

A hydrofoil is a type of boat with a wing underneath it. At sufficiently high speeds, the wing will generate enough lift to raise the hull out of the water, thus greatly reducing the drag. Thus a hydrofoil is more efficient than similarly sized boats at high speeds (above about 30 knots).

Unfortunately, it is impossible to make a scale model of a boat and test it in a water tunnel that maintains both Froude and Reynolds number scaling. For example, if the model boat is denoted by 1 and the full-scale boat by 2, then for a test made in a water tunnel, the fluid viscosities are equal, so $\nu_1 = \nu_2$. Of course the value of gravity will not change from the model test to the full-scale test, so $g_1 = g_2 = g$. Reynolds number scaling then requires that $V_2/V_1 = L_1/L_2$, which is the model scale factor, and Froude scaling requires that $V_2/V_1 = \sqrt{L_2/L_1}$, so both cannot be done at the same time (except for the trivial case of $L_1/L_2 = 1$). If a different fluid were used, such as an oil, it might be possible to match both the Froude and the Reynolds numbers simultaneously, but this would be an expensive test.

6.5.4 Supersonic Drag

The speed of sound is the speed at which sound waves travel through a medium. It can be shown from thermodynamics that the speed of sound, a, is

$$a = \sqrt{\left.\frac{dP}{d\rho}\right|_s} \tag{6.31}$$

where the subscript, s, means constant entropy. For a vehicle, the Mach number, Ma, is defined as the ratio of the speed of the vehicle to the local speed of sound:

$$\text{Ma} = \frac{V}{a} \tag{6.32}$$

For vehicles moving through liquids, the speed of sound is so high that the Mach number will be very small and compressibility effects can be safely ignored. Thus it is only for vehicles (usually airplanes) moving through gases (usually air) that the Mach number need be calculated. The only time another gas might be encountered is when an interplanetary probe is sent to the surface of another planet and has to navigate the atmosphere of that planet. But for aircraft traveling through Earth's atmosphere, the speed of sound in air is of interest.

For ideal gases—and air under atmospheric conditions is well modeled as an ideal gas—the speed of sound is

$$a = \sqrt{k(R/M)T} \tag{6.33}$$

For air, the average molecular weight is about $M = 29$ kg/kmol, and $k = 1.4$. (Actually, k varies with temperature, but not very rapidly.) So with $R = 8314$ J/kmol-K, the speed of sound in air as a function of temperature is

$$a = 20.0 \frac{m}{s\sqrt{K}} \sqrt{T}$$

for the temperature expressed in absolute units of degrees Kelvin. Notice that the speed of sound does not depend on pressure, only on the temperature. In Earth's atmosphere, the temperature decreases with increasing elevation.

Supersonic "Wave" Drag

When an airplane travels below the speed of sound its presence can be heard before the plane arrives because sound waves emanating from the plane travel faster than the plane itself. As the airplane increases its speed, the sound waves generated by the plane get closer and closer to each other. When the plane travels exactly at the speed

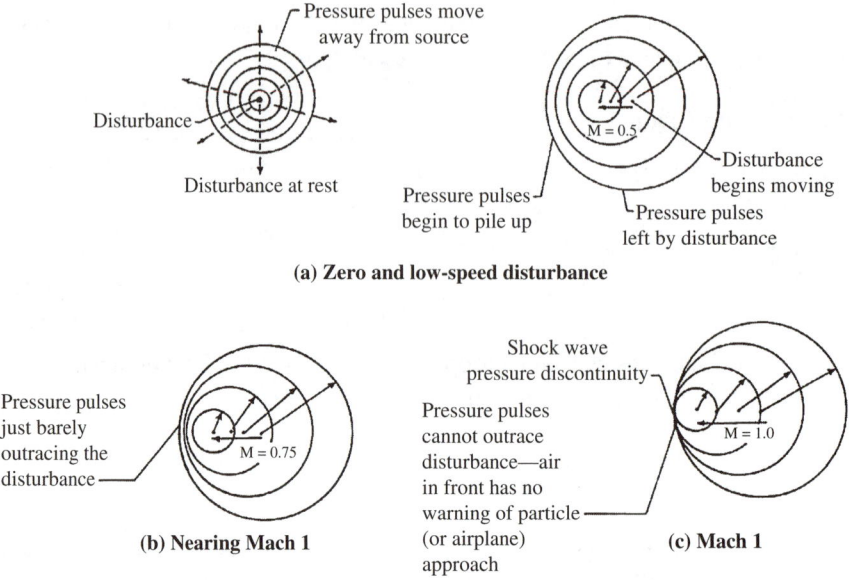

FIGURE 6.30 Propagation of sound waves from aircraft at different speeds. From [NASA75].

of sound, the sound waves travel at the same rate as the plane. When the plane travels at supersonic speeds, it travels faster than the sound waves and will actually be well past a listener on the ground before the sound waves are heard. (See Figure 6.30.)

The increase in drag at transonic and supersonic velocities is due to wave drag. The shock waves formed in supersonic flight (see Figure 6.32) have energy, and that energy comes at the expense of propulsion power of the aircraft. In the transonic range, there is a rapid increase in drag until Mach number = 1 is reached, and then the drag coefficient decreases, although it remains at a value higher than in the low subsonic range. In transonic flow, a small local shock wave forms over the wings where there is a local pocket of supersonic flow, and then typically the flow decelerates again to subsonic speeds before passing over the wing. In supersonic flow, the entire wing structure sees subsonic velocities, assuming the wings are swept back behind the Mach cone, and there is smoother flow around the wings and less drag than in the transonic regime [NASA75]. The Mach angle, α, is the angle of the Mach cone around a supersonic flying object. The value of the Mach angle can be determined from

$$\sin \alpha = \frac{a}{V} = \frac{1}{\text{Ma}} \tag{6.34}$$

Figure 6.31 shows the general trend in the variation of the drag coefficient of an airplane with Mach number.

6.5 Vehicle Aerodynamics 309

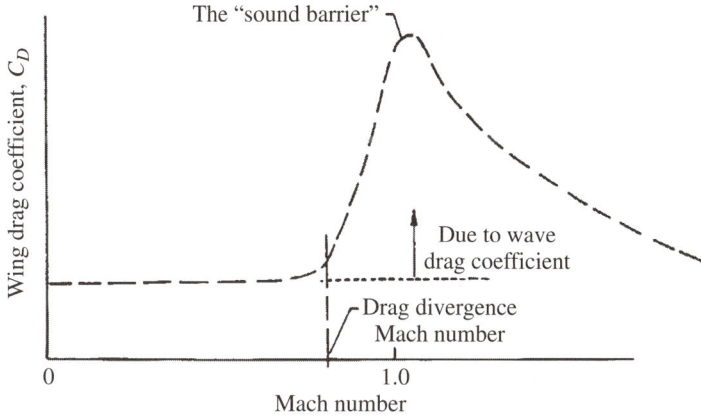

FIGURE 6.31 Typical trend of drag coefficient with Mach number for an airplane. Adapted from [NASA75].

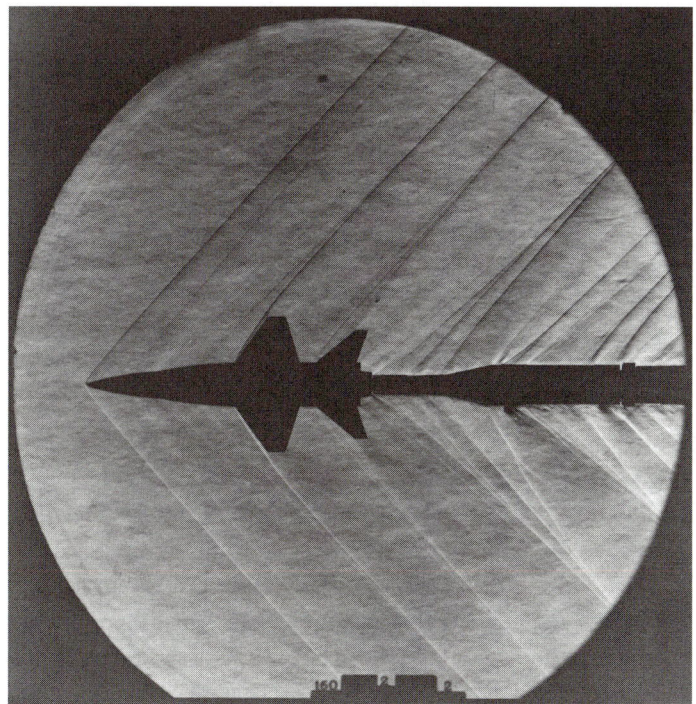

FIGURE 6.32 Schlieren photograph of shock waves off an X-15 model in a wind tunnel at supersonic speeds. Courtesy of NASA.

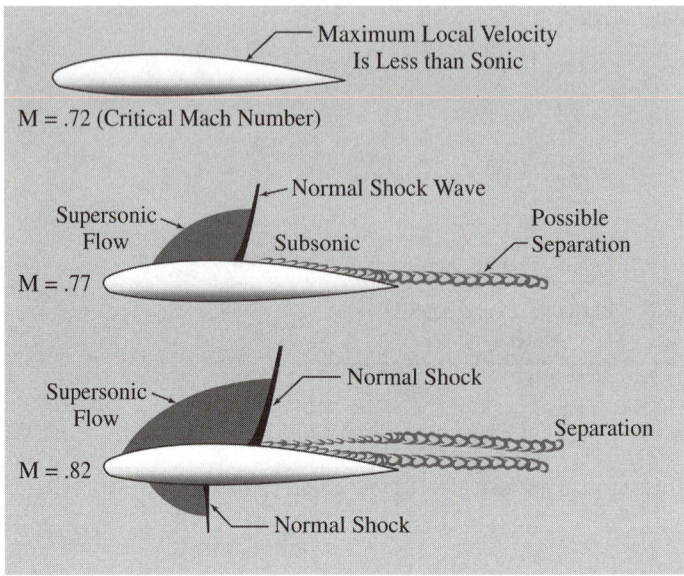

FIGURE 6.33   Schematic of transonic flow over wings. From [FAA04].

6.5.5 Transonic Drag

Why is the transonic drag coefficient higher than the supersonic drag coefficient? In transonic flow, the local shock wave formed over the wing interacts with the boundary layer so that separation of the boundary layer occurs immediately behind this shock. (See Figure 6.33.) This results in a large increase in the pressure drag, which is also called shock-induced (boundary layer) separation [NASA75]. The Mach number at which shock waves start to form over the wings of an aircraft is called the *drag divergence Mach number*, or sometimes the critical Mach number. Because of the rapid increase in drag coefficient above the critical Mach number, large increases in engine thrust are needed to achieve further increases in speed.

The critical Mach number can be increased, and the transonic drag reduced, by angling the wings backward from the fuselage, referred to as swept-back wings. Swept-back wings will delay the formation of the shock waves at transonic conditions to a higher Mach number, resulting in lower wave drag over all transonic and supersonic Mach numbers. The disadvantages of swept-back wings are that there is high induced drag and reduced effectiveness of trailing-edge flaps, as well as the added structural weight required to support the wing [NASA75]. That is why planes that travel at low speeds where transonic effects are not an issue typically have straight wings with no sweepback. Sweeping the wings forward instead of backward achieves the same wave drag reduction effect, but moves the center of lift forward, which makes

6.5 Vehicle Aerodynamics

FIGURE 6.34 Photo of the NASA X-29. Courtesy of NASA-Dryden Flight Research Center.

the plane more unstable. NASA built an experimental airplane, the X-29, that had forward-swept wings and required digital computer fly-by-wire controls to keep the plane flying. (See Figure 6.34 and 6.35.)

One possible way to create an aircraft that is efficient at both low and high speeds is to use variably swept wings, such as on the F-14 Tomcat (Figure 6.36). However the extra weight, volume, and complexity (maintenance) make this design a losing proposition. No aircraft currently in production have variably swept wings.

One major problem facing any potential supersonic transport aircraft is the noise associated with the sonic boom from the aircraft. In addition to the discomfort caused to listeners on the ground, sonic booms can cause minor structural damage such as broken windows and cracked plaster. This limited previous commercial supersonic transport aircraft to overseas routes, which in turn limits the profitability of the airplane. Recent research at NASA into reducing the noise from supersonic aircraft has included reshaping the nose of an F-5 aircraft (Figure 6.37) and adding a long nose-boom on an F-15 in the "Quiet Spike" project (Figure 6.38).

6.5.6 Wind-Tunnel Testing

To determine the aerodynamic performance of a vehicle, wind-tunnel testing offers the advantage of relatively cheap testing with less setup and turnaround time than full-scale testing in flight (or driving) with an actual vehicle. As we will discuss in Chapter 9, for a wind-tunnel test with a scale model to be an accurate representation

FIGURE 6.35 Photo of a NASA X-29 model in a water tunnel. Courtesy of NASA-Dryden Flight Research Center.

FIGURE 6.36 Photo of the of F-14 Tomcat. Courtesy of NASA-Dryden Flight Research Center.

6.5 Vehicle Aerodynamics

FIGURE 6.37 Photo of the NASA F-5. Courtesy of NASA-Dryden Flight Research Center.

of the aerodynamics around a full-scale vehicle, conditions of dynamic similarity must be met. This means that the Reynolds numbers of the scale model and the actual vehicle should match, and if compressibility effects are important, the Mach numbers should also match. A brief history of wind-tunnel development is given in the following paragraphs, summarized from [Day03] and [NASA81].

FIGURE 6.38 Photo of the NASA Quiet Spike. Courtesy of NASA-Dryden Flight Research Center.

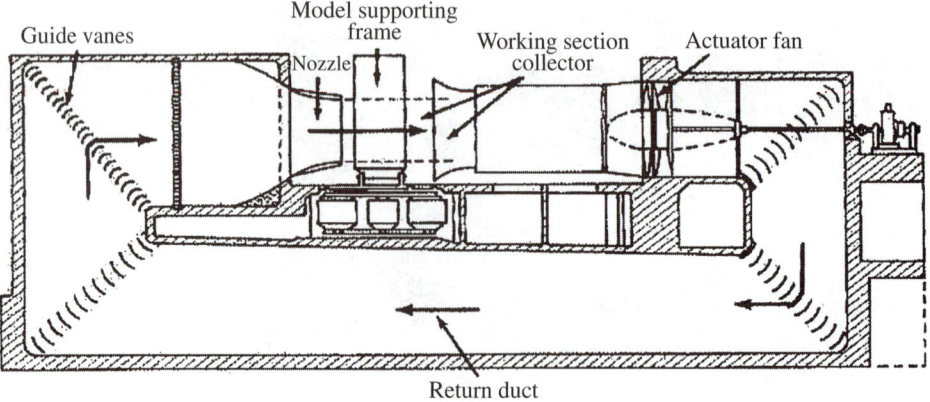

FIGURE 6.39 Schematic of an early wind tunnel of closed-circuit design. Courtesy of NASA.

From the first wind tunnel built by Frank Wenham in 1871 through the Wright Brothers' 1902 wind tunnel, most of the early wind tunnels were little more than long tubes or boxes with a fan to generate the air flow. The Wright Brothers' wind tunnel had a cross-sectional area of 16 in^2 and operated up to 35 mph. In Europe, Gustave Eiffel built wind tunnels after experimenting by dropping objects from his famous tower. Many of the most significant advances took place under Ludwig Prandtl's leadership in Germany. He built the first closed-circuit wind tunnel in 1908 (see Figures 39–40). Figure 6.39 shows a schematic of an early closed-circuit wind tunnel, similar to what Prandtl would have used, and Figure 6.40 shows a picture of another of the early wind tunnels.

FIGURE 6.40 Photo of an early NASA recirculating wind tunnel. Courtesy of NASA.

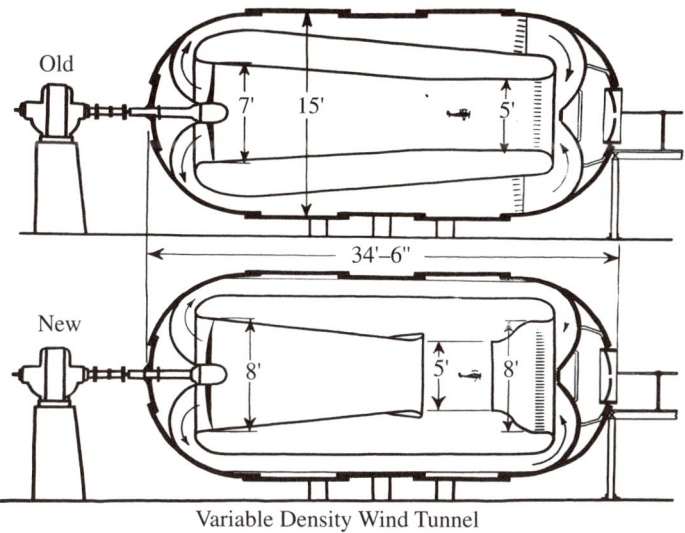

FIGURE 6.41 Schematic of the NACA Variable Density Wind Tunnel. Courtesy of NASA.

The NACA Variable Density Tunnel (VDT) (Figure 6.41), built at the Langley Laboratory in 1921–1923, was the first wind tunnel that could operate at pressures higher than atmospheric. Such pressures increase the density, so that higher Reynolds numbers could be achieved at lower velocities. This was accomplished by placing the entire wind tunnel inside a pressure vessel. In the 1940s supersonic wind tunnels were in use. As we will discuss in Chapter 8, for internal supersonic flow, there must be a converging section followed by a diverging section to accelerate the flow past Mach 1, with the airplane model placed in the diverging section. Because of the large power requirements to operate a supersonic wind tunnel in steady state, the blow-down wind tunnel was developed. In this design a large chamber is filled with compressed air, which is then released to provide a few seconds of supersonic flow. As the air expands it cools, so condensation of water in the air can be an issue.

In 1972 a cryogenic wind tunnel was built at NASA Langley. Liquid nitrogen was injected into the wind tunnel to cool the gas, which lowered the viscosity and increased the Reynolds number for the same pumping power. This tunnel had a greater capability to match Reynolds and Mach numbers simultaneously and could operate up to Mach 1.2. Today the largest wind tunnel in the world is in the National Full-Scale Aerodynamics Complex at NASA's Ames Research Center. This wind tunnel has a test section with cross sectional area of 80 ft by 100 ft (24 m by 31 m). (See Figure 6.42.)

While most commercial and military planes operate in the transonic regime, wind tunnels in which speeds are kept low enough that compressibility effects can be ignored are still commonly used for ground-vehicle studies and as educational tools.

FIGURE 6.42 Photo of the Ames Full-Scale Wind Tunnel. Courtesy of NASA.

Many auto makers have built full-scale wind tunnels for automobile testing. One challenge with wind-tunnel testing of ground vehicles is how to simulate the relative motion between the vehicle and the ground. One solution has been to incorporate a moving belt on the floor of the wind tunnel underneath the vehicle.

Over the past few decades, wind-tunnel testing has been used less and less as CFD tools have become more mature and computers faster. CFD analysis is used more often, particularly in the early stages of development of a project, with wind-tunnel testing reserved for final prototypes. However, due to the difficulty in modeling turbulent flows, it seems likely that wind-tunnel testing will continue to be used for the foreseeable future.

The two basic types of wind tunnel are open circuit and closed circuit. The advantage of open-circuit wind tunnels is that they are usually simpler and cheaper to build, and because of this low cost they are often found as educational tools in universities. Closed-circuit wind tunnels use less energy for a given size and flow velocity, because some of the kinetic energy of the exhaust stream is recovered as it returns. They also tend to be quieter. The disadvantages are the higher initial cost of construction, and the need to purge the wind tunnel if smoke is used for visualization.

How much power does it take to run a wind tunnel? By neglecting losses, we can calculate the minimum power required for an open-circuit wind tunnel as

$$\dot{W} = \frac{1}{2}\rho A V^3$$

For air at STP, the minimum power required to run a small wind tunnel of 25 cm by 25 cm cross section at an air speed of 50 m/s would be

$$\dot{W} = \frac{1}{2}\left(1.2\ \frac{\text{kg}}{\text{m}^3}\right)(0.25\ \text{m} \times 0.25\ \text{m})\left(50\ \frac{\text{m}}{\text{s}}\right)^3 = 5{,}860\ \text{W}$$

which is equal to 7.8 hp. At a speed of 100 m/s, the power increases by a factor of 8 to 46,900 W (62.8 hp). Note that this calculation does not account for inefficiencies in the fan-motor combination or the friction flow losses through the tunnel.

Experimental measurement techniques in fluids are discussed in Chapter 9. Among the instruments commonly used in wind tunnels are boundary layer rakes, tufts, pitot tubes, pressure-sensitive paint, smoke, and static pressure taps. A full six-component balance allows for all three forces and all three moments on the model to be measured.

One of the potential pitfalls of wind-tunnel testing is the possible aerodynamic interference between the support structure and the model. One should always measure the drag of the support tare or sting with no model on it first, and then subtract that from the total drag measured with the model in place. However, there are effects of the air flow on the model on the air flow on the sting, and vice-versa, so there is always some unknown interference drag component not accounted for. For supersonic tests the sting should always be at the rear of the aircraft. In cases where the model Re is lower than the anticipated full-scale Re, a trip strip can be placed at the front of the model to induce transition to turbulent flow in the boundary layer. The trip strip may be as simple as a piece of double-sided tape with a rough grit applied to one side. If the test object is large compared the wind tunnel cross section (say, larger than about 5% on a cross-sectional area basis), the blockage effect causes the air to accelerate as it goes past the test object, and there is no longer a uniform velocity of air.

6.5.7 Aerodynamic Stability

Stability is potentially an issue in any of the three axes about which a plane can rotate—roll, pitch, and yaw. In the roll axis, dihedral is commonly used to put inherent stability into a plane design. Dihedral refers to the upward angle above a horizontal plane at which the wings are tilted. Most commercial planes have some dihedral to improve the stability in roll. How does this work? If a plane were to roll slightly to one side, there would be a restoring moment to bring it back to center, as shown in Figure 6.43. Some planes, mostly military planes, have anhedral. The loss of lift associated with small angles of dihedral is very small. For example, a dihedral angle of 5° corresponds to a vertical component of the lift vector that is $\cos(5°) = 0.996$ of the overall lift, so only 0.4% of the lift is lost by putting the wings at a dihedral angle of 5°.

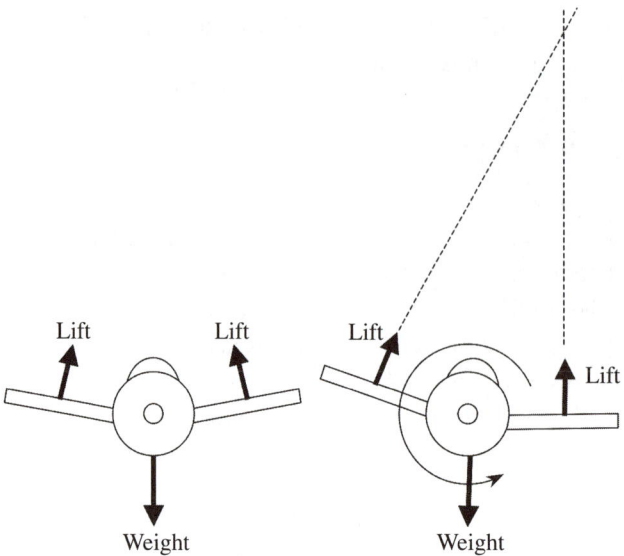

FIGURE 6.43 Sketch showing how dihedral enhances the stability of a plane in roll.

In the pitch axis, the relative locations of the center of mass and the center of lift are important in determining the stability characteristics of the airplane in pitch (see Figure 6.44). If the center of mass is in front of the center of lift, then the plane will be stable in pitch. If the center of mass is behind the center of lift, then the plane will be unstable. Yaw control is usually obtained with a large vertical tail.

6.5.8 Lifting Bodies

Although most commercial passenger aircraft produced today have the same basic shape (long cylindrical fuselage, swept-back wings, one engine mounted under each wing), there are many other ways to fly. One of the more interesting methods is

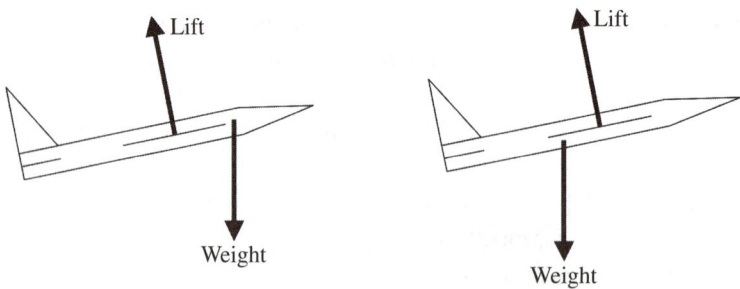

FIGURE 6.44 Sketch showing how the relative locations of the center of mass and center of lift are important for stability in pitch.

through lifting bodies. The history of lifting body development covered in this section is adapted from [Hallion84], [Reed97], and [Thompson99].

The initial motivation for creating a lifting body aircraft—that is, one that would generate lift through the shape of its body and hence would not need wings—was that NASA wanted its astronauts to be able to control their spacecraft so that they could land at a specific spot when returning to Earth from space. While wings are excellent at generating lift for airplanes, it is much more difficult to design spacecraft wings that can withstand the high stresses involved at the high velocities of reentry into Earth's atmosphere.

The first lifting body aircraft was the M2-F1, built in 1962 out of wood. The M2-F1 had no engine and was tested by being towed behind a C-47 transport plane, as shown in Figure 6.45. It was nicknamed the Flying Bathtub because of its shape.

The M2-F2 was a similar aircraft that was capable of higher speeds than the M2-F1. Another lifting body was also developed by engineers at NASA's Langley Research Center in Virginia (all of the previous research had been at NASA's Dryden Flight Research Center in California) and was designated the HL-10. Both the M2-F2 and HL-10 were dropped from a B-52 mothership, and once safely away from the mothership would ignite their rocket engines to achieve high speeds. The final part of the test flight was an unpowered landing (also called a dead-stick landing) on the dry lakebed at Dryden. Some of the problems the lifting bodies had were instability and flow separation. In May of 1967, the M2-F2 crashed, severely injuring its pilot. The

FIGURE 6.45 Photo of an M2-F1 being towed in flight. Courtesy of NASA-Dryden Flight Research Center.

FIGURE 6.46 Photo of three NASA lifting bodies. Courtesy of NASA-Dryden Flight Research Center.

M2-F2 was later rebuilt as the M2-F3, with an additional vertical fin to increase the roll stability. The HL-10 was eventually flown to a maximum speed of Mach 1.86. Another wingless lifting body aircraft, the X-24A, was also built and tested, and in 1972 it was further modified into the dart-shaped X-24B, which eventually reached a speed of Mach 1.75. The X-24A, M2-F3, and HL-10 are shown in Figure 6.46, and streamlines around a water tunnel model of the X-24A are shown in Figure 6.47.

While the research by NASA in the 1960s demonstrated that lifting bodies could be flown at high speeds and landed, the research also indicated some limitations of

FIGURE 6.47 Photo of a flow past a lifting body in a water tunnel. Courtesy of NASA-Dryden Flight Research Center.

aircraft without wings. In addition to the issues with stability and flow separation mentioned earlier, another major concern was the high landing speeds of the aircraft. Although NASA decided to go with a winged-vehicle design for the space shuttle, the lifting body idea received renewed interest in the 1990s when NASA was designing a crew return vehicle for the International Space Station to replace or supplement the Soyuz capsules. A prototype vehicle was built and designated the X-38 (Figure 6.48). The X-38, which was very similar to the X-24A, was designed to deploy a parachute about the size of an American football field during the final descent to reduce the landing speed.

NASA chose to go with a capsule geometry, similar to Apollo and Gemini, for its new space shuttle return vehicle. But the lifting body idea is still being researched,

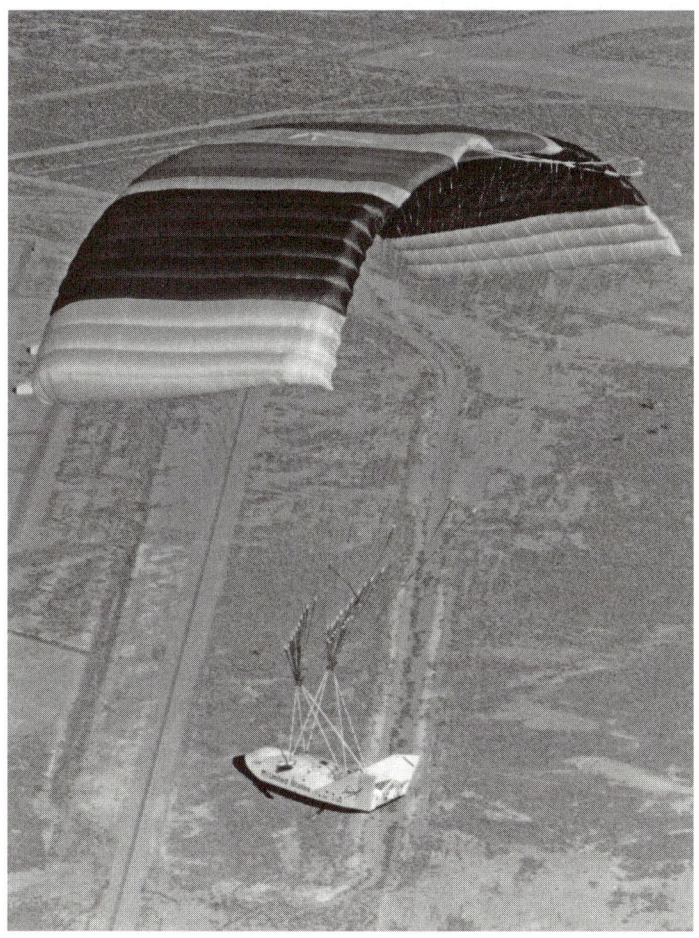

FIGURE 6.48 Photo of the X-38. Courtesy of NASA-Dryden Flight Research Center.

FIGURE 6.49 Photo of the X-48. Courtesy of NASA-Dryden Flight Research Center.

most prominently in the X-48 blended-wing-body aircraft, which combines wings with a lifting body. The X-48 (Figure 6.49) is a scale model of a potential commercial passenger aircraft. It has the potential to be more aerodynamically efficient, and thus use less fuel, than conventional airliners.

6.6 Transient Drag

When an object is started from rest, the drag coefficient will be higher than the steady-state drag coefficient due to transient effects. Specifically, the fluid around the object must be accelerated from rest as well. Thus,

$$F = \frac{1}{2}C_D \rho A V^2 + ma + (C_a \rho \forall)a = \frac{1}{2}C_D \rho A V^2 + (m + C_a \rho \forall)a$$

where C_a is the added mass coefficient. The quantity $C_a \rho \forall$ is the added mass, and the sum $(m + C_a \rho \forall)$ is the virtual mass. For a sphere $C_a = 0.5$, and for a cylinder $C_a = 1.0$. This transient drag is important for a bubble rising through a liquid, for example.

EXAMPLE 6.12

A helium balloon is released from rest at sea level. (See Example 2.24 for the data.) What is the initial acceleration calculated by (a) neglecting the virtual mass effect and (b) including the virtual mass?

SOLUTION: Neglecting the virtual mass effect, the initial acceleration would be equal to the buoyant force divided by the mass of the balloon. The buoyant force is equal to the weight of air displaced:

$$F_B = \rho_a g V = \left(1.21 \frac{\text{kg}}{\text{m}^3}\right)\left(9.8 \frac{\text{m}}{\text{s}^2}\right)\left(\frac{\pi}{6}(2 \text{ m})^3\right) = 49.7 \text{ N}$$

and the total weight of the balloon is:

$$W = \rho_{\text{He}} g V = W_{\text{skin}} = \left(0.17 \frac{\text{kg}}{\text{m}^3}\right)\left(9.8 \frac{\text{m}}{\text{s}^2}\right)\left(\frac{\pi}{6}(2 \text{ m})^3\right) + 10.5 \text{ N} = 17.4 \text{ N}$$

So the mass of the balloon is $m = W/g = 17.4 \text{ N}/9.8 \text{ m/s}^2 = 1.78 \text{ kg}$. Thus the initial acceleration is

$$a = \frac{F_B - W}{m} = \frac{49.7 \text{ N} - 17.4 \text{ N}}{1.78 \text{ kg}} = 18.1 \frac{\text{m}}{\text{s}^2}$$

With the effects of virtual mass included, the initial acceleration should actually be

$$a = \frac{F_B - W}{m + C_a \rho V} = \frac{49.7 \text{ N} - 17.4 \text{ N}}{1.78 \text{ kg} + (0.5)(0.17 \frac{\text{kg}}{\text{m}^3})(\frac{\pi}{6}(2 \text{ m})^3)} = 15.1 \frac{\text{m}}{\text{s}^2}$$

Hence the initial upward acceleration is overestimated by 3 m/s when the added mass is not taken into account. The "added mass" is 0.36 kg for this problem.

Summary

After reading this chapter and working through the problems, you should understand the basic principles of aerodynamics and be able to calculate lift and drag forces for an object. You also should be able to write a short computer program to simulate the two-dimensional trajectory of an object if the initial conditions are known. We presented a brief introduction to a wide variety of topics on aerodynamic and hydrodynamic drag and lift. You can consult the references or the reading list in Appendix E for more detailed information on a specific subject.

References

[Munk21] Munk, M. 1921. *The Minimum Induced Drag of Aerofoils*. NACA Report 121.

[Prandtl21] Prandtl, L. 1921. *Applications of Modern Hydrodynamics to Aeronautics*. NACA Report 116.

[NACA33] Jacobs, E., K. Ward, and R. Pinkerton. 1933. *The Characteristics of 78 Related Airfoil Sections from Tests in the Variable-Density Wind Tunnel.* NACA Report 460.

[NACA34] Abbott, I. 1934. *The Drag of Two Streamline Bodies as Affected by Protuberances and Appendages.* NACA Report 451.

[NACA36] Brevoort, M., and U. Joyner. 1936. *Experimental Investigation of the Robinson-type Cup Anemometer.* NACA Report 513.

[NACA38] Lindesy, W. 1938. *Drag of Cylinders of Simple Shapes.* NACA Report 619.

[NACA49] Scher, S., and L. Gale. 1949. *Wind-Tunnel Investigation of the Opening Characteristics, Drag, and Stability of Several Hemispherical Parachutes.* NACA Technical Note 1869.

[Jones50] Jones, R. 1950. *The Spanwise Distribution of Lift for Minimum Induced Drag of Wings Having a Given Lift and a Given Bending Moment.* NACA Technical Note 2249.

[NACA53] Delany, N., and N. Sorensen. 1953. *Low-Speed Drag of Cylinders of Various Shapes.* NACA Technical Note 3038.

[Goldstein65] Goldstein, S. 1965. *Modern Developments in Fluid Dynamics.* Dover.

[Hoerner65] Hoerner, S. 1965. *Fluid Dynamic Drag.* S. Hoerner (self-published).

[Schlicting68] Schlicting, H. 1968. *Boundary Layer Theory.* McGraw-Hill.

[NASA71] Davenport, E., and E. Lee. 1971. *Effects of Retrorocket Exhaust on Drag of 120° Cone at Subsonic Speeds.* NASA Technical Memorandum X-2275.

[Achenbach74] Achenbach, E. 1974. "The Effects of Surface Roughness and Tunnel Blockage on the Flow Past Spheres." *Journal of Fluid Mechanics*, vol. 65, pp. 113–125.

[NASA75] Talay, T. 1975. *Introduction to the Aerodynamics of Flight.* NASA SP-367.

[Bearman76] Bearman, P., and J. Harvey. 1976. "Golf Ball Aerodynamics." *Aeronautics Quarterly*, vol. 27, pp. 112–122.

[NASA81] Baals, D., and W. Corliss. 1981. *Wind Tunnels of NASA.* NASA SP-440.

[Hinds82] Hinds, W. 1982. *Aerosol Technology.* Wiley.

[Harvald83] Harvald, A. 1983. *Resistance and Propulsion of Ships.* Wiley.

[Hallion84] Hallion, R. 1984. *On the Frontier—Flight Research at Dryden, 1946–1981.* NASA SP-4303.

[Sprackling85] Sprackling, M. 1985. *Liquids and Solids.* Routledge & Kegan Paul.

[Mehta85] Mehta, R. 1985. "Aerodynamics of Sports Balls." *Annual Reviews in Fluid Mechanics*, vol. 17, 151–189.

[JSME88] Japan Society of Mechanical Engineers. 1988. *Visualized Flow.* Pergamon Press.

[Blevins92] Blevins, R. 1992. *Applied Fluid Dynamics Handbook*. Krieger.

[Reed97] Reed, D. R. 1997. *Wingless Flight: The Lifting Body Story*. NASA SP-4220.

[Crowe98] Crowe, C., M. Sommerfeld, and Y. Tsuji. 1998. *Multiphase Flows with Droplets and Particles*. CRC Press.

[Thompson99] Thompson, M., and C. Peebles. 1999. *Flying Without Wings: NASA Lifting Bodies and the Birth of the Space Shuttle*. Smithsonian Institution Press.

[NASA01] Whitmore, S., S. Sprague, and J. Naughton. 2001. *Wind-Tunnel Investigations of Blunt-Body Drag Reduction Using Forebody Surface Roughness*. NASA TM-2001-210390.

[Mehta01] Mehta, R., and J. Pallis. 2001. "The Aerodynamics of a Tennis Ball." *Sports Engineering*, vol. 4, pp. 177–189.

[Adair02] Adair, R. 2002. *The Physics of Baseball*. Harper.

[Bray03] Bray, K., and D. Kerwin. 2003."Modelling the Flight of a Soccer Ball in a Direct Free Kick." *Journal of Sports Sciences*, vol. 21, pp. 75–85.

[Watts03] Watts, R., and G. Moore. 2003."The Drag Force on an American Football." *American Journal of Physics*, vol. 71, pp. 791–793.

[Hubbard03] Sawicki, G., M. Hubbard, and W. Stronge. 2003. "How to Hit Home Runs: Optimum Baseball Bat Swing Parameters for Maximum Range Trajectories." *American Journal of Physics*, vol. 71, pp. 1151–1162.

[Day03] Day, D. "The Evolution of the Wind Tunnel." Centennial of Flight Commission. http://www.centennialofflight.gov/.

[Cooper04] Cooper, K. 2004. *Pickup Truck Aerodynamics—Keep Your Tailgate Up*. SAE Paper 2004-01-1146.

[FAA04] *Airplane Flying Handbook*. 2004. Federal Aviation Administration. FAA-H-8083-3A.

[Fontanella06] Fontanella, J. 2006. *The Physics of Basketball*. John Hopkins University Press.

Problems

1. Find the drag coefficient for the car you own (or, if you do not own a car, for a car you would like to own). How much power does it take for your car to overcome aerodynamic drag when traveling at 60 mph at sea level?

2. Find the drag coefficient for a Boeing 757. How much power is required to overcome aerodynamic resistance when the plane is traveling at 500 mph at an altitude of 30,000 ft?

3. A nuclear power plant has a cooling tower of height 80 m and average diameter of 60 m. Compute the drag force on the tower in a wind of 15 m/s by modeling the tower as a circular cylinder and assuming a uniform velocity across the tower.

4. If a baseball pitcher throws a ball at 85 mph, at what speed will it cross the plate?

5. Calculate the terminal velocity of a tennis ball. How does this compare to the speeds seen in professional men's tennis matches?

6. Calculate the terminal velocity of a golf ball. How does this compare to the speeds seen in professional men's golf?

7. Based on current diesel fuel prices, how much money would the average truck driver save in a year if the drag on her tractor-trailer was reduced by 1%?

8. If your school has a rapid prototyper, use it to build a scale model of a mass-produced car. Put it in a wind tunnel and measure the drag coefficient. How does the measured value compare to the manufacturer's published data?

9. If an airplane with an $L/D = 16$ runs out of fuel at an altitude of 10,000 ft, how far can it glide before it will have to land?

10. Explain how the Flettner rotor works as a means of propulsion.

11. What is the rising speed of a 1-mm air bubble in water?

12. What is the terminal velocity of a 1-mm raindrop?

13. What is the terminal velocity of a whiffle ball?

14. How can rolling resistance on a vehicle be measured?

15. Repeat the rocket problem in Example 3.14 in Chapter 3. Include aerodynamic drag, with a rocket diameter of 3 m and drag coefficient of 0.6.

16. Make a plot of drag coefficient versus vehicle weight for current-year vehicles. Group the results by type of vehicle: SUV, pickup truck, sedan, and so on.

17. Estimate the change in highway mileage of a pickup truck if the drag coefficient is reduced by 10%. If the vehicle is driven for 15,000 miles in one year, how much money would be saved by reducing the drag coefficient by 50% at today's fuel prices?

18. Do the stitches on a baseball cause an increase or decrease in drag compared to a smooth sphere?

19. What is the difference in drag coefficient between a golf ball and a smooth sphere traveling at the same Reynolds number?

20. If your university has a wind tunnel, measure the drag coefficient of a sphere as a function of Reynolds number. How do your results compare to published data? Are the differences within the experimental uncertainty of your measurements?

21. If a wind tunnel has a cross section of 20 cm by 30 cm, what is the largest diameter sphere that can be placed inside the wind tunnel?

22. What upward vertical velocity of air is required to suspend a 1-mm drop of diesel fuel in place at STP? Repeat for a drop of size 100 μm and 10 μm. Plot velocity versus drop size for your results. The density of diesel fuel can be taken as approximately 750 kg/m^3.

23. A tennis ball, a golf ball, and a baseball are dropped simultaneously off the edge of a building 10 m high. What is the order in which they will hit the ground? Does the height of the building factor into this problem at all?

24. A typical squash ball is a smooth sphere with diameter of 4 cm and mass of 24 g. If the maximum velocity of a squash ball is 70 m/s, what is the drag coefficient at that speed?

25. What is the terminal velocity of a squash ball, if it were dropped off the side of a tall building?

26. A chemical lab fume hood induces an upward velocity of 5 m/s in the air in the hood. What is the largest diameter of water drop that can be carried away by the fume hood?

27. A large aircraft carrier displaces 90,000 metric tons of water, has a length of 330 m, a width at the waterline of 40 m, and a waterline depth of 15 m. The ship engines produce 200 MW of power, and the top speed is 55 km/hr. Estimate the drag coefficient from this data. What percentage of the drag do you think is wave drag? What are the Reynolds and Froude numbers for this ship?

28. Estimate the percentage of drag on a tractor-trailer that comes from skin friction by modeling the sides, top, and bottom of the trailer as a flat plates parallel to the flow. Take the vehicle speed to be 30 m/s.

29. After lift-off, the 82-ton solid rocket boosters are recovered from the space shuttle (see Figure 6.50). Three parachutes of 115 ft diameter are used for each booster. What is the terminal velocity of the boosters?

30. A baseball, a golf ball, a soccer ball, and a basketball are all dropped at the same time from a height of 10 m into air at sea-level conditions. What is the order in which the balls will hit the ground?

31. If a pitcher throws a curveball with a release velocity of 80 mph and spin of 1800 rpm about a vertical axis, how much sideways deflection will there be in the trajectory of the pitch due to the lift generated due to spin?

FIGURE 6.50 Photo of a solid rocket booster with parachutes deployed. Courtesy of NASA.

32. A soccer player kicks a soccer ball off the ground toward the goal 10 m away at an angle of inclination of 20°, with an initial velocity of 30 m/s. Will the flow over the ball be laminar or turbulent during its flight path?

33. For a football launched at a height 2 m above the ground, thrown at an initial speed of 25 m/s, what angle of inclination should be used to maximize the horizontal distance traveled?

34. For a golf ball spinning at 3600 rpm with initial speed of 135 mph, what launch angle should be used to maximize the driving distance?

35. The army wants to drop a vehicle from a plane weighing 500 lbf and have it hit the ground with a speed of no more than 20 mph. What size of dome-shaped parachute must be used?

36. Estimate the lift force of a helicopter with blades of an NACA 0012 profile 10 ft long and 0.5 ft wide spinning at 1000 rpm with an angle of attack of 5°.
37. If the drag coefficient of a tractor-trailer could be reduced from 0.6 to 0.25 by using aerodynamic devices, what is the reduction in the required engine power for steady driving at 60 mph on level ground?
38. A zeppelin airship (Figure 6.51) travels at an elevation of 300 m at a speed of 60 km/hr. The length of the zeppelin is 75 m and the diameter is 18 m. Using the drag coefficient of a streamlined body from Table 6.2, calculate the drag force and the power required to overcome drag.
39. A passenger airplane has a maximum L/D ratio of 16 and is designed to cruise at that point at an altitude of 30,000 ft at a speed of Ma = 0.85. If the plane weighs 120,000 lbf (including cargo), what is the required engine thrust force and power?
40. For the conditions of the previous problem, if the engines use a fuel of chemical energy content of 42,000 kJ/kg, and the engine has a thermodynamic efficiency of 30%, what is the rate at which fuel is consumed?
41. In the United States, the Department of Transportation has established vehicle limits of 80,000 lbs gross weight for tractor-trailers. What percentage of engine power goes to overcoming aerodynamic resistance at steady level driving on a highway?
42. An engineer wants to perform a test of a surface-going boat in a towing tank filled with oil instead of water. If a 1/10 scale model is to be used (real vehicle length of 10 m, model length of 1 m) and the designed vehicle speed is 15 m/s, is it

FIGURE 6.51 Picture of a zeppelin. Courtesy of ZeppelinNT.

possible to match both the Reynolds and the Froude numbers with the scale model in the oil tank? If so, what type of oil should be used?

43. A submarine is to be tested in a wind tunnel, as shown in Figure 6.52. If the submarine is a 1/15 scale model, and the design speed for the full-scale submarine is 5 m/s, what wind tunnel velocity should be used?

44. An aircraft manufacturer is willing to go to any expense to match both Reynolds number and Mach number for a wind-tunnel model. For a plane of length 10 m designed to operate at Ma = 0.9 at sea level, and a model scale of 1/10, how could the wind tunnel be made to match both numbers simultaneously? Be specific and quantitative in your answer.

45. How would the drag coefficient of a well-used tennis ball differ from that of a brand new one? Why?

46. Perform an experiment to measure the drag coefficient of a whiffle ball. What is your measured value of C_D? How does this compare to other sports balls?

FIGURE 6.52 Photo of a submarine in a wind tunnel. Courtesy of NASA.

47. Perform an experiment to measure the lift and drag coefficients of a Frisbee. What is the lift-to-drag ratio for the Frisbee you tested?
48. For the car you own and drive, what percentage of the fuel chemical energy is used to overcome drag in steady highway driving?
49. Calculate the buoyant force at sea level for a baseball, golf ball, soccer ball, and basketball. Compare this force to the weight of the ball for each.
50. Calculate the terminal velocities at sea level for a baseball, golf ball, soccer ball, and basketball.
51. Find the optimum trajectory for a free throw in basketball, based on the size and location of the basket rim, and finite launch velocity.
52. What is the minimum launch velocity needed for a three-point shot in basketball? How does this compare to the minimum velocity needed for a free throw?
53. Measure the surface roughness of a standard basketball. Compute the velocity needed to shoot a free throw if the ball is launched at an initial angle of 45°, and drag is neglected in this initial calculation. Use this velocity to compute a Reynolds number, and based on the data of Achenbach, determine whether the boundary layer around a basketball is laminar or turbulent.
54. Calculate the design speed (also called hull speed) of a boat that is 500 ft long. Express your answer in units of ft/s, m/s, and knots.
55. If a boat displacing 100,000 tons of water with a waterline length of 1000 ft uses 250,000 hp to travel at a speed of 35 knots, what is the total resistance coefficient ($C_F + C_W$) of the boat? What are the Reynolds and Froude numbers for the boat?
56. For a pickup truck with a cross-sectional area of 3 m² and a drag coefficient of 0.5, how much power is required to overcome aerodynamic drag at a speed of 30 m/s? How does this compare to a typical passenger car?
57. During a professional game, a radar gun captured a baseball leaving the bat at 105 mph. A standard baseball has a diameter of 7.32 cm and a weight of 150 g, and the effective drag coefficient for a baseball around 100 mph has been found experimentally to be $C_D = 0.31$. The game was played inside a dome at a city near sea level, so that the conditions inside the stadium could be taken as standard atmospheric conditions with no wind. Perform the following calculations: (a) To the nearest 1°, compute the initial launch angle of the ball from the bat that gives the farthest horizontal distance traveled for the ball. (b) For the initial velocity and trajectory above, how much farther would the ball have traveled if it were hit at Mile High Stadium in Denver (assume a standard atmosphere)?
58. A plane travels at a speed of 2300 km/hr in an atmosphere at 10°C. Find the Mach angle.

7 Rotating Machinery

In This Chapter
- Conservation of Angular Momentum
- Pumps, Compressors, Fans, and Propellers
- Turbines
- Wind Turbines

The objective of this chapter is to reexamine the principle of conservation of angular momentum (first introduced in Chapter 3) with reference to simple rotating systems, to apply the principle to devices that either add energy to or extract energy from fluids, and to survey the use of wind turbines for generating electrical power.

■ 7.1 Conservation of Angular Momentum

Angular momentum was introduced in Chapter 3, but the principle of conservation of angular momentum is presented again here, as we apply it to simple rotating machinery, also commonly called turbomachinery.

When the fluid thrust force does not act through the center of gravity of a system, it can generate a *torque* on the system. A torque acts to put a system into rotational motion, unless there is another opposing torque or moment. Torque is usually expressed in units of N-m or ft-lbf, or occasionally in-lbf, and a torque is equivalent to a moment.

The angular momentum of a system must be defined relative to a coordinate origin. Whereas the linear momentum of a rigid object is the mass times the velocity, the angular momentum of a rigid object is the product of the linear momentum times the length of a lever arm from the object to the coordinate origin. In vector terms this is expressed as

$$\vec{H} = \vec{r} \times (m\vec{V}) \tag{3.27}$$

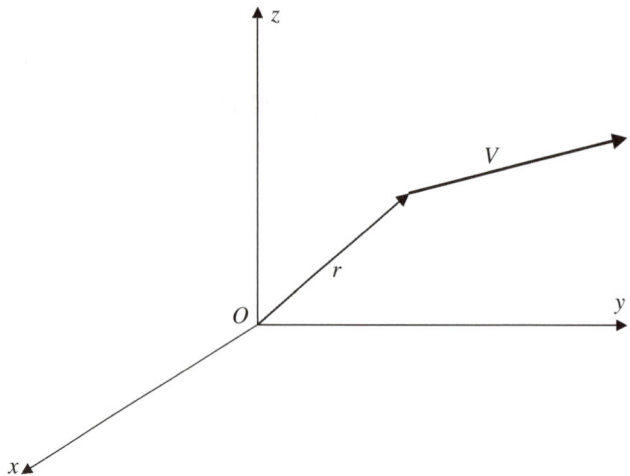

FIGURE 7.1 A right-hand coordinate system for angular momentum.

where $\vec{H}$ is the angular momentum vector. For a fluid system defined within a control volume, V, the total angular momentum is

$$\vec{H} = \int_V \rho(\vec{r} \times \vec{V}) \tag{3.28}$$

The time rate of change of the angular momentum of a fluid system is equal to the rate of change of fluid angular momentum within the control volume, plus any changes due to convective fluxes of angular momentum into or out of the control volume:

$$\frac{d\vec{H}}{dt} = \frac{\partial}{\partial t}\int_V \rho(\vec{r} \times \vec{V}) + \int_S \rho(\vec{r} \times \vec{V})(\vec{V} \bullet d\vec{A}) \tag{3.29}$$

Just as a force produces a change in linear momentum, a torque produces a change in angular momentum. For a right-hand coordinate system defined about an origin point, O (see Figure 7.1), a general equation for the conservation of angular momentum is

$$\sum \vec{M}_O = \frac{d\vec{H}}{dt} = \frac{\partial}{\partial t}\int_V \rho(\vec{r} \times \vec{V}) + \int_S \rho(\vec{r} \times \vec{V})(\vec{V} \bullet d\vec{A}) \tag{3.30}$$

So the sum of any external moments or torques on the system is equal to the total change in angular momentum.

For the special case of steady-state flow with uniform velocity profiles at inlet and outlets, we have

$$\sum \vec{M}_O = \sum \dot{m}_{\text{out}} (\vec{r} \times \vec{V}) - \sum \dot{m}_{\text{in}} (\vec{r} \times \vec{V}) \tag{3.31}$$

Equation 3.31 is the form of the conservation of angular momentum equation we will attempt to apply to most problems. Depending on the problem, it may be more convenient to have the control volume rotate with the device. This relative velocity is defined relative to the control volume. For a problem with a water sprinkler, the control volume moves with the sprinkler arm. If the rotational speed and water velocity are measured, the frictional torque on the sprinkler can be deduced from

$$T = \dot{m} R V_{\text{rel}} \tag{3.34}$$

Conversely, if the torque and flow rate are known, the predicted rotational speed will be

$$\omega = \frac{V}{R} - \frac{T}{\dot{m} R^2} \tag{3.35}$$

Also note the power associated with a torque is

$$\dot{W} = T\omega \tag{3.36}$$

It is common engineering practice in situations involving rotating shafts to write the equation for shaft power as

$$\dot{W} = 2\pi N T \tag{3.37}$$

where the factor of 2π is included to account for the conversion of revolutions to radians. Note that since rotational speed is usually expressed in rpm, the cycle time in minutes must be converted to seconds to put the power in standard units, such as watts. Systems where angular momentum is important include turbomachinery such as compressors and turbines, water sprinklers, dishwasher arms, automotive torque converters, and so on.

7.2 Pumps, Compressors, Fans, and Propellers

Pumps and compressors are used to increase the pressure of a fluid. When the fluid is a liquid, the device is referred to as a *pump*; when the fluid is a gas or vapor, the device is called a *compressor*. Thermodynamically we say the device adds energy to the fluid.

A compressor will increase the pressure of the gas, while a fan generally does not cause a significant pressure rise.

There are many different types of pumps. *Displacement pumps* work by trapping a fluid in a confined volume, compressing it, and then releasing it. Examples include reciprocating piston pumps, rotary gear or vane pumps, and peristaltic pumps. A *peristaltic pump* is one in which a rolling action is used to compress and move a flexible surface to move the fluid. Peristaltic pumps are found in biological systems. *Rotary pumps* spin a fluid to increase its momentum. Rotary pumps can be centrifugal pumps, axial pumps, or mixed axial and radial flow pumps. In this section, we will focus on rotating centrifugal pumps and compressors (see Figure 7.2).

Figure 7.3 shows a drawing of an old displacement pump, and Figure 7.4 shows the four strokes of the pump processes for this type of reciprocating piston pump.

A simple control volume shows that the rate at which a pump adds energy to a fluid is

$$\dot{W} = Q \, \Delta P$$

The efficiency of a pump is simply defined as the ratio of the useful power output to the required power input:

$$\eta_{\text{pump}} = \frac{Q \, \Delta P}{\dot{W}_{\text{input}}}$$

FIGURE 7.2 Photo of a centrifugal compressor. Courtesy of NASA Glenn Research Center, Cleveland, Ohio.

7.2 Pumps, Compressors, Fans, and Propellers

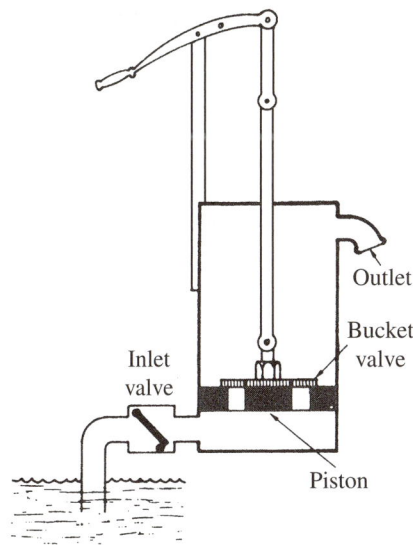

FIGURE 7.3 Diagram of a lift pump, from [Audel70]. Used with permission of Howard W. Sams & Co.

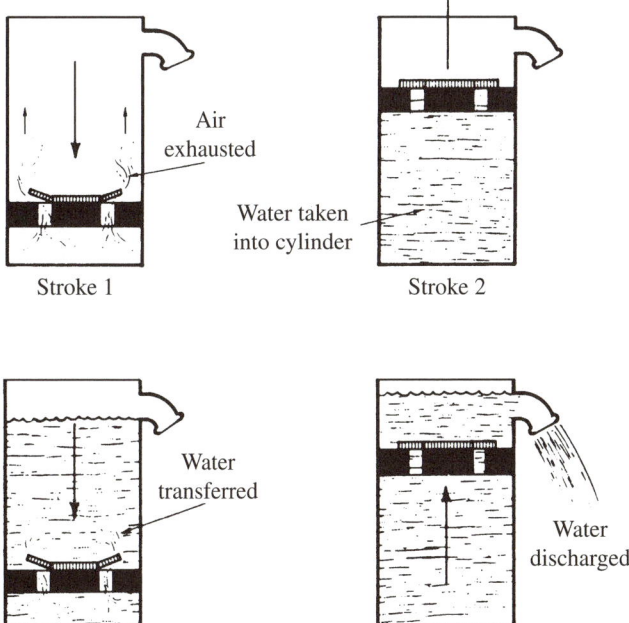

FIGURE 7.4 Diagram of the strokes in a lift pump cycle, from [Audel70]. Used with permission of Howard W. Sams & Co.

Most pumps are driven by an electric motor. If the rotational speed, ω, of the pump-motor shaft is known, and the torque, T, applied by the motor is known as well, then the power input is

$$\dot{W}_{input} = T\omega$$

Note that many references prefer to refer to the pump head, h_{pump}, instead of the pressure increase generated, ΔP. The relation between the two is

$$\rho g h_{pump} = \Delta P$$

7.2.1 Ideal Centrifugal Pump

A *centrifugal pump* (Figure 7.5) is a piece of turbomachinery that uses a rotating impeller to add energy to the fluid. Flow enters axially from one end along the pump axis, and leaves the impeller flowing radially outward into the volute chamber, where it is collected and funneled into the downstream piping. A centrifugal pump is also a variable displacement pump, in that the actual flow rate achieved depends on the total dynamic head the pump must work against.

A designer can alter the flow capacity of a centrifugal pump by changing the impeller diameter, the design of the overall shape of the pump housing and its components, and changing the pump speed. For inflow of flow rate Q along the axis with no angular momentum, and circumferential outflow through an exit of width w, if the

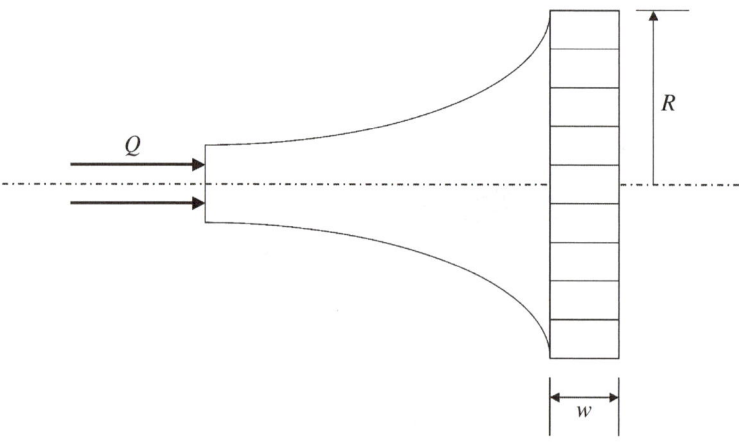

FIGURE 7.5 Sketch of an ideal centrifugal pump.

outlet radius is R, then the outlet area is $2\pi Rw$, and the average outflow velocity (normal to the outlet area) is

$$V_n = \frac{Q}{2\pi Rw} \tag{7.1}$$

If the angle of the blades at the exit of the pump is θ, and we assume that the flow stays attached to the blades as it exits, then the circumferential outlet velocity relative to the rotor is $V_{\theta,\text{rel}} = V_n \tan(\theta)$. But the rotor is moving with a speed of ωR at its periphery, so the velocity seen by a stationary observer is

$$V_\theta = -V_n \tan(\theta) + \omega R \tag{7.2}$$

The angular momentum balance for the system is that the angular momentum generated by the fluid leaving the pump must be balanced by the torque, T, on the pump:

$$T = \dot{m}(\vec{R} \times \vec{V}) = \rho Q R V_\theta = \rho Q R(-V_n \tan(\theta) + \omega R)$$

Substituting in the value for V_n from Equation 7.1 gives an expression for T in terms of the flow rate, rotational speed, and geometry of the pump, as specified by R, ω, and θ:

$$T = \rho Q R \left(-\frac{Q}{2\pi Rw} \tan(\theta) + \omega R \right) \tag{7.3}$$

The ideal power to run the pump is the product of the torque and the rotational speed:

$$\dot{W} = T\omega = \rho Q R \omega \left(-\frac{Q}{2\pi Rw} \tan(\theta) + \omega R \right) \tag{7.4}$$

Euler's pump/turbine formula is given in Equation 7.5, where the subscript t denotes tangential velocity, and it is a more general formula for the ideal torque from a piece of turbomachinery. Equations 7.3 and 7.5 are equivalent under the assumption of negligible rotational motion of the incoming flow and applying the definition of the outflow angle, θ.

$$T = \dot{m}(r_2 V_{2t} - r_1 V_{1t}) \tag{7.5}$$

Based on Euler's turbine formula for torque, a general expression for pump efficiency can be derived:

$$\eta_{\text{pump}} = \frac{\rho Q (U_2 V_2 - U_1 V_1)}{T\omega} \tag{7.6}$$

where U is radial velocity and V is circumferential velocity.

A centrifugal pump is like a turbine operating in reverse (or vice versa). The pump power required (or the turbine power output) is proportional to flow velocity (Q/A) cubed, and linearly proportional to rotational speed. The performance of the centrifugal pump depends on the flow rate Q, the pressure rise ΔP, the rotational speed ω, and the diameter D. Most references use the pump head, h_{pump}, instead of the pressure drop. The two are related simply by the specific weight of the fluid.

$$h_{\text{pump}} = \frac{\Delta P}{\rho g} \tag{7.7}$$

The specific speed of a pump is commonly written as follows (this is not dimensionally homogeneous):

$$N_s = \frac{\omega \sqrt{Q}}{h_{\text{pump}}^{-3/4}} \tag{7.8}$$

Note that this is different from the definition of specific speed for a fan, where the pressure rise/fan head is usually small. While not as commonly used in practice, a proper nondimensional form of the specific speed is

$$N_s = \frac{\omega \sqrt{Q}}{(gh_{\text{pump}})^{-3/4}} \tag{7.9}$$

This form of the specific speed has the advantage that you do not have to remember what units are used.

The efficiency of a pump can usually be well characterized by the specific speed and the flow rate. The maximum pressure rise, or head, is achieved at zero flow rate, and the pressure rise decreases with increasing flow rate until it reaches zero at the maximum flow rate of the pump. That is, as the pressure the pump is pushing against is decreased, more flow will result. From dimensional analysis, it can be shown that the flow rate is linearly proportional to the angular speed of the pump, and the pump head, or pressure rise, is proportional to the square of the rotational speed. If the efficiency of the pump does not change much, the power is proportional to the rotational speed cubed.

The *critical speed* of a pump is the speed at which vibrations, which increase with increasing operating speed, become so severe that there is a risk of damage to the pump. A general rule of thumb is that this critical speed should be 20% higher than rated speed [Wahren97]. Causes of pump vibrations include misalignment, collection of foreign material nonuniformly inside the pump on the impeller, the pump being not rigidly supported, and mechanical defects [Audel70]. Figure 7.6 shows a cross section of a centrifugal pump with the mechanical parts marked.

7.2 Pumps, Compressors, Fans, and Propellers 341

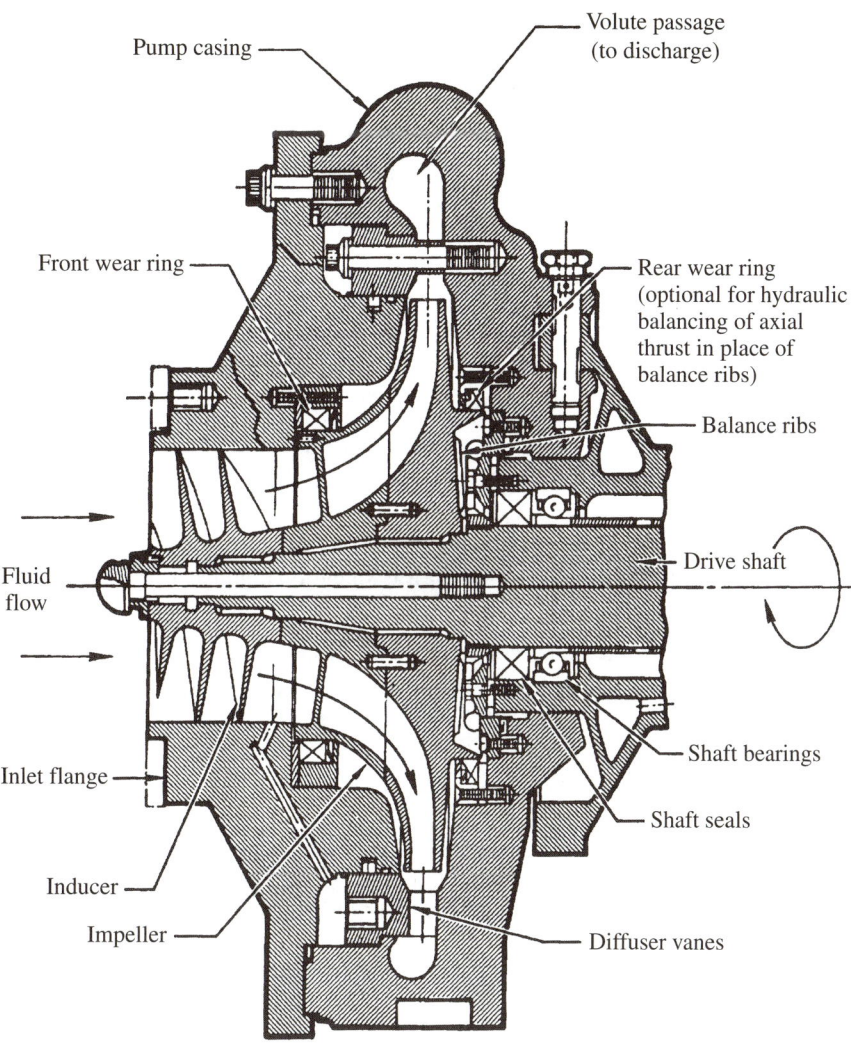

FIGURE 7.6 Cross-sectional diagram of a centrifugal pump, from [NASA73].

In the pump industry, the viscosity of the working fluid is cited at the operating temperature of the pump, and not at the ambient or initial temperature. Common reference temperatures used are 100°F (38°C) and 140°F (60°C) [Wahren97]. As the temperature increases, the viscosity of a liquid decreases, as seen in Figure A.1 in Appendix A.

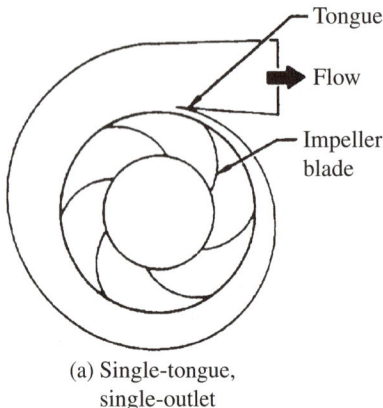

(a) Single-tongue, single-outlet

FIGURE 7.7 Diagram of a centrifgual pump volute, from [NASA73].

The *volute*, shown in Figure 7.7, is the spiral shape of the curve of the housing around the impeller blades from the center of the pump to its periphery. Centrifugal pumps need to be primed before they are operated to make sure they are full of liquid when started. Otherwise the internal parts of the pump could be damaged.

7.2.2 Pump Scaling Laws

The similarity rules for pumps can be derived by dimensional analysis, which is covered in detail in Chapter 9. For two or more pumps of the same design having geometric similarity, the following scaling laws hold true for the flow rate

$$Q \approx \omega D^3 \qquad (7.10)$$

the pressure rise produced

$$\Delta P \approx \rho \omega^2 D^2 \qquad (7.11)$$

and the pump power

$$\dot{W} \approx \rho \omega^3 D^5 \qquad (7.12)$$

So for a given pump housing geometry and design, the results of a test at one size can be used to predict the performance at other sizes, either larger or smaller. One caveat is that we assume cavitation does not occur. If cavitation is an issue, then an additional scaling parameter such as the cavitation number should be added. A general trend is that large pumps have higher efficiency (due to higher Reynolds numbers and smaller roughness ratios), as well as greater economic motivation to maximize the efficiency. Another general trend is that the efficiency of a centrifugal pump decreases as the fluid viscosity increases, as when oil is pumped instead of water.

7.2.3 Net Positive Suction Head (NPSH) and Cavitation

The maximum possible net suction head of a pump is

$$\text{NPSH} = h_{max} = \frac{P - P_{vap}}{\rho g}$$

For water at 1 atm pressure, this head, expressed in units of length, is

$$h_{max} = \frac{101{,}325 \text{ Pa} - 2{,}315 \text{ Pa}}{(1{,}000 \text{ kg/m}^3)(9.8 \text{ m/s}^2)} = 10.1 \text{ m}$$

So if the inlet pipe to a pump had water coming in at atmospheric pressure, the pump could be no more than 10.1 m above the inlet, or else the water would vaporize. Of course in practice the distance must be lower than this to account for losses and pump inefficiencies, and to provide a factor of safety. The pump also needs to be primed before being started, to ensure that the impellers are completely immersed. The pump when running can create partial vacuum in the intake so that the fluid, usually water, can be at pressures less than atmospheric. However, cavitation does not occur exactly at the vapor pressure, but will occur at higher pressures due to impurities in the liquid—including dissolved air—and the dynamic changes in pressure around the impeller blades. The pressure required to prevent cavitation increases as the pump speed and flow rate increase.

NPSHA is the net positive suction head available, which is calculated as follows:

$$\text{NPSHA} = \frac{P}{\rho g} - \frac{P_{vap}}{\rho g} - h_f \pm \Delta z \qquad (7.13)$$

where P is the local atmospheric pressure, and Δz is the static suction. If the liquid surface level is below the pump, Δz has a negative value. P_{vap} is the liquid vapor pressure, and h_f is the friction loss, expressed as a head in units of length. NPSHA must exceed NPSHR to allow pump operation without cavitation. NPSHR is the net positive suction head required by the pump, and can be obtained from the pump performance curve for a given pump (refer to Figure 7.8 for an example). It is recommended to allow some margin of error between NPSHA and NPSHR. For many applications, a few feet or a meter or so will do. It is also important to remember that water will boil at temperatures less than 100°C if the pressure is less than standard atmospheric pressure, so the temperature of the water is also important.

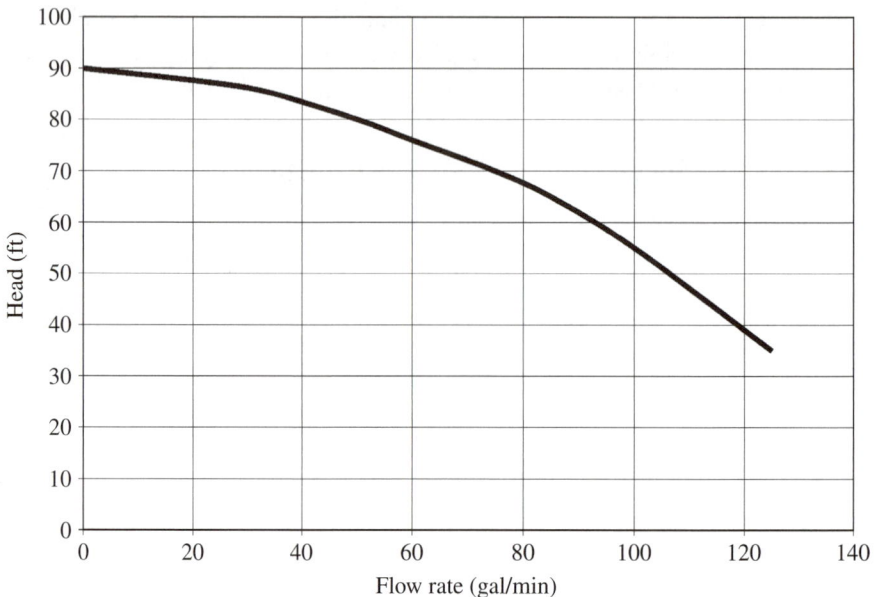

FIGURE 7.8 Generic pump performance curve for a 2-hp swimming pool pump.

Cavitation occurs when the local pressure falls below the thermodynamic vapor pressure and gas bubbles are formed. Cavitation can result in oscillations in machinery performance and in damage to the blades if the bubbles collapse. In many pumps, the pressure at the inlet may be subatmospheric (negative gage pressure). The net positive suction head (NPSH) is the difference between the absolute stagnation pressure in the fluid and the liquid vapor pressure. As stated above, the net positive suction head available (NPSHA) should be greater than the net positive suction head required (NPSHR) to avoid cavitation. NPSHR comes from measured data. A correlation given by Wislicenus [Wislicenus65] is that a pump is in danger of cavitation when

$$N_s = \frac{\omega \sqrt{Q}}{(g \times \text{NPSH})^{3/4}} \geq 2.98 \tag{7.14}$$

7.2.4 Pump Performance Curves

For a given pump operational speed and pipe size, the relationship between the flow rate provided by the pump and the pressure or head it has to work against is determined from experimental data, provided by the manufacturer as a pump performance curve. The obvious trend is that the lower the pressure the pump has to push against, the higher the flow rate it provides.

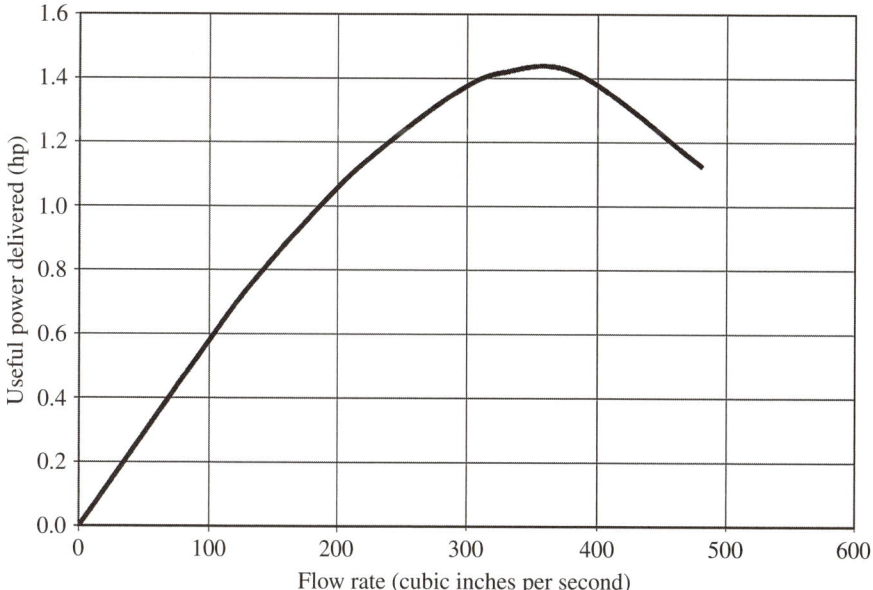

I FIGURE 7.9 Flow power curve calculated from the pump curve data in Figure 7.8.

Figure 7.8 shows a generic pump curve that would be representative of a 2-hp pump used for personal swimming pools, and Figure 7.9 shows the power delivered by the same pump as a function of flow rate. The actual flow rate this pump will provide in a given installation is determined by generating a system pressure curve, as will be illustrated in the next example. (See Figure 7.10.) For a fixed piping system, the head loss will increase with the flow rate squared, as the friction loss increases with increasing flow rate. If the pump has been properly sized, there will be a point where the pump curve and the system curve cross, which will be the operating point of the system.

EXAMPLE 7.1

If the swimming pool pump of Figure 7.8 has to pump water through 40 ft of 1.5-in steel pipe, and through elbows and a filter with a total cumulative resistance of $K = 2.0$, what is the flow rate of water through the system?

SOLUTION The total resistance, or the total head loss, will increase with flow rate in a nonlinear fashion, so there is no explicit solution for this problem. The head loss as a function of flow rate for this system has been calculated in Excel

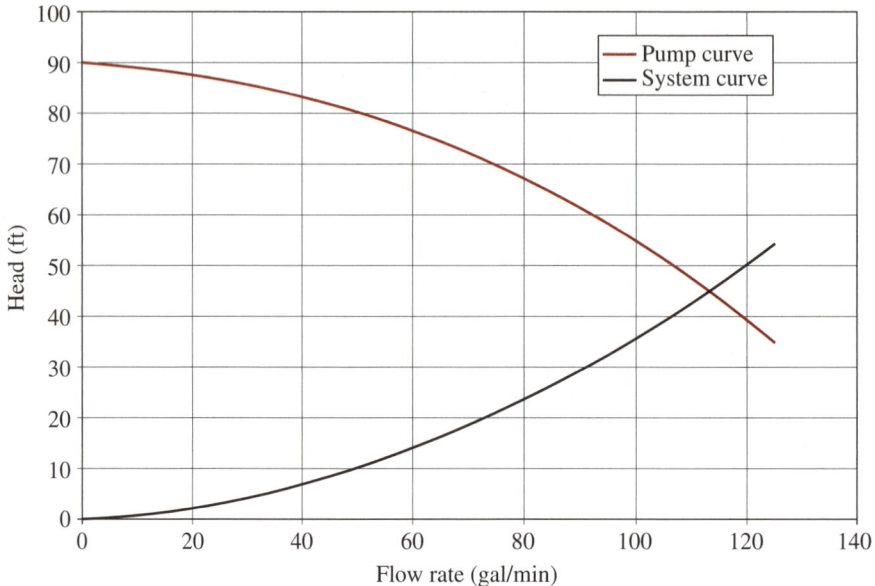

FIGURE 7.10 Pump performance curve of Figure 7.8 shown along with the system curve for Example 7.1.

and plotted along with the pump performance curve in Figure 7.10. The point where the two lines cross will be the operating point of the system for the given conditions. For this problem the flow rate will be about 110 gal/min.

Most manufacturers provide more information on their pump curves than a single performance curve at constant speed. Figure 7.11 shows a graph from a manufacturer for a specific model that shows performance curves for three different sizes of piping, along with the efficiency of the pump at different points and the NPSH of the pump.

EXAMPLE 7.2

For the Split Flow™ pump performance curve shown in Figure 7.11, what is the maximum flow rate of water that could be obtained when working against a pressure head of 125 psi? What is the NPSH and the efficiency of the pump at this point?

SOLUTION Since the pump curve is given in units of feet, the pressure must be converted to head:

$$h_{\text{pump}} = \frac{\Delta P}{\rho g} = \frac{125 \text{ lbf/in}^2}{62.4 \text{ lbf/ft}^3} \times \frac{144 \text{ in}^2}{1 \text{ ft}^2} = 288 \text{ ft}$$

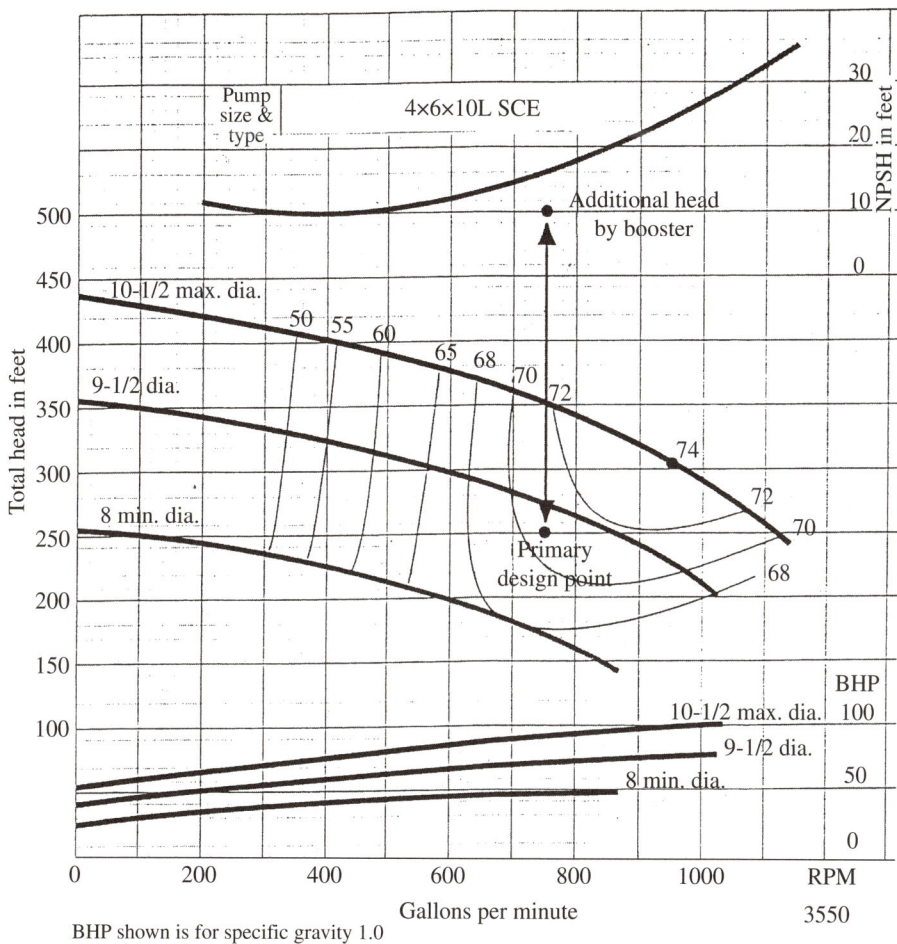

FIGURE 7.11 Pump performance curves. Courtesy of Split Flow™ Pumps.

For a given total head the pump is working against, the maximum flow rate will be achieved with the maximum pipe size that can be used with the pump, which for this pump is $D = 10.5$ in. Reading across the curve, the corresponding flow rate is 1,000 gal/min. Reading off the curve at the top, for a flow rate of 1000 gal/min, the NPSH is about 27 ft. This is the maximum amount of suction head the pump can draw before cavitating. From the curve at the bottom of Figure 7.11, the power required by the pump at 1,000 gal/min using 10.5-in pipe is close to 100 hp.

To find the efficiency, we must calculate how much of that power is converted to useful energy. The flow work of the pump is

$$\dot{W} = Q\,\Delta P = \left(1{,}000\,\frac{\text{gal}}{\text{min}}\right)\left(125\,\frac{\text{lbf}}{\text{in}^2}\right) \times \frac{144\,\text{in}^2}{1\,\text{ft}^2}\,\frac{1\,\text{min}}{60\,\text{s}}\,\frac{0.1337\,\text{ft}^3}{1\,\text{gal}} = 40{,}100\,\frac{\text{ft}\cdot\text{lbf}}{\text{s}}$$

The conversion factor to hp is 1 hp = 550 ft-lbf/s, so the useful power output of the pump is 40,100/550 = 72.9 hp. We then calculate the efficiency as

$$\eta = \frac{72.9\,\text{hp}}{100\,\text{hp}} = 72.9\%$$

Note that this value falls between the markings for 72% and 74% indicated on the pump curve.

7.2.5 Fans

Fans are used to move gases, most often air, with very small pressure rises when compared to pumps. Impellers in leaf blowers and vacuums and other devices come in various shapes, and some have the blades curved forward, while others have the blades curved backward. The scaling in fans (see Chapter 9 on dimensional analysis) is as follows:

$$Q = Q_0 \frac{\omega}{\omega_0}\left(\frac{D}{D_0}\right)^3 \tag{7.15}$$

$$\dot{W} = \dot{W}_0 \left(\frac{\rho}{\rho_0}\right)\left(\frac{\omega}{\omega_0}\right)^3\left(\frac{D}{D_0}\right)^5 \tag{7.16}$$

If, as is often the case, two fans operate with air at standard conditions, then the density ratio is 1.0. The nondimensional *specific speed* of a fan, N_s, is defined as

$$N_s = \frac{\omega Q^{1/2} \rho^{3/4}}{\dot{W}^{3/4}} \tag{7.17}$$

Compressors and fans are similar in design and operation, except that there is normally a significant pressure rise associated with compressors, and often the goal of a compressor is to increase the density of the gas. Automotive turbochargers, which are discussed later, are centrifugal flow compressors, while the compressors used in the large jet engines that propel large passenger planes are axial flow compressors.

7.2.6 Propellers

Propellers are used for both air and sea/maritime propulsion. Most prop-driven aircraft have variable-pitch propellers. The typical range is between 10° to 25° pitch for light aircraft, and 10° to 50° for larger, faster aircraft. Detailed blade theory is quite complex and will not be covered here. CFD or measurements are normally used in the analysis of a specific design. General trends may be obtained from an overall control volume analysis.

In the control volume shown in Figure 7.12, air at velocity V approaches the propeller, where it is thrust backward with an increase in velocity ΔV. The streamlines are drawn parallel to the airflow, so that no mass crosses the streamlines. By applying conservation of momentum to the control volume, we can calculate the thrust force, F, on the propeller as

$$F = \dot{m}\,\Delta V \tag{7.18}$$

The rate at which work is done by the propeller is the force times the velocity of the air, so that the useful power output of the propeller is

$$\dot{W}_{\text{output}} = FV = (\dot{m}\,\Delta V)V \tag{7.19}$$

This is also called the *propulsive power* of the propeller.

If conservation of energy is applied to the control volume, we can find the maximum amount of power that can be extracted from a fluid going through a velocity

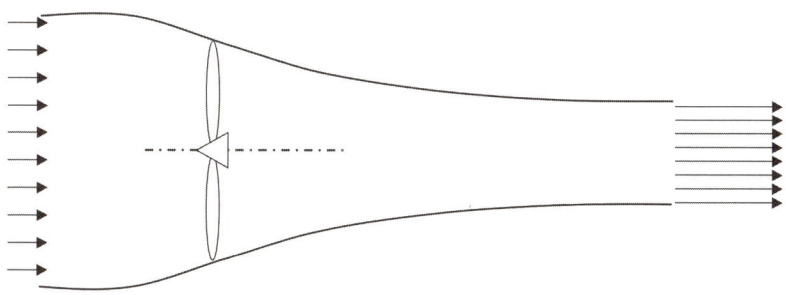

FIGURE 7.12 Control volume sketch for propeller analysis.

change ΔV. Assuming no heat transfer or friction losses or changes in potential energy, the steady-state conservation of mass equation will be

$$\dot{W}_{theo} = \dot{m}\left(\frac{V_{out}^2}{2} - \frac{V_{in}^2}{2}\right) = \dot{m}\frac{((V+\Delta V)^2 - V^2)}{2} = \dot{m}\left(\frac{2V\Delta V + \Delta V^2}{2}\right) \quad (7.20)$$

Thus the efficiency of the propeller can also be defined as the ratio of the power of the propeller divided by the change in kinetic energy of the fluid:

$$\eta = \frac{\dot{W}_{output}}{\dot{W}_{input}} = \frac{\dot{m}V\Delta V}{\dot{m}(V\Delta V + \frac{\Delta V^2}{2})} = \frac{1}{1 + \frac{\Delta V}{2V}} \quad (7.21)$$

So, qualitatively, a propeller that moves more mass, but moves it through a smaller velocity change, will be more efficient. Increasing the air speed velocity also increases the efficiency, but practical considerations—such as the onset of compressible flow effects for aircraft propellers and cavitation for marine propellers—limit how fast a propeller can spin. A brief discussion of the relative merits of propeller and jet propulsion is included in Chapter 8.

Some of the important scaling parameters are listed here. The *blade advance ratio*, J, is ratio of the air speed to the speed at the tips of the propeller blades:

$$J = \frac{V}{\omega D} \quad (7.22)$$

Propeller efficiency is the ratio of the propulsion power produced divided by the mechanical power input to the propeller from the engine:

$$\eta = \frac{FV}{\omega T} \quad (7.23)$$

The power input to the propeller is the product of the engine output torque, T, and the rotational speed at which it spins, ω. Aircraft propeller speeds are limited by the need to keep the flow at the blade tips from becoming supersonic. In general, $\eta = \eta(J)$.

In marine propellers, cavitation must also be accounted for. The nondimensional *cavitation number* is

$$Ca = \frac{P - P_{vap}}{\frac{1}{2}\rho V^2} \quad (7.24)$$

Cavitation was discussed in the previous section.

7.3 Turbines

7.3.1 Francis, Pelton, and Kaplan Turbines

Hydraulic turbines, also called hydraulic motors, use liquids instead of gases to generate power. To increase the efficiency, the liquid should be discharged at as low of a speed and pressure as possible. In many applications this low pressure will be the ambient atmospheric pressure. The three main types of hydraulic turbines used in hydroelectric power plants are reaction turbines (Francis), impulse turbines (Pelton water wheel), and propeller turbines (Kaplan).

Francis Turbine

In a Francis turbine (see Figure 7.13) the flow comes in tangentially and radially and spins the blades around, and then exits axially from the bottom center of the turbine. Figure 7.14 shows a center plane cutaway schematic of a Francis turbine. Francis turbines are commonly used in applications with high flow rates of water.

Pelton Turbine

As discussed in Chapter 3, the power generated by a Pelton water wheel is given by

$$\dot{W} = \rho \omega R Q (V_{jet} - \omega R)(1 - \cos\beta) \qquad (3.38)$$

where V_{jet} is the velocity of the jet striking the wheel, and thus $(V_{jet} - \omega R)$ is the relative velocity between the water jet and the wheel. The power calculated by Equation 3.38 is plotted as a function of rotational speed in Figure 7.15. Impulse turbines, such as

FIGURE 7.13 Photo of a Francis turbine. Courtesy of Voith Siemens.

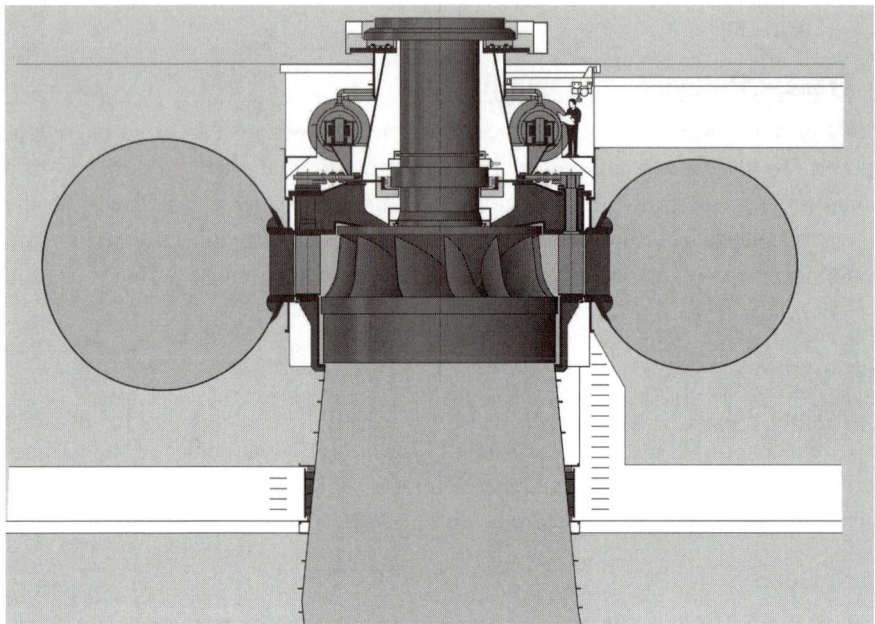

FIGURE 7.14 Schematic of a Francis turbine. Courtesy of Voith Siemens.

the Pelton turbine, are generally only used in applications with relatively high head (pressure) but low flow rate. Reaction turbines are used for other cases. The maximum speed of an ideal Pelton water wheel was found in Example 3.19 (the actual optimum speed is slightly slower due to nozzle losses and friction). Assuming that the rotational speed of the water wheel can be varied by the designer, if not the user, the maximum power can be found by taking the derivative of the power (Equation 3.38) with respect to the rotational speed and setting it equal to zero. The derivative is

$$\frac{d\dot{W}}{d\omega} = \rho R Q (1 - \cos\beta)(V_{\text{jet}} - 2\omega R)$$

Setting the derivative equal to zero and solving for ω yields

$$\omega_{\text{MaxPower}} = \frac{V_{\text{jet}}}{2R}$$

Now we substitute this value for ω back into the equation for power to find the maximum power:

$$\dot{W}_{\max} = \rho\omega RQ\left[V_{\text{jet}} - \left(\frac{V_{\text{jet}}}{2R}\right)R\right](1 - \cos\beta) = \frac{1}{2}\rho\omega RQV_{\text{jet}}(1 - \cos\beta) \quad (7.25)$$

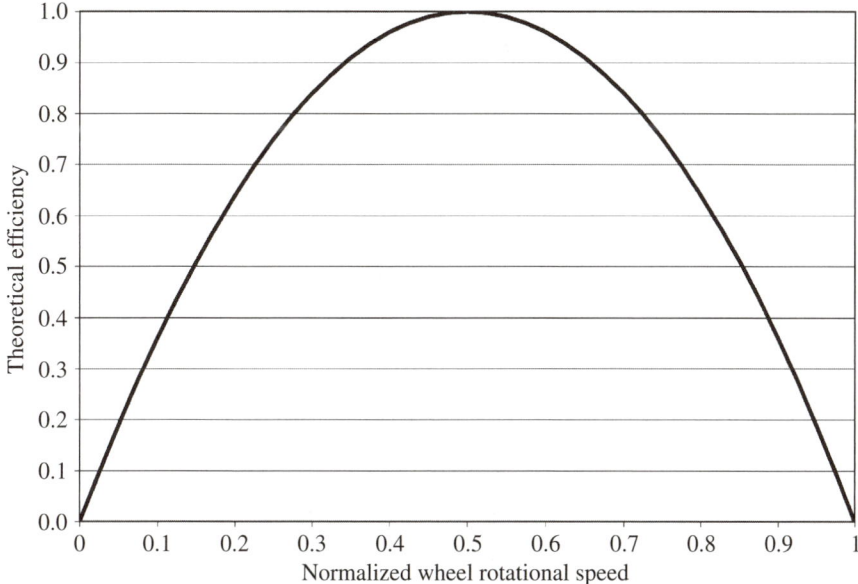

FIGURE 7.15 Plot of the efficiency of an ideal Pelton water wheel versus the nondimensional wheel rotational speed ($\omega R/V_{jet}$) of the water wheel.

So the maximum power occurs at half the maximum rotational speed for an ideal Pelton turbine. Figure 7.16 shows a picture of a typical Pelton turbine.

Kaplan Turbine

The Kaplan turbine (Figures 7.17 and 7.18) was developed by Victor Kaplan in the early 1900s. The unique feature of Kaplan turbines is that the blades have adjustable pitch. Because of this, the turbine is efficient over a wide range of loads/flow rates. A Kaplan turbine is somewhat similar in appearance to an airplane propeller. Typical efficiencies are around 90%. Kaplan turbines are very expensive and typically used only in large installations.

7.3.2 Centrifugal Turbines

For an ideal centrifugal turbine the maximum possible rotational speed when no torque is applied (called the *free-wheeling speed*) is

$$\omega_{max} = \frac{Q \tan(\theta)}{2\pi R^2 w} \tag{7.26}$$

FIGURE 7.16 Photo of a Pelton turbine. Courtesy of Voith Siemens.

The derivation of the equations for the ideal centrifugal turbine is the same as that for the ideal centrifugal pump presented earlier, except that the device extracts energy from the liquid rather than adding energy to it. The ideal turbine efficiency can be calculated as

$$\eta_{\text{pump}} = \frac{T\omega}{\rho Q(U_2 V_2 - U_1 V_1)} \tag{7.27}$$

FIGURE 7.17 Photo of a Kaplan turbine. Courtesy of Voith Siemens.

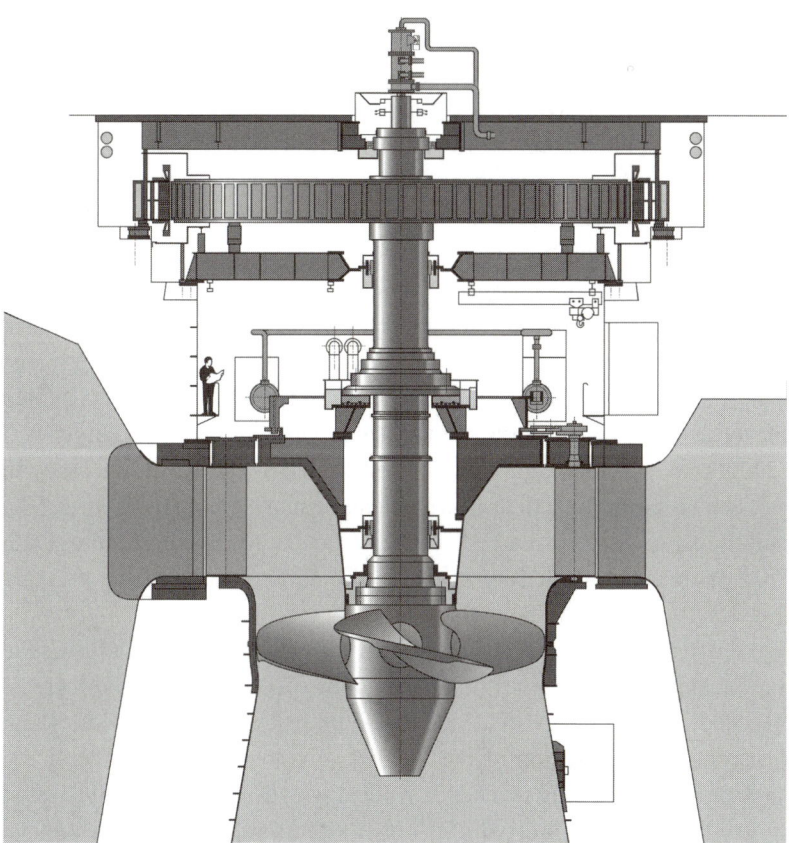

FIGURE 7.18 Schematic of a Kaplan turbine. Courtesy of Voith Siemens.

7.3.3 Applications

Two applications of small turbines are automotive turbochargers and torque converters. Both are connected to energy-extracting turbomachinery. Turbochargers are connected to an air compressor, while torque converters have a pump.

Turbochargers

In naturally aspirated engines, air at local atmospheric conditions is sucked into the engine. In gasoline engines, the density of the intake air is usually below local atmospheric pressure due to pressure losses over the throttle valve. The amount of power of an engine is proportional to the amount of fuel in the engine, and the amount of fuel that can be injected is limited by the amount of air in the engine cylinders (called the *stoichiometric condition* when there is just enough air to completely burn all the fuel). Simply making the cylinders larger (or adding more cylinders) means making the engine block larger, which greatly increases the size and weight of the engine. However, a turbocharger can increase the peak power of an engine without adding as much weight and volume as adding extra cylinders would.

Turbochargers and superchargers are forced induction systems. They compress the air coming in through the intake manifold before it enters the engine, raising the air's pressure and density so that a greater mass of air may enter the engine. Hence more fuel can be burned and a higher peak power obtained compared to an engine of the same displacement with no turbochargers. Turbochargers also increase the thermodynamic efficiency of an engine because they extract waste energy from the exhaust that would otherwise be expelled to the environment as heat. The energy of the exhaust is extracted by means of a turbine at the end of the exhaust manifold, which is coupled by a direct shaft to the compressor in the intake. An engine with a turbocharger will typically have a higher power-to-weight ratio than an engine of similar design but with no turbocharger. The speed of the exhaust turbine can be as fast as 150,000 rpm. The typical maximum boost pressure of an automotive turbocharger is around 6 to 8 psi, for roughly a 50% increase in air density and potential power, but the boost pressure in those used for diesel engines can be higher.

Some design challenges of turbochargers are that the exhaust turbine increases the backpressure on the engine, the turbine shaft needs high-speed bearings, and increasing the pressure inside the engine cylinder increases the risk of the engine knocking. *Knocking* is the premature detonation of the fuel–air mixture in the engine cylinder before the flame from the spark plug reaches it. Knocking causes damage to the piston and should be avoided. Using a fuel with a higher octane rating can help prevent knock. Reducing the compression ratio in a turbocharged engine relative to a naturally aspirated engine of similar design can also avoid knock.

Superchargers

Superchargers, like turbochargers, are used to increase the mass flow rate of air into an engine and hence increase the peak power. While the compressor in a turbocharger is driven by a turbine in the exhaust, the compressor in a supercharger is driven off the engine drive shaft, through a chain or belt, like other engine accessories. So adding a turbocharger increases the thermodynamic efficiency of an engine, but adding a supercharger does not. If that is the case, why would an engineer choose to use a supercharger rather than a turbocharger? The main reason is turbocharger lag: There is sometimes a significant delay between when the accelerator is pressed and the turbocharger becoming fully active, because it takes some time for the exhaust gas to reach sufficient velocity to increase the intake air compression. Superchargers respond much more quickly because they are driven directly off the drive shaft. There are also fewer moving parts with superchargers, so they are easier to manufacture and less maintenance is required.

Superchargers typically spin around 50,000 rpm and give pressure boosts of around a 50% increase in air flow rate. In both super- and turbochargers, the compressor causes the air to heat up, which reduces its density, so intercoolers are usually used to cool the air and increase the intake air charge density. The air exits the supercharger and goes into the intake manifold, and then to the individual engine cylinders. Superchargers are commonly used on piston-powered airplanes, where performance degrades with increasing altitude due to the reduced ambient air density. There are three types of superchargers that have been in common use: roots type, twin screw, and centrifugal superchargers. The centrifugal type uses an impeller, while the other two use meshing lobes.

The volumetric efficiency of a supercharger can be over 90%, depending on the speed and boost pressure. The benefits of a supercharger outweigh the costs, in that the power lost to run the supercharger will be less than the power gained by having extra air to burn more fuel. However, because the engine has to burn some extra fuel to power the supercharger, although the peak power of the engine is increased, its specific fuel consumption is increased as well. Thus a plane can fly faster and higher, but it will not be able fly as far (or a car can have a higher top speed and get better acceleration, but will get worse mileage).

Torque Converters

In cars and trucks with manual transmissions, the clutch—which is located between the engine and the transmission—allows the vehicle to come to a complete stop without killing the engine and making it go to 0 rpm. In cars with automatic transmission, there is no clutch, but the torque converter performs the same function of allowing the engine to rotate while the wheels are stationary, and it performs other functions as well. The clutch in manual transmissions disconnects the engine from the transmission. In this state, all of the power generated by the engine is dissipated in internal engine friction (or used via the alternator to charge the battery and power accessories). A torque converter translates power from the rotating engine shaft to the engine transmission. Thus, it is a type of fluid coupler. It is used in automatic transmissions in place of the clutch to separate the engine from the rest of the drivetrain.

Inside the torque converter housing, there are three main elements that contain arrays of blades that direct the fluid (see Figure 7.19):

1. Pump. Mechanically driven via a shaft connected to the engine.
2. Turbine. Delivers the load output to the transmission.
3. Stator. Does not rotate, located between the pump and turbine.

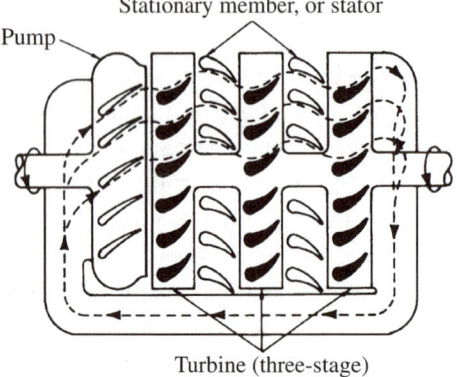

FIGURE 7.19 Schematic of a torque converter.

There are three common modes of torque converter operation:

1. Stall. This is when the pump is spinning but the turbine is not, as when the car has come to a stop and the engine is still spinning at idle, but the wheels are locked in place by the brakes.

2. Acceleration. During times when the vehicle is accelerating, such as getting up to speed coming out of a stop, the engine and pump will be spinning at higher speeds than the transmission and turbine. The torque converter works as a torque multiplier to increase the effective gear ratio to the wheels to provide the torque needed for acceleration. The power (torque times rotational speed) is approximately the same for the pump and turbine (assuming only small losses).

3. Lock Up. When the vehicle reaches steady cruising speed—and the pump and turbine inside the torque converter are spinning at close to the same speed—the mechanical lock-up clutch engages, so that the transmission input shaft spins at the same speed as the engine. The car would still drive essentially the same without the lock-up clutch, but the drivetrain efficiency would be less due to fluid friction losses in the torque converter.

7.4 Wind Turbines

Wind turbines are used to extract energy from wind to generate electric power. Moving wind has kinetic energy that can be captured and converted to mechanical power, and from mechanical power into electrical power, much the same way that the potential energy of water in a high reservoir can be converted into mechanical power in a Pelton, Kaplan, or Francis turbine, and then from this into electrical power in a generator. Ultimately almost all forms of power on Earth, except for nuclear power, are driven by the sun. Photovoltaic and solar power plants take their energy directly from the sun; wind energy is more indirect and is the result of some areas of Earth's surface becoming hotter than others due to solar exposure, generating convection currents and wind. Figure 7.20 shows how the availability of wind power varies in the United States.

Wind turbines for electrical power generation can be grouped into two classes—horizontal-axis wind turbines (HAWTs) (see Figure 7.21) and vertical-axis wind turbines (VAWTs). Most wind turbines in use are horizontal-axis wind turbines. For the vertical-axis wind turbines, the Darrieus wind turbine—invented in France in the 1920s by G.J.M. Darrieus—is the most prominent example. VAWTs do not require yaw control to orient them into the wind, but Darrieus turbines cannot start themselves—they require an external motor to begin operation. In addition to the Darrieus rotor, other types of VAWTs include the Savonius rotor, Madaras rotor, and Flettner rotor. While the Savonius rotor is self-starting, it has low efficiency.

In the early part of the 20th century in the United States, windmills with small power output were commonly used on farms for water pumping and to provide local electricity. Today utility wind turbines commonly produce more than 1 MW of power each. In the United States, most individual turbines are sized 2 MW and smaller.

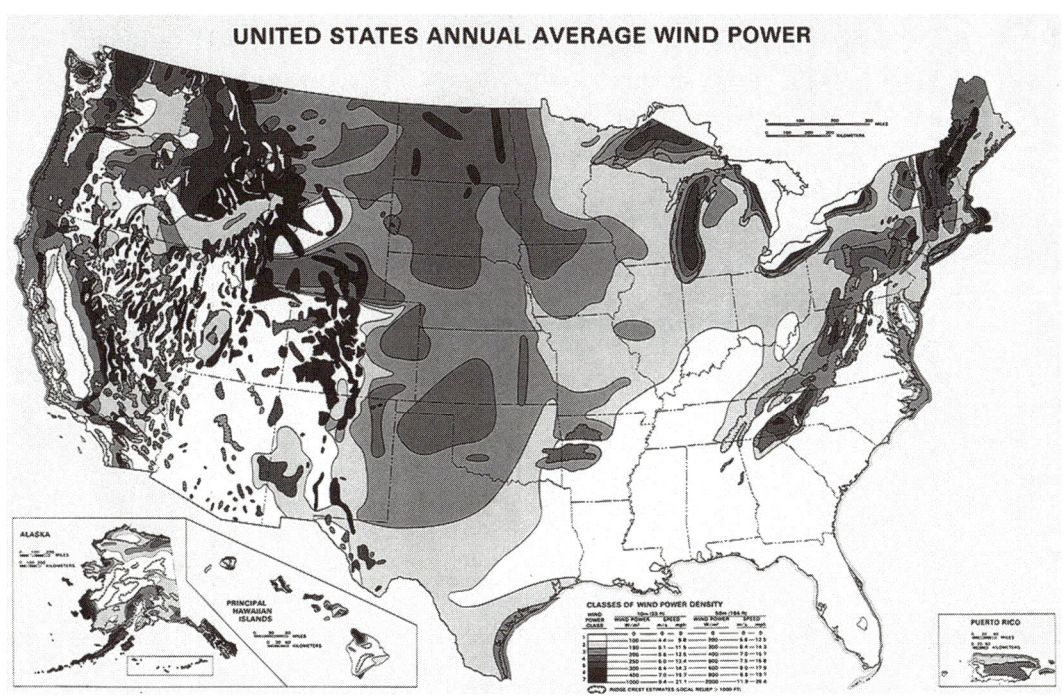

FIGURE 7.20 Wind resource map of the United States. Courtesy of the Department of Energy/National Renewable Energy Lab.

FIGURE 7.21 Photo of horizontal-axis wind turbines. Courtesy of NASA.

Table 7.1 lists several countries and their installed wind power. Note that in the United States, the wind power generation capacity represents a little over 1% of the nationwide electricity consumption. The country with the highest percentage of its power generated by wind is Denmark, with about 20% [DOE07].

Table 7.1 Installed Wind Power by Country

Country	Amount (in units of MW)
Germany	22,277
United States	16,904
Spain	14,714
India	7,845
China	5,875
Denmark	3,088
Italy	2,721
France	2,471
United Kingdom	2,394
Portugal	2,150
Rest of World	13,591
Total	94,030

From [DOE07].

The efficiency of a wind turbine can be defined as

$$\eta = \frac{\dot{W}}{\frac{1}{2}\rho V^3 \pi R^2} \quad (7.28)$$

Some references call the efficiency defined this way the *coefficient of performance*, C_p. This is the fraction of the wind's power that is extracted and converted to mechanical power. For a given wind turbine, this efficiency or coefficient of performance is not constant, but varies with speed (see Figure 7.22). It will also vary with pitch angle if the turbine has a variable pitch angle, which most HAWTs do.

The *ideal Rankine analysis* is a one-dimensional analysis that neglects any pressure or velocity changes in the radial and azimuthal directions. A horizontal-axis wind turbine is essentially a propeller in reverse, so that the analysis of the ideal wind turbine is quite similar to the analysis of an ideal propeller presented earlier.

As the wind approaches the turbine, its velocity steadily decreases as it gets closer and closer to the turbine, while the air pressure steadily rises, so that the total energy of the airstream is conserved via the Bernoulli equation. As the air travels

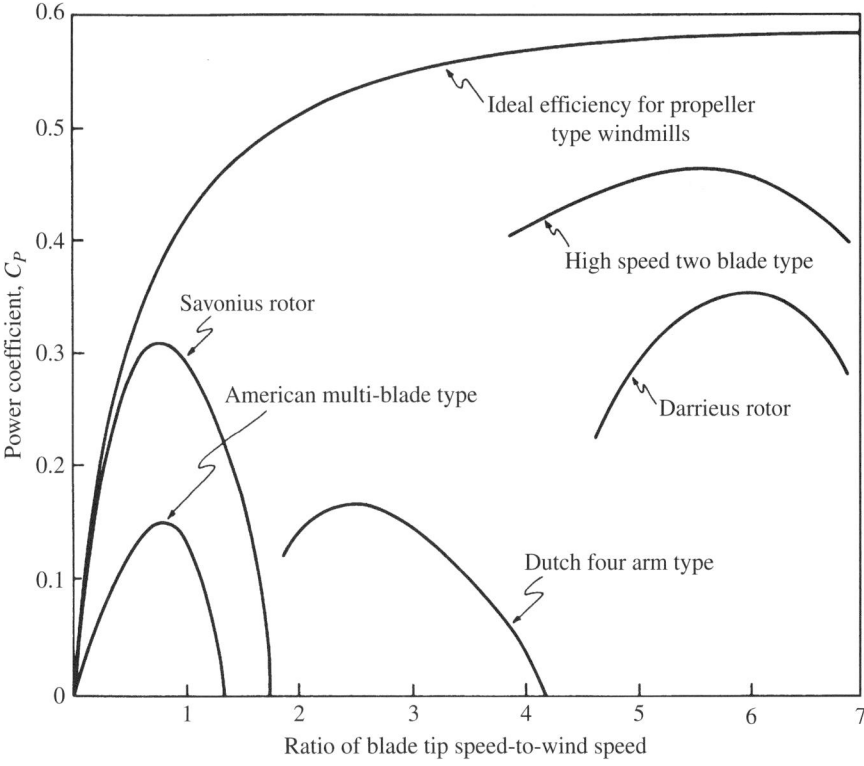

FIGURE 7.22 Efficiency of different wind turbine designs as a function of tip speed ratio, from [NSF75].

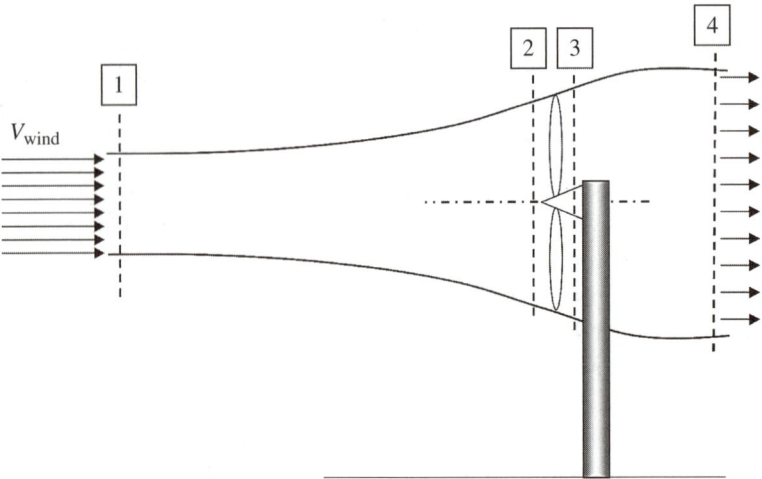

FIGURE 7.23 Sketch of streamlines for ideal Rankine analysis of a wind turbine.

through the turbine itself, it imparts kinetic energy to the turbine, losing its pressure energy in the process. The pressure will actually drop below atmospheric immediately behind the turbine. As the air travels away from the turbine, its pressure will rise back to atmospheric and, consistent with the Bernoulli equation, its velocity will continue to decrease until atmospheric pressure is reached.

For the ideal wind turbine shown in Figure 7.23, the wind approaches the turbine at station 1 at atmospheric pressure and velocity V_1, which is the wind velocity. As shown in Figure 7.24, in the far wake (4) the pressure again returns to atmospheric,

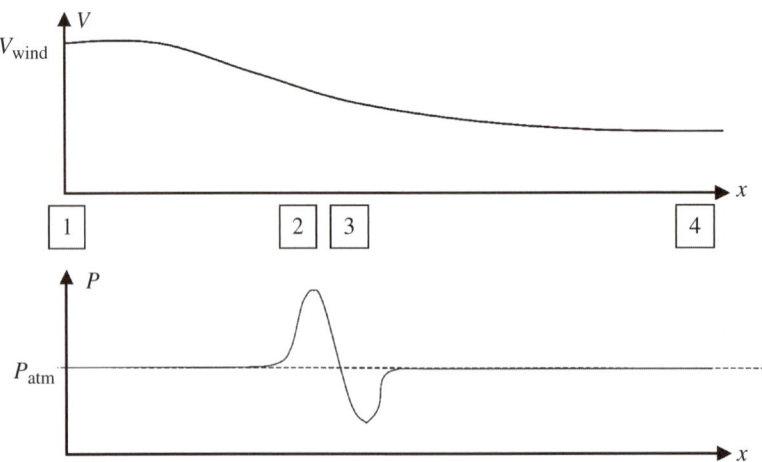

FIGURE 7.24 Trends of velocity and pressure in ideal wind turbine analysis.

but immediately in front of the wind turbine (2) the pressure rises above atmospheric, while immediately behind it (3), it drops below atmospheric. If the streamlines have been drawn correctly, so that no flow occurs across the streamlines, then by conservation of mass we have

$$\dot{m}_1 = \dot{m}_2 = \dot{m}_3 = \dot{m}_4$$

By the Bernoulli equation for the regions outside the turbine,

$$P_{atm} + \frac{1}{2}\rho V_1^2 = P_2 + \frac{1}{2}\rho V_2^2$$

and

$$P_3 + \frac{1}{2}\rho V_3^2 = P_{atm} + \frac{1}{2}\rho V_4^2$$

States 2 and 3 are located immediately in front of and behind the turbine so that $A_2 = A_3$, and then by conservation of mass $V_2 = V_3$. Thus for a control volume drawn around 2 and 3, the force on the turbine is

$$F = (P_2 - P_3)A$$

However, for a large control volume drawn from states 1 to 4, the force on the turbine is

$$F = \dot{m}(V_1 - V_4)$$

By combining the above four equations, we can show that

$$V_2 = \frac{1}{2}(V_1 + V_4) \tag{7.29}$$

Defining a as the *interference factor*, which is the proportion of the oncoming wind that is stopped at the turbine to do work, then by this definition it is clear that the incoming air flow is decelerated from the free-stream velocity of V_1 to a velocity of $V_2 = V_1(1 - a)$ at the plane of the turbine blades and, by combining with the previous equation, $V_4 = V_1(1 - 2a)$. The fluid force on the blades is

$$F = \dot{m}(V_1 - V_4) = \left(\rho \frac{\pi}{4}D^2 V_2\right)(V_1 - V_4) = \rho \frac{\pi}{2}D^2 V_1^2 a(1 - a) \tag{7.30}$$

Thus the power extracted from the wind can be written as

$$\dot{W} = FV = \rho \frac{\pi}{2} D^2 V^3 a (1-a)^2 \tag{7.31}$$

which is the product of the thrust force times the wind velocity. The power output can be maximized by taking the derivative $dW/da = 0$. This is obtained when $a = 1/3$. This maximum theoretical efficiency of a wind turbine obtained when $a = 1/3$ is $\eta = 0.593$, or 59.3%, which is known as the *Betz limit*. Figure 7.25 plots this theoretical efficiency.

The *tip speed ratio*, X, is defined as

$$X = \frac{\omega R}{V} \tag{7.32}$$

As was shown in Figure 7.22, generally the higher the tip speed ratio, the higher the efficiency, though the optimum tip speed ratio will vary for specific designs. At very high tip speeds, vibrations, centrifugal loading, and noise become concerns.

Solidity, σ, is the ratio of the blade area to the area swept out by the blades as they spin. There is a trade-off here. A turbine with low solidity, or low total blade area, will have a high flow rate of air going through, but a low pressure drop. A turbine with high solidity will generate a high pressure drop, but at the cost of a lower flow rate. If the rotational speed of the turbine is too low compared to the wind speed, the blade will stall, with a resultant loss of power.

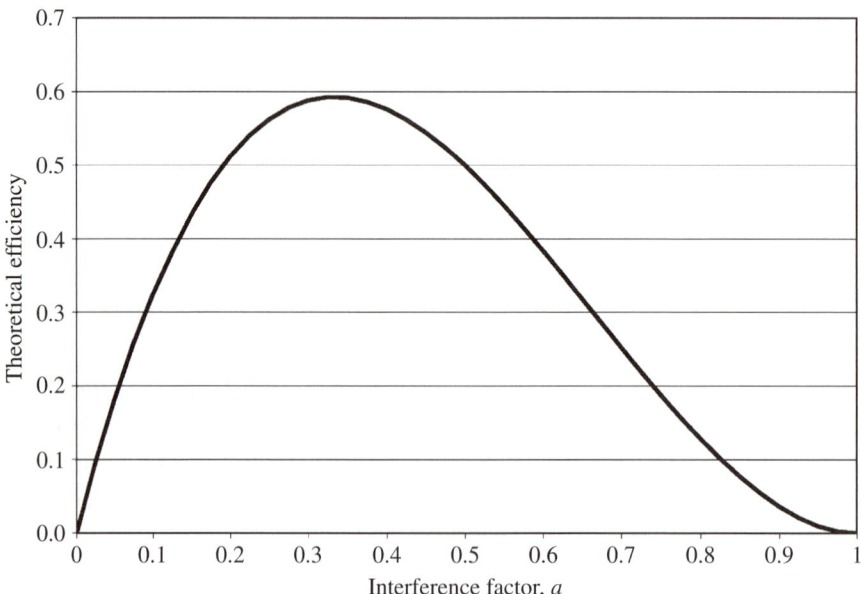

FIGURE 7.25 Plot of the theoretical efficiency of a wind turbine as a function of the interference factor, *a*.

EXAMPLE 7.3

Consider a wind turbine of diameter 20 m operating in a wind of 10 m/s at sea level in standard atmospheric conditions. What is the maximum power that the wind turbine could produce?

SOLUTION If the wind turbine could extract all of the energy from the wind with 100% efficiency, then its power output would be equal to the initial power of the air going through a disk of the same diameter as the turbine. Thus,

$$\dot{W} = \frac{1}{2}\rho A V^3 = \frac{1}{2}\rho\left(\frac{\pi}{4}D^2\right)V^3$$

Substituting in the numerical values yields

$$\dot{W} = 0.5\left(1.2 \frac{\text{kg}}{\text{m}^3}\right)\left[\frac{\pi}{4}(20 \text{ m})^2\right]\left(10 \frac{\text{m}}{\text{s}}\right)^3 = 235{,}600 \text{ W}$$

Of course the turbine cannot extract this much power from the wind for several reasons. They include three-dimensional flow effects (including induced drag at the propeller tips), friction and pressure losses along the blades, and internal friction in the turbine-generator assembly. Applying the Betz limit for efficiency of 59.3%, the maximum power that could theoretically be obtained is (0.593)(235,600 W) = 139,700 W.

One of the earliest research wind turbines constructed in the United States was the NASA MOD-1 wind turbine (Figure 7.26), installed on Howard's Knob near Boone, NC, in 1979. It was rated for 2 MW, had rotor diameter of 61 m, and a rated rotational speed of 34.7 rpm [NASA79]. With the rotor diameter of 61 m, the blade-swept area was 2,820 m². The electrical generator was an AC synchronous generator operating at 1,800 rpm. The site was at an elevation of 1,347 m, and the tower height was 43 m, so the hub was at an elevation of 1,390 m, where the density is about 14% less than sea level. The turbine was designed to achieve rated power at speeds above 16 m/s. When the wind was above 20 m/s, it shut down to avoid overstressing the blades. The total mass of the MOD-1, excluding foundation, was 317,000 kg (700,000 lbf); the breakdown of weight distribution is given in Table 7.2.

Mean wind speeds are in the range of 3 to 7 m/s for much of the United States [Johnson85]. A turbine's height above ground is important because wind speed

FIGURE 7.26 Photo of the NASA MOD-1 wind turbine. Courtesy of NASA Glenn Research Center; photo by Martin Brown.

generally increases with increasing height (think of a big boundary layer). A general correlation for wind velocity as a function of height above ground is given by Patel as

$$V = V_{ref}\left(\frac{h}{h_{ref}}\right)^\alpha \quad (7.33)$$

where V_{ref} is the velocity measured at a reference height h_{ref}. The exponent α depends on the type of terrain, has a value of $\alpha = 0.10$ for smooth surfaces, such as a lake or

Table 7.2 Weight Distribution in the NASA MOD-1 Wind Turbine

Assembly	Weight (%)
Tower	45.1
Nacelle (includes generator)	24.2
Rotor	15.3
Yaw and brakes	7.8
Controls	7.6

Adapted from [NASA79].

ocean, and takes on higher values for terrain with more obstacles, such as plants, trees, and buildings [Patel99].

To prevent excessively high rotational speeds, which would create stresses in the blades and eventual failure, HAWTs employ a braking mechanism. Wind turbines generally have a yaw control that keeps the turbine aligned with the oncoming wind. In the old straight-bladed windmills used on American farms, this was accomplished with a simple tail vane. The optimum spacing of towers in a large wind farm depends on the terrain, but is usually around 1.5 to 3 rotor diameters apart perpendicular to the wind, and 8 to 12 rotor diameters apart along the prevailing wind direction [Patel99].

■ Summary

After reading this chapter and working through the problems, you should understand and be able to analyze and perform calculations for simple turbomachinery systems, including centrifugal pumps, fans, propellers, and wind turbines, such as the NASA MOD-2 shown in Figure 7.27.

I Figure 7.27 Photo of the NASA MOD-2 wind turbine. Courtesy of NASA.

References

[Audel70] Black, Perry. 1970. *Audel Pump Handbook*. Howard W. Sams & Co.

[NASA73] *Liquid Rocket Engine Centrifugal Flow Turbopumps*. 1973. NASA SP-8109.

[Wahren97] Wahren, U. 1997. *Practical Introduction to Pumping Technology*. Gulf Publishing Company.

[Wislicenus65] Wislicenus, George. 1965. *Fluid Mechanics of Turbomachinery*. Dover.

[DOE07] Annual Report on U.S. Wind Power Installation, Cost, and Performance Trends: 2007. Published by DOE in 2008.

[NSF75] Eldridge, Frank. 1975. *Wind Machines*. National Science Foundation.

[NASA79] Spera, D., L. Vitema, T. Richards, and H. Neustadter. 1979. *Preliminary Analysis of Performance and Loads Data from the 2-Megawatt MOD-1 Wind Turbine Generator*. NASA TM-81408.

[Johnson85] Johnson, Gary. 1985. *Wind Energy Systems*. Prentice-Hall.

[Patel99] Patel, Mukund. 1999. *Wind and Solar Power Systems*. CRC Press.

Problems

1. Prepare a report on the economics of wind turbines for generating electric power. What is the cost of installation per kW of electrical production? How long would it take to recoup the investment costs? How does this compare to conventional (fossil-fuel) sources of electricity?

2. Look at a pump catalog from a supplier of your choice. What is the most power you can get for less than $100? What is the maximum pressure rise for the pump you selected?

3. How are pressure and flow rate and power related for a centrifugal water pump?

4. Is it more economical to build a few very large wind turbines, or a large number of smaller ones?

5. A turbine has a diameter of 20 m and is spinning at 40 rpm in wind at atmospheric conditions of 20°C and 90 kPa. Plot the tip speed ratio as a function of wind velocity for velocities from 5 to 20 m/s.

6. If a particular wind turbine generates 1.0 MW at sea level, how much power would it produce in Denver at the same wind speed and same air temperature?

7. What is the limit on the percentage of its electrical power that a nation could obtain from wind power?

8. If a wind turbine operating at sea level has an efficiency of 30% in a 15 m/s wind, a transmission and generation efficiency of 90%, and a rotor diameter of 25 m,

how much power can it generate? What is the horizontal force on the support structure? Repeat the calculations for a diameter of 50 m.

9. A piston pump is to be built with a stroke length of 20 cm and is to run at an operating speed of 1 cycle per 2 s. If the flow rate to be provided is 50 L/min, what is the minimum diameter of piston that could be used?

10. A particular centrifugal pump has an efficiency of 70% at a flow rate of 125 gal/min, has a pump head of 500 ft, and spins at 3500 rpm. What is the power required to run the pump, in units of hp? What is the torque, in ft-lbf, on the pump shaft? Which numbers do you need to select an electrical motor to power the pump?

11. A particular centrifugal pump has an efficiency of 70% at a flow rate of 500 L/min, has a pump head of 130 m, and spins at 3500 rpm. What is the power required to run the pump, in units of kW? What is the torque, in N-m, on the pump shaft?

12. What are typical efficiencies of an automotive torque converter? What parameters does the efficiency depend on?

13. A centrifugal pump delivers 500 gal/min of water at standard conditions. If the input power to the pump is 20 hp and the efficiency of the pump is 75%, determine the pressure rise of the water across the pump, in units of psi.

14. If a pump can deliver a flow rate of 400 gal/min of water with a pressure rise of 40 psi, and it takes 12 hp to run, what is the efficiency of the pump?

15. A gasoline piston-cylinder engine of displacement 2.0 L produces 125 hp at maximum capacity (wide-open throttle). If a supercharger is added to the engine that draws 25 hp from the engine and is 90% efficient, how much more power could the engine produce, assuming it runs at a stoichiometric condition of 15-to-1 air-to-fuel ratio and uses a fuel with heating value of 42,000 kJ/kg. Assume also that the engine is 35% efficient at converting heat into power. State any other assumptions you have to make to solve this problem.

16. If your electrical bill is $100 per month and electricity costs $0.10/kW-hr, what is the rate at which you consume electrical power, assuming constant usage? If you live in a place where you have fairly constant winds with an average speed of 10 m/s, and the average air density is 1.1 kg/m³, how large of a wind turbine would you have to build to supply your household energy needs, assuming a turbine efficiency of 35%?

17. A designer wants to build a small, ducted wind turbine to protect the birds from the turbine blades. How will the presence of a shroud around the wind turbine affect the aerodynamics and efficiency, and what shape should be chosen for the profile of the duct?

18. What is the typical cost of building a new wind turbine, in dollars per installed megawatt? What is the projected lifetime of a new wind turbine? What ongoing costs are associated with operating a wind turbine? (Although there are no fuel

costs, there are upkeep and maintenance costs to consider.) How does the total cost of wind power compare with other sources of power, such as coal power plants?

19. Using the correlation given in Equation 7.33, if the wind velocity is measured to be 10 m/s at a height 100 m above the surface of flat level ground, what do you expect the velocity to be at a height of 250 m? How much would the power generated by a wind turbine increase if the height of the turbine were increased from 100 m to 250 m above ground?

20. What braking mechanisms are employed in wind turbines to limit the maximum speed?

21. Given that wind speed is variable, what energy storage mechanisms could you think of that would allow a wind turbine to store its excess energy produced in times of high winds, and then discharge it at times of low wind, so that a constant amount of power would be output?

22. Write a short report on the effects of wind turbines on birds and what can done to minimize the casualty rate of birds colliding with turbines.

23. If the instantaneous wind velocity at a particular location follows a log-normal distribution with $m = 10$ m/s and $\sigma = 5$ m/s, find the average power produced by a turbine of efficiency 40% and diameter 50 m with an air density of 1.1 kg/m^3. How does this power compare to the power produced if the wind was constant at 10 m/s?

24. For the pump in Example 7.1, if the pool size is 10 ft by 20 ft and has water to a depth of 5 ft, how long will it take to completely cycle the water of the pool? A reasonable goal is to recirculate pool water in 8 h. If this is not achievable for a given pump, what could you do to reduce the water turnover time?

8 Additional Applications

In This Chapter
- Rockets
- Jet Engines
- Liquid Sprays
- Flow for Electronics Cooling
- Flows in Biological Systems

The objective of this chapter is to survey a wide range of applications in the field of fluid mechanics and to discuss compressible flow as applied to rockets. For many of the topics only a basic introduction is given; Appendix E lists references for further reading for more detailed information.

■ 8.1 Rockets

Rockets (Figure 8.1) use the combustion of a fuel and oxidizer to produce hot rapidly expanding product gases. One end of the rocket chamber is open, and the flow of the exhaust gases out this opening gives the rocket its thrust. To increase the thrust, the rocket exit has a nozzle or constriction at its end, which partially blocks the flow of gas and increases the pressure of the gas inside the rocket chamber as well. Both of these effects produce more thrust. The amount of thrust produced by the rocket depends on the mass flow rate through the engine, the exit velocity of the exhaust, and the pressure at the nozzle exit:

$$F_{\text{thrust}} = P_e A_e + \rho A_e V_e^2 \tag{8.1}$$

Although for compressible subsonic and supersonic flows it cannot be assumed that the pressure at the exit is equal to atmospheric, in general the second term is much larger than the first. To increase the thrust, the exit velocity should be made as high as possible. A fuel should be chosen that releases large amounts of chemical energy that can be converted to the kinetic energy of the escaping exhaust gases.

FIGURE 8.1 Picture of a rocket. Courtesy of NASA.

8.3.1 Rocket Fuels

Commonly used rocket fuels can be classified into two groups: solid propellant and liquid propellant. The solid fuel rockets must combine the fuel and oxidizing agent into a single solid fuel that is then ignited. Once a solid fuel rocket is ignited, it cannot be turned off. In liquid fuel rockets, the fuel and oxidizer are kept in separate tanks, and then must be mixed and ignited to produce thrust. Some liquid fuel rockets can be turned off and restarted, and some can be throttled to regulate the thrust. The oxidizing agent is usually liquid oxygen, but there are other possibilities.

Liquid fuels are preferred to gas fuels because of their greater density. Compressed gas fuels require large and heavy pressure vessels to hold them. Liquid oxygen must be kept at or below a temperature of −300°F (−184°C), while liquid hydrogen is even more challenging to handle, requiring a storage temperature of −423°F (−253°C) [Heppenheimer03]. Hydrogen and oxygen propellants give the highest energy and the best exhaust velocity. However, they cannot be stored for long periods because they evaporate readily. Hydrogen is a preferred rocket fuel because of its high fuel energy and the high exhaust velocity of its combustion products in the exhaust. The heating value of hydrogen is 120,000 kJ/kg.

The chemical reaction for hydrogen combustion is

$$2H_2 + O_2 \rightarrow H_2O$$

For a hydrocarbon fuel such as kerosene, the balanced combustion reaction is

$$C_xH_y + O_2 \rightarrow xCO_2 + (y/2)H_2O$$

Kerosene is a mixture of components formed during the refining of petroleum. The compounds that make up kerosene have carbon chains 12 to 15 atoms long—this is heavier than gasoline. Kerosene has a heating value of 43,000 kJ/kg.

The standard military jet fuel in the United States, JP-8, is kerosene-based and contains additives to improve lubrication and corrosion performance. Jet-A is a commonly used commercial fuel for jet engines. Figure 8.2 shows a schematic of NASA's X-15 rocket plane, which used anhydrous ammonia and liquid oxygen in the main engine, and hydrogen peroxide in the thrusters for the reaction control system. Figure 8.3 shows a picture of the main engine being fired in a ground test.

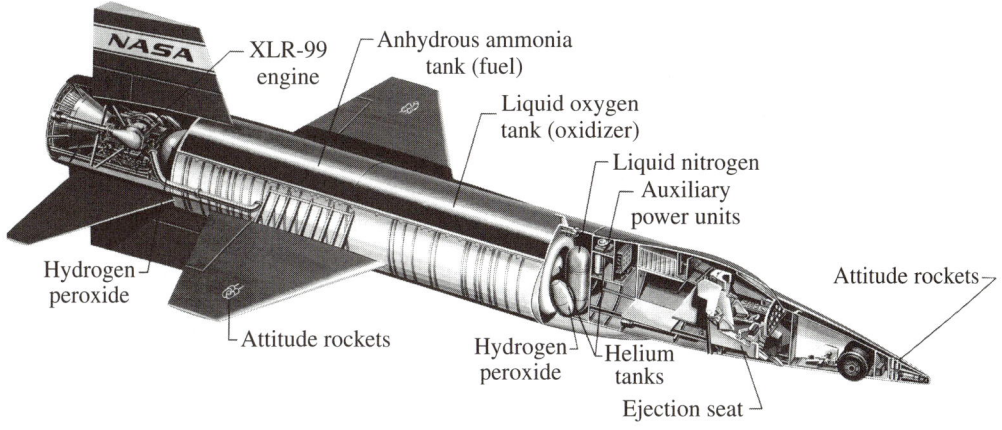

FIGURE 8.2 Diagram of the components of an X-15 rocket plane. Courtesy of NASA-Dryden Flight Research Center.

FIGURE 8.3 Picture of an X-15 rocket plane on a test stand undergoing an engine test. Courtesy of NASA-Dryden Flight Research Center.

Mono-methyl hydrazine (MMH) is a *hypergolic* rocket fuel that is liquid at standard conditions. When mixed with the oxidizing agent nitrogen tetroxide, it reacts readily without the need for an ignition source. The balanced chemical reaction is

$$CH_3NHNH_2 + N_2O_4 \to CO_2 + 2H_2O + 2N_2$$

The fuel in the maneuvering thrusters of the space shuttle is a form of hydrazine, and the oxidizer is nitrogen tetroxide. These substances are liquid at room temperature, but hydrazine is very toxic.

Solid fuel rockets generally have lower exhaust velocities than liquid fuel rockets, which result in less thrust as well as lower efficiency. The advantage of solid fuel rockets is that they can be stored for a long time. The solid rocket boosters (SRBs) of the space shuttle (see Figures 8.4 and 8.5) have an exhaust velocity around 8,400 ft/s (2,600 m/s), while the liquid fuel main engines have an exhaust velocity of around 14,600 ft/s (4,400 m/s). These main engines use liquid hydrogen and liquid oxygen, which are cryogenic fuels that must be cooled to keep them in a liquid state. The space shuttle has very large nozzles, 14 ft (4 m) in diameter, which are gimbaled, so that the direction of thrust can be changed [Heppenheimer03].

8.1 Rockets 375

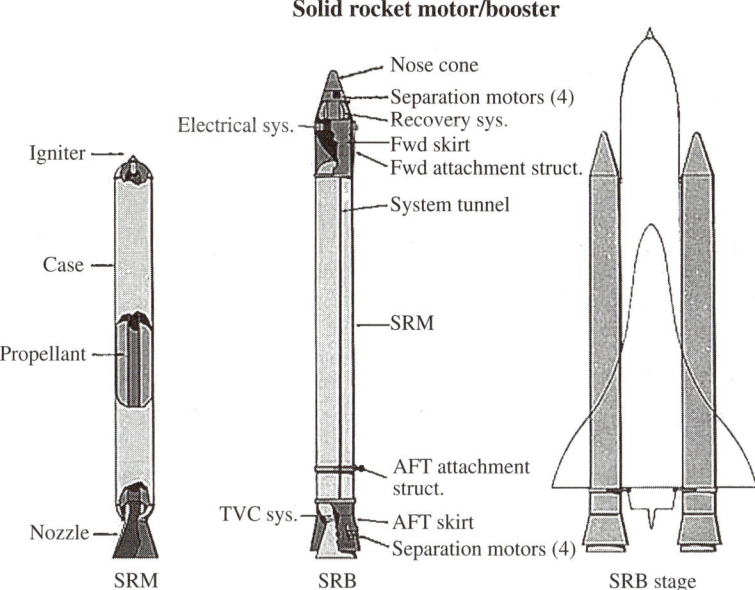

FIGURE 8.4 Schematic of space shuttle solid fuel rocket boosters. Courtesy of NASA Dryden Flight Research Center.

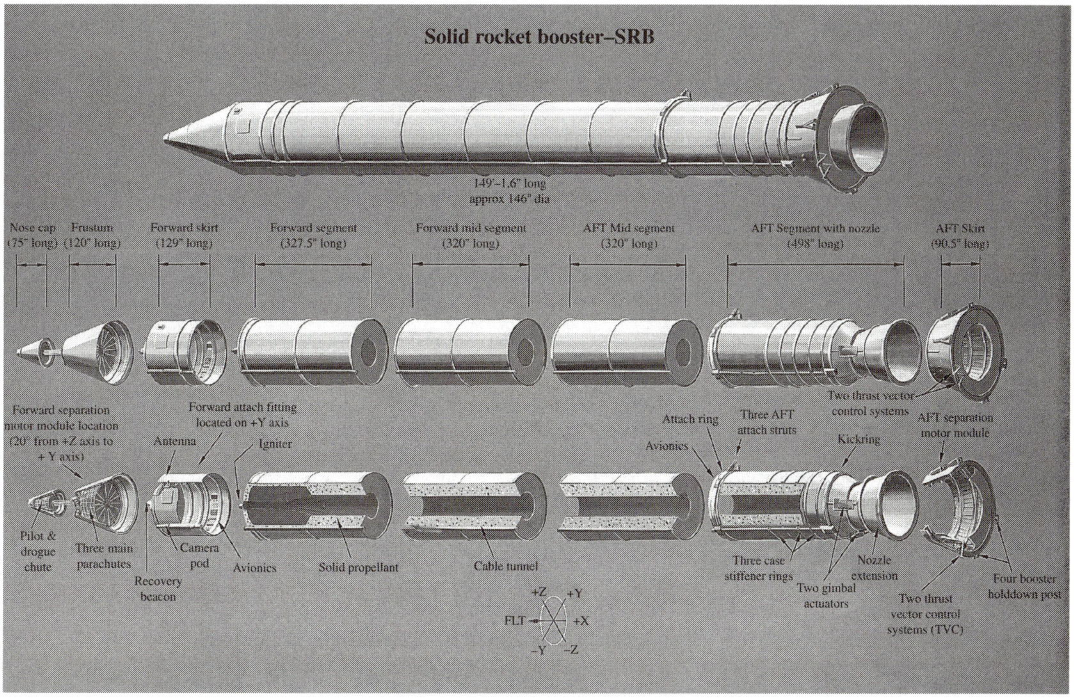

FIGURE 8.5 Detailed schematic of the internal parts of the solid fuel rocket boosters of the space shuttle. Courtesy of NASA/Marshall Space Flight Center Collection.

The pumping requirements for these rocket motors is also significant. The Saturn V rocket that propelled the Apollo capsules had five main engines, which had a cumulative mass flow rate of 15 tons per second of liquid propellants. Each of the five engines had its own set of pumps, and those pumps used as much as 60,000 hp (44,700 kW) of pumping power. The main fuel pumps of the Space Shuttle are also rated at 60,000 hp [Heppenheimer03].

In addition to the traditional liquid fuel and solid fuel rockets, there has been recent interest in hybrid rockets. These would use a liquid–solid mixture, typically a rubber-based solid fuel along with liquid oxygen. Such a mixture needs to be cast within a strong casing [Heppenheimer03].

8.3.2 Transient Rocket Equation

A brief introduction to analysis of rocket motion was presented in Chapter 3, and a one-dimensional equation for the motion of a rocket was derived for the case of constant fuel burn rate and negligible drag. We solved this equation by both analytical and numerical methods in the examples in Chapter 3. Aerodynamic drag was covered in Chapter 6, along with the numerical solution technique for the two-dimensional motion of a projectile. Here we present Algorithm 8.1, which gives the numerical method for a two-dimensional rocket trajectory, including drag and variable engine thrust force as sketched in Figure 8.6. We will illustrate by way of an example.

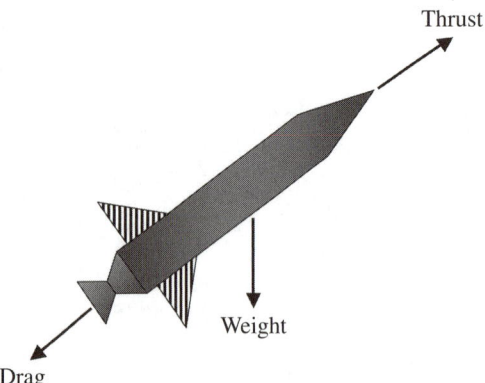

FIGURE 8.6 Free-body diagram for a rocket in two-dimensional motion.

8.1 Rockets

EXAMPLE 8.1

A popular college project is to have students build model rockets and predict their performance, and then launch the rockets and see how close the actual performance comes to the predicted. Suppose you have built a model rocket of mass 50 g, length 30 cm, tube diameter of 2.5 cm, and drag coefficient of 0.75, to be used with a small motor that produces a total of 4 N of thrust over 0.5 s of firing time. If the rocket is launched at an angle of 60° to the horizontal, how far will it travel horizontally? Assume there is no significant wind at the time of launch.

SOLUTION To solve this problem, we could use Algorithm 6.1, but it must be modified to include a thrust force. This is done in Algorithm 8.1. For these conditions, the numerical solution calculated in MATLAB is that the rocket will travel 110 m horizontally, with a maximum altitude of 15 m.

ALGORITHM 8.1

1. Set the initial conditions x_0, y_0, V_{x0}, V_{y0}. If the magnitude, V_0, and angle, θ, of the initial velocity are specified instead, then $V_{x0} = V_0 \cos\theta$ and $V_{y0} = V_0 \sin(\theta)$.
2. Set $x(0) = x_0$, $y(0) = y_0$, $V_x(0) = V_{x0}$, $V_y(0) = V_{y0}$.
3. Set the initial time $t = 0$, and select the numerical time step, Δt.
4. Specify the mass, m, and cross-sectional area, A, of the projectile and the density, ρ, and viscosity, ν, of air for local atmospheric conditions. If the rocket is going to high altitudes, include a correlation or look-up table for density as a function of height.
5. Specify the thrust curve $F_T(t)$ for the rocket motor.
6. LOOP: For the current time value t

    Compute the magnitude of velocity: $V(t) = \sqrt{V_x^2 + V_y^2}$
    Compute the Reynolds number: $Re = V\,d/\nu$
    Compute the drag coefficient from the appropriate correlation: $C_D = C_D(Re)$
    Compute the drag force: $F = \frac{1}{2}C_D\rho A V^2$
    Compute the x and y components of the drag force:

    $F_x = F * V_x(t)/V$

    $F_y = F * V_y(t)/V$

Look up the magnitude of the thrust force from the stored thrust curve:

```
F_T = F_T(t)
```

Compute the *x* and *y* components of the thrust force:

```
F_T,x = F_T * V_x(t)/V

F_T,y = F_T * V_y(t)/V
```

Perform integrations of the four variables:

```
V_x(t + Δt) = V_x(t) + (-F_D,x + F_T,x)/m * Δt

V_y(t + Δt) = V_y(t) + (-g - F_D,y/m + F_T,y/m) * Δt

x(t + Δt) = x(t) + V_x(t) * Δt

y(t + Δt) = y(t) + V_y(t) * Δt
```

Increment the current time by time step: `t = t + Δt`
7. END LOOP when y < 0

8.1.3 Compressible Flow

Reviewing Equation 8.1, the thrust force produced by a rocket is

$$F_{\text{thrust}} = P_e A_e + \rho A_e V_e^2$$

and the mass flow rate of propellants through the rocket nozzle is

$$\dot{m} = \rho A_e V_e$$

So for a given mass flow rate the thrust force can be increased by increasing the velocity, or, alternatively, for a fixed rocket geometry and nozzle exit area A_e, it would appear that we could increase the thrust by increasing the rate at which propellants are burned to increase the mass flow rate and velocity. Unfortunately, the speed at which a fluid can be forced through a fixed area is limited by compressibility effects.

The maximum mass flow rate that can be achieved is called the *choked flow condition*. As will be shown in the next section, the flow in a converging nozzle cannot be accelerated past Ma = 1.0, no matter how much energy is added to the fluid. However, if a diffuser is added to the nozzle after the flow is choked, then the flow can be accelerated to supersonic velocities. So supersonic rocket nozzles are made of a converging section followed by a diverging section. The connection between them, called the *throat*, is the location of the minimum cross-sectional area.

The goal of the analysis in the next subsection is to develop correlations that will allow us to calculate the mass flow rate at the choked throat of a converging–diverging nozzle, and then, with the mass flow rate known, to calculate the thrust of the rocket based on the expansion ratio of the diffuser section. The equations and analysis of Chapter 3 are not sufficient here because the Mach numbers will be above 0.3, and the density will change as the Mach number of the flow changes.

Speed of Sound

From conservation of mass and momentum in an isentropic process, it can be shown that

$$a^2 = \frac{dP}{d\rho} \quad \text{or} \quad a = \sqrt{\left.\frac{dP}{d\rho}\right|_s} \tag{8.2}$$

where a is the speed of sound. For liquids, the speed of sound can be related to the modulus of elastisicity,

$$a = \sqrt{\frac{E}{\rho}} \tag{8.3}$$

For gases, if an ideal gas can be assumed, then an expression may be derived using the ideal gas law and assuming an isentropic process. Thus,

$$\frac{P}{\rho^k} = C \tag{8.4}$$

For an isentropic process, Equation 8.4 can be written as $P = C\rho^k$. Taking the derivative with respect to ρ of both sides yields

$$\frac{dP}{d\rho} = Ck\rho^{k-1}$$

Substituting in the value of C from Equation 8.4 to eliminate the constant, C, gives

$$\frac{dP}{d\rho} = \left(\frac{P}{\rho^k}\right)k\rho^{k-1} = k\frac{P}{\rho} \tag{8.5}$$

Substituting Equation 8.5 into Equation 8.2 gives the speed of sound in an ideal gas

$$a = \sqrt{k\frac{P}{\rho}} \tag{8.6}$$

The ideal gas law ($P/\rho = (R/M)T$) can then be used to express the speed of sound as a function of temperature:

$$a = \sqrt{k(R/M)T} \qquad (8.7)$$

where $k = c_P/c_V$ is the ratio of specific heats for the gas, R is the universal gas constant, M is the molecular weight of the gas, and T is the absolute temperature in units of Kelvin or degrees Rankine. For air at standard atmospheric conditions, $k = 1.4$. Note that $c_P - c_V = R/M$. Recall that enthalpy is defined as $h = u + Pv$. If a fluid were truly completely incompressible, the speed of sound would be infinite.

Stagnation Properties

The stagnation properties are those that would be obtained if the fluid was decelerated to a complete stop in an isentropic (adiabatic) process. For an isentropic process of an ideal gas,

$$P\rho^{-k} = C \qquad (8.8)$$

The stagnation pressure is

$$P_0 = P\left(1 + \frac{k-1}{2}\text{Ma}^2\right)^{\frac{k}{k-1}} \qquad (8.9)$$

the stagnation temperature is

$$T_0 = T\left(1 + \frac{k-1}{2}\text{Ma}^2\right) \qquad (8.10)$$

the stagnation enthalpy is

$$h_0 = h + \frac{V^2}{2} \qquad (8.11)$$

and the stagnation density is

$$\rho_0 = \rho\left(1 + \frac{k-1}{2}\text{Ma}^2\right)^{\frac{1}{k-1}} \qquad (8.12)$$

Critical Conditions

The critical speed is defined as Ma = 1. By convention, critical conditions are denoted by an asterisk, so $V^* = a^*$. Since the speed of sound, a, is a function of temperature, which can vary, there is a critical value of a for a given problem. For air at standard conditions with $k = 1.4$, and at the critical velocity Ma = 1, the critical values of stagnation temperature, pressure, and density are

$$T_0^* = T^*\left(1 + \frac{0.4}{2}1^2\right) = 1.20T^*$$

$$P_0^* = P^*\left(1 + \frac{0.4}{2}1^2\right)^{1.4/0.4} = 1.89P^*$$

$$\rho_0^* = \rho^*\left(1 + \frac{0.4}{2}1^2\right)^{1/0.4} = 1.58\rho^*$$

One-Dimensional Isentropic Flow

For one-dimensional isentropic flow, conservation of mass, momentum, and energy apply. Also, the second law of thermodynamics leads to $s_1 = s_2$. If we allow the area of the duct to change, there are three possibilities: The area can increase, stay the same, or decrease. So we can write that the area $A = A(x)$, the properties of the fluid that vary are $P = P(x)$ and $T = T(x)$, and the velocity is $V = V(x)$. For one-dimensional flow along the x-axis, conservation of mass is

$$\frac{\partial}{\partial x}(\rho A V) = 0 \qquad (8.13)$$

and the energy equation is

$$\frac{\partial}{\partial x}\left(u + \frac{P}{\rho} + \frac{V^2}{2}\right) = 0 \qquad (8.14)$$

assuming frictionless flow (no wall shear stress). With the further assumption that the specific heats of the gas are constant, then $u = c_V T$ and the energy equation can be rewritten as

$$\frac{\partial}{\partial x}\left(c_V T + \frac{P}{\rho} + \frac{V^2}{2}\right) = \frac{\partial}{\partial x}\left(c_P T + \frac{V^2}{2}\right) = 0 \qquad (8.15)$$

For one-dimensional steady flow with no viscous forces, the x-momentum equation of the Navier-Stokes equations (see Chapter 4) simplifies to

$$\rho V \frac{dV}{dx} + \frac{dP}{dx} = 0 \tag{8.16}$$

Multiplying by dx yields $dP + \rho V\, dV = 0$. It will be convenient later to have this equation divided by ρV^2, which will result in

$$\frac{dV}{V} = -\frac{dP}{\rho V^2} \tag{8.17}$$

Expanding out the conservation of mass equation (Equation 8.13) gives

$$\frac{d}{dx}(\rho A V) = \frac{d\rho}{dx} AV + \rho \frac{d}{dx}(AV) = \frac{d\rho}{dx} AV + \rho A \frac{dV}{dx} + \rho V \frac{dA}{dx} = 0$$

Dividing both sides of the equation by ρAV and multiplying by dx yields

$$\frac{d\rho}{\rho} + \frac{dA}{A} + \frac{dV}{V} = 0 \tag{8.18}$$

This is a differential form of the continuity equation. Now Equation 8.17 can be substituted into Equation 8.18 to eliminate dV/V.

$$\frac{d\rho}{\rho} + \frac{dA}{A} - \frac{dP}{\rho V^2} = 0$$

This can be rearranged to

$$\frac{dA}{A} = \frac{dP}{\rho V^2} - \frac{d\rho}{\rho} = \frac{dP}{\rho V^2} - V^2 \frac{d\rho}{\rho V^2} \frac{dP}{dP} = \frac{dP}{\rho V^2}\left(1 - V^2 \frac{1}{dP/d\rho}\right)$$

Since the speed of sound is $a^2 = dP/d\rho$ (Equation 8.2) and $\mathrm{Ma} = V/c$, then

$$\frac{dA}{A} = \frac{dP}{\rho V^2}(1 - \mathrm{Ma}^2) \tag{8.19}$$

So if $\mathrm{Ma} < 1$, an increase in area (diffuser) results in an increase in pressure (as was shown in Chapter 3), and a decrease in area (nozzle) results in a decrease in pressure.

However, when Ma > 1 the opposite occurs. How does the velocity change with area in a compressible flow? This can be found by substituting Equation 8.17 into Equation 8.19:

$$\frac{dA}{A} = -\frac{dV}{V}(1 - \text{Ma}^2) \qquad (8.20)$$

So when Ma < 1, a nozzle, which has a decrease in area (negative dA), will result in an increase in velocity, again agreeing with results from Chapter 3 for incompressible flow. However, when Ma > 1, a decrease in area will result in a decrease in velocity, but an increase in area will result in an increase in velocity! (For supersonic flows sometimes things are the opposite of what you might expect.) Note that when Ma = 1, $dV/dA = 0$, which implies it is impossible to maintain sonic flow in a constant area duct. Internal sonic flow can only be reached when there is a change in sign of dA, such as at the throat of a converging–diverging nozzle.

The four equations needed to describe one-dimensional isentropic flow of an ideal gas through a passageway of varying area are listed as follows [NACA53]. The three relations for stagnation properties are

$$P_0 = P\left(1 + \frac{k-1}{2}\text{Ma}^2\right)^{\frac{k}{k-1}}$$

$$T_0 = T\left(1 + \frac{k-1}{2}\text{Ma}^2\right)$$

$$\rho_0 = \rho\left(1 + \frac{k-1}{2}\text{Ma}^2\right)^{\frac{1}{k-1}}$$

To this must be added a relationship between the local cross-sectional area to the cross-sectional area at the throat (critical condition) as a function of Mach number.

$$\frac{A}{A^*} = \frac{1}{\text{Ma}}\left[\frac{1 + (\frac{k-1}{2})\text{Ma}^2}{1 + \frac{k-1}{2}}\right]^{\frac{k+1}{2(k-1)}} \qquad (8.21)$$

These four ratios have been tabulated as a function of Mach number in Table B.1 of Appendix B for an ideal gas of specific heat ratio $k = 1.4$.

The isentropic flow equations are useful in calculations of flow through an intake restrictor, such as in the intake of a Formula SAE racecar. There are two other types of compressible flows for which simple relations can be derived. Flow in a constant area duct or pipe with heat addition is known as *Rayleigh flow*, and compressible flow with friction is *Fanno flow*. For further information and flow tables, refer to a dedicated text such as Zucrow and Hoffman [Zucrow76].

Shock Waves

Shock waves are one of the few discontinuities that exist in nature (a liquid interface being another). The thickness of a shockwave is about 5 times the mean free path. For air at STP, the mean free path is $\lambda = 0.1$ μm, so the thickness of a shock wave at standard conditions is about 0.5 μm (0.00002 in.). Shock waves are *not* isentropic—they are a one-way process.

For a converging nozzle, when the flow accelerates to Ma $= 1$, the flow is *choked* and cannot accelerate any further. A detonation is a supersonic combustion wave. Isentropic flow through a variable-area duct has applications to rocket nozzles, ramjets, Venturis, and restrictors in engines.

There are two types of shocks. A shock that is perpendicular to the flow direction is a *normal shock*, whereas a shock that is inclined at an angle with respect to the main flow direction is an *oblique shock* (Figure 8.7). The Mach number and velocity decrease across a shock as the flow decelerates from supersonic to subsonic speeds, while the pressure and density of the gas rise across the shock. A shock does no work, and there is no external heat transfer to or from a shock, so by conservation of energy the total temperature and total enthalpy across a shock are constant.

Normal Shocks

For a normal shock wave, let point 1 denote the supersonic flow coming into the shock wave, and point 2 the subsonic flow exiting the shock wave, as shown in Figure 8.8. Note that all velocities are relative to the shock wave. The conservation of mass equation across a normal shock wave is

$$\rho_1 V_1 = \rho_2 V_2 \tag{8.22}$$

the conservation of momentum equation is

$$P_1 + \rho_1 V_1^2 = P_2 + \rho_2 V_2^2 \tag{8.23}$$

and the conservation of energy equation is

$$h_1 + \frac{V_1^2}{2} = h_2 + \frac{V_2^2}{2} \tag{8.24}$$

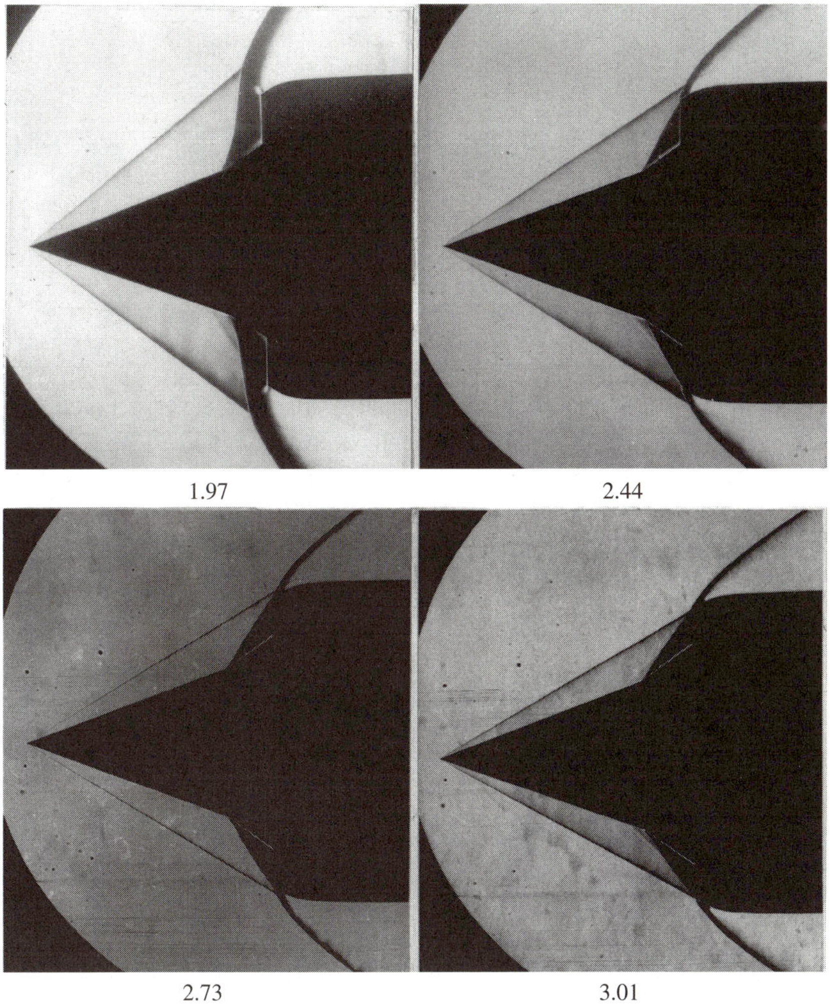

1.97

2.44

2.73

3.01

FIGURE 8.7 Oblique shocks on an engine inlet at different Mach numbers. Courtesy of NASA.

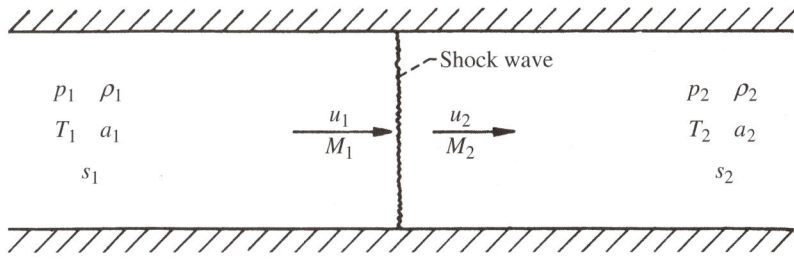

FIGURE 8.8 Diagram for a normal shock wave, from [NACA53].

With the additional assumption of a "perfect gas" with constant specific heats, the energy equation becomes

$$c_P T_1 + \frac{V_1^2}{2} = c_P T_2 + \frac{V_2^2}{2} \tag{8.25}$$

The flow across a shock can be considered adiabatic, but not isentropic, so that the second law of thermodynamics for the shock wave is

$$s_2 > s_1 \tag{8.26}$$

Assuming the flow conditions at the initial state, 1, are completely specified, this gives us a system of four equations (mass, momentum, energy, and ideal gas law) and four unknowns (P_2, ρ_2, T_2, V_2). So for a given set of completely specified upstream conditions, there will be one unique state downstream. Theoretically we can solve these equations simultaneously, perhaps aided by software such as EES, but it is often convenient to use shock flow tables with tabulated values. Normal shock flow tables are included in Appendix B for a value of $k = 1.4$. The ratio of properties across the shock is tabulated as a function of the incoming Mach number, Ma_1. With judicious combination of the four equations and the definition of the Mach number, we can calculate the Mach number of the gas exiting the shock, Ma_2, as

$$Ma_2 = \sqrt{\frac{Ma_1^2 + \frac{2}{k-1}}{\frac{2k}{k-1} Ma_1^2 - 1}} \tag{8.27}$$

Across the normal shock wave the Mach number decreases to the value specified as Ma_2, which is always subsonic. Since the stagnation temperature is constant across the shock (energy is conserved), the temperature ratio is

$$\frac{T_2}{T_1} = \frac{1 + \frac{k-1}{2} Ma_1^2}{1 + \frac{k-1}{2} Ma_2^2} \tag{8.28}$$

With the temperatures and Mach numbers known, the value of the velocity can be calculated as

$$\frac{V_2}{V_1} = \frac{Ma_2}{Ma_1} \sqrt{\frac{T_1}{T_2}} \tag{8.29}$$

With the velocities known, conservation of mass can be used to find the ratio of densities (since area is constant):

$$\frac{\rho_2}{\rho_1} = \frac{\text{Ma}_1}{\text{Ma}_2} \sqrt{\frac{1 + \frac{k-1}{2}\text{Ma}_2^2}{1 + \frac{k-1}{2}\text{Ma}_1^2}} \qquad (8.30)$$

And then the pressure ratio can be obtained from the momentum equation:

$$\frac{P_2}{P_1} = \frac{1 + k\text{Ma}_1^2}{1 + k\text{Ma}_2^2} \qquad (8.31)$$

The right-hand side of all these equations depends only on the free-stream Mach number. So knowing the Mach number, we can determine all the conditions associated with the normal shock. The equations described are also listed in reference [NACA53].

EXAMPLE 8.2

The conditions upstream of a normal shock in an internal flow of air are 100 kPa (absolute), 20°C, and 720 m/s. Find the downstream conditions.

SOLUTION We calculate the initial density from the ideal gas law, and we can compute the initial speed of sound as well:

$$\rho_1 = \frac{PM}{RT} = \frac{(100{,}000 \text{ Pa})(29 \text{ kg/kmol})}{(8{,}314 \text{ J/kmol} \cdot \text{K})(293 \text{ K})} = 1.19 \text{ kg/m}^3$$

$$a_1 = \sqrt{(1.4)\left(\frac{8{,}314 \text{ J/kmol} \cdot \text{K}}{29 \text{ kg/kmol}}\right)(293 \text{ K})} = 343 \text{ m/s}$$

$\text{Ma}_1 = V_1/a_1 = (720 \text{ m/s})/(343 \text{ m/s}) = 2.10$. From Table A.9, the outgoing Mach number is 0.547, the temperature ratio is 1.85695, the pressure ratio is 4.9783, and the density ratio is 2.8119. Thus the outlet temperature, pressure, and density are $T_2 = 1.85695\ T_1 = 1.85695\ (293\text{ K}) = 544\text{ K} = 271°\text{C}$; $P_2 = 4.9783\ P_1 = 4.9783\ (100\text{ kPa}) = 497.8\text{ kPa}$; and $\rho_2 = 2.8119\ \rho_1 = 2.8119\ (1.19 \text{ kg/m}^3) = 3.346 \text{ kg/m}^3$. By conservation of mass the outlet velocity V_2 is $V_1 \rho_1/\rho_2 = (720 \text{ m/s})/2.8119 = 256 \text{ m/s}$. Note that if the value for Ma_1 did not fall on the table entries, one could either interpolate or use Equations 8.27–8.31 directly. Use of the equations involves considerably more work, unless they are pre-programmed in a computer program.

Rocket Efficiency

The *specific impulse* is the thrust of an engine divided by the *weight flow rate* of fuel into the engine. The weight flow rate of fuel is

$$\dot{w} = \dot{m}g = \rho g A V \qquad (8.32)$$

and the specific impulse is:

$$I_{\text{SP}} = \frac{F_T}{\dot{m}g} \qquad (8.33)$$

since specific impulse is the thrust divided by the total mass flow of propellant, specific impulse is the inverse of specific fuel consumption used in discussing the performance of other types of propulsion systems. For a rocket, if we assume that the pressure at the rocket exit is equal to the atmospheric pressure, then thrust is generated solely due to the momentum of the exiting hot gasses, and the thrust force is $F_T = \dot{m}V_e$. Under this assumption, the specific impulse becomes

$$I_{\text{SP}} = = \frac{F_T}{\dot{m}g} = \frac{\dot{m}V_e}{\dot{m}g} = \frac{V_e}{g} \qquad (8.34)$$

According to Anderson, the specific impulse from a kerosene-oxygen rocket is about 240 s, and from a hydrogen-oxygen rocket about 360 s [Anderson07]. The specific impulse of a rocket in general is typically 100 to 450 s, which is low compared to jet engines. But rockets can be used in space since they carry their own oxidizer, and are efficient at high speeds. The general thrust equation, from conservation of momentum, for any engine operating in steady state is

$$F_T = \dot{m}_{\text{out}}V_{\text{out}} - \dot{m}_{\text{in}}V_{\text{in}} + (P_e - P_\infty)A_e \qquad (8.35)$$

where the subscript *e* denotes the exit conditions. For a rocket there is no inlet stream of fluid (air), so

$$F_{T,\text{rocket}} = \dot{m}_e V_e + (P_e - P_\infty)A_e \qquad (8.36)$$

where by convention the subscript e is used for the exit conditions. For jet engines, fuel carried on board the vehicle is burned in the combustion chamber, so the mass of exhaust gas exiting is greater than the mass of air coming in to the intake of the jet engine. For a jet engine the thrust equation is

$$F_{T,\text{jet}} = (\dot{m}_{\text{fuel}} + \dot{m}_{\text{air}})V_e - \dot{m}_{\text{air}}V_\infty + (P_e - P_\infty)A_e \tag{8.37}$$

The stoichiometric air-to-fuel ratio for most hydrocarbon fuels is around 15:1. Jet engines typically run lean, so neglecting the mass flow rate of the fuel in Equation 8.36 will lead to an error of 5% or less. The momentum term is usually much larger than the pressure term.

The *nozzle area ratio* is the ratio of the nozzle exit area to the throat area. The nozzle area ratio determines the amount of expansion of the exhaust gases through the nozzle and is related to exhaust gas pressures. If a rocket designer is asked to provide a nozzle that is to be operated only on the ground—as in the case of an experimental rocket engine—he chooses a nozzle area ratio such that the exhaust gas pressure at the nozzle exit is equal to ambient pressure, usually considered as sea-level pressure or 1 atmosphere. If the combustion pressure is 20 atm, the gases undergo an expansion ratio of 20, and this corresponds roughly to a nozzle area ratio of 4. If, for some reason, the designer provides a nozzle area ratio less than that needed for complete expansion, the exhaust gases emerge from the nozzle exit at greater than ambient pressure. In this case the gases are said to be *underexpanded*, for they have to expand farther to reach ambient pressure. On the other hand, if the designer provides a larger nozzle area ratio than that needed for complete expansion, the exhaust gases reach a pressure equal to ambient pressure while still in the nozzle. In some cases the gases will continue to follow the nozzle walls and expand to a pressure lower than ambient. In this case the gases are said to be *overexpanded*. Sooner or later, overexpanded gas must be reconciled with the ambient pressure, and nature provides for this adjustment by means of a shock wave. The ideal nozzle is one that provides for complete gas expansion, neither more nor less, for theory shows that this yields maximum exhaust velocity. This poses a problem to the designer of a launch vehicle, because as soon as the vehicle is launched, the ambient pressure begins to fall, approaching zero at very high altitudes.

Figure 8.9 shows one design to deal with this challenge. The linear aerospike engine has a variable expansion ratio, unlike conventional rocket engines that have a fixed expansion ratio. The diamond-shaped shock pattern in the exhaust of an SR-71 shown in Figure 8.10 is typical of the exhaust plumes from supersonic jets and rockets.

FIGURE 8.9 Linear aerospike rocket nozzle, designed to give the proper expansion of exhaust gases at a range of altitudes. Courtesy of NASA Dryden Flight Research Center; Photo by: Carla Thomas.

FIGURE 8.10 Diamond pattern in exhaust of an SR-71 due to shock waves. Courtesy of NASA.

8.2 Jet Engines

The first jet engines were developed independently by Hans von Ohain in Germany and Frank Whittle in England during WWII. The basic principles of jet propulsion had been understood for some time before then, but jet engines require speeds of at least several hundred mph to be effective.

The three most critical parts of a jet engine are the compressor, combustor, and turbine. The inlet nozzle and exhaust nozzle are also important. The compressor rotates rapidly, compressing the air. The compressed air flows into a combustor where fuel is injected, mixed with the compressed air, and burned. The hot, high-pressure exhaust gases then pass through a turbine, forcing it to spin rapidly. The turbine draws power from this hot air flow. A long shaft connects the turbine and compressor; the spinning turbine uses its power to turn the compressor.

8.2.1 Turbojets and Turbofans

Early turbojet engines (Figure 8.11), such as those of the Me 262, had high fuel consumption rates. Thus, an initial challenge for early jet engine designers was to design and build an engine that could give high thrust with lower fuel consumption. A key breakthrough was the concept of the turbofan engine (Figure 8.12). A large fan, somewhat similar to an airplane propeller but with more long blades set closely together, is added in front of the compressor in the engine nacelle. The fan adds its thrust to that

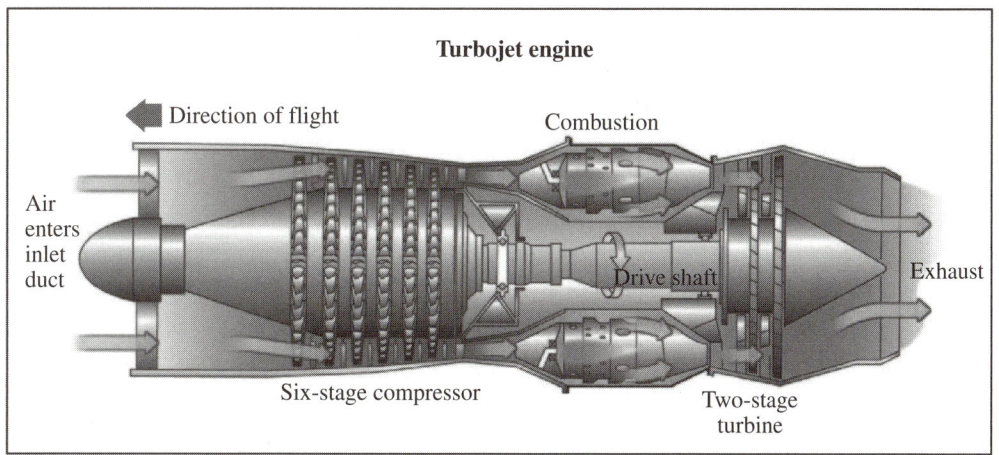

FIGURE 8.11 Diagram of a turbojet engine. Courtesy of FAA.

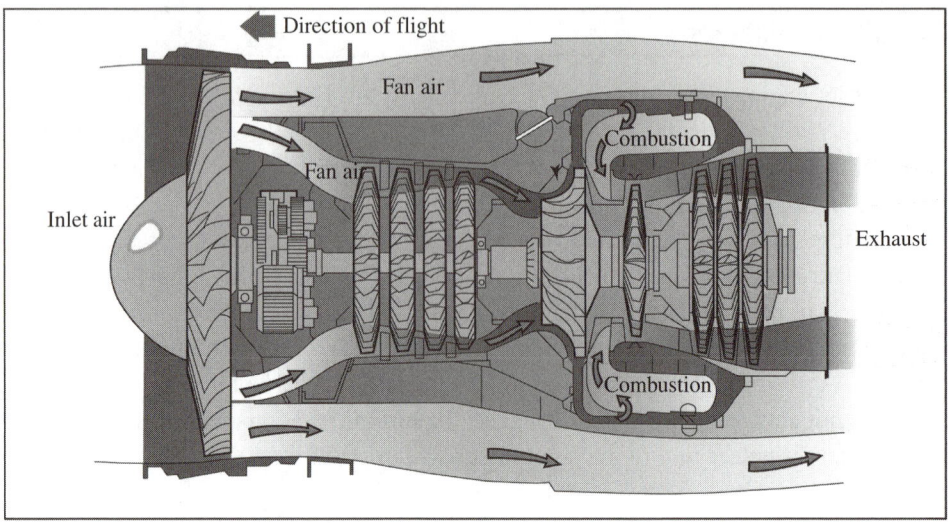

FIGURE 8.12 Diagram of a turbofan engine. Courtesy of FAA.

of the exhaust jet exiting the turbine and is driven off the main shaft. This design more than doubled the thrust of earlier engines, and it also further improved fuel economy. In addition, turbofan engines were relatively quiet compared to turbojet engines. GE and Pratt & Whitney both built turbofans after 1965. The engine for the Me 262, the Jumo 004, delivered 2,000 lbf (8,900 N) of thrust. The J-57 turbojet was rated at 13,500 lbf (60,000 N) of thrust. The early turbofans of the 1970s produced around 40,000 lbf (180,000 N) of thrust, and GE's recent GE 90 turbofan is rated at 90,000 lbf (400,000 N) of thrust [Heppenheimer 03].

Various measures of quantitatively assessing the efficiency of a jet engine include specific thrust, specific impulse, thrust specific fuel consumption (TSFC), and thermodynamic efficiency. TSFC is the mass of fuel burned by an engine per unit of time divided by the thrust the engine produces. The units of this efficiency factor are mass per time divided by force (in English units, pounds mass per hour per pound; in metric units, kilograms per hour per Newton). Mathematically, TSFC is a ratio of the engine fuel mass flow rate to the thrust F produced by the engine:

$$\text{TSFC} = \frac{\dot{m}_f}{F_{\text{thrust}}} \qquad (8.38)$$

If we compare the TSFC for two engines, the engine with the lower TSFC is the more fuel-efficient engine.

EXAMPLE 8.3

One design for an engine calls for a thrust of 2000 lbf, while consuming 2000 lbm/hr of fuel. Another design uses the same basic engine with an afterburner added. With the afterburner, the engine can now produce 3000 lbf of thrust, but now consumes 4500 lbm/hr of fuel. Which engine is more efficient?

SOLUTION The first engine has a TSFC of 2000/2000 = 1, and the engine with the afterburner has a TSFC of 4500/3000 = 1.5, so the first engine is more efficient since it has the lower TSFC value. This says that adding the afterburner produces more thrust, but costs much more fuel for each pound of thrust added. Afterburners are good when a sudden burst of speed is needed, but are poor in terms of fuel efficiency. Thus they are used on military fighter aircraft but not on commercial aircraft.

A turbofan typically has a bypass ratio of around 5, where the *bypass ratio* is defined as the ratio of air moving through the fan to air going to the combustion chamber. The maximum speed of turbofan-powered aircraft is limited by the possibility of shock waves forming and damaging intake and engine systems. A turboprop airplane has very high efficiency at low speeds and high power-to-weight ratio, but top speed is limited by propeller blade speeds needing to be subsonic. Figure 8.13 shows a

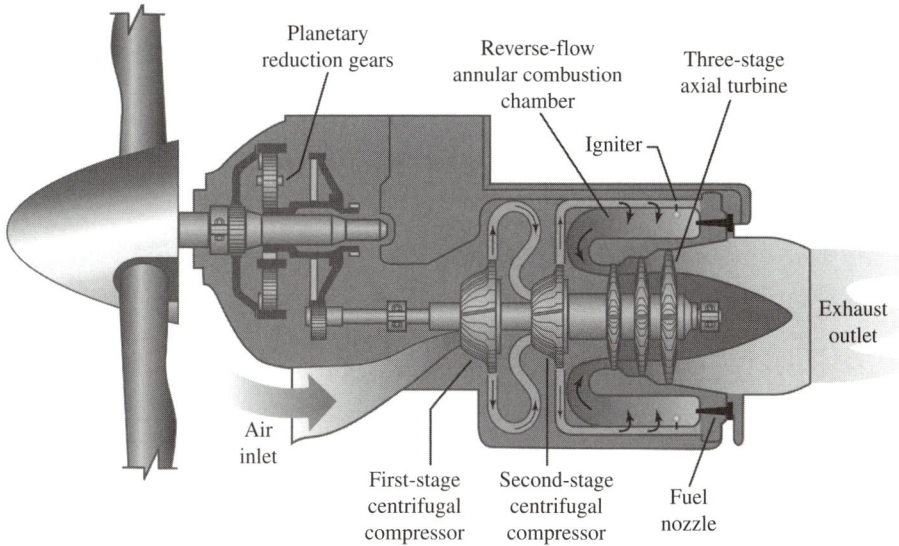

FIGURE 8.13 Schematic of a turboprop engine. Courtesy FAA.

schematic of turboprop engine design. A turbofan can be used up to about Mach 2, with specific impulses above 5000 s for subsonic speeds. With an afterburner, turbofans can be used up to Mach 3, but with a decrease in efficiency. Figure 8.14 shows a general comparison of efficiency of different options for propulsion.

Performance Metrics

The same parameters used to characterize rocket engines can be used for jet engines. The *propulsive efficiency* is defined as the ratio of useful power output to the rate of energy input, which can be quantified as

$$E_{\text{prop}} = \frac{TV}{\Delta KE} \tag{8.39}$$

where T is the thrust, V is the flight speed, and ΔKE is the kinetic energy increase of the fluid. The principal use of the propulsive efficiency parameter for flight is to indicate that the various propulsion systems operate best in different speed ranges. This is shown qualitatively by Figures 8.15 and 8.16. The propeller is the most efficient

The following is from NASA SP-4404 Liquid Hydrogen as a Propulsion Fuel, 1945–1959 by John L. Sloop, 1978.

I FIGURE 8.14 Comparison of different options for propulsion, from [NASA75].

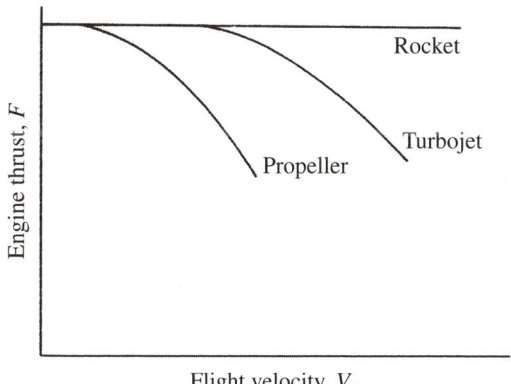

FIGURE 8.15 Thrust generated as a function of speed for different propulsion options, from [NASA75].

propulsive method at low speeds, and the jet engine achieves best efficiency only at relatively high flight speeds. The very high exhaust velocities of the rocket make its propulsive efficiency high only at very high flight speeds.

Since thrust is the mass flow rate change in fluid velocity, propulsive efficiency can also be expressed as

$$\eta = \frac{2}{1 + V_j/V_o} \qquad (8.40)$$

where V_j is exhaust jet velocity, and V_o is flight velocity.

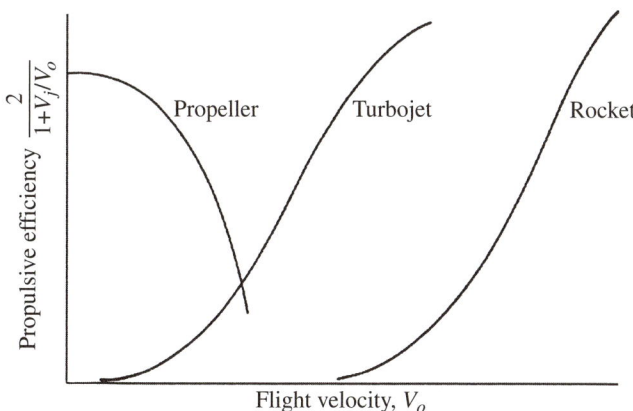

FIGURE 8.16 Propulsion efficiency as a function of speed, from [NASA75].

8.2.2 Ramjets

The simplest of all jet engines for propulsion, the ramjet has no moving parts. The front of the engine is a simple inlet, followed by a diffuser that decelerates the air to subsonic speeds and increases the pressure. After the diffuser, fuel is injected and then burns just downstream of the flame holder, which is a grid or ring structure used to provide a stable place for the flame location. The hot exhaust gases then pass through the exit nozzle, generating thrust. Ramjets can be used at higher speeds than turbojet or turbofan engines, up to about Mach 5, but they have lower efficiencies than turbofan engines at Ma < 3. Ramjets are used as afterburners on fighter planes. Ramjets cannot be started at rest since the flame would go out the intake; in fact, they need a high initial speed to get started. They are also inefficient at low speeds since they rely on ram air compression to compress the air. The specific impulse of ramjets burning hydrocarbon fuels is usually in the range of 1000 to 1500 s.

8.2.3 Scramjets

The term *scramjet* comes from the name **S**upersonic **C**ombustion ramjet. Thus a scramjet is a ramjet that is operated in the hypersonic regime. While in a ramjet the air is decelerated to subsonic speeds for combustion, in a scramjet the flow is supersonic throughout the engine. Unlike the ramjet, there is no diffuser section in a scramjet. Scramjets should be able to operate in the range from Ma = 5 to Ma = 15. One challenge in scramjet design and operation is to keep flame from quenching at the high operating speeds.

Scramjets cannot completely replace traditional rockets because, like ramjets, they cannot start from rest but must be accelerated to supersonic speeds to be started. Scramjets also require air to run, and so cannot work in the near vacuum of space. However, scramjets do provide promise in reducing the total weight of a multistaged rocket. If used as the second stage of a three-stage rocket, for example, the scramjet stage would not need to carry liquid oxygen since it can burn the oxygen in the air. This would reduce the weight of propellants in the rocket and allow for a cheaper launch system, or perhaps for more payload to be carried into orbit.

The first successful flight of a scramjet, reaching speeds of Mach 7, was achieved in 2002 by the University of Queensland in Australia. NASA's Hyper-X program built the 12-foot long X-43 scramjet-powered research vehicle, which was flown on top of a modified Pegasus rocket. A picture and schematic of the X-43 are shown in Figures 8.17 and 8.18, respectively. The Pegaus–X-43 combination was dropped from a modified B-52 aircraft. The Pegasus rocket then took the X-43 to an altitude of over 90,000 ft and speeds high enough for the scramjet to operate. Then the X-43 was released and flew under its own power. In the final flight of the X-43 in 2004, it achieved a speed of Mach 9.6, or nearly 7,000 mph, and provided thrust for a duration of 13 s.

8.2 Jet Engines 397

FIGURE 8.17 Photo of an X-43 scramjet. Courtesy of NASA Dryden Flight Research Center; Photo by: Tom Tschida.

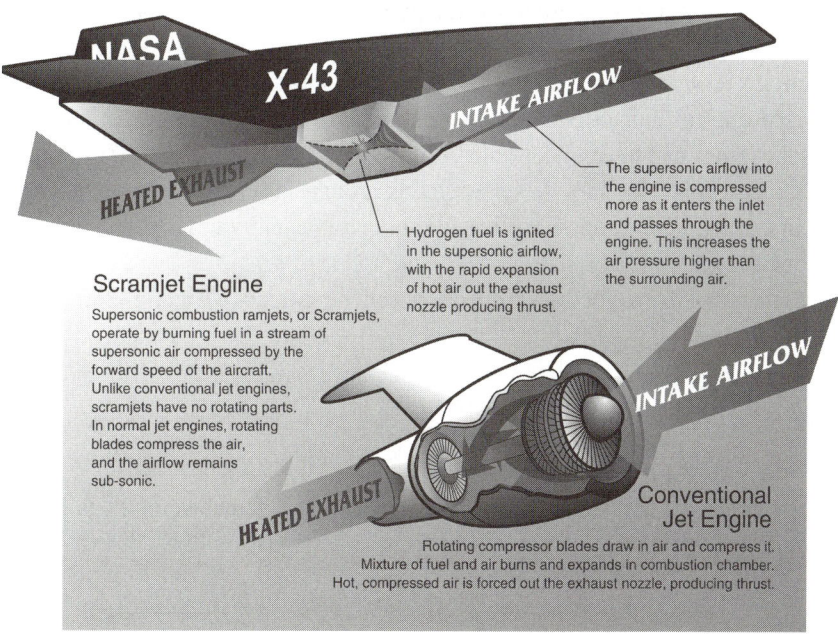

FIGURE 8.18 Schematic of an X-43 scramjet. Courtesy of NASA Dryden Flight Research Center; by: Dave Faust.

8.2.4 Pulse Detonation Engines

Pulse detonation engines (PDEs) are supersonic engines that operate on a transient, or cyclical, basis rather than in steady state. A *detonation* is a supersonic combustion wave. Most flames, such as those inside gasoline or diesel engines, are *deflagrations*, which are subsonic combustion waves. A basic PDE design is a tube with valves at one end that open to let in the fuel and air, and with the other end open to the atmosphere. After the fuel and air are introduced into the engine, the valves are closed and the fuel–air mixture is ignited. Since there is only one opening in the chamber, the hot exhaust gases travel back toward the exit. If enough energy is released in combustion for the flame velocity to reach supersonic speeds before it exits, then a detonation is achieved.

PDEs have high *theoretical* efficiency and could be used at high speeds, which would make them particularly well suited for use in missiles. They have fewer moving parts and thus would have lower cost than turbine engines. Practically, however, they have not been successful, and they are very noisy and hard to start.

8.3 Liquid Sprays

8.3.1 Fuel Injectors

Typical automotive fuel injectors are in the range of 40 to 60 psi (275 to 410 kPa). For diesel fuel injectors, the trend is for ever-increasing injection pressures. While increasing the injection pressure does decrease the drop size, the higher injection pressures are mainly used due to their effects on combustion and emissions in diesel engines. Modern diesel engines can have injection pressures of 20,000 psi (140 MPa) or higher. The typical injector hole size in a diesel engine is around 200 μm, with either 6 or 8 holes on an injector tip common. Gasoline fuel injectors usually have 1 or 2 holes of diameter about 1 mm.

For fuel injectors the velocity of injection can be calculated using Bernoulli's equation:

$$V = \sqrt{\frac{2 \Delta P}{\rho}} \tag{8.41}$$

In reality there are losses, frictional and otherwise, so the actual velocity is lower. Engineers prefer to keep the same functional form as in Equation 8.41, but we add a *discharge coefficient*, C_d, to account for these losses.

$$V = C_d \sqrt{\frac{2 \Delta P}{\rho}} \tag{8.42}$$

When most of the losses in flow are due to a *vena contracta* effect, an alternative viewpoint may be taken. That viewpoint is that the velocity is approximately what would be predicted by the frictionless Bernoulli equation, but that the *effective area* of flow has been reduced below the geometric area of the nozzle orifice. The definition of the effective flow area of the injector nozzle is

$$A_{\text{eff}} = C_d A \tag{8.43}$$

EXAMPLE 8.4

An automotive fuel injector at 60 psi delivers ethanol fuel at room temperature. The injector is a single-hole nozzle with a diameter of 1 mm, $C_d = 0.75$. If the engine is running at 1800 rpm, and the injector is open for 90 crank angles (CA), how much mass of fuel is injected? What is the effective flow area of the injector?

SOLUTION First, we calculate how long the injector is open:

$$90 \text{ CA} \times \frac{1 \text{ rev}}{360 \text{ CA}} \times \frac{1 \text{ min}}{1800 \text{ rev}} \times \frac{60 \text{ s}}{1 \text{ min}} = 0.00833 \text{ s}$$

Next, we calculate the velocity of the injected fuel:

$$V = 0.75 \sqrt{\frac{2(60 \text{ lbf/in}^2)}{(49.3 \text{ lbm/ft}^3)} \frac{144 \text{ in}^2}{1 \text{ ft}^2} \frac{32.2 \text{ lbm} \cdot \text{ft/s}^2}{1 \text{ lbf}}} = 79.7 \text{ ft/s}$$

In English units the flow area is

$$A = \frac{\pi}{4}d^2 = \frac{\pi}{4}(1 \text{ mm})^2 \left(\frac{0.00328 \text{ ft}}{1 \text{ mm}}\right)^2 = 0.00000845 \text{ ft}^2$$

and the mass injected is

$$m = \rho A V \Delta t = \left(49.3 \frac{\text{lbm}}{\text{ft}^3}\right)(0.00000845 \text{ ft}^2)\left(79.7 \frac{\text{ft}}{\text{s}}\right)(0.00833 \text{ s}) = 0.000277 \text{ lbm}$$

This is equivalent to 0.126 g, or 126 mg, in metric units. The effective flow area of the injector is

$$A_{\text{eff}} = C_d A = 0.75 \frac{\pi}{4}(1 \text{ mm})^2 = 0.589 \text{ mm}^2$$

Physical Processes in Liquid Sprays

There are four main physical processes that occur in liquid sprays: atomization, vaporization, secondary breakup, and collisions and coalescence. Other effects that are important are the entrainment of air into the spray and the turbulent dispersion of the liquid drops.

Atomization

Atomization is the process of breaking up a continuous stream of liquid into a discontinuous stream of drops. Models of atomization include Taylor analogy breakup, Kelvin-Helmholtz instability, Rayleigh-Taylor instability, and Information theory, among many others. Regimes of atomization range from Rayleigh breakup at very low Weber numbers to fully atomized at very high Weber numbers, which are of relevance to diesel sprays.

Rayleigh breakup (Figure 8.19) is a slow-speed breakup. It can be observed by carefully setting the flow on your kitchen faucet so that a continuous stream of liquid just barely comes out of the faucet (not a dripping flow). The Rayleigh breakup is a varicose instability. A liquid surface is unstable if the surface area created by breaking it into drops is less than the undeformed area of the liquid cylinder issuing out of the orifice. That is, a liquid stream is unstable if a lower energy state exists that it can reach.

Consider a ligament of cylindrical shape of length L coming out of an orifice of diameter D. The surface area of the ligament is πDL, and its volume is $(\pi/4)D^2L$. If an instability of wavelength λ were to grow on the surface of the ligament until at such point it pinches off in the middle and breaks the cylinder, then two drops would be formed. If the drops formed have diameter d, then the surface area of the two drops would be $2\pi d^2$, and the volume of the two drops would be $2(\pi/6)d^3$. By conservation of mass, the diameter of the two drops formed would be

$$d = 2\sqrt[3]{\frac{3}{8 \cdot 4}D^2L} = 0.9086\sqrt[3]{D^2L}$$

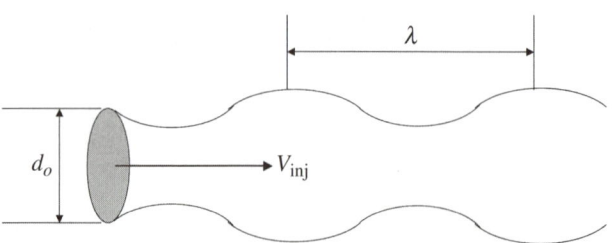

FIGURE 8.19 Schematic of Rayleigh breakup parameters on a deformed surface of a liquid cylinder.

The liquid cylinder is unstable if it is possible for the surface area of the drops to be less than the surface area of the cylinder. This is true when $2\pi d^2 < \pi DL$. Substituting in the relation for d derived from conservation of mass yields

$$8\pi \left(\frac{3}{32}D^2 L\right)^{2/3} < \pi DL$$

Solving for L yields the criterion for breakup of the cylinder:

$$L > 4.5D \tag{8.44}$$

So if a liquid cylinder has a length less than 4.5 times it diameter, it may be stable. But if the length is greater than 4.5 times the diameter, it will break up into spherical drops. What is the size of the drops that will be formed? Assuming the shortest wavelength is the one that will dominate and applying conservation of mass, we can calculate the drop size. If we assume each wavelength breaks up into one drop, then conservation of mass can be applied to find the diameter of the drop:

$$\frac{\pi}{4}D_0^2 \lambda = \frac{\pi}{6}d_{\text{drop}}^3 \tag{8.45}$$

For low-speed Rayleigh breakup, the optimum wavelength is $\lambda = 4.5D$, which results in $d = 1.9D$. The drop size prediction shown here is only valid when the flow rate is low enough that kinetic effects in the breakup process can be ignored, and surface tension forces dominate the breakup of the liquid. Figure 8.20 shows the breakup of a liquid sheet into drops.

FIGURE 8.20 Photo of the atomization of a liquid sheet breaking up due to instabilities on a liquid surface.

At the other end of the regime of breakup at very high Weber numbers is the fully atomized regime, which is of most relevance to fuel injectors. The diameter of drops formed in the atomization regime is much smaller than the injector hole size. The drops shear off right at the orifice. In fact, the liquid stream may already be broken up due to turbulence, cavitation, and other internal flow effects inside the fuel injector.

There is no truly fundamental theory to predict the drop size in the atomization regime, but there are several models. One is the Taylor analogy breakup (TAB) model. In this model the equation for mean injected drop size is

$$d = B_d \frac{2\pi\sigma}{\rho_{air} V^2} \lambda_m(\text{Ta}) \qquad (8.46)$$

where B is a nondimensional model constant of order (1), and λ_m is a nondimensional function of the Taylor number:

$$\text{Ta} = \frac{\rho_l}{\rho_{air}} \left(\frac{\text{Re}}{\text{We}_l} \right)^2 \qquad (8.47)$$

as described in reference [O'Rourke80].

Vaporization

Vaporization is the change of phase of the injected fluid from a liquid to a gas vapor, due to heat transfer from the gas (usually air) to the liquid. The D^2-law governs the lifetime of an isolated drop that is vaporizing:

$$D^2 = D_o^2 - k_{vap} t$$

where

$$k_{vap} = \frac{2k}{\rho c_P} \ln(1 + B)$$

and the thermal transfer number, B, is defined as

$$B = \frac{c_P(T_{air} - T_{drop})}{h_{fg}}$$

Setting $D(t) = 0$, the drop lifetime is calculated as

$$t = \frac{D_0^2}{k_{vap}}$$

which shows that if the size of a drop is doubled, it will take four times as long to vaporize.

Secondary Breakup

Secondary breakup is the further breakup of the spray drops after they have been injected due to aerodynamic forces. Note that surface tension acts to bring the atomized drops into a spherical shape, which minimizes their surface area, while aerodynamic forces can stretch the drops into different shapes. The key parameter in secondary breakup is the aerodynamic Weber number:

$$\text{We} = \frac{\rho_{\text{air}} V^2 D}{\sigma} \tag{8.48}$$

Of secondary importance is the Ohnesorge number:

$$\text{Oh} = \frac{\mu_l}{\sqrt{\rho_l \sigma D}} \tag{8.49}$$

In high relative velocity systems (diesel engines, gas turbines, and air-assist atomizers used on viscous fluids) the drops formed by primary atomization are subjected to aerodynamic shear forces, which lead to deformation and breakup. Most of the experimental studies on secondary breakup have been done by placing a single drop in a shock tube. The nondimensional timescale of drop breakup is

$$t^* = t \frac{V}{D} \sqrt{\frac{\rho_{\text{air}}}{\rho_l}} \tag{8.50}$$

For We < 12 the drop breakup process is very slow, and for engineering purposes it can be assumed that no breakup occurs in this case. So defining $\text{We}_{\text{crit}} = 12$, then the largest size of drops that can expected can be deduced:

$$d_{\max} = \frac{\text{We}_{\text{crit}} \sigma}{\rho_{\text{air}} V^2} \tag{8.51}$$

where V is the relative velocity between the drop and the gas. Note that viscosity will act to resist any shearing motion and breakup. To account for the effects of viscosity one popular correlation defines a "corrected Weber number" that can then be substituted into the previous correlation [Pilch87]:

$$\text{We}_{\text{crit,cor}} = 12(1 + 0.077 \, \text{Oh}^{1.6}) \tag{8.52}$$

EXAMPLE 8.5

Estimate the largest possible size of a rain drop at sea level under atmospheric conditions.

SOLUTION The maximum velocity a rain drop will reach is its terminal velocity, which is reached when the weight equals the drag force:

$$\frac{\pi}{6}d^3\rho_l g = \frac{1}{2}C_D\rho_a\frac{\pi}{4}d^2V^2$$

Rearranging to solve for V yields

$$V = \sqrt{\frac{4}{3}\frac{d\rho_l g}{C_D\rho_a}}$$

The criterion for the largest drop that can survive aerodynamic breakup was given previously as

$$d_{max} = \frac{We_{crit}\sigma}{\rho_{air}V^2}$$

This gives us two equations with two unknowns (d, V). We can substitute to eliminate V from the second equation, yielding

$$d = \frac{We_{crit}\sigma}{\rho_{air}}\frac{3C_D\rho_{air}}{4d\rho_l g}$$

And solving this equation for d, we have

$$d = \sqrt{\frac{3}{4}\frac{We_{crit}\sigma C_D}{\rho_l g}}$$

Now in general $C_D = C_D(d, V)$, so this problem will likely require an iterative solution. Making an initial assumption of $C_D = 0.5$, the first guess for d will be

$$d = \sqrt{0.75\frac{12(0.072 \text{ N/m})(0.5)}{(1000 \text{ kg/m}^3)(9.8 \text{ m/s}^2)}} = 0.00575 \text{ m} = 5.75 \text{ mm}$$

Now the value of C_D needs to be checked by computing the velocity and the Reynolds number:

$$V = \sqrt{\frac{4(0.00575 \text{ m})(1000 \text{ kg/m}^3)(9.8 \text{ m/s}^2)}{3(0.5)(1.2 \text{ kg/m}^3)}} = 11.2 \text{ m/s}$$

$$Re = \frac{Vd}{\nu} = \frac{(11.2 \text{ m/s})(0.00575 \text{ m})}{(1.5 \times 10^{-5} \text{ m}^2/\text{s})} = 4290$$

which is indeed in the flat part of the drag curve for a sphere, so the initial guess for C_D was correct. This also agrees with experimental measurements that the largest raindrops are about 5 mm in diameter. Two minor effects that were not taken into account in this problem are that there will be internal circulation in the drop, which will reduce the drag force slightly, and the drop will deform so that the front will flatten out somewhat, increasing the effective cross-sectional area.

Collisions and Coalescene

Since most sprays are turbulent, the turbulent dispersion of the drops can introduce relative velocities between the injected drops, which can lead to collisions between drops. Once drops collide, there are two likely outcomes—coalescence into one large drop, or a separation or glancing collision that maintains the original two drops (at very high collision kinetic energies it is possible for the drops to shatter into many small drops). The easiest way to model the collision rate between the drops is to make an analogy to the kinetic theory of ideal gases, in which molecules are modeled as hard spheres that take on straight-line trajectories between collisions and have random directions. In sprays, the assumption also must be made that the turbulence in the spray is *isotropic*, so that it scatters the spray drops in all directions.

A single drop of radius R will sweep out a volume $\pi R^2 V \Delta t$ over a time period Δt. If the drops are uniformly distributed in space with a number density ($n/\mathcal{V}$), then the expected number of drops to be hit by one drop would be $n\pi R^2 V \Delta t/\mathcal{V}$. The total number of collisions for all drops is then calculated by multiplying by n, and dividing by 2 to avoid double-counting possible collision partners. The resulting collision rate between drops per unit time is then

$$\nu_{\text{coll}} = \frac{n^2 \pi R^2 V_{\text{rel}}}{\mathcal{V}} \tag{8.52}$$

Once it is determined that spray drops have collided, the outcome of the collision has to be determined. For example, they may coalesce and form a larger drop. This is important because it takes longer for larger drops to vaporize, according to the D^2-law. In diesel engines longer vaporization times result in slower combustion, which results in lower thermodynamic efficiency and worse emissions.

The standard approach in CFD codes is to break the process into two steps: First, how often do the drops collide? Second, what happens when they collide? Note that sprays are complex, turbulent, three-dimensional processes, so analytical or exact solutions to the Navier-Stokes equations are not possible. The approaches to deal with sprays are either empirical correlations or CFD. Since correlations are only valid for the type of spray from which the data came from, CFD modeling is very common, especially for new designs.

For drops to collide, there must be a relative velocity between them. The main cause is turbulence, but areas of acceleration or deceleration in the flow can also be responsible. If drops touch they will coalesce, at least temporarily, due to surface tension. Computing drop–drop collisions can be a numerically taxing chore. If we do this deterministically, and want to consider all possible collision partners, then if there are $n = 1$ million drops in a spray, there will be $O(n^2)$ calculations to perform. For a computer capable of 100 MLOPS, it will take approximately 10 hours to do the computation. So stochastic procedures are normally used; we group the drops.

When two drops collide and form one new drop, conservation of mass can be employed:

$$\rho \frac{4}{3}\pi R_1^3 + \rho \frac{4}{3}\pi R_2^3 = \rho \frac{4}{3}\pi R_{new}^3$$

If both of the initial drops were the same size, so that $R_1 = R_2$, the new drop will have a size given by

$$R_{new} = R_1 \sqrt[3]{2} = 1.26 R_1$$

If the D^2-law holds true, the new drop will take $(1.26)^2 = 1.59$ times as long to vaporize as the original drops. Note that if two drops of greatly unequal size collide and coalesce, say that $R_1 = 10 R_2$, the size of the new drop is about that of the larger:

$$R_{new} = \sqrt[3]{1^3 + \left(\frac{1}{10}\right)^3} R_1 = 1.0003 R_1$$

So in this case because the mass of drop R_1 was so much greater than drop R_2, the new drop is not much different in size.

8.3 Liquid Sprays

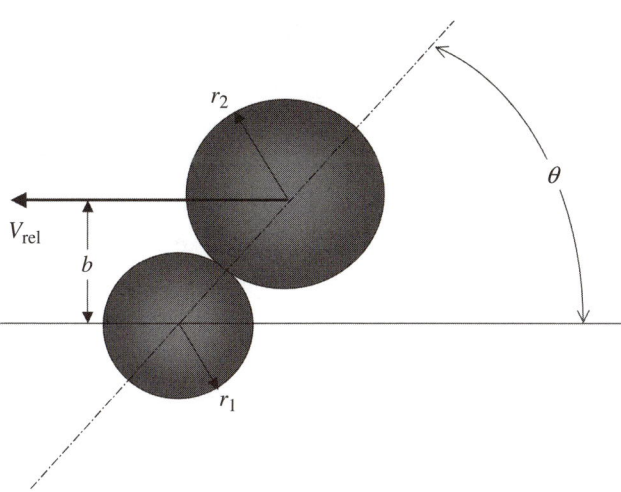

FIGURE 8.21 Schematic of a drop–drop collision. The offset between the centerlines of the drops and the relative velocity vector is $b = (r_1 + r_2)\sin(\theta)$.

The model of Brazier-Smith et al. [Post02] is commonly used to predict the outcome of a drop–drop collision:

1. First, assume the two drops temporarily coalesce and form one spherical drop.
2. Then compute the rotational kinetic energy of the temporary, coalesced drop.
3. If the rotational kinetic energy is greater than the change in kinetic energy required to form the original two drops from the current one, then the drops will split apart. Otherwise they will stay coalesced.

The equations needed to perform these steps follow.

Figure 8.21 shows a schematic of two drops colliding. The moment of inertia for a sphere rotating about its center is

$$I = \frac{8\pi R^5 \rho}{15} \tag{8.54}$$

The angular momentum about the center of gravity of the system is

$$\Omega = \frac{4\pi \rho V b r_1^3 r_2^3}{3(r_1^3 + r_2^3)} \tag{8.55}$$

The rotational kinetic energy can be related to the angular momentum and the moment of inertia by

$$\mathrm{KE}_{\mathrm{rotational}} = \frac{\Omega^2}{2I} \tag{8.56}$$

Substituting in for the angular momentum and the moment of inertia, the rotational kinetic energy is

$$KE_{rotational} = \frac{5\pi \rho V^2 b^2 r_1^6 r_2^6}{3(r_1^3 + r_2^3)^{11/3}}$$

The change in surface energy from one drop to two drops is equal to the surface tension times the change in area for the process:

$$\Delta SE = \sigma \Delta A = \sigma 4\pi \left[(r_1^2 + r_2^2) - R^2\right] \tag{8.57}$$

The radius of the coalesced drop, R, can be related to the radii of the initial two drops by conservation of mass:

$$R = \sqrt[3]{r_1^3 + r_2^3} \tag{8.58}$$

Setting the change in surface energy equal to the rotational kinetic energy gives

$$\sigma 4\pi \left[(r_1^2 + r_2^2) - (r_1^3 + r_2^3)^{2/3}\right] = \frac{5\pi \rho V^2 b^2 r_1^6 r_2^6}{3(r_1^3 + r_2^3)^{11/3}}$$

With some rearranging, we have

$$\frac{\rho V^2}{\sigma} = \frac{12\pi}{5\pi} \left[(r_1^2 + r_2^2) - (r_1^3 + r_2^3)^{2/3}\right] \frac{(r_1^3 + r_2^3)^{11/3}}{b^2 r_1^6 r_2^6}$$

With the introduction of the nondimensional parameters we obtain

$$B = \frac{b}{r_1 + r_2} \tag{8.59}$$

$$\gamma = \frac{r_2}{r_1} \tag{8.60}$$

where $r_2 > r_1$, and

$$We = \frac{\rho V^2 r_1}{\sigma} \tag{8.61}$$

Then the energy balance can be simplified to

$$We_{crit} = \frac{2.4}{B^2} f(\gamma) \tag{8.62}$$

where $f(\gamma)$ is a rather complex function. For $\gamma = 1$, $f(\gamma) = 1.3$. A good curve fit to $f(\gamma)$ is $f(\gamma) = \gamma^3 - 2.4\gamma^2 + 2.7\gamma$. If We $>$ We$_{crit}$, then the drop breaks back into the original two drops after a collision, and if We $<$ We$_{crit}$, the drops permanently coalesce into a larger drop after colliding [Post02].

Drops from fuel injectors are not all the same size but have a range of sizes due to turbulence in the fuel and turbulence in the air, among other factors. The mean drop size is normally defined as the ratio of the mass of the drops to the surface area of the drops, which is termed the *Sauter mean diameter* (SMD):

$$d_{32} = \frac{\Sigma d^3}{\Sigma d^2} \tag{8.63}$$

8.3.2 Other Applications

Another application of liquid sprays, besides fuel injectors, is metered dose inhalers for asthma patients. These inhalers are designed to produce drops of approximately 5 μm that can be taken deep into the lungs to administer the medicine where it is needed. Larger drops would be filtered out in the upper respiratory tract.

Liquid sprays are also used in inkjet printers. Of the two types of printers—thermal/bubble jet and piezoelectric—the thermal jet is by far the most common. Inkjet technology is also called *drop on demand* printing. The diameter of the ink drops is proportional to the hole size in the printhead. As the inkjet printer has evolved over time, the hole size has gotten smaller and smaller (leading to higher DPI resolution). Of course, smaller drops deliver less ink, so to compensate and increase the mass flow rate without the print speed getting too slow, more nozzles (holes) have been added. Modern printers have $O(1000)$ nozzles on each printhead. For multicolor printers, there is a separate printhead for each color. For a hole size of 60 μm, the volume of a single drop is $O(10$ picoliters$)$. Ejection velocities are $O(10$ m/s$)$.

EXAMPLE 8.6

In the TAB atomization model, the predicted drop size is given by

$$d = C_1 \frac{2\pi\sigma}{\rho_{air} U_0^2}$$

where U_0 is the injection velocity and C_1 is a parameter that depends on the fluid viscosity and surface tension as well as both ambient and fluid densities and injection velocities. If $C_1 = 19.5$ for the conditions of interest, what is the predicted drop size for tetradecane ($C_{14}H_{30}$) fuel at an injection velocity of 100 m/s, into air at atmospheric conditions?

SOLUTION The surface tension of $C_{14}H_{30}$ at room temperature is 0.022 N/m, and the density of air is about 1.2 kg/m^3, so the predicted drop size is

$$d = 19.5 \frac{2\pi(0.022 \text{ N/m})}{(1.2 \text{ kg/m}^3)(100 \text{ m/s})^2} = 0.000225 \text{ m} = 0.225 \text{ mm}$$

Note that in a diesel engine, fuel injection takes place near top dead center, so for a naturally aspirated engine with a compression ratio of 18, the air density would be 21.6 kg/m^3, and the predicted drop size would be 12.5 μm.

EXAMPLE 8.7

A small hole is accidentally poked in the bottom of an ethanol fuel tank mounted to the side wall of an engine lab. Estimate the size of the ethanol drops that will drip out through the hole.

SOLUTION There is no pressure applied to the fuel and no driving force other than gravity to generate a high mass flow rate. Also, the problem states that the hole is small. Thus we can assume that the flow out the hole will drip in a slow process rather than be atomized. A force balance between a surface tension force resisting the fluid leaving the tank and a gravity force trying to pull the ethanol out yields

$$\sigma \pi d_{\text{hole}} = \frac{\pi}{6} \rho d_{\text{drop}}^3 g$$

Rearranging to solve for d_{drop} gives

$$d_{\text{drop}} = \sqrt[3]{\frac{6\sigma d_{\text{hole}}}{\rho g}}$$

So for a hole size of 1.0 mm, the expected drop size is

$$d_{\text{drop}} = \sqrt[3]{\frac{6(0.023 \text{ N/m})(0.001 \text{ m})}{(791 \text{ kg/m}^3)(9.8 \text{ m/s}^2)}} = 0.0026 \text{ m} = 2.6 \text{ mm}$$

Note that for such slow dripping processes, where the velocity is so small as to be negligible (different from Rayleigh breakup), the relevant nondimensional parameter is the Bond number:

$$\text{Bo} = \frac{\rho g d^2}{\sigma} \tag{8.64}$$

and so the predicted drop size can be expressed in terms of the Bond number as

$$\frac{d_{\text{drop}}}{d_{\text{hole}}} = \sqrt[3]{\frac{6}{\text{Bo}}}$$

8.4 Flow for Electronics Cooling

In many consumer electronic devices, the computer processors will be damaged if they become too hot. According to one government report, the failure rate of electronic components increases by a factor of 10 when the temperature is increased from 75°C to 140°C [MIL91]. Figure 8.22 shows the increase in the failure rate as a function of temperature.

In an electronic component, the electrical heat generated is calculated as

$$\dot{W} = I^2 R \tag{8.65}$$

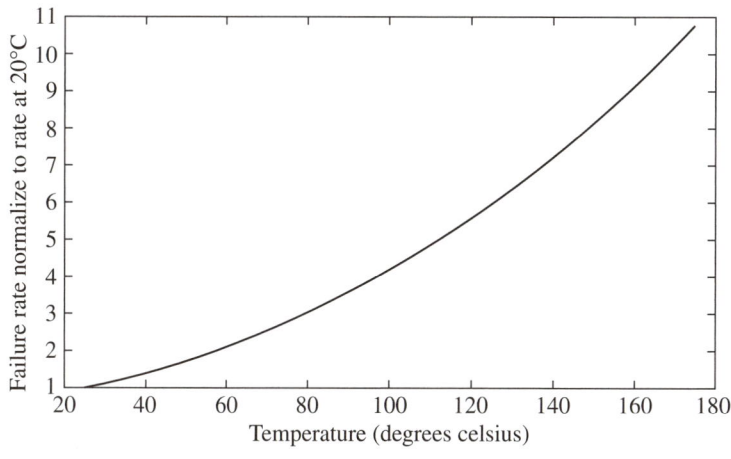

FIGURE 8.22 Failure rate of electronic components as a function of temperature, from [MIL91].

where I is the current and R the resistance. In steady state, the required cooling load, $\dot{Q}$, is equal to the electric power input. So if a device is using 10 W of electrical energy, that 10 W must be removed by the cooling system. As with all engineering designs, we want to be conservative and add a factor of safety to oversize the cooling system. However, providing too much cooling leads to unnecessary weight, cost, parasitic power consumption, and noise.

To be able to engineer cooling systems for electronics, an understanding of fans (Chapter 7) and boundary layers (Chapter 4) is required, as well as a basic understanding of heat transfer. The first law of thermodynamics, from Chapter 3, is

$$\frac{dE}{dt} = \dot{Q} - \dot{W} + \sum_{\text{in}} \dot{m}\left(h + \frac{V^2}{2} + gz\right) - \sum_{\text{out}} \dot{m}\left(h + \frac{V^2}{2} + gz\right) \tag{3.39}$$

Under the assumptions of steady state, no work, negligible changes in kinetic and potential energies, and constant specific heats, this reduces to

$$\dot{Q} = \dot{m}_{\text{air}} c_P (T_{\text{out}} - T_{\text{in}}) \tag{8.66}$$

Here $\dot{Q}$ is used for the heat transfer rate in watts. We could also have derived this equation by performing an energy balance on the whole system.

Equation 8.66 shows that the total heat generated by the electronics, $\dot{Q}$, is balanced by the change of thermal energy of the cooling air from the point it enters the system to when it leaves. But what is the mechanism by which $\dot{Q}$ is transmitted from the electronics to the fluid? The heat is transmitted by one of the three modes of heat transfer:

1. Conduction. Heat transfer by direct contact through a continuous substance.
2. Convection. Heat transfer by a fluid moving over a solid surface.
3. Radiation. Heat transfer by electromagnetic radiation emitted from the surface of a material.

For the temperature range considered here for electronics cooling, we can safely ignore radiation. So let's consider conduction first. Fourier's law of conduction is

$$\dot{Q} = kA \frac{dT}{dx} \tag{8.67}$$

where k is the thermal conductivity of the substance. The parameter k is high for metals and low for most fluids. For aluminum, k is particularly high, which is why it is popular for use in heat sinks. A is the cross-sectional area across which heat flows, and dT/dx is the change in temperature over distance.

Now let us consider convection. The heat transfer due to fluid convection is governed by Newton's law of cooling:

$$\dot{Q} = hA(T_{\text{surface}} - T_{\text{fluid}}) \tag{8.68}$$

where A is the surface area exposed to the fluid and h is the convection coefficient. Here, $h = h(\text{Nu}) = h(\text{Re}, \text{Pr})$ are relevant nondimensional parameters. The Nusselt number, Nu, is a nondimensional heat transfer coefficient,

$$\text{Nu} = \frac{hL}{k} \tag{8.69}$$

where k is the thermal conductivity of the substance and L is a relevant length scale. For flow over a cylinder of diameter D, the length scale is $L = D$. The Prandtl number, Pr, is a property of the fluid and is the ratio of momentum transfer to thermal transfer within the fluid:

$$\text{Pr} = \frac{\nu}{\alpha} = \frac{c_P \mu}{k} \tag{8.70}$$

where α is the *thermal diffusivity*, which is related to other properties of the fluid by

$$\alpha = \frac{k}{\rho c_P} \tag{8.71}$$

Note that Pr = 0.7 for air, and about 7 for water. The thermal boundary layer looks very similar to the velocity boundary layer.

8.4.1 Air Cooling

Low-power systems may be cooled adequately by *natural convection*. Natural convection is the cooling due to the buoyancy-induced motion of a fluid when temperature differences cause density differences and the fluid near the hot surface rises. One advantage of natural convection cooling systems is that there are no moving parts, so it is quiet, no power is wasted in a fan, and there is no possibility of fan failure. Given the small distances in electronics cooling and low velocities of natural convection systems, the boundary layers are laminar. An important nondimensional number in natural convection cooling is the Grashof number, Gr:

$$\text{Gr} = \frac{g\beta(T_s - T_\infty)L^3}{\nu^2} \tag{8.72}$$

where β is the coefficient of thermal expansion, defined as

$$\beta = \frac{1}{V}\frac{dV}{dT} \tag{8.73}$$

For an ideal gas such as air, $V = NRT/P$, so that

$$\beta = \frac{P}{NRT}\left(\frac{NR}{P}\right) = \frac{1}{T} \tag{8.74}$$

is true for ideal gases. Note that the temperature must be expressed in absolute units, such as K. The value of β for water at 20°C is 0.000088 K^{-1}; for air at 20°C it is 1/293 K = 0.0034 K^{-1}.

For higher cooling rates, *forced convection* must be used. Increasing the velocity with fans in forced convection increases the heat transfer coefficient, h, so that more heat may be removed. In forced convection the flow could be laminar or turbulent. Axial fans—which are simple, cheap, light, and small—are commonly used in computer cooling. However, they can only be used for small pressure resistance, and they are noisy. Turbulent flow enhances thermal mixing and heat transfer, just as it enhances fluid mixing, but it also causes more frictional resistance, which requires more fan power. Correlations for the Nusselt number are almost always expressed in the following functional form:

$$Nu = C \cdot Re^m Pr^n \tag{8.75}$$

Such correlations can be found in a heat transfer book for a variety of geometries. For example, for flow inside a cylinder, one correlation is [NACA54]

$$Nu = 0.031 Re^{0.8} Pr^{0.4} \tag{8.76}$$

For external flow over a cylindrical wire, a correlation is [NACA52]

$$Nu = 0.478 Re^{0.5} Pr^{0.3} \tag{8.77}$$

and for a flat plate in laminar free-convection flow, a correlation is [NACA53b]

$$Nu = 0.548 Gr^{1/2} Pr^{1/2} \tag{8.78}$$

where Gr is the Grashof number. This correlation is for a vertical plate.

8.4.2 Liquid Cooling

In applications where air cooling cannot provide a sufficient heat transfer rate, liquid cooling can be used. Liquids have higher thermal conductivities and heat capacities and much higher densities than gases. Challenges with using liquids include leakage and corrosion. If the electrical components are completely immersed in the liquid, the system can be designed either so the liquid remains in the liquid state, or so the liquid boils so that the effect of the latent heat of vaporization can be used to further enhance the cooling rate. Fluids used in such direct contact cooling must be *dielectric* (water is not, fluorocarbon fluids are).

Liquid cooling systems for PCs are similar to the cooling systems in cars. A liquid cooling system needs a pump to move liquid over the hot components, and a radiator to transfer heat from the liquid to the air, and a fan to move air over the radiator. As transistors switch on and off, electricity is moving through wires and other components, generating heat. In conventional systems, the heat is transferred to a heat sink on top of the processor. The heat sink has a lot of fins that provide the surface area for air to flow over to remove the heat.

■ 8.5 Flows in Biological Systems

Our discussion of applications of fluid mechanics to biological systems will be broken into two parts. We will first discuss internal flows, such as the flow of blood and air inside living animals. Then we will deal with external flows, especially the drag on flying and swimming creatures.

8.5.1 Internal Flows

Human beings and other animals have internal flows of air through the respiratory system, blood through the circulatory system, and water through the digestive and urinary systems. Most internal biological flows are laminar [Vogel96].

The lungs of the average adult human have surface area of 30 m^2 and process 10 to 20 m^3 of air per day. At resting condition, about 700 cc of air is inhaled and exhaled with each breath, at a rate of about 12 breaths per minute. Under typical conditions there is about 1100 cc of "reserve air" left in the lungs after exhaling [Hinds82]. The surfaces of the upper respiratory system—defined as the nose to the trachea—are covered with a mucus, which is moved by the cilia upward to the pharynx, where it is periodically swallowed, thus removing any particles that lodge on those surfaces. Particle removal from the lower respiratory system is slower, and particles can remain in the lungs for months. Large particles impact the walls of the upper respiratory system due to *inertial impaction*. Particles larger than 10 μm generally always deposit on wall surfaces.

Fluid movement in the respiratory system is important to health. Particulate matter in the atmosphere has been linked to bronchitis, asthma, pneumonia, and other upper respiratory infections. PM_{10} is defined as inhalable particles (larger deposits in the upper respiratory tract) and $PM_{2.5}$ is defined as fine particles. National Ambient Air Quality Standards for fine particulates in the ambient environment is 65 μg/m^3, based on a 24-hour average [deNevers00].

For a human at resting conditions, the blood outflow of each side of the heart is about 6 L/min. The aorta has an internal diameter of about 2.5 cm, and so by conservation of mass the velocity of blood in the aorta is about 0.2 m/s. Capillaries have a diameter of about 6 μm, with a velocity of 1 mm/s. By conservation of mass there are about 3,000,000,000 capillaries in parallel flow with each other. The sum total cross-sectional area of all the capillaries in a human being is about 0.1 m^2 [Vogel96]. Why is the cross-sectional area of the capillaries so much higher than that of the aorta, since they can be considered as pipes in series? The answer is that the frictional hydraulic resistance changes with pipe diameter. Assuming laminar flow, the pressure drop through a pipe is given by

$$Q = \frac{\pi \Delta P D^4}{128 L \mu}$$

Thus for a constant flow rate Q, the resistance to flow ΔP must be proportional to D^{-4}. Murray's law [LaBarbera90] states that at each branch, the sum of the cube of the diameter of each parallel pipe vessel should be the same:

$$\sum_i D_i^3 = \sum_j D_j^3$$

for any two levels of branching, i and j. Murray's law was derived by including the fluid mechanics costs of flow resistance along with the biomechanical cost of building and maintaining the system of vessels and support structure. See Table 8.1 to see how well Murray's law stacks up to actual data.

Most mammals have an aortic blood pressure around 13,000 Pa (1.89 psi) [Vogel96]. The aortic pressure in giraffes is about three times this value. Why do you think this is so? Murray's law and Poiseuille flow assume fully developed laminar flow, which is valid when Re < 2,000 and L/D > 10. Here $L/D \approx 1$ for most of the passageways, so the flow is not fully developed. Note that the fact that the flow is not fully developed probably enhances transport at the walls, since fully developed laminar flow has very low velocities near the wall.

Table 8.1 Fluid Exchange Transport Systems in Mammals

Vessel	Average radius	Number	$\sum r^2$	$\sum r^3$	$\sum r^4$
Homo sapiens					
Aorta	1.25	1	1.56	1.95	2.44
Arteries	0.2	159	6.36	1.27	0.25
Arterioles	0.003	1.4×10^7	127.4	0.382	0.0011
Capillaries	0.0006	3.9×10^9	1432	0.86	0.0005
Venules	0.004	3.2×10^8	1273	2.55	0.0051
Veins	0.25	200	12.9	3.18	0.80
Vena cava	1.5	1	2.25	3.38	5.06

| **Canis familiaris** | | | | | |

Vessel	Average radius	Number	$\sum r^3$
Aorta	0.5	1	1.25
Large arteries	0.15	40	1.35
Main arterial branches	0.05	600	0.75
Terminal branches	0.03	1800	0.49
Arterioles	0.001	4.0×10^7	0.40
Capillaries	0.0004	1.2×10^9	0.77
Venules	0.0015	8.0×10^7	2.7
Terminal veins	0.075	1800	7.59
Main venous branches	0.12	600	10.37
Large veins	0.30	40	10.8
Vena cava	0.625	1	1.53

Hamster cheek pouch retractor muscle arteriolar network

Vessel order	Average radius	Number	$\sum r^3$
0	1.8	476	2.78
1	2.85	144	3.33
2	4.2	41	3.03
3	7.6	12	5.27
4	13.65	2	5.09

From [LaBarbera90]. Used with permission.

8.5.2 External Flows

Not all birds fly, and not every creature that flies is a bird. Insects and bats also fly, and fish and aquatic mammals swim—or "fly" through water. There are around 10,000 species of bird, of which about 8,500 fly [Videler06]. Not all birds fly the same way.

Hummingbird flight is very different from falcon flight. Also, bird flight is quite different from airplane flight: Birds are capable of changing the airfoil profile of their wings during flight.

It appears that most bird body shapes have been optimized to provide the lowest drag for the highest volume [Videler06], even within the wide range of bird sizes (bee hummingbirds weigh just a few grams and are 7 cm in length, whereas the wandering albatross has a wingspan over 3 m long and a mass over 8 kg). Birds have been seen flying at altitudes as high as commercial airliners fly, and alabatrosses have been known to fly for thousands of miles over the ocean.

The Great Flight Diagram of Tennekes, shown in Figure 8.23, shows the range of sizes and airspeeds for birds, along with the same information for insects and man-made aircraft for comparison.

For birds and bats and insects, their wings are not only wings but propellers as well. During the downstroke of a wingbeat, the outer half of a bird's wings functions almost exactly like a propeller, with the primary feathers angled forward, much like the pitch on an airplane propeller. Figure 8.24 shows a bald eagle with the primary feathers at the end of each wing angled down and forward to produce thrust during the downstroke. The lift force on the wing causes the feathers at the end to bend upward. Figure 8.25 shows the same effect on an osprey. Figure 8.26 illustrates the many different parts of the wingstroke. Birds fold their wings in during the upstroke to minimize negative lift. At extremely small Reynolds numbers, twist and camber do not seem to affect the performance of a wing, which is likely why many insects have straight wings.

Just as for an airplane, there are four forces acting on a bird in flight: thrust, drag, lift, and weight. In a bird the wings provide both the thrust and the lift, although for most birds in flapping flight the inner part of the wing provides most of the lift while the outer part of the wing provides most of the thrust. For a bird flying forward in level flight at a constant velocity V, the four forces must balance, so that lift equals weight and thrust equals drag. By scaling arguments it can be shown that the order of magnitude of the power required for bird flight is

$$\dot{W}_{\text{flight}} \approx \frac{W^2}{\rho b^2 V} + \rho A_{\text{body}} V^3 C_D \qquad (8.79)$$

where the first term represents the power to overcome the induced drag (see Chapter 6 for an explanation of induced drag) and the second term is the power to overcome the body drag of the bird. Here W is the weight and b is the wingspan. Since the first term is inversely proportional to the velocity, and the second term is directly proportional to the velocity cubed, the power required for flight cannot be minimized by either flying extremely slow or extremely fast, but rather by flying at some finite intermediate

8.5 Flows in Biological Systems 419

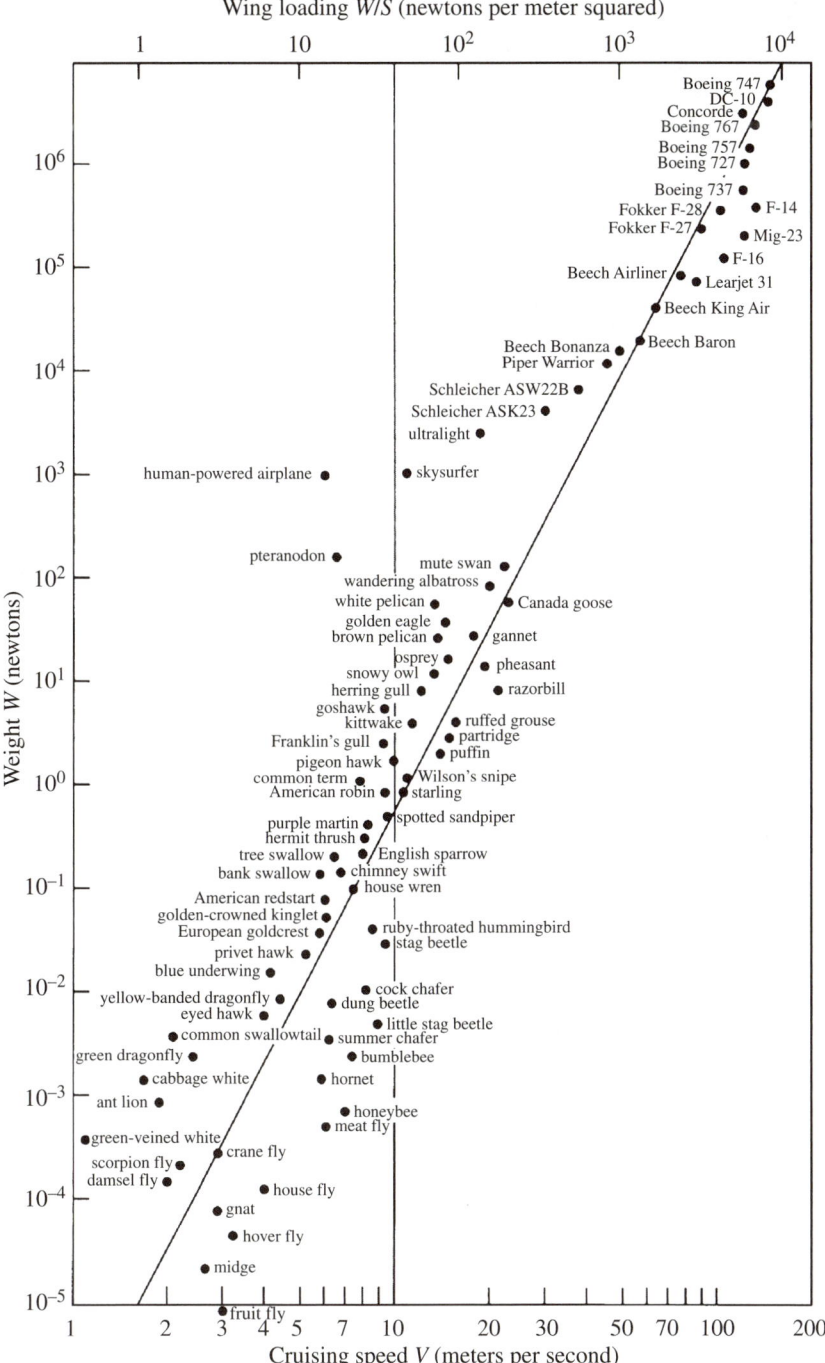

FIGURE 8.23 Great Flight Diagram of Tennekes, from [Tennekes97]. Used with permission of MIT Press.

FIGURE 8.24 Bald eagle in flight. Courtesy of NASA Kennedy Space Center.

FIGURE 8.25 Osprey in flight. Courtesy of NASA Kennedy Space Center.

FIGURE 8.26 White pelicans in flight. Courtesy of NASA Kennedy Space Center.

speed, called the *speed for minimum power*. Note that this is not the same as the speed for maximum range. The speed for maximum range is slightly higher than the speed for minimum power. Videler reports the maximum speed of a peregrine falcon has been measured by researchers at 51 m/s in a dive [Videler06].

Bird flight can take several modes, including gliding, soaring, flapping, and bounding. Some birds are capable of flapping in a hovering flight. Figure 8.27 gives an illustration of bounding flight, in which the finch has folded his wings in next to his body to minimize drag as he glides to a landing on the feeder. Shortly before the picture was taken, the finch must have been flapping to reach that position and velocity. Figure 8.28 shows an example of gliding flight.

In fish, the nature of their swimming, where the whole body moves, is such that the thrust-producing motions and the drag-inducing motions are inseparable [Vogel96]. However, just as some birds use bounding flight to reduce energy expenditures, fish also glide through water for short periods as well. Vogel notes that the drag coefficient of a California sea lion and a seal has been measured to be around 0.004, almost the same as that for turbulent flow over a flat plate of the same area at the same Reynolds number. Table 8.2 shows the drag coefficients for a variety of animals.

FIGURE 8.27 Finch in bounding flight, approaching a landing.

FIGURE 8.28 Gull in gliding flight.

Table 8.2 Drag Coefficients for Animals That Move Through Fluid Media. Reference areas are defined as F = frontal, W = wetted, P = planform, V = (volume)$^{2/3}$

Animal	Reynolds number	Reference area	Drag coefficient
Flea	65–205	F	0.96–1.02
Fruit fly	300	F	1.16
Tadpole	1,000–2,500	F	0.36–0.74
Frog	1,500–8,000	W	0.11–0.24
Locust	8,000	F	1.47
Dytiscid beetle	8,600–15,000	F	0.28–0.33
Crab	10,000	P	0.30–0.35
Cephalopod	100,000	V	0.48
Trout	50,000–200,000	W	0.015
Mackerel	100,000–175,000	W	0.0043–0.0052
Falcon	380,000	F	0.24
Duck (underwater)	420,000	W	0.028
Saithe	500,000	W	0.005
Penguin	1,000,000	W	0.0044
Seal	1,600,000	W	0.004
Human, swimming	1,600,000	W	0.035
Sea lion	2,000,000	W	0.0041

Adapted from [Vogel96]. Used with permission of Princeton University Press.

Summary

After reading this chapter and working through the problems, you should have an understanding of the basics of jet and rocket propulsion, including simple one-dimensional compressible flows. Other topics introduced include fuel injectors and liquid sprays, convective cooling of electronics, and flows in biological systems.

References

[JSME88] Japan Society of Mechanical Engineers. 1988. *Visualized Flow: Fluid Motion in Basic and Engineering Situations Revealed by Flow Visualisation.* Pergamon Press.

[NACA52] Scadron, M. and I. Warshasky. 1952. *Experimental Determination of Time Constants and Nusselt Numbers for Bare Wire Thermocouples in High-Velocity Air Streams and Analytic Approximation of Conduction and Radiation Errors.* NACA-TN-2599.

[NACA53] *Equations, Tables, and Charts for Compressible Flow.* 1953. NACA Report 1135.

[NACA53b] Ostrach, S.1953. *Analysis of Laminar Free-Convection Flow and Heat Transfer about a Flat Plate Parallel to the Direction of the Generating Body Force.* NACA Report 1111.

[NACA54] Aladyev, I. 1954. *Experimental Determination of Local and Mean Coefficients of Heat Transfer for Turbulent Flow in Pipes.* NACA TM 1356.

[NASA75] Talay, T. 1975. *Introduction to the Aerodynamics of Flight.* NASA SP-367.

[O'Rourke80] O'Rourke, P. and F. Bracco. 1980. "Modelling of Drop Interactions in Thick Sprays and a Comparison with Experiments." *Proceedings of the Institution of Mechanical Engineers*, Vol. 9, pp. 101–106.

[Hinds82] Hinds, William. 1982. *Aerosol Technology.* Wiley.

[Pilch87] Pilch, M. and C. Erdman. 1987. "Use of Breakup Time Data and Velocity History Data to Predict the Maximum Size of Stable Fragments for Acceleration-Induced Breakup of a Liquid Drop." *International Journal of Multiphase Flow*, Vol. 13, pp. 741–757, 1987.

[LaBarbera90] LaBarbera, M. 1990. "Principles of Design of Fluid Transport Systems in Zoology." *Science*, Vol. 249, pp. 992–1000.

[MIL91] MIL-HDBK-217F. 1991. Reliability Prediction of Electronic Equipment. Department of Defense.

[Vogel96] Vogel, S. 1996. *Life in Moving Fluids.* Princeton University Press.

[Tennekes97] Tennekes, H. 1997. *The Simple Science of Flight.* MIT Press.

[deNevers00] deNevers, N. 2000. *Air Pollution Control Engineering.* McGraw-Hill.

[Post02] Post, S. and J. Abraham. 2002. "Modelling the Outcome of Drop-Drop Collisions in Diesel Sprays." *International Journal of Multiphase Flow*, Vol. 28, pp. 997–1016.

[Heppenheimer03] Heppenheimer, T. 2003. *Principles of Rocketry*. Centennial of Flight Commision. http://www.centennialofflight.gov/

[Videler06] Videler, J. 2006. *Avian Flight*. Oxford University Press.

[Anderson07] Anderson, J. 2007. *Introduction to Flight*. McGraw-Hill.

[Zucrow76] Zucrow, M. and J. Hoffman. 1976. *Gas Dynamics*. Wiley.

■ Problems

1. What limits the maximum flight speed of a plane using turbofan engines?
2. What is the efficiency of a fuel injector?
3. Based on the Great Flight Diagram of Tennekes, what would you expect the cruising speed of a micro air vehicle weighing 5 lbf to be?
4. Repeat Example 8.1, but assume a 10 mph headwind.
5. Suppose a computer generates 150 W of heat, and takes in air at 20°C. If the maximum allowable outlet air temperature is 45°C, what is the required flow rate of air? If the space available for a fan is a circle of diameter 7 cm, what average velocity of air is required? Estimate the minimum amount of power required to run the fan.
6. An engineer is told to build a methane–liquid oxygen rocket for a particular mission. What temperature will the methane have to be cooled to, in order to liquefy it at atmospheric pressure? How does this compare to the condensation temperatures of other cryogenic fuels? What is the density of liquid methane?
7. Two rain drops, one of diameter 1 mm and another of diameter 0.5 mm, are falling from a cloud. Assuming both are traveling at their terminal velocities, what is the relative velocity between the two drops?
8. If the two drops of Problem 7 collide, what is the Weber number associated with the collision? If the centers of the two drops are offset by 0.3 mm when they collide, does the Brazier-Smith model predict coalescence or separation for the two drops?
9. For Example 8.1, if the rocket is launched at an angle to the horizontal of 45°, what is the horizontal distance the rocket will traverse?
10. For Example 8.1, what initial launch angle should be chosen to maximize the horizontal distance traversed?
11. A ramjet is flying at Mach 1.1 and burns fuel at a rate of 1.0 kg/s. If the flow is assumed to be compressible, how much thrust is generated?

Problems

12. Repeat Problem 11, but include the compressibility effects.
13. Repeat Example 8.7 using water as the substance in the tank.
14. Repeat Example 8.7 with a hole size of 10 μm.
15. A computer takes in air at 20°C and 100 kPa and generates 100 W of heat. If the maximum allowed exit temperature of the air is 42°C, what is the minimum flow rate of air needed to cool the computer?
16. If the surface area of the heat sink over a computer processor is 25 cm², and the convective cooling coefficient for flow over a flat plate can be used, what velocity is needed to provide this cooling rate? The processor uses 50 W of power and has a maximum allowable surface temperature of 65°C, and the incoming air is at 25°C.
17. For a jet flying at Ma = 3.0 at an altitude of 80,000 ft, calculate the temperature at a stagnation point on the front of the aircraft.
18. An airplane travels at a speed of 600 mph in air at pressure $P = 0.75$ atm and temperature $T = -5°F$. Calculate the pressure, temperature, and density at a stagnation point on the plane.
19. A tank of volume 100 ft³ is evacuated to a pressure of 0.001 atm and allowed to settle to room temperature. It is connected to the outside environment at 1 atm and 65°F through a converging–diverging nozzle with throat diameter of 1 in, and outer bell mouth diameter of 5 in. If a valve is suddenly opened to allow the outside air to enter the evacuated tank, what will be the Mach number of the incoming air flow at the throat?
20. For Problem 19, for how long will the flow remain choked at the throat after the valve is opened?
21. Find the speed of sound in gaseous nitrogen when it is at $-100°C$, with $k = 1.4$ and molecular weight $M = 32$.
22. Explain the fundamental differences between jet propulsion and rocket propulsion.
23. At a point in the flow of an ideal gas, the gas is at $P = 5$ bar, with a density of $\rho = 5.6$ kg/m³ and velocity $V = 150$ m/s. Estimate the stagnation properties of the gas (P and T).
24. A stream of air flows in a duct of diameter $D = 15$ cm at the rate of 1 L/s. The stagnation temperature is 35°C. At one location in the duct the static temperature is $T = 8°C$. Calculate the Mach number, Ma, and velocity, V, at this point.
25. Air is kept in a tank at a pressure of 100 psi and a temperature of 75°F. If the air is allowed to issue out of the tank in a one-dimensional isentropic flow through a duct, what is the maximum possible flow rate per unit area of the duct? If you used the incompressible Bernoulli equation for this problem, what flow rate per unit area would you calculate?

26. A converging–diverging nozzle is designed to expand air from a large chamber in which the pressure is 800 kPa and the temperature is 40°C to a design Mach number of 2.7. The throat area of the nozzle is 0.08 m². Find the exit area of the nozzle and the mass flow rate through the nozzle under design conditions.

27. Air is kept at a pressure of 10 bar and temperature of 300°C in a large vessel. A valve is suddenly opened and the air exits through a converging–diverging nozzle to a pressure of 1 bar. Assuming frictionless adiabatic flow, for a desired flow rate of 1 kg/s of the air, calculate the Mach number at the exit of the nozzle, the velocity at the exit of the nozzle, and the required cross-sectional area at the throat.

28. Air flows from a large reservoir in which the pressure is 300 kPa and the temperature is 40°C through a nozzle. If the pressure at some section of the nozzle is measured as 200 kPa, find the temperature and velocity at this section. The Mach number at a different section is Ma = 1.5. Find the temperature, pressure, and velocity at this second section.

29. Compare the specific fuel consumption of a turbojet and a ramjet that are being designed for flight at Ma = 2.0 and z = 10,000 m altitude. The turbojet compressor pressure ratio is 10 and the maximum allowable temperature in the Brayton cycle for this engine is 1,200 K. For the ramjet the maximum allowable temperature is 2,400 K. Conventional hydrocarbon fuels are to be used (heating value 44,000 kJ/kg). Assume k = 1.4. Which engine will be more efficient? How sensitive are your answers to the magnitude of the maximum allowable temperature?

30. Air at a temperature of −10°C flows through a supersonic wind tunnel. Imperfections along the wall generate weak Mach lines that extend downstream. These waves are at an angle of 40° to the flow. Find the Mach number and the velocity in the wind tunnel.

31. A converging–diverging nozzle has an exit area to throat area ratio of 3.0. The upstream stagnation pressure is 1.0 MPa. The upstream stagnation temperature is 300 K and the fluid is air. What is the velocity of the fluid in the throat of the nozzle?

32. Air flows steadily through a horizontal nozzle discharging to the atmosphere. The area at the nozzle inlet is 0.1 m². The area at the nozzle exit is 0.02 m². Determine the pressure required at the inlet to the nozzle to produce an outlet speed of 50 m/s.

33. A static thrust stand for jet engine testing has an intake air velocity of 200 m/s and an exhaust gas velocity of 500 m/s. The intake area is 1.0 m². At the inlet, the static pressure is 78.5 kPa and the static temperature is 286 K. The exhaust static pressure is 101 kPa. Assume an ideal gas with R = 0.2869 kJ/kg-K. Calculate the mass flow rate and the thrust of the engine.

9 Fluid Measurement Techniques

In This Chapter
- Velocity Probes
- Flow Rate Measurements
- Pressure Transducers
- Nondimensionalization of Flow Data

The objective of this chapter is to gain familiarity with the different options for flow measurement devices on the market, including the strengths and weaknesses of each.

■ 9.1 Velocity Probes

Flow meters can measure either local velocity or the total flow rate. Of course, the total flow rate can be deduced from pointwise velocity measurements, if enough data points are taken so that the velocity profile can be integrated to sufficient accuracy. Alternatively, if the shape of a velocity profile is known, such as for laminar flow in a round pipe, then a single velocity measurement will suffice to deduce the flow rate.

For a local velocity measurement at a point, the available velocity probes include Pitot tubes, hotwire anemometers, and laser-based measurements such as LDV and PIV. LDV and PIV are expensive modern techniques. Pitot tubes can only measure one component of the velocity; hot-wire anemometers and LDVs can measure two or even all three components of the velocity at a point for the most sophisticated systems.

One requirement for velocity probes is that the probe itself be small enough to resolve the spatial gradients in the velocity profile (i.e., the velocity of the fluid should not change significantly across the diameter of the probe). Also, for transient flow fields, including turbulent flows, the frequency response of the probe should be fast enough to resolve the temporal gradient in the flow field. Of course, in some steady turbulent flow fields, only the mean velocity needs to be measured.

In conducting flow measurements, the engineer must ask what potential sources of errors are present, apart from the measurement technique. (See Appendix D for information on quantifying experimental error.) Also, for discrete measurements, the effects of sample size must be considered. The larger the number of measurements, the higher can be the confidence in the accuracy of the measurements. A good instrument should have a large dynamic range factor—at least 10, and preferably 30 or higher—and should be fast and easy to use.

9.1.1 Pitot Tube

The simplest and cheapest way to measure velocity at a point is with a Pitot tube. A Pitot tube can be connected to a simple U-tube manometer filled with water or oil, so that no expensive electronic transducers need be used. The Pitot tube also was discussed in Chapter 3. Figures 9.1–9.4 show different configurations of Pitot tubes.

The disadvantages of a Pitot tube are that it has a poor dynamic response, and it does not measure transient velocities very well. It also is a rather intrusive device that perturbs the flow around it. For internal flows, a tap must be inserted into the wall to allow the Pitot tube to be moved, and this restricts where the velocity measurements can be made. Care must also be taken to ensure that the Pitot tube is aligned with the flow. Placing multiple holes around the periphery for the static pressure measurements helps to correct for off-axis alignment errors.

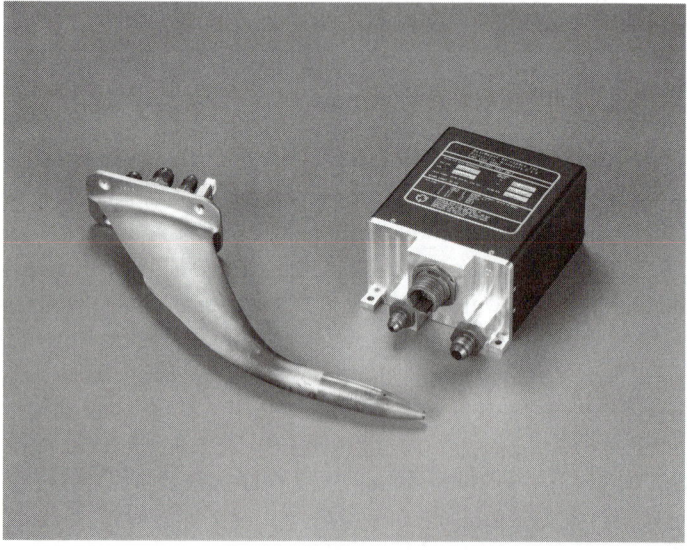

I FIGURE 9.1 Picture of Pitot tube for aircraft application. Courtesy NASA-Dryden Flight Research Center.

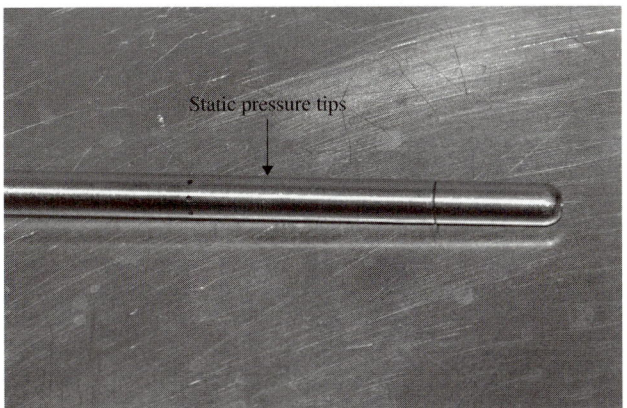

FIGURE 9.2 Close-up picture of the tip of a Pitot tube probe. The static pressure taps are visible.

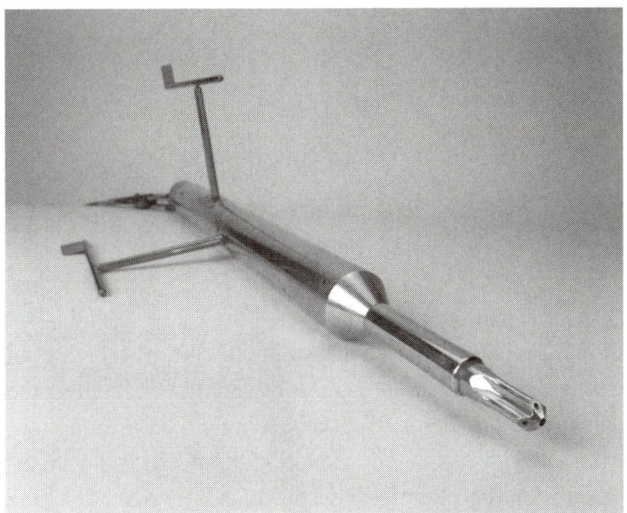

FIGURE 9.3 Nose-boom Pitot tube for an aircraft indicated velocity sensor. Courtesy NASA-Dryden Flight Research Center.

The equation used in flow velocity measurements using a Pitot tube is

$$V = \sqrt{\frac{2\Delta P}{\rho}} \qquad (9.1)$$

Here ΔP is the difference in pressure between the stagnation port at the tip of the probe and the static ports on the side, so that $\Delta P = P_{stagnation} - P_{static} = \frac{1}{2}\rho V^2$. This differential

FIGURE 9.4 Boundary layer rake, which is a collection of Pitot tubes. Courtesy NASA-Dryden Flight Research Center.

pressure can be measured with a differential manometer, or more commonly with an electronic differential pressure transducer. Note that in gases at very low velocities, the pressure rise will be small and it may be difficult to get an accurate measurement.

For subsonic flow, in the range $0.3 < \text{Ma} < 1.0$, Equation 9.1 needs to be modified because the fluid must be considered a compressible fluid, and the density of the gas in the stagnation and static ports may not be the same. For isentropic flow (see Chapter 8) the compressible equation for a Pitot tube is

$$V = \sqrt{\frac{2k}{k-1}\frac{P_{\text{static}}}{\rho_{\text{static}}}\left[\left(\frac{P_{\text{stagnation}}}{P_{\text{static}}}\right)^{\frac{k-1}{k}} - 1\right]} \tag{9.2}$$

where ρ_{static} is the density of the fluid as it flows around the Pitot tube, and $k = c_P/c_V$ is the ratio of specific heats for the gas, where $k = 1.4$ for air at standard conditions. Note that for the Pitot tube we assume that the Reynolds number is high enough that we may neglect viscous losses.

EXAMPLE 9.1

A Pitot tube is used to measure the velocity of water inside a pipe of diameter 4 cm. If the reading on the total pressure port is 250 kPa (gage) and the reading on the static pressure port is 5.26 kPa (gage), determine the velocity in the pipe. If you can assume the velocity is nearly uniform across the diameter, what is the flow rate through the pipe, in units of liters per minute?

SOLUTION From the Bernoulli equation $P_{total} = P_{static} + \frac{1}{2}\rho V_2$, so that $\frac{1}{2}\rho V^2 = \Delta P$, and the velocity is calculated from

$$V = \sqrt{\frac{2\,\Delta P}{\rho}}$$

Substituting in the numerical values yields

$$V = \sqrt{\frac{2(250{,}000\text{ Pa} - 5{,}260\text{ Pa})}{1{,}000\text{ kg/m}^3}} = 22.1\text{ m/s}$$

Recall that $1\text{ Pa} = 1\text{ N/m}^2 = 1\text{ kg/m-s}^2$. The volume flow rate is the product of the velocity and the area, so

$$Q = AV = \frac{\pi}{4}D^2 V = \frac{\pi}{4}(0.04\text{ m})^2(22.1\text{ m/s}) = 0.02777\text{ m}^3/\text{s}$$

The conversion factor from cubic meters to liters is $1\text{ m}^3 = 1000\text{ L}$, so the flow rate is

$$Q = 0.02777\,\frac{\text{m}^3}{\text{s}} \times \frac{1{,}000\text{ L}}{1\text{ m}^3} \times \frac{60\text{ s}}{1\text{ min}} = 1{,}666\text{ L/min}$$

9.1.2 Hot-Wire Anemometer

The next step up from the Pitot tube, and historically the next velocity probe invented, is the hot-wire anemometer. The idea behind the hot-wire probe is that the convective cooling rate increases with increasing velocity. For cylinders, heat transfer correlations are well established, so cylindrical wires are used. A voltage is applied across the wire, which causes an electrical current to flow through the wire, which causes electrical resistive heating. The heat generated is taken away by the fluid flowing over the wire.

The hot-wire anemometer can be operated in one of three modes: constant current, constant temperature, or constant voltage. In constant-current mode the change in the wire temperature is measured. In constant-temperature mode, the current required to maintain constant temperature is measured, and in constant-voltage mode the change in wire resistance is measured. Then convective heat transfer correlations (see Chapter 8) are used to determine the fluid velocity.

Multiple wires can be crossed in a hot-wire anemometer to measure more than one velocity component. One advantage of a hot-wire anemometer is that the probe can be made relatively small to reduce the disturbance to the flow field. Also, this device has a fast temporal response, so that transient velocities—including turbulent velocities—can be measured.

One disadvantage is that the wire is fragile. The diameter of the wire in a hot-wire probe is usually around 5 μm, which contributes to its fragility. Common materials used in the wires are platinum or tungsten.

9.1.3 Laser-Based Measurements

The most modern (and most expensive) velocity measurements are laser-based measurements, which include LDV (laser Doppler velocimetry) and PIV (particle image velocimetry). LDV measures velocities at a point as a function of time, and PIV measures velocities across a plane in space at a point in time. Both methods rely on tracer particles that are seeded into a flow. Those particles then interact with a laser beam shown through the flow. Both LDV and PIV can be used in either gases or liquids, provided the fluid is transparent.

In PIV, the displacement, Δx, of particle images between two exposures of known time separation, Δt, is used to calculate the velocity:

$$V = \frac{\Delta x}{\Delta t} \tag{9.3}$$

Rather than attempting to track each individual particle, PIV takes a statistical average of the particle motion over a given area. A typical interrogation spot is around 32 by 32 pixels. Two two-dimensional fast Fourier transforms (FFTs) are applied to the interrogation spot to obtain the average displacement.

Figure 9.5 shows a schematic of a basic PIV setup. Recent advances in PIV include stereoscopic PIV, in which two cameras are used simultaneously to get all three velocity components. Figure 9.6 shows an example PIV image from an autocorrelation system with two exposures on a single image. Current PIV systems use cross-correlation, in which two separate images are taken and compared to each other.

One limitation of PIV measurement systems is the reality that spurious vectors often arise when interrogating an image. The spurious vectors can arise if there is too

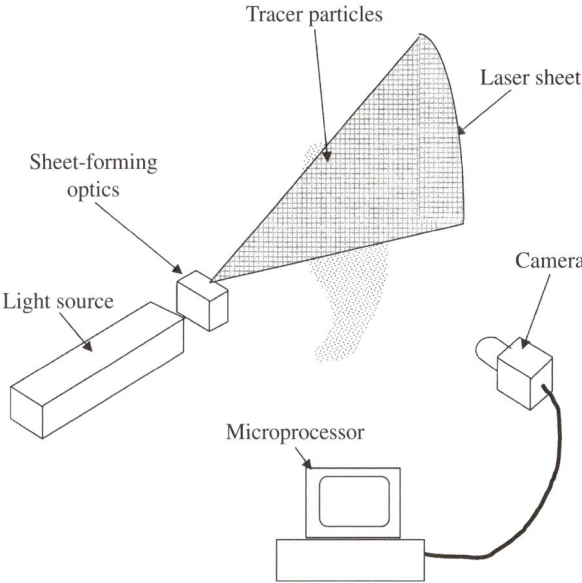

FIGURE 9.5 Schematic of a basic PIV system.

much out-of-plane motion, so that there are not enough image pairs to get a suitable signal-to-noise ratio. Spurious vectors can also be caused by blooming in the CCD sensor when a large particle (or a clump of particles) scatters so much light that a pixel is saturated. In some CCD cameras, that extra energy will spill over into the surrounding pixels.

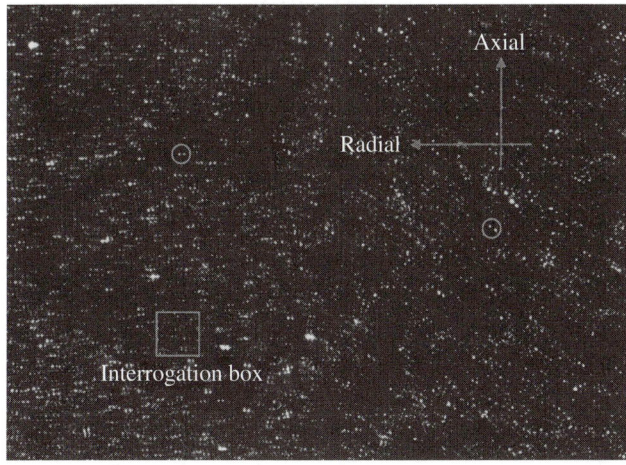

FIGURE 9.6 Example of a double-exposed PIV image.

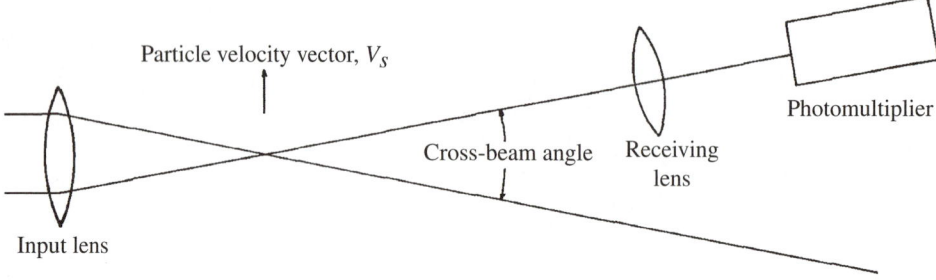

FIGURE 9.7 Schematic of an LDV system, from [NASA71].

In laser Doppler velocimetry (LDV) [also called phase Doppler anemometry (PDA) or Laser Doppler anemometry (LDA)], two laser beams are crossed to create interference fringes. Drops moving through these fringes scatter light to a detector. Figure 9.7 shows an LDV system. For a known interference fringe spacing (calculated from the laser wavelength, λ, and the angle of beams crossing, θ), the velocity of a particle moving through the probe volume is calculated as

$$V = \lambda \frac{f_D}{2 \sin(\theta/2)} \qquad (9.4)$$

where f_D is the Doppler frequency, in units of hertz, as measured by a photodetector [NASA71].

The advantages of LDV are that it is based on proven technology and that it has fine-scale temporal and spatial resolution. Also, it is a nearly nonintrusive measurement, except when the seed particles are introduced. In flows that already contain dust or drops or bubbles that can be used as seed particles, it is completely nonintrusive. LDV works for any index of refraction or drop shape (i.e., nonspherical is okay). The standard system provides two-dimensional measurements, and a two-color system can provide three-dimensional velocity vector measurements.

One disadvantage of LDV is that there is a slow data acquisition rate if the concentration of seed particles is too dilute. But the seed particle concentration cannot be too high, because there can be only one drop or particle in the probe volume at a time. There also is an inherent velocity bias with LDV—slower drops cause more "counts," as they stay in the probe volume longer than faster ones. Software has been written to try to account for this. Also, the beam crossing angle must be known accurately for the velocity to be calculated.

Continuous-wave (CW) lasers are typically used for LDV, whereas pulsed lasers are used for PIV. As with any system using lasers, safety is a concern. For the high-powered lasers used in LDV and PIV even reflections of the laser beam off a wall or other object can cause eye damage. Because one should never look directly into a laser beam or its reflections, safety goggles that are specified for the wavelength of the laser light being used should be worn whenever the laser is on.

A crude way to measure velocity is with flow tracers, an illumination source, and a camera. Either a double exposure or a long single exposure (streak images) can be used. The length the tracer particle moved is divided by the time of exposure for a streak, or the time between illuminations for a double-pulsed exposure, to get the approximate velocity. This technique relies on the same assumption as PIV and LDV: that the tracer particles have the same velocity as the fluid. The quality of the tracer particles can be determined by calculating the Stokes number:

$$\text{St} = \frac{\rho_{\text{particle}} DV}{18\mu} \tag{9.5}$$

Note that this is different from the Reynolds number because the density used is the particle density and not the surrounding fluid density.

EXAMPLE 9.2

For the conditions of Example 9.1, if a PIV system is to be used with a 1-megapixel camera (1000 by 1000 pixels), and the time between exposures is 0.0002 s, what is the size of the flow field that can be imaged?

SOLUTION A general rule of thumb is that the maximum displacement of the PIV particles should be no more than $\frac{1}{4}$ of the interrogation spot size. So if an interrogation box is 32 by 32 pixels, the particle image should move no more than 8 pixels between exposures. Since the maximum velocity to be measured is 10 m/s and the time between exposures is 0.0002 s, the maximum physical displacement of a particle is (10 m/s)(0.0002 s) = 0.002 m = 2 mm. Equating 8 pixels with 1 cm gives the scale factor of the image: 1 pixel = 0.25 mm. So with a total size of 1000 by 1000 pixels, the physical space that can be imaged is 25 cm by 25 cm.

For sprays and other multiphase flows (covered in Chapter 8), drop sizing is also important. Optical techniques of photography—light scattering and diffraction—are most commonly used, although the older mechanical cascade impactor sometimes still finds use. For two-dimensional imaging, one needs to use a fast shutter speed or to leave the shutter open and use a short-duration light burst (as from a pulsed laser) to image the drops sharply without streaking or blurring. Manual sizing from a picture takes a long time, and automated methods still are not completely robust—that is, they still require user input. There is also a tradeoff in the aperture setting of the camera. A smaller aperture will cause a greater depth of field, which will result in fewer out-of-focus drops, but at the expense of less light making it to the camera, which will require either a longer exposure time or a brighter light source to get a decent exposure.

Other techniques for drop and particle sizing include laser diffraction and interference refraction. A forward light scattering (diffraction) device uses coherent, collimated light, and it works for either transparent or opaque drops. The technique is based on Fraunhofer diffraction of a parallel beam of monochromatic light by a moving drop. Just as light passing through a small aperture creates a diffraction pattern, light passing by a small drop also creates a diffraction pattern. In fact, the scattering from a drop differs from the scattering through an orifice only by the shadow of the drop. The smaller the drop, the wider the diffraction pattern. The scattering angle of the diffracted light, β, scales as

$$\beta \approx \frac{1}{d^2} \tag{9.6}$$

where d is the diameter of the drop.

Laser diffraction systems have the advantage of being able to work with non-spherical particles and a large dynamic range (with the use of different lenses, sizes from 1 µm to 3 mm can be measured). Among the disadvantages are that it is a line-of-sight measurement, where data are averaged over the length of the laser beam.

The interference refraction method can be used only with transparent drops. These transparent drops act as a spherical lens to the passing laser light, with an effective focal length of

$$f = \frac{R}{2(n-1)} \tag{9.7}$$

where R is the drop radius and n is the liquid index of refraction. In this method, two laser beams are crossed to create interference fringes for the drops (lenses) to magnify. The change in fringe spacing from the magnified light is measured to get R. The interference refraction method also has the advantage that it can be combined with LDV to measure drop size and velocity simultaneously.

9.1.4 Flow Visualization

It some cases it may be possible to estimate velocities in a fluid field from a photograph or sequence of photographs. While often more qualitative than quantitative, flow visualization is a frequently used tool that can give valuable insights into the flow. Methods of flow visualization include tufts, smoke, dye injection, oil injection, and various optical techniques including schlieren and shadowgraph techniques. For multiphase flows, basic photography and holography are also used. Figures 9.8 through 9.18 show examples of flow visualization techniques.

9.1 Velocity Probes 437

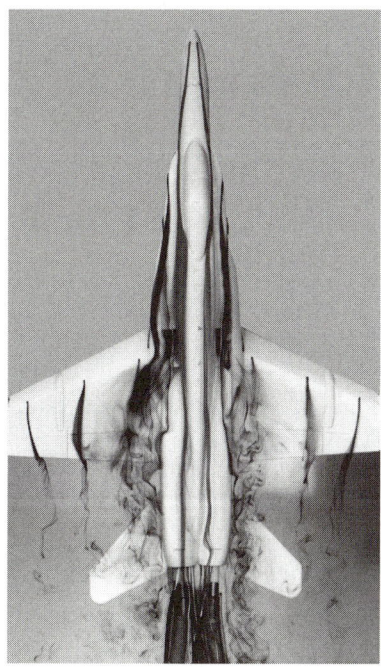

FIGURE 9.8 Dye used to visualize flow over a model of an F-18 in a water tunnel. Courtesy of NASA-Dryden Flight Research Center.

FIGURE 9.9 Dye used to visualize flow over a model of an F-117 in a water tunnel. Courtesy of NASA-Dryden Flight Research Center.

FIGURE 9.10 Smoke flow from the nose of an X-29. Courtesy of NASA-Dryden Flight Research Center.

FIGURE 9.11 Smoke flow over an F-18 in flight. Courtesy of NASA-Dryden Flight Research Center.

9.1 Velocity Probes 439

FIGURE 9.12 Red liquid tracer flow over the fuselage of an F-18 High Angle of Attack Research Vehicle (HARV). Courtesy of NASA-Dryden Flight Research Center.

FIGURE 9.13 An F-18 HARV showing the release of glycol-based liquid from very small holes in the nose of the aircraft. Courtesy of NASA-Dryden Flight Research Center.

FIGURE 9.14 An F-18 HARV showing the release of glycol-based liquid from very small holes in the nose of the aircraft. Courtesy of NASA-Dryden Flight Research Center.

FIGURE 9.15 Liquid tracer flow near the wing leading edge of an F-18 HARV. Courtesy of NASA-Dryden Flight Research Center.

9.1 Velocity Probes

FIGURE 9.16 Tufts used to visualize the surface flow patterns over an F-18 HARV. Courtesy of NASA-Dryden Flight Research Center.

FIGURE 9.17 Tufts used to visualize the surface flow patterns over an F-18 HARV. Courtesy of NASA-Dryden Flight Research Center.

FIGURE 9.18 Heat-sensitive paint on the wing of an X-15 from testing in 1964. Courtesy of NASA-Dryden Flight Research Center.

Schlieren and shadowgraph images make use of the fact that the index of refraction of a gas changes with the density of a fluid. A shadowgram is a monochromatic image, in which the local intensity of the image is proportional to the second derivative of the refractive index of the fluid. In schlieren images, changes in the image are proportional to the first derivative of density of the fluid. Figures 9.19 and 9.20 are examples of schlieren images.

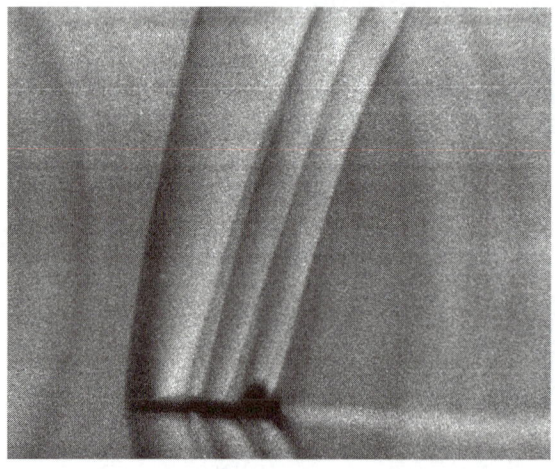

FIGURE 9.19 Schlieren photograph of T-38 shock waves at Ma = 1.1, flying at an altitude of 13,000 ft. Courtesy NASA-Dryden Flight Research Center.

FIGURE 9.20 Schlieren photograph of a shock wave at Mach 7. Courtesy of NASA.

A basic schlieren system can be constructed at low cost from a point light source, two identical convex lenses, and a knife edge. The light from the point source is directed through the first lens to create a beam of light that passes through the test section. If there are variations in density of the fluid in the test section, this will cause distortions in the transmitted light beam. The light is then focused by the second lens. A knife edge is placed at the focal point of the second lens in such a way as to block out about half of the light. The remaining light is collected on an image plane, which is usually the sensor of a CCD camera in modern systems [Settles06].

9.2 Flow Rate Measurements

Common flow rate measurement devices are Venturi meters, rotameters, orifice plates, laminar flow elements, turbine meters, ultrasonic flow meters, and Coriolis meters. The orifice plate is the cheapest to install, although none of these devices are very expensive. Some common examples of flow rate measurements are for flow of petroleum and petroleum distillates, flow of gases such as natural gas and air, and wastewater discharge rates [Goldstein96].

9.2.1 Venturi Meter

The derivation of the equations for Venturi meters (Figures 9.21, 9.22, and 9.23) was presented in Chapter 3. A Venturi meter is installed as a converging–diverging section in a section of straight pipe, with the largest diameter of the Venturi equal to the pipe diameter. Pressure taps are placed at the inlet to the Venturi and at the throat (or just

444 CHAPTER 9 FLUID MEASUREMENT TECHNIQUES

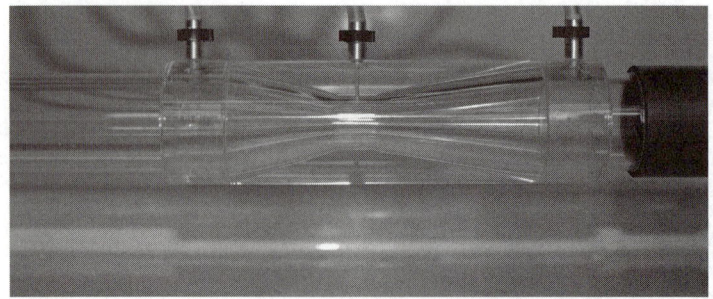

FIGURE 9.21 Picture of a Venturi flow meter. Courtesy of Armfield.

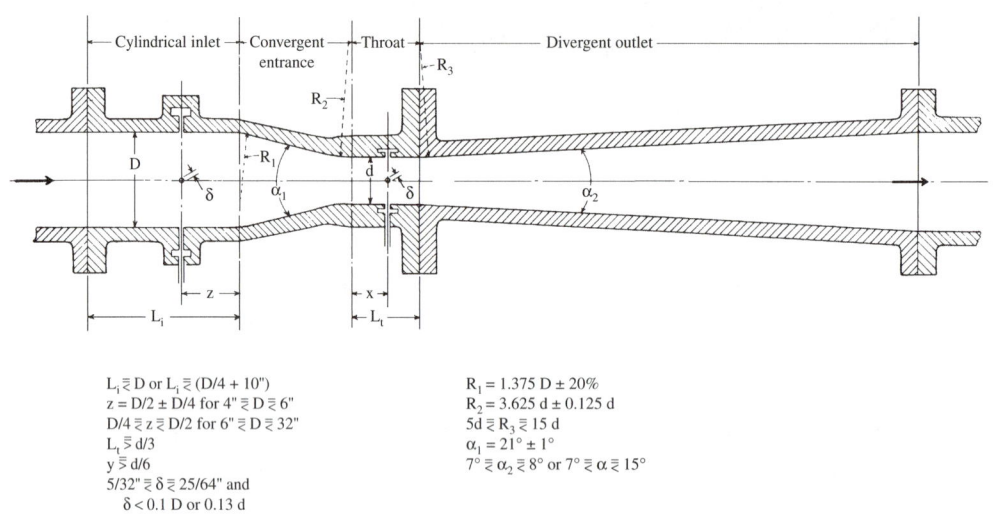

$L_i \lesssim D$ or $L_i \lesssim (D/4 + 10")$
$z = D/2 \pm D/4$ for $4" \lesssim D \lesssim 6"$
$D/4 \lesssim z \lesssim D/2$ for $6" \lesssim D \lesssim 32"$
$L_t \lesssim d/3$
$y \lesssim d/6$
$5/32" \lesssim \delta \lesssim 25/64"$ and
$\delta < 0.1\,D$ or $0.13\,d$

$R_1 = 1.375\,D \pm 20\%$
$R_2 = 3.625\,d \pm 0.125\,d$
$5d \lesssim R_3 \lesssim 15\,d$
$\alpha_1 = 21° \pm 1°$
$7° \lesssim \alpha_2 \lesssim 8°$ or $7° \lesssim \alpha \lesssim 15°$

FIGURE 9.22 Schematic of a classical Hershel Venturi, from [Bean83]. Used with permission of ASME.

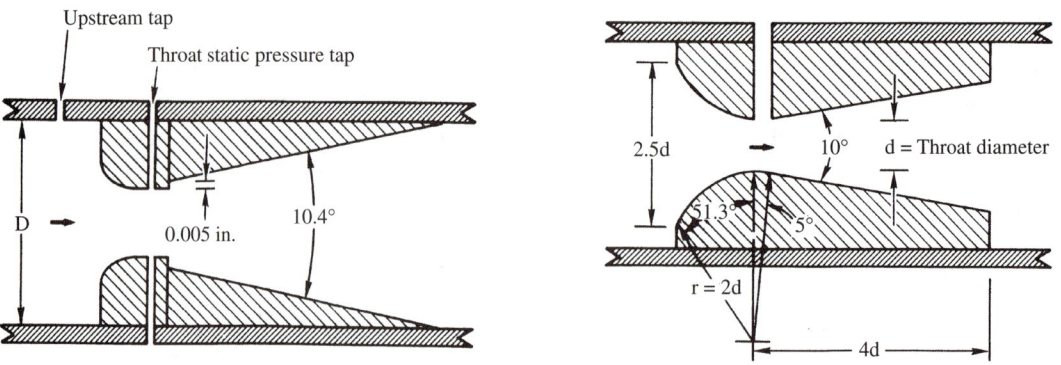

FIGURE 9.23 Schematic of two different Venturi designs, from [Blevins92]. Used with permission of Krieger.

slightly beyond the throat). From the pressure measurements, the average velocity in the pipe at the inlet is

$$V = \frac{(d/D)^2}{\sqrt{1-(d/D)^4}}\sqrt{\frac{2\Delta P}{\rho}} \tag{9.8}$$

where D is the pipe diameter and d is the throat diameter. The volume flow rate, Q, is then calculated as

$$Q = D^2 \frac{(d/D)^2}{\sqrt{1-(d/D)^4}}\sqrt{\frac{2\Delta P}{\rho}} \tag{9.9}$$

The discharge coefficients for a Venturi are generally in the range of 0.985 to 0.995 [Baker02]. Venturis have a low pressure drop and are less affected by upstream flow distortions than other meters, but have higher costs and a longer length to install than the orifice plate. They are good in applications with very high flow rates because of their low pressure drop.

9.2.2 Rotameter

Rotameters (Figure 9.24) are often calibrated in SCFM. Because rotameters must be installed vertically, they generally cannot be installed inline like orifice plates and Venturis can. The rotameter has a conical shape, with the area increasing upward. The tube is usually made of glass so that the position of the float may be observed visually. Baker reports that the typical accuracy of a rotameter is ±2% of full-scale reading, with a dynamic range ratio of 10 to 1 [Baker02]. The flow rate can be easily obtained visually without calculations. Rotameters are easy to install and use, but they have low accuracy, typically provide no electrical readout, and are sensitive to changes in the density and viscosity of the fluid. In addition, rotameters must be calibrated for fluid type and temperature.

Unlike orifice plate meters, which produce a pressure drop in the system that increases with increasing flow rate, the pressure drop across a rotameter is nearly constant with flow rate [Hinds82]. Based on a simple balance between the weight of the float and the drag force the fluid exerts on the float to hold it in place, we have

$$m_f g = \frac{1}{2} C_D \rho \frac{\pi}{4} d_f^2 V^2$$

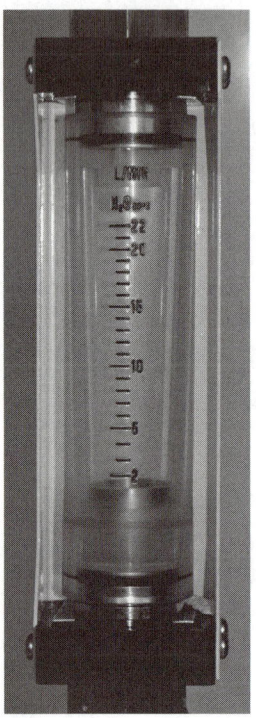

FIGURE 9.24 Picture of a rotameter. Courtesy of Armfield.

The flow rate, Q, is equal to the velocity times the area of the rotameter at the float location, A_O:

$$Q = A_O V = A_O \sqrt{\frac{2 m_f g}{C_D \rho \frac{\pi}{4} d_f^2}}$$

The drag coefficient is not that for a simple sphere (or whatever shape of float is used) because of the influence of the walls. The drag coefficient can be extracted out and replaced with a rotameter coefficient, C_R, that is specific to the given rotameter float combination. The flow rate can then be calculated as

$$Q = C_R A_O \sqrt{\frac{2 m_f g}{\rho (\frac{\pi}{4} d_f^2)}} \tag{9.10}$$

The rotameter coefficient is usually in the range of 0.6 to 0.8 [Hinds82]. Note that the cross-sectional area of the rotameter, A_O, is not constant but changes with the float location.

9.2 Flow Rate Measurements

In practice Equation 9.10 is not used to calculate flow rate since manufacturers provide a calibration curve. If the rotameter is to be used with a fluid, such as water, of the same density for which it is calibrated, then no density correction need be made. For gases, however, where the density will change with operating temperature, it is necessary to correct the manufacturer's calibration for changes in density. As can be seen from Equation 9.10, the flow rate varies with the square root of density, so a simple correction is

$$Q = Q_{STP}\sqrt{\frac{\rho_{STP}}{\rho}} \tag{9.11}$$

If the user has an accurate test meter available, it is recommended to use that to recalibrate the flow meter for actual operating conditions.

9.2.3 Orifice Plate Meter

Orifice plate meters (Figure 9.25) are usually made to ASME standards. At sufficiently high Reynolds number, the discharge coefficient, C_d, is constant for a given orifice. The flow rate is measured indirectly by measuring the pressure drop ΔP across the orifice. The standards usually call for the upstream pressure tap to be placed a distance of one diameter, D_1, upstream of the orifice plate, and the downstream pressure tap to be at a distance $D/2$ downstream of the orifice plate. This will typically put the

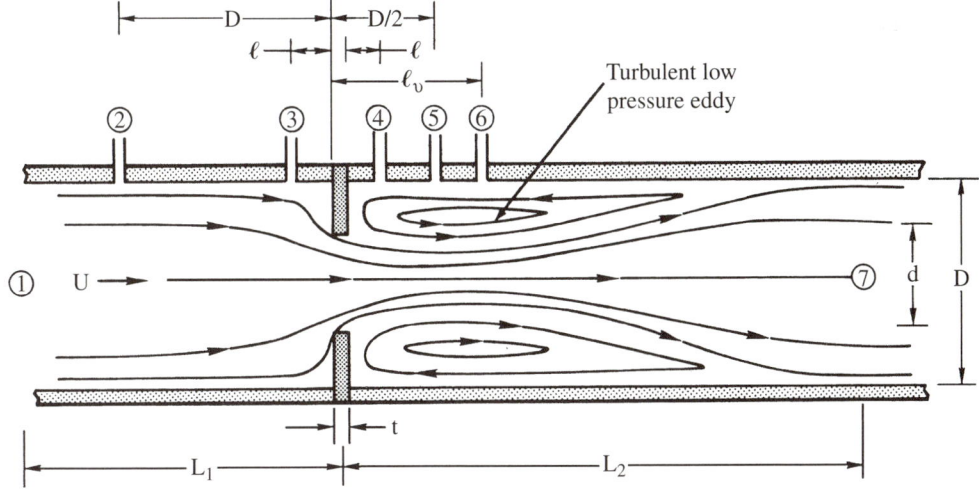

FIGURE 9.25 Sketch of an orifice plate meter, illustrating the typical turbulent flow through a square-edged orifice plate. Locations 3 and 4 are for flange taps, and locations 2 and 5 are for vena contracta taps. From [Blevins92]. Used with permission of Krieger.

downstream pressure tap in the recirculation zone, before the fluid has completely expanded back to the walls and the fully developed velocity profile is recovered. Orifice plates are cheap and easy to install, but have nonlinear response, high pressure drop, and small dynamic range.

The discharge coefficient, C_d, is defined as the ratio of the actual mass flow rate through a device to the theoretical flow rate if there were no losses. Thus,

$$C_d = \frac{\dot{m}_{actual}}{\dot{m}_{theoretical}} \quad (9.12)$$

The theoretical mass flow rate through a constriction (such as Venturi) with a measured pressure drop ΔP from the large entrance to the small throat is

$$\dot{m}_{theoretical} = \rho A V = \rho A_2 \sqrt{\frac{2\,\Delta P}{\rho[1 - (\frac{D_2}{D_1})^4]}} = \frac{A_2}{\sqrt{1 - (\frac{D_2}{D_1})^4}} \sqrt{2\rho\,\Delta P} \quad (9.13)$$

In many references, to make the equations more compact, the diameter ratio, β, is introduced:

$$\beta = \frac{D_2}{D_1} \quad (9.14)$$

where D_2 is the throat diameter, so $D_2 < D_1$ and $\beta < 1.0$. Then the area ratio between the main pipe and the constriction is

$$\frac{A_2}{A_1} = \left(\frac{D_2}{D_1}\right)^2 = \beta^2$$

The values of C_d do not change much with Reynolds number for $Re > 100{,}000$ and $\beta \geq 0.5$. In other words, at high Reynolds numbers, C_d is approximately constant for a given orifice. The total pressure lost through the use of an orifice plate meter is given by

$$P_{loss} = \left[1 - \left(\frac{d}{D}\right)^{1.9}\right] \Delta P_{meas} \quad (9.15)$$

[Baker02]. Using the correlation of Equation 9.15, we can relate the value of the discharge coefficient, C_d, to the minor loss coefficient, K, by

$$K = (1 - \beta^{1.9})\left(\frac{1 - \beta^4}{C_d^2}\right) \quad (9.16)$$

9.2 Flow Rate Measurements

EXAMPLE 9.3

An orifice plate meter has an internal diameter of 20 mm and is used inside a pipe of internal diameter 51 mm. The discharge coefficient of the orifice plate is 0.601. For a measured pressure drop across the meter of 30 mm of water, find the mass and volume flow rate of water and estimate the unrecoverable pressure loss associated with the use of the meter. The area of the plate is $(\pi/4)d^2 = (\pi/4)(0.02 \text{ m})^2 = 0.000314 \text{ m}^2$.

SOLUTION The pressure drop from the manometer can be converted to Pa by using the specific weight of water: $\Delta P = \rho g \Delta h = (1000 \text{ kg/m}^3)(9.8 \text{ m/s}^2)(0.030 \text{ m}) = 294$ Pa.

$$Q = C_d A_2 \sqrt{\frac{2 \Delta P}{\rho [1 - (\frac{D_2}{D_1})^4]}}$$

$$= 0.601(0.000314 \text{ m}^2)\sqrt{\frac{2(294 \text{ Pa})}{(1000 \frac{\text{kg}}{\text{m}^3})[1 - (\frac{20 \text{ mm}}{51 \text{ mm}})^4]}} = 0.000146 \frac{\text{m}^3}{\text{s}}$$

The average velocity in the pipe is $V = Q/A = 0.000146 \text{ m}^3/\text{s}/0.00204 \text{ m}^2 = 0.0715$ m/s, and the velocity at the orifice is $V = Q/A = 0.000146 \text{ m}^3/\text{s}/0.000314 \text{ m}^2 = 0.465$ m/s. One method to find the unrecoverable pressure drop associated with the flow meter is to use Equation 9.16 to find the value of K. For this meter $\beta = 20 \text{ mm}/51 \text{ mm} = 0.392$, so

$$K = [1 - (0.392)^{1.9}]\left[\frac{1 - (0.392)^4}{(0.601)^2}\right] = 2.25$$

Then with K known, the pressure drop is

$$\Delta P = K \frac{1}{2} \rho V^2 = (2.25)\frac{1}{2}\left(1000 \frac{\text{kg}}{\text{m}^3}\right)\left(0.465 \frac{\text{m}}{\text{s}}\right)^2 = 243 \text{ Pa}$$

The unrecoverable pressure loss is about 82% of the measured drop across the meter.

9.2.4 Laminar Flow Element

In a laminar flow element (LFE), the flow is divided into an array of smaller passages, with diameters small enough to ensure that the Reynolds number is less than 2000.

Since the flow is laminar, the flow rate can be calculated from the measured pressure drop using Equation 5.8 from Chapter 5:

$$V = \frac{d^2 \Delta P}{32 \mu L}$$

where d is the diameter of the laminar tubes. The flow rate is the product of velocity times total area:

$$Q = AV = \left(\frac{\pi}{4}D\right)^2 \left(\frac{d^2 \Delta P}{32 \mu L}\right) = \frac{\pi}{128} \frac{D^2 d^2}{\mu L} \Delta P \tag{9.17}$$

where D is the diameter of the main pipe. So the flow rate is linear with respect to the measured pressure drop.

EXAMPLE 9.4

An LFE is used to measure the flow rate through a pipe. Calculate the total flow rate, Q, of air, knowing that the measured pressure loss $\Delta P = 20$ Pa over the LFE with a length of $L = 5$ cm. The diameter of the main pipe, D, is 10 cm, and the diameter of the flow elements, d, is 3 mm. You may assume that the total cross-sectional area of the LFE is essentially the same as the cross-sectional area of the main flow pipe.

SOLUTION The velocity in the laminar flow element is

$$V = \frac{(0.003 \text{ m})^2 (20 \text{ N/m}^2)}{32 (1.8 \times 10^{-5} \text{ kg/m} \cdot \text{s})(0.05 \text{ m})} = 6.25 \text{ m/s}$$

so the volume flow rate is

$$Q = AV = \frac{\pi}{4}(0.10 \text{ m})^2 (6.25 \text{ m/s}) = 0.0490 \frac{\text{m}^3}{\text{s}}$$

9.2.5 Turbine Meters

The flow rate through a turbine is related to the rotational speed of the turbine, in a nearly linear relationship. Baker reports accuracy is usually around ±0.5% of the measured flow rate [Baker02]. A turbine meter is of course sensitive to swirl in the flow, so it should not be placed close to a pipe elbow. Overall, turbine meters have high accuracy and a relatively large linear range. They can be used with either liquids or gases, but tend to be more accurate with liquids. Turbine meters work best with an upstream flow straightener, and they may need to be recalibrated over time as their bearings wear.

9.2.6 Ultrasonic Flow Meter

One type of ultrasonic flow meter uses the Doppler effect and requires particles, drops, or bubbles in the flow that reflect the sound waves. The Doppler shift equation is

$$\Delta f = 2f\left(\frac{V}{a}\right)\cos\theta \qquad (9.18)$$

where f is the frequency of the transmitted signal, a is the speed of sound, V is the flow velocity, and θ is the angle of the detector with respect to the flow direction. This type of ultrasonic flow meter is easy to install, but requires the presence of a second phase and relies on the assumption that the second phase travels at the same velocity as the carrier fluid.

The other type is a time-of-flight measurement device that relies on the sound waves being transmitted at an angle across the flow. This system uses at least two transmitters, one facing upstream and the other downstream. The difference in the time between the two is used—both flows go the same distance. For this device, the speed of sound in the fluid must be known. This device is nonintrusive (no obstruction or disturbance to the flow), and in some cases can be added to existing piping systems. Another advantage is that it can be used to measure wind speed outdoors.

9.2.7 Coriolis Meter

The Coriolis meter is different from the other flow meters presented here in that it measures the mass flow rate of the fluid rather than the volume flow rate. Coriolis flow meters are not sensitive to upstream disturbances and require little maintenance. However, they can be heavy and bulky, and they are sensitive to vibrations.

9.2.8 Accuracy of Flow Meters

The accuracy of a measurement can be quantified in several ways. The *repeatability* of a measurement with a particular device is the variation of obtained values over multiple trials in which the values of the independent variables (i.e., the flow conditions) are not changed. The *uncertainty* is the range of possible values represented by the readout on the instrument. No instrument is infinitely precise. (A more detailed discussion on experimental uncertainty—including a method to calculate the effects of propagations of errors into calculations from measurements—is presented in

Appendix D.) The *linearity* of a device is the closeness of the output to a linear fit to the input. Many devices are designed to be linear. The *range* is the span from the minimum to the maximum values that can be measured with a particular device, and the *turndown ratio* is the ratio of the maximum to the minimum values that can be measured. (This is also referred to as the *dynamic range* of the instrument.) In many devices the error is specified as a percentage of the full-scale reading of the device, so using a meter that is oversized for a particular application will result in increased experimental uncertainty and unnecessary cost.

Errors are usually classified as systematic or random. Random errors are those due to the limitations of the precision and accuracy of the device and are usually considered acceptable, as long as their magnitude is within the manufacturer's specifications. Systematic errors result in an offset or other nonrandom error that is generally undesirable. Sources of systematic errors include invalid calibration, misalignment, and so on.

For maximum accuracy and reliability, most manufacturers specify that their flow meters should be placed in a section of straight pipe far away from any disturbances to the flow. Sources of disturbances include elbows, expansions, and contractions in the pipe system. When a fluid flows through a pipe elbow, a three-dimensional swirling motion results. Expansions and contractions also can provide variant velocity profiles.

A fluid meter is usually calibrated by comparing it to a more accurate flow meter. Perhaps the most accurate way to measure flow rate is to collect the fluid flowing through a system over a period of time and weigh the mass of fluid collected, or measure the volume of fluid collected. According to Baker, the best accuracy that can typically be achieved with flow meters is around ±0.1% for liquid flows and ±0.2% for gas flows [Baker02].

9.3 Pressure Transducers

The older, analog methods for measuring pressure include manometers (Chapter 2) and Bourdon tubes. In most modern applications an electromechanical device is desired so that the measurements can be automatically acquired, processed, and stored to a computer without the need for a technician to make visual observations of the pressure and record the values by hand in a notebook, as was done in the past.

The different types of electromechanical pressure transducers include capacitance, piezoelectric, and piezo-resistive [Baker02]. Sensors can also be classified according to the type of pressure they measure, such as absolute, gage, vacuum, or differential. An absolute pressure sensor, such as barometer, measures the pressure relative to a perfect vacuum of zero pressure. The more common gage pressure sensor measures pressure relative to the local atmospheric pressure. A vacuum pressure system measures the amount the pressure is below the local atmospheric, while a differential pressure transducer measures the difference in pressure across two points. The simple analog tire pressure gage measures the pressure in the tires with a mechanical spring that compresses linearly with the force applied on it according to Hooke's law. Some pressure gages also use strain gages.

The piezoelectric effect refers to the behavior of certain types of crystals that change capacitance under mechanical loading. With proper electronic circuitry and an amplifier, a measurable output can be obtained that has a linear response with respect to a change in pressure. These sensors can measure very high pressures, on the order of 10,000 psi. They exhibit fairly good temperature resistance, although for applications in internal combustion engines, when they are used to measure in-cylinder pressures, they must be water cooled to avoid damaging the sensor. Quartz is a common material for the crystal in the sensor.

It is quite common to measure pressure and temperature simultaneously in the same location. This is especially desirable in gases, where the ideal gas law can be used to compute the gas density from measurements of pressure and temperature if the composition (molecular weight) of the gas is known. By far the most common device used to measure temperatures in fluid flows is the thermocouple. Thermocouples are cheap, simple, easy to install, and give an electronic output that can be recorded by a computerized data acquisition system. Thermocouples generally work in a range of temperatures from −200 to 1700°C [Baker02], which is a wide range. The upper end of the range is limited by the melting point of the metals used in the thermocouple, which is unfortunately below the flame temperatures encountered in many combustion systems. Thermistors are more accurate (and precise) than thermocouples, but have a smaller temperature range, and are generally only useable from −100 to 300°C [Baker02].

9.4 Nondimensionalization of Flow Data

In 1883 Reynolds wrote, "As there is no such thing as absolute space and time recognized in mechanical philosophy, to suppose that the character of motion of fluids in any way depended on absolute size or absolute velocity, would be to suppose such motion without the pale of the laws of motion. If then fluids in their motion are subject to these laws, what appears to be the dependence of the character of the motion on the absolute size of the tube, and on the absolute velocity of the immersed body, must in reality be a dependence on the size of the tube as compared with the size of some other object, and on the velocity of the body as compared with some 'other velocity'" [Reynolds83].

9.4.1 Common Nondimensional Numbers

Engineers often find it more efficient to communicate with other engineers in terms of nondimensional variables rather than in dimensioned units when discussing fluid problems. The goals of dimensional analysis are to present data as clearly as possible, identify parameters that are important to a particular problem, and allow meaningful modeling studies. Among the reasons that engineers nondimensionalize flow data is so they can compare data with other flows of greatly different sizes, to make results easier to graph and present to other engineers, and even to hide actual data values from competitors.

One of the earliest examples of the use of nondimensional parameters is when Reynolds realized that the transition from laminar to turbulent flow did not depend on the pipe diameter, the flow velocity, or the fluid viscosity individually, but on the nondimensional grouping of those three parameters that we now call the Reynolds number. In fact, in fluid mechanics the most commonly used nondimensional number is the Reynolds number. Since the time of Reynolds, many other groupings of variables to form nondimensional parameters have been discovered to describe various fluid phenomena. A list of some of the more common ones is presented in Table 9.1.

The *principle of dimensional homogeneity* states that any physically based equation must be dimensionally homogeneous. That is, each of the additive terms of the equation must have the same dimensions (units). An example is Bernoulli's equation:

$$P + \frac{1}{2}\rho V^2 + \rho g z = C$$

Each of the three terms in Bernoulli's equation has the same units, of pressure. Mathematical manipulations, including integration and differentiation, cannot change dimensional homogeneity.

The basic units in any unit system are mass, length, time, and temperature. When angles are expressed in radians they are nondimensional. The choice of a scaling parameter for length is usually obvious, such as the diameter of a round object or the length of a flat one. Oftentimes the maximum or the average velocity in a flow is known, and that can be used to scale the velocity. With these two parameters known, the time scale can be taken as the ratio of length scale to velocity scale.

The simplest possible nondimensional scaling parameters for distance, velocity, and time are as follows, where the superscript * is used to denote a nondimensional variable.

$$x^* = \frac{x}{L} \tag{9.19}$$

$$V^* = \frac{V}{U_0} \tag{9.20}$$

$$t^* = t\frac{U_0}{L} \tag{9.21}$$

Table 9.1 Common Nondimensional Numbers in Fluid Mechanics

Symbol	Name	Equation	Notes
Reynolds	Re	$\dfrac{\rho V L}{\mu}$	Ratio of inertial to viscous forces; most commonly used
Mach	Ma	$\dfrac{V}{a}$	Compressible and supersonic flow
Froude	Fr	$\dfrac{V}{\sqrt{gL}}$	Surface waves
Strouhal	St	$\dfrac{fL}{V}$	Vortex shedding frequency
Stokes	St	$\dfrac{\rho_{particle} D V}{18\mu}$	Response of suspended drop or particle
Prandtl	Pr	$\dfrac{\nu}{\alpha}$	Heat transfer
Weber	We	$\dfrac{\rho V^2 D}{\sigma}$	Atomization and drop breakup
Capillary	Ca	$\dfrac{\mu V}{\sigma}$	Shear force to surface tension
Bond	Bo	$\dfrac{\rho g L^2}{\sigma}$	Ratio of gravity to surface tension forces
Cavitation	Ca	$\dfrac{P - P_{vap}}{\frac{1}{2}\rho V^2}$	Cavitation in hydrodynamics
Friction factor	f	$\dfrac{\tau}{\frac{1}{2}\rho V^2}$	Friction loss in internal flow
Drag coefficient	C_D	$\dfrac{F_D}{\frac{1}{2}\rho A V^2}$	Friction drag in external flow
Pressure coefficient	C_P	$\dfrac{P - P_\infty}{\frac{1}{2}\rho V^2}$	Used in aerodynamics for surface profiles
Grashof number	Gr	$\dfrac{g\beta(\Delta T)L^3}{\nu^2}$	Natural convection

While this scaling commonly works, there may be some problems where other scaling is more important, such as the time scale of gravity-driven surface waves of wavelength λ. For flow of free-surface waves of a liquid, a better choice for the time scale might be

$$t^* = \sqrt{\dfrac{\lambda}{g}} \qquad (9.22)$$

Note that there is only one possible combination of the wavelength (m) and the gravitational acceleration (m/s²) that would have units of time (s). While the nondimensional parameters presented here will be useful to the engineer for a great many fluids problems, obviously they do not cover every possible range of phenomena that can be encountered in practice. The Buckingham pi theorem, presented in the next section, provides guidance for how to identify the important nondimensional parameters for a given problem.

9.4.2 Buckingham Pi Theorem

The pi theorem allows us to reduce the number of independent parameters in a problem if their units overlap. This works because the output will vary as a ratio or product of variables, rather than all the variables separately when their dimensions overlap. The pi theorem states that the number of independent nondimensional parameters needed to characterize the system is equal to the number of variables and parameters that influence the system, minus the number of fundamental units contained within those variables that are repeated, and are repeated in a way that is independent of each other. The nondimensional parameters are formed by multiplying the variables together in such a way that all the units cancel. The rules of a pi analysis are that all the variables and parameters must be used in the construction of the nondimensional parameters—none can be left over. Thus each nondimensional parameter will be a product of variables, called a pi group. The notation for pi multiplication is

$$\prod_{i=1}^{n} a_i = a_1 \times a_2 \times a_3 \times \cdots \times a_n$$

In a pi group, the variables need not all be raised to the power 1; they could be raised to the power $\frac{1}{2}$ or -1, for example.

Procedure for Pi Analysis

First, we must identify the independent variables (inputs) and the dependent variables (outputs). The dependent variable(s) should be identified first. In many problems, there is only one dependent variable. The first pi group formed is always linearly proportional to the dependent variable. The second and any other pi groups formed cannot use the dependent variable but must be made from combinations of the independent variables. There is usually more than one way to perform a pi analysis for a particular problem. Thus we have some flexibility—there is some art in performing a pi analysis, beyond just following rote instructions. We are also guided by previous experience, and should look for relevant nondimensional parameters identified by previous researchers in the area, such as those listed in Table 9.1.

9.4 Nondimensionalization of Flow Data

EXAMPLE 9.5

Logically it makes sense that the aerodynamic drag force, F, on an object should depend on how large the object is (quantified by its length L), how fast it is moving (V), the mass of air it has to move (quantified by the density, ρ), and the frictional resistance of the fluid (quantified by the viscosity, μ). To fully characterize the drag force by testing 10 different models of different lengths L and 10 different speeds V each, with 10 different values each of the fluid density and viscosity (changed by pressurizing and heating the air), would require $10 \times 10 \times 10 \times 10 = 10{,}000$ tests, which would take a very long time to perform and be quite expensive as well. In addition, presentation of the data would be quite challenging. Two variables can easily be plotted on a two-dimensional graph, and three variables can be shown by a surface plot or by plotting multiple curves on the same two-dimensional plot, but to show the results of changes in five different variables will require several graphs (or else one extremely messy graph). Fortunately, the pi theorem allows us to simplify the problem.

SOLUTION We start by listing all the variables and their units.

$$F: \frac{\text{kg} \cdot \text{m}}{\text{s}^2}, \quad L: \text{m}, \quad V: \frac{\text{m}}{\text{s}}, \quad \rho: \frac{\text{kg}}{\text{m}^3}, \quad \mu: \frac{\text{kg}}{\text{m} \cdot \text{s}}$$

In this problem, there are three independently repeated fundamental units, mass (kg), length (m), and time (s). The pi theorem says that the complexity of the problem can be reduced, without any loss of generality, from five variables (F, L, V, ρ, μ) to $5 - 3 = 2$ nondimensional parameters. A nondimensional analysis always starts with the *dependent variable*, which in this case is the force, F. The other four variables are *independent variables* that can be controlled by the designing engineer or the operator, while the force, F, depends on the value of the other variables. To nondimensionalize the force, we create a new parameter, F^*, that is linearly proportional to F:

$$F^* = CF$$

Now the task is to arrange the other variables into a coefficient, C, that makes F^* nondimensional. The units of kg, m, and s^{-2} must be eliminated. There is one more than one way to proceed, but in this example we will start by eliminating the units of mass (kg). This leaves two choices: divide F by ρ or by μ. In this case, we choose to divide by the density, giving

$$F^* = C'\frac{F}{\rho}$$

When F (kg m/s^2) is divided by ρ (kg/m^3), the remaining units are m^4/s^2. We will save the units of length (m) until last because that can easily be taken care of by L.

Now, there are two variables that contain time (s): velocity and viscosity. Using the viscosity would reintroduce mass, so velocity is the wise choice here. Thus, F should be divided by V^2 to eliminate time:

$$F^* = C'' \frac{F}{\rho V^2}$$

The units remaining on the coefficient C'' are m^2, so dividing by L^2 will make F^* nondimensional. So the final form of the nondimensional drag force is

$$F^* = \frac{F}{\rho V^2 L^2}$$

We now have one of the two nondimensional parameters needed to characterize the problem. The only variable that has not been used is the viscosity, μ, so that will be the basis of the second nondimensional parameter. We start forming a nondimensionalized viscosity in the same way we nondimensionalized the force:

$$\mu^* = C\mu$$

Any of the other variables can be used to nondimensionalize μ except F, which is the dependent variable. To eliminate the mass (kg) the only choice is the density, ρ. When μ (kg/m-s) is divided by ρ (kg/m^3) the remaining units are m^2/s. The only variable left that contains time (s) is the velocity, V. Dividing μ/ρ by the velocity, V, will leave units of length (m), which can be eliminated simply by dividing by L. So the final form of the nondimensionalized viscosity is

$$\mu^* = \frac{\mu}{\rho V L}$$

Now that we have formed the two nondimensional parameters, we can write the functional relationship between them as

$$F^* = f(\mu^*)$$

The dimensional analysis cannot tell us exactly what the function f is, but it tells us that if the value of μ^* is known, we have all that is needed to determine the value of F^*. The function f will vary with the geometry of the object, but for a given shape, it is only necessary to perform enough wind tunnel tests to cover the appropriate range of μ^*. Then the values of F^* can be determined from those tests. Then the results can be plotted on a simple two-dimensional graph that can be shared with any engineer.

In practice, aerodynamic engineers use slightly modified forms of the parameters from Example 9.5. The Reynolds number is actually the inverse of μ^*, and the commonly used drag coefficient, C_D, is twice F^*. So the functional relationship used in common practice is

$$C_D = f(\text{Re})$$

For example, suppose it is proposed to build a plane of length $L = 10$ m to fly at 300 m/s in air at sea level ($\nu = 1.5 \times 10^{-5}$ m²/s). A geometrically similar 1/10 scale model is constructed to be flown in a wind tunnel. If air at standard conditions is used in the wind tunnel, then in order to match the Reynolds numbers, $\text{Re}_{\text{model}} = \text{Re}_{\text{actual}}$, the velocity would be

$$\frac{(V)(1 \text{ m})}{1.5 \times 10^{-5} \text{ m}^2/\text{s}} = \frac{(300 \text{ m/s})(10 \text{ m})}{1.5 \times 10^{-5} \text{ m}^2/\text{s}}$$

This equation gives $V = 3000$ m/s. The Mach number of the actual plane to be built would be $\text{Ma} = V/a = 300/343 = 0.87$, while the model Mach number is $\text{Ma} = 3000/343 = 8.7$. Thus exact similarity is not achieved. What is the solution to this problem? A change of fluid could be used, or else we could settle for partial similarity.

EXAMPLE 9.6

Even though the valve trains in engines operate transiently, the flow over the valves may be analyzed using steady-state head loss by invoking a quasi-steady assumption. The mass flow rate $\dot{m}$ is believed to be a function of the gas density ρ, the viscosity μ, the throat area of the valve A, and the pressure drop over the valve, $\Delta P = P_1 - P_2$.

SOLUTION First we write the units for each variable:

$\dot{m}$	ρ	μ	A	ΔP
$\dfrac{\text{kg}}{\text{s}}$	$\dfrac{\text{kg}}{\text{m}^3}$	$\dfrac{\text{kg}}{\text{m} \cdot \text{s}}$	m^2	$\dfrac{\text{kg}}{\text{m} \cdot \text{s}^2}$

There are five variables and three repeated units (kg, m, s), so using the pi theorem we can reduce this problem to $5 - 3 = 2$ nondimensional parameters, such as $\dot{m}^* = f(\mu^*)$. First let's find the nondimensional forms (pi groups) of the mass flow rate and viscosity. There are three different choices of parameters to eliminate kg from $\dot{m}^*$, but only one choice, ΔP, to eliminate s (we exclude μ to be used for the other nondimensional parameter). Thus $\dot{m}^* = \frac{\dot{m}}{\sqrt{\Delta P}}$ leaves us with units of $\sqrt{\text{kg} \cdot \text{m}}$. Dividing by the square root of density will eliminate kg, and then dividing by A will eliminate m, so that the final form of $\dot{m}^*$ is

$$\dot{m}^* = \frac{\dot{m}}{A\sqrt{\rho\,\Delta P}}$$

The viscosity μ can be nondimensionalized using ρ, A, and ΔP:

$$\mu^* = \frac{\mu}{\sqrt{\rho A\,\Delta P}}$$

For some engine and valve configurations, the $\dot{m}^*$ is constant for a wide range of Reynolds numbers. Assuming this holds true for our engine, if the mass flow rate of air is measured to be 0.86 kg/s when $A = 0.004$ m², $\rho = 1.2$ kg/m³, and $\Delta P = 40{,}000$ Pa, we can predict the mass flow rate for the same conditions and a pressure drop of 10,000 Pa in the actual engine. If $\dot{m}_1^* = \dot{m}_2^*$, then $\frac{\dot{m}_1}{A_1\sqrt{\rho_1 \Delta P_1}} = \frac{\dot{m}_2}{A_2\sqrt{\rho_2 \Delta P_2}}$. Since the areas and densities are constant between the two cases, the mass flow rate scaling is $\dot{m}_2 = \dot{m}_1 \sqrt{\frac{\Delta P_2}{\Delta P_1}} = 0.86\,\text{kg/s}\sqrt{\frac{10{,}000\,\text{Pa}}{40{,}000\,\text{Pa}}} = 0.43\,\text{kg/s}$.

EXAMPLE 9.7

We need to know the force F that will be experienced by a dam that holds back water in a channel of depth H and width L. The dam will have a complex three-dimensional curved eggshell shape. The force will depend on the depth (H), the width (L), the density (ρ), and the acceleration due to gravity (g).

SOLUTION First we write down all important variables and parameters in this problem and their units:

F	H	L	ρ	g
$\text{kg}\frac{\text{m}}{\text{s}^2}$	m	m	$\frac{\text{kg}}{\text{m}^3}$	$\frac{\text{m}}{\text{s}^2}$

There are five independent variables and three repeated fundamental units (kg, m, s), so $5 - 3 = 2$ nondimensional parameters are needed. We start with the dependent variable F, which has all three units (kg, m, s) in it. Density, ρ, is only other variable that contains mass, so we divide F by ρ to eliminate kg. Similarly, g is only other variable that contains time, so we divide F by g to eliminates. Only the units of length are left, which could be eliminated by either H or L. For this problem, we choose to divide F by H^3 to eliminate the units of length. The only independent variable left is L, which can be easily nondimensionalized by dividing by H. The resulting two nondimensional parameters are

$$F^* = \frac{F}{\rho g H^3} \quad \text{and} \quad L^* = \frac{L}{H}$$

Thus the functional relationship is of the form $F^* = f(L^*)$.

Suppose a scale-model dam is built with $L = 10$ cm and $H = 5$ cm. If the measured force on this model dam is 1.3 N, what is the expected force on the actual dam with $L = 40$ and $F = 20$ m? When $L^*_{model} = L^*_{actual}$, then $F^*_{model} = F^*_{actual}$. Then the actual force is

$$F_{actual} = F_{model}\left(\frac{H_{actual}}{H_{model}}\right)^3 = 1.3 \text{ N}\left(\frac{20 \text{ m}}{0.05 \text{ m}}\right)^3 = 83.2 \text{ MN}$$

EXAMPLE 9.8

In Chapter 5, we stated that the pressure drop, ΔP, through a length L of straight pipe depends on the velocity through the pipe, the density and viscosity of the fluid, the diameter of the pipe, and the surface roughness of the pipe. Determine how many nondimensional parameters are needed to characterize the system and what they are.

SOLUTION Again, we first write down all important variables and parameters in this problem and their units:

ΔP	L	D	V	ρ	μ	ε
$\frac{\text{kg}}{\text{m} \cdot \text{s}^2}$	m	m	$\frac{\text{m}}{\text{s}}$	$\frac{\text{kg}}{\text{m}^3}$	$\frac{\text{kg}}{\text{m} \cdot \text{s}}$	m

There are seven independent variables and three repeated units (kg, m, s) so there will be $7 - 3 = 4$ nondimensional parameters. The pressure drop can be nondimensionalized as

$$\Delta P^* = \frac{\Delta P}{\rho V^2}$$

The viscosity can be nondimensionalized by

$$\mu^* = \frac{\mu}{\rho V D}$$

and the surface roughness can be nondimensionalized by

$$\varepsilon^* = \frac{\varepsilon}{D}$$

Finally, the length can be nondimensionalized simply by

$$L^* = \frac{L}{D}$$

EXAMPLE 9.9

For a horizontal-axis wind turbine, the power produced is a function of the diameter of the turbine, the density of the working fluid (air), the wind velocity, the rotational speed of the turbine, and the number of turbine blades. Determine the important nondimensional parameters to use for analysis of the wind turbine.

SOLUTION We first write down all important variables and their units:

$\dot{W}$	D	V	ρ	Ω	n
$kg\frac{m^2}{s^3}$	m	$\frac{m}{s}$	$\frac{kg}{m^3}$	$\frac{1}{s}$	—

There are six variables and three repeated units (kg, m, s) so there are $6 - 3 = 3$ nondimensional parameters. There are two logical choices for functional relationships: either $\dot{W}^* = f(n, V^*)$ or $\dot{W}^* = f(n, \Omega^*)$. Choosing the first, the nondimensional power is first formed by dividing $\dot{W}$ by ρ to eliminate mass, then dividing by ω^3 to eliminate time, and finally multiplying by D^5 to eliminate length, yielding

$$\dot{W}^* = \frac{\dot{W} D^5}{\rho \Omega^3}$$

The nondimensional velocity is easier to find:

$$V^* = \frac{V}{\Omega D}$$

■ Summary

After reading this chapter and working through the problems, you should be able to select an appropriate measurement device for a given application. To make the proper selection, you need to ask several questions. First, what is the quantity that needs to be measured? Is it velocity, volume flow rate, mass flow rate, momentum flow rate, force, pressure, temperature, or something else? Second, how much accuracy is required? The more accurate the transducer, the more costly it is likely to be. Third, what is the likely range over which measurements will be taken? In general, higher flow rates will require larger flow meters, which are more expensive. Additional considerations include maintenance and lifetime of the device, availability, technical support, warranty, reliability, calibration required, and so on.

■ References

[NASA71] Meyers, J. 1971. Investigation and Calculations of Basic Parameters for the Application of the Laser Doppler Velocimeter. NASA Technical Note D-6125.

[Settles06] Settles, G. 2006. *Schlieren and Shadowgraph Techniques.* Springer.

[Goldstein96] Goldstein, R. 1996. *Fluid Mechanics Measurements.* CRC Press.

[Bean83] Bean, H. 1983. *Fluid Meters.* ASME.

[Blevins92] Blevins, R. 1992. *Applied Fluid Dynamics Handbook.* Krieger.

[Hinds82] Hinds, William. 1982. *Aerosol Technology.* Wiley.

[Baker02] Baker, R. 2002. *An Introductory Guide to Flow Measurement.* Wiley.

[Reynolds83] Reynolds, O. 1883. "An Experimental Investigation of the Circumstances which Determine whether the Motion of Water Shall be Direct or Sinuous, and of the Law of Resistance in Parallel Channels." *Philosophical Transactions of the Royal Society of London*, Vol. 174, pp. 935–982.

■ Problems

1. You need to measure the flow rate of air through an HVAC duct. What are your options?
2. What is the definition of SCFM? If your flow meter is calibrated in SCFM, what does this mean for your use of it?
3. How do you calibrate a pressure transducer?
4. What is a dead-weight tester?

5. Why did Mr. Fahrenheit choose to make the freezing and boiling points of water at 32° and 212°, respectively, in his temperature scale?

6. How could you make measurements of the gas temperature in a flame that is over 2000°C?

7. Compare the costs of a rotameter, turbine meter, and Coriolis meter all set to measure water flow rates up to 50 gal/min?

8. What is the discharge coefficient for the valve in Example 9.6?

9. The pressure difference for air flow is measured with a Pitot tube to be 2,000 Pa, with an error of ±25 Pa, the pressure of the air at the test area is measured to be 99,580 Pa ± 10 Pa, and the temperature of the air is measured to be 21°C ± 0.5°C. What is the measured velocity, and what is the range of uncertainty in the measured velocity?

10. In a PIV system, the displacement of the fluid at a particular location is measured to be 4.1 pixels in the CCD camera, with a magnification factor of 10 (1 cm on the image corresponds to 10 cm in the flow). The camera has a 2/3 size CCD sensor (16 by 24 mm) and is a 6-megapixel camera (2000 by 3000). The time between exposures is 10 ms. What is the measured velocity? If the uncertainty in the PIV software is ±0.1 pixel, what is the uncertainty in the measured velocity?

11. List at least three different ways you could measure the drag coefficient of a baseball.

12. The speed of a baseball is measured as a function of time with a radar gun that has an accuracy of ±0.5 mph, and the time is known to ±0.02 s. If the average speed of the baseball is 90 mph and the average drag coefficient is 0.3, what is the uncertainty in the calculated value of the drag coefficient? (See Appendix D for information on propagation of errors in calculations.) You can assume that errors in measuring the air density and the diameter of the baseball are negligible.

13. The flow rate of water is to be measured continuously through a flow meter in a system that operates 24/7. The pipe where the flowmeter is to be located has an internal diameter of 5 cm, and the typical flow velocity at that point is 0.5 m/s. Your boss wants you to compare the operating costs of a Venturi flow meter and an orifice flow meter, both with the same contraction ratio of $\beta = 0.5$. The discharge coefficient for the Venturi is 0.98 and for the orifice is 0.62. Electricity costs $0.10/kW-hr, and the pump is 70% efficient. How much will it cost to operate each flow meter over the projected 20-year life of the facility?

14. Repeat the previous problem assuming that electricity rates go up 5% over the previous year's rates each year.

15. An interference-refraction drop-sizing instrument is used to measure the size of water drops. If a water drop has a diameter of 0.5 mm, what is its effective focal length?

16. What is the diffraction limit for visible light? What is the size of the smallest object that can be accurately seen using visible light?

17. What are the relative advantages and disadvantages of the use of a laminar flow element compared to an orifice meter to measure air flow rates?

18. Air flows with a velocity of 10 m/s through a pipe of 10 cm diameter. What is the unrecoverable pressure loss associated with the use of a Venturi meter, an orifice meter, and a laminar flow element?

19. Air flows through a pipe of 20 cm diameter. A laminar flow element is installed, with the hydraulic diameter of the flow elements equal to 1 mm. The cross-sectional area of the laminar flow element is 90% of that of the main pipe due to the finite thickness of the flow element walls. If the measured pressure drop in the laminar flow element is 2.1 kPa over 15 cm of length, what is the flow rate of air in the pipe?

20. If the flow in a round pipe is laminar, can you use a Pitot tube to measure the volume flow rate?

21. A pipe of 10 cm internal diameter has a Venturi flow meter with a throat diameter of 4 cm installed. If the measured pressure difference is 42.0 kPa, find the volume flow rate of water in liters per minute.

22. A pipe of 10 cm internal diameter has an orifice flow meter with a throat diameter of 5 cm installed. If the measured pressure difference is 42.0 kPa, and the discharge coefficient of the orifice meter is $C_d = 0.61$ m, find the volume flow rate of water in liters per minute.

23. Air flows through a Venturi meter in a pipe of 8 cm diameter, with a Venturi throat diameter of 4 cm. What is the maximum velocity of air in the pipe at which the Venturi meter can still be used without account for changes in the density of the air due to the change in static pressure at the throat? What is the maximum velocity that can be used until compressibility effects in the throat become significant?

24. The vertical velocity of a Penaud helicopter depends on the density of the air, the weight of the helicopter, the length of the blades, and the rotational speed of the blades. Perform a pi analysis for this problem. If the helicopter geometry is kept to scale and doubled in size (length), how do you expect the velocity to change? Will it get bigger or smaller?

25. A $\frac{1}{4}$-scale model of an automobile is tested in a wind tunnel at an air speed of 120 m/s. The cross-sectional area of the model is 0.14 m², and the measured drag force on the model is 25.7 N. Calculate the power to overcome drag on the full-scale car traveling at typical highway speeds.

26. A series of experimental tests is going to be performed to measure the aerodynamic characteristics of a Frisbee. List all the variables that should be measured for this testing. Perform a pi analysis to reduce this to the minimum number of nondimensional parameters to characterize the flight of a Frisbee.

27. For a centrifugal pump, the variables are Q, ω, D, $\dot{W}$, and ΔP, and the density and viscosity of the fluid are also important. Perform a pi analysis for this problem. How many nondimensional parameters are required?

28. Based on the results of the previous problem, if a given pump of size D_1 is tested and provides a flow rate Q_1 and takes power $\dot{W}_1$ to perform the task, how would a pump of the same design and geometrically similar construction perform if the size is $D_2 = 2D_1$? What flow rate would be provided and what input power would be required?

29. A series of experimental tests is going to be performed to analyze the performance of a marine propeller in a water tunnel. List all the variables that need to be measured and quantified for these experiments. Perform a pi analysis to reduce this to the minimum number of nondimensional parameters needed to characterize the propeller performance.

30. A cylinder of radius R and length L is placed inside a slightly larger cylinder with a small gap of thickness H between them. If a fluid of density ρ and viscosity μ is placed between the two cylinders and the outer cylinder is fixed in place, find the nondimensional form of the torque T required to turn the cylinder at rotational speed ω as a function of the nondimensional rotational speed and any other needed nondimensional parameters.

31. The steady forward flight speed of a bird depends on the weight of the bird, the density and viscosity of the air, the area of the wings, and the length of the wing span. Perform a pi analysis for this problem.

32. A $\frac{1}{4}$-scale model of a tractor–trailer is to be tested in a wind tunnel. If the maximum highway speed of the vehicle is 65 mph, what is the maximum speed that should be used in a wind tunnel test? If the wind tunnel is at sea level and the actual truck is to be used in Denver, how does that change the velocity that should be used in the wind tunnel?

33. The steady level flight speed, V, of a tubular rocket plane, such as the X-1 or X-15, depends on the exit velocity of the rocket exhaust relative to the plane, V_e, the weight of the plane, W, the diameter of the fuselage, D, the area of the wings, A, the density of the air, ρ, the altitude at which it will fly, h, and the lift and drag coefficients, C_L and C_D, for the plane. How many nondimensional parameters are needed to characterize this problem? Give one possible grouping.

34. The power generated by a wind turbine depends on the velocity of the wind, the length of the turbine blades, the density of the air, and the efficiency of the electric generator and drive train. Perform a pi analysis for this problem.

35. The rate at which an inkjet printer can print (pages/s) depends on the area of the paper, A, the size of the inkjet holes, D, the number of printhead holes, N, the density of the ink, ρ, the electrical power supplied to the printhead, $\dot{W}$, the heat of vaporization of the ink, h_{fg}, and the surface tension of the ink, σ. Perform a pi analysis for this problem. What does your pi analysis tell you about the best way to increase the printing rate of a printer? What other factors not considered here might also need to be included?

36. The roll rate (degrees per second) of an airplane depends on the speed at which the airplane is flying, the mass of the plane, the density of the air, the area of the ailerons, and the distance of the ailerons to the central axis of the plane. Perform a pi analysis for this problem.

37. What is the Deborah number, De?

38. What is the Ohnesorge number, Oh? What is the relationship between the Ohnesorge number, Reynolds number, and Weber number?

39. The rate at which chocolate candies can be made (number/hour) depends on the volume of the molds, the diameter of the pouring spout, the density of molten chocolate, the acceleration due to gravity, and the viscosity of molten chocolate. Perform a pi analysis for this problem.

# Properties of Common Fluids

A.1 Properties of Fluids

If the density and the absolute viscosity are known, the kinematic viscosity is calculated according to

$$\nu = \frac{\mu}{\rho} \tag{A.1}$$

Alternatively, if the kinematic viscosity and density are known, the absolute viscosity can be calculated as

$$\mu = \rho\nu \tag{A.2}$$

The viscosity of gases increases as a function of temperature, while the viscosity of liquids decreases with increasing temperature. The viscosity of air and other gases is almost independent of pressure. To good engineering approximation, the pressure dependency of viscosity can be neglected, leaving the viscosity as a function of temperature solely [NACA53]. For gases, an empirical correlation known as *Sutherlands' formula* is often used to approximate changes in absolute viscosity as a function of temperature:

$$\mu = \frac{\beta T^{3/2}}{T + S} \tag{A.3}$$

where β and S are constants that depend on the particular gas. For air the values are $\beta = 1.458 \times 10^{-6}$ kg/(m-s) and $S = 110.4$ K [USATM76].

Property values are contained in the following tables. Table A.1 has the properties for liquid water, Table A.2 the properties for air, Table A.3 the properties for steam vapor, Table A.4 properties for common gases, and Table A.5 has data for common liquids.

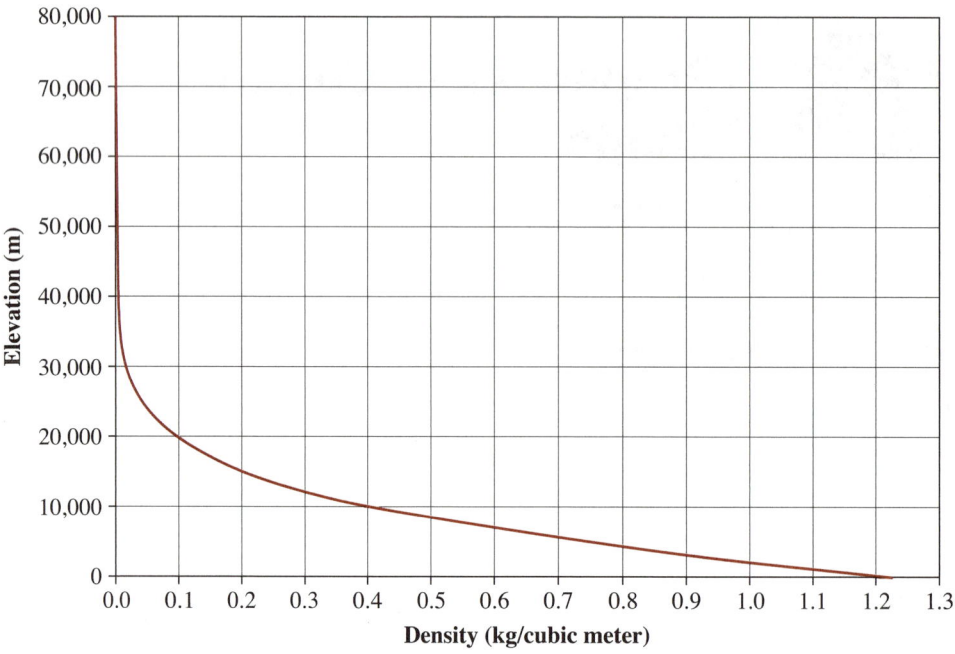

FIGURE A.1 Profile of density as a function of altitude in the standard atmosphere.

Table A.1 Properties of Water as a Function of Temperature at 1 atm

Temperature (°C)	Density (kg/m³)	Absolute viscosity (kg/m-s)	Kinematic viscosity (m²/s)	Vapor presure (atm)
0	999.8	1.79×10^{-3}	1.79×10^{-6}	0.0060
10	999.7	1.31×10^{-3}	1.31×10^{-6}	0.0120
20	998.2	1.00×10^{-3}	1.00×10^{-6}	0.0228
30	995.7	7.97×10^{-4}	8.01×10^{-7}	0.0415
40	992.2	6.53×10^{-4}	6.58×10^{-7}	0.0722
50	988.0	5.47×10^{-4}	5.53×10^{-7}	0.1208
60	983.2	4.66×10^{-4}	4.74×10^{-7}	0.1951
70	977.8	4.04×10^{-4}	4.13×10^{-7}	0.3054
80	971.8	3.54×10^{-4}	3.65×10^{-7}	0.4644
90	965.3	3.14×10^{-4}	3.26×10^{-7}	0.6877
100	958.4	2.82×10^{-4}	2.94×10^{-7}	0.9943

Data from [NIST05] and [NBS55].

Table A.2 Properties of Air as a Function of Temperature at 1 atm

Temperature (°C)	Density (kg/m³)	Absolute viscosity (kg/m-s)	Kinematic viscosity (m²/s)	Specific heat ratio, k	Speed of sound, a (m/s)
−20	1.396	1.62×10^{-5}	1.16×10^{-5}	1.4035	319.1
−10	1.343	1.67×10^{-5}	1.24×10^{-5}	1.4031	325.3
0	1.293	1.72×10^{-5}	1.33×10^{-5}	1.4027	331.5
10	1.248	1.77×10^{-5}	1.42×10^{-5}	1.4023	337.4
20	1.205	1.81×10^{-5}	1.51×10^{-5}	1.4019	343.3
30	1.165	1.86×10^{-5}	1.60×10^{-5}	1.4016	349.1
40	1.128	1.91×10^{-5}	1.69×10^{-5}	1.4011	354.8
50	1.093	1.95×10^{-5}	1.79×10^{-5}	1.4007	360.4
60	1.060	2.00×10^{-5}	1.89×10^{-5}	1.4002	365.9
70	1.029	2.04×10^{-5}	1.99×10^{-5}	1.3997	371.3
80	1.000	2.09×10^{-5}	2.09×10^{-5}	1.3991	376.6
90	0.972	2.13×10^{-5}	2.19×10^{-5}	1.3985	381.8
100	0.946	2.17×10^{-5}	2.30×10^{-5}	1.3979	387.0
150	0.837	2.38×10^{-5}	2.84×10^{-5}	1.3942	411.4
200	0.748	2.57×10^{-5}	3.44×10^{-5}	1.3897	434.5
250	0.676	2.75×10^{-5}	4.07×10^{-5}	1.3846	456.0
300	0.617	2.93×10^{-5}	4.74×10^{-5}	1.3791	476.4
350	0.567	3.09×10^{-5}	5.45×10^{-5}	1.3733	495.6
400	0.525	3.25×10^{-5}	6.19×10^{-5}	1.3676	514.1
450	0.489	3.40×10^{-5}	6.96×10^{-5}	1.3621	531.8
500	0.457	3.55×10^{-5}	7.77×10^{-5}	1.3567	548.9

Data from [NBS55].

Table A.3 Properties of Steam at 1 atm

Temperature (°C)	Density (kg/m³)	Absolute viscosity (kg/m-s)	Kinematic viscosity (m²/s)
100	0.596	1.25×10^{-5}	2.09×10^{-5}
150	0.524	1.43×10^{-5}	2.73×10^{-5}
200	0.467	1.64×10^{-5}	3.51×10^{-5}
250	0.421	1.79×10^{-5}	4.24×10^{-5}
300	0.384	2.00×10^{-5}	5.21×10^{-5}
350	0.353	2.15×10^{-5}	6.09×10^{-5}
400	0.327	2.36×10^{-5}	7.23×10^{-5}
450	0.304	2.51×10^{-5}	8.26×10^{-5}
500	0.284	2.73×10^{-5}	9.60×10^{-5}

Data from [NBS55].

Table A.4 Properties of Common Gases at 20°C and 1 atm

Gas	Formula	Molecular weight (kg/kmol)	Density (kg/m³)	Absolute viscosity (kg/m-s)	Kinematic viscosity (m²/s)
Hydrogen	H_2	2.02	0.084	8.82×10^{-6}	1.05×10^{-4}
Nitrogen	N_2	28.01	1.164	1.78×10^{-5}	1.53×10^{-5}
Oxygen	O_2	32.00	1.330	2.03×10^{-5}	1.52×10^{-5}
Carbon dioxide	CO_2	44.01	1.830	1.50×10^{-5}	8.21×10^{-6}
Methane	CH_4	16.04	0.667	1.13×10^{-5}	1.69×10^{-5}
Propane	C_3H_8	44.09	1.833	8.39×10^{-6}	4.58×10^{-6}
Helium	He	4.00	0.166	1.96×10^{-5}	1.18×10^{-4}
Argon	Ar	39.95	1.661	2.28×10^{-5}	1.37×10^{-5}

Data from [NASA65], [NBS55], and [NIST05].

The surface tension of a liquid decreases with increasing temperature until the critical point is reached, at which point the surface tension is no longer defined. A correlation for the surface tension of water as a function of temperature is

$$\sigma = 235.8 \frac{J}{m^2} \left(\frac{647.15 - T}{647.15} \right)^{1.256} \left[1 - 0.625 \left(\frac{647.15 - T}{647.15} \right) \right] \tag{A.4}$$

Table A.5 Properties of Selected Liquids at 20°C and 1 atm

Liquid	Formula	Density (kg/m³)	Absolute viscosity (kg/m-s)	Kinematic vescosity (m²/s)	Boiling point (°C)
Ethanol	C_2H_5OH	790.0	1.20×10^{-3}	1.52×10^{-6}	78.3
Water	H_2O	1000.0	1.00×10^{-3}	1.00×10^{-6}	100.0
Hydrogen peroxide	H_2O_2	1449.9	1.25×10^{-3}	8.62×10^{-7}	150.0
Hydrazine	N_2H_4	1006.5	9.70×10^{-4}	9.64×10^{-7}	113.3
Ammonia*	NH_3	683.0	2.60×10^{-4}	3.81×10^{-7}	−33.3
Liquid oxygen*	O_2	1141.6	1.90×10^{-4}	1.66×10^{-7}	−183.3
Liquid hydrogen*	H_2	70.7	1.40×10^{-5}	1.98×10^{-7}	−252.8

*Properties at boiling point. Data from [NASA65].

for the temperature, T, in units of Kelvin, and where 647.15 K is the critical temperature for water [Vargaftik83].

A.2 Properties of the Atmosphere

Table A.6 lists the relative concentrations of the different gases in the atmosphere for dry air. The concentration of water vapor (H_2O) depends on the relative humidity.

Properties of Air as a Function of Elevation

The ideal gas law applies to air in the atmosphere as long as the air is dense enough for the continuum assumption to apply. Table A.7 lists the properties of the standard atmosphere, and Figures A.1 and A.2 show plots of density and pressure as a function of elevation.

Table A.6 Constituent Gases in the Atmosphere
for clean, dry atmospheric air near sea level

Constituent gas	Formula	Content, percent by volume
Nitrogen	N_2	78.084
Oxygen	O_2	20.948
Argon	Ar	0.934
Carbon dioxide	CO_2	0.031
Trace gases*		0.003

*Trace gases include neon (Ne), helium (He), krypton (Kr), hydrogen (H_2), xenon (Xe), methane (CH_4), nitrogen oxide (N_2O), ozone (O_3), sulfur dioxide (NO_2), ammonia (NH_3), carbon monoxide (CO), and iodine (I_2). From [USATM76].

Table A.7 Properties of the Atmosphere as a Function of Elevation Above Sea Level

Elevation (m)	Density (kg/m³)	Pressure (bar)	Temperature (°C)	Speed of sound (m/s)	Absolute viscosity (kg/m-s)	g (m/s²)
0	1.225	1.01325	15.00	340.3	1.788×10^{-5}	9.807
100	1.213	1.0012	14.35	339.9	1.786×10^{-5}	9.806
200	1.208	0.9895	13.70	339.5	1.783×10^{-5}	9.806
300	1.190	0.9777	13.05	339.1	1.780×10^{-5}	9.806
400	1.179	0.9661	12.40	338.8	1.778×10^{-5}	9.805
500	1.167	0.9546	11.75	338.4	1.774×10^{-5}	9.805
750	1.139	0.9263	10.13	337.4	1.767×10^{-5}	9.804
1,000	1.112	0.8988	8.50	336.4	1.758×10^{-5}	9.804
2,000	1.007	0.7950	2.00	332.5	1.726×10^{-5}	9.800
3,000	0.909	0.7012	−4.49	328.6	1.694×10^{-5}	9.797
4,000	0.819	0.6166	−10.98	324.6	1.661×10^{-5}	9.794
5,000	0.736	0.5405	−17.47	320.6	1.628×10^{-5}	9.791
6,000	0.660	0.4722	−23.96	316.5	1.595×10^{-5}	9.788
7,000	0.590	0.4110	−30.45	312.3	1.561×10^{-5}	9.785
8,000	0.526	0.3565	−36.94	308.1	1.527×10^{-5}	9.782
9,000	0.467	0.3080	−43.42	303.9	1.493×10^{-5}	9.779
10,000	0.414	0.2650	−49.90	299.5	1.458×10^{-5}	9.776
15,000	0.195	0.1211	−56.50	295.1	1.422×10^{-5}	9.761
20,000	0.089	0.0553	−56.50	295.1	1.422×10^{-5}	9.745
25,000	0.040	0.0255	−51.60	298.4	1.448×10^{-5}	9.730
30,000	0.018	0.0120	−46.64	301.7	1.475×10^{-5}	9.715
35,000	0.0085	0.0057	−36.64	308.3	1.529×10^{-5}	9.700
40,000	0.0040	0.0029	−22.80	317.2	1.601×10^{-5}	9.684
50,000	0.0010	0.00080	−2.50	329.8	1.704×10^{-5}	9.654
60,000	0.00031	0.00022	−26.13	315.1	1.584×10^{-5}	9.624
70,000	0.000083	0.000052	−53.57	297.1	1.438×10^{-5}	9.594
80,000	0.000018	0.000010	−74.51	282.5	1.321×10^{-5}	9.564

Data from [USATM76].

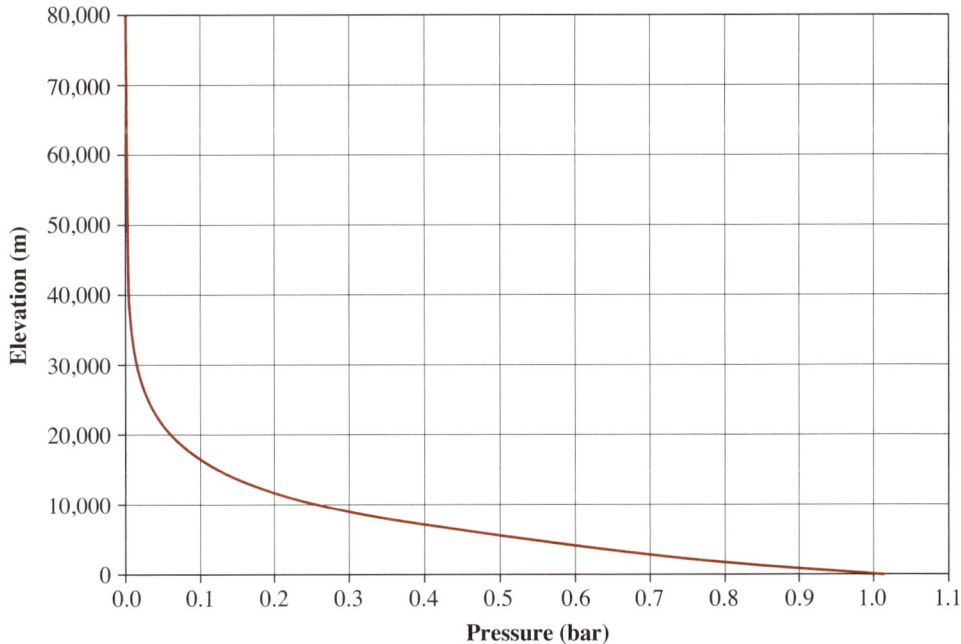

FIGURE A.2 Profile of pressure as a function of altitude in the standard atmosphere.

References

[NACA53] NACA. 1953. *Equations, Tables, and Charts for Compressible Flow.* NACA Report 1135.

[USATM76] National Oceanic and Atmospheric Administration, National Aeronautics and Space Administration, and U.S. Air Force. 1976. *U.S. Standard Atmosphere.*

[NIST05] Linstrom, P. and W. Mallard. 2005. NIST Chemistry WebBook. NIST Standard Reference Database Number 69.

[NBS55] U.S. Department of Commerce. 1955. *Tables of Thermal Properties of Gases.* National Bureau of Standards Circular 564.

[NASA65] Carter, G. 1965. *Liquid Propellants Safety Handbook.* NASA TM-X-56611.

[Vargaftik83] Vargaftik, N., B. Volkov, and L. Voljak. 1983. "International Tables of the Surface Tension of Water." *Journal of Physical Chemistry Reference Data*, Vol. 12, pp. 817–820.

Compressible Flow Tables

This section presents compressible flow tables used when solving compressible flow problems. If a general equation solver such as EES is available, then it may be used instead of these tables to solve the equations. Compressible flow problems arise in flow through rocket nozzles and around supersonic projectiles, but they are also seen in internal flows of gases if the velocity becomes high enough. Table B.1 contains data for one-dimensional isentropic flow, and Table B.2 the relations for flow across a normal shock wave. The data in Table B.1 is plotted in Figure B.1, and the data from Table B.2 is plotted in Figure B.4.

We stated in the text that compressibility effects can generally be ignored when Ma < 0.3, and for flows of low Mach number a gas can be treated as incompressible, or as having a constant density that will not change with velocity. Figure B.2 shows a plot of the ratio of the density to the stagnation density as a function of Mach number for isentropic flow. It can be seen that the change in density from the stagnation value is less than 5% when the Mach number is less than 0.3. Similarly, Figure B.3 shows the predicted stagnation pressure as a function of Mach number for the pressure being predicted using the incompressible Bernoulli equation, and the more correct compressible flow relationship.

Table B.1 Relations for Isentropic Flow of an Ideal Gas with a Specific Heat Ratio $k = 1.4$ as a Function of Mach Number

Mach number	T/T_0	P/P_0	ρ/ρ_0	A/A^*
0.00	1.0000	1.0000	1.0000	
0.10	0.9980	0.9930	0.9950	5.8218
0.20	0.9921	0.9725	0.9803	2.9635
0.30	0.9823	0.9395	0.9564	2.0351
0.40	0.9690	0.8956	0.9243	1.5901
0.50	0.9524	0.8430	0.8852	1.3398
0.60	0.9328	0.7840	0.8405	1.1882
0.70	0.9107	0.7209	0.7916	1.0944
0.80	0.8865	0.6560	0.7400	1.0382
0.90	0.8606	0.5913	0.6870	1.0089
1.00	0.8333	0.5283	0.6339	1.0000
1.10	0.8052	0.4684	0.5817	1.0079
1.20	0.7764	0.4124	0.5311	1.0304
1.30	0.7474	0.3609	0.4829	1.0663
1.40	0.7184	0.3142	0.4374	1.1149
1.50	0.6897	0.2724	0.3950	1.1762
1.60	0.6614	0.2353	0.3557	1.2502
1.70	0.6337	0.2026	0.3197	1.3376
1.80	0.6068	0.1740	0.2868	1.4390
1.90	0.5807	0.1492	0.2570	1.5553
2.00	0.5556	0.1278	0.2300	1.6875
2.50	0.4444	0.0585	0.1317	2.6367
3.00	0.3571	0.0272	0.0762	4.2346
3.50	0.2899	0.0131	0.0452	6.7896
4.00	0.2381	0.0066	0.0277	10.7188
4.50	0.1980	0.0035	0.0174	16.5622
5.00	0.1667	0.0019	0.0113	25.0000

Adapted from [NACA53].

Table B.2 Relations for Flow Across a Normal Shock for an Ideal Gas with a Specific Heat Ratio $k = 1.4$

M_1	M_2	P_{02}/P_{01}	T_2/T_1	P_2/P_1	ρ_2/ρ_1
1.00	1.0000	1.0000	1.0000	1.0000	1.0000
1.10	0.9118	0.9989	1.0649	1.2450	1.1691
1.20	0.8422	0.9928	1.1280	1.5133	1.3416
1.30	0.7860	0.9794	1.1909	1.8050	1.5157
1.40	0.7397	0.9582	1.2547	2.1200	1.6897
1.50	0.7011	0.9298	1.3202	2.4583	1.8621
1.60	0.6684	0.8952	1.3880	2.8200	2.0317
1.70	0.6405	0.8557	1.4583	3.2050	2.1977
1.80	0.6165	0.8127	1.5316	3.6133	2.3592
1.90	0.5956	0.7674	1.6079	4.0450	2.5157
2.00	0.5774	0.7209	1.6875	4.5000	2.6667
2.10	0.5613	0.6742	1.7705	4.9783	2.8119
2.20	0.5471	0.6281	1.8569	5.4800	2.9512
2.30	0.5344	0.5833	1.9468	6.0050	3.0845
2.40	0.5231	0.5401	2.0403	6.5533	3.2119
2.50	0.5130	0.4990	2.1375	7.1250	3.3333
2.60	0.5039	0.4601	2.2383	7.7200	3.4490
2.70	0.4956	0.4236	2.3429	8.3383	3.5590
2.80	0.4882	0.3895	2.4512	8.9800	3.6636
2.90	0.4814	0.3577	2.5632	9.6450	3.7629
3.00	0.4752	0.3283	2.6790	10.3333	3.8571
3.50	0.4512	0.2129	3.3151	14.1250	4.2609
4.00	0.4350	0.1388	4.0469	18.5000	4.5714
4.50	0.4236	0.0917	4.8751	23.4583	4.8119
5.00	0.4152	0.0617	5.8000	29.0000	5.0000

APPENDIX B COMPRESSIBLE FLOW TABLES

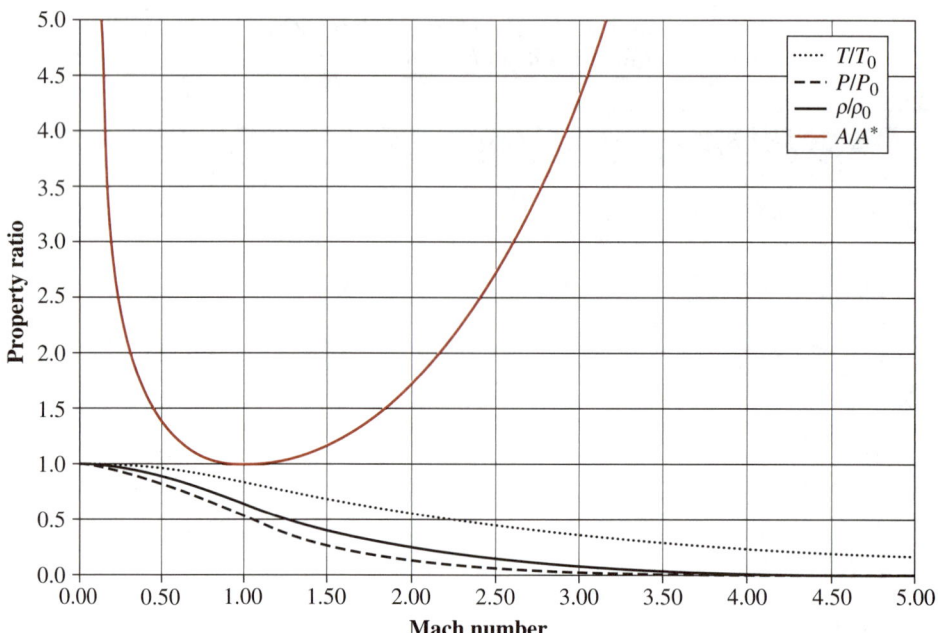

FIGURE B.1 Plot of isentropic flow relations as a function of the Mach number.

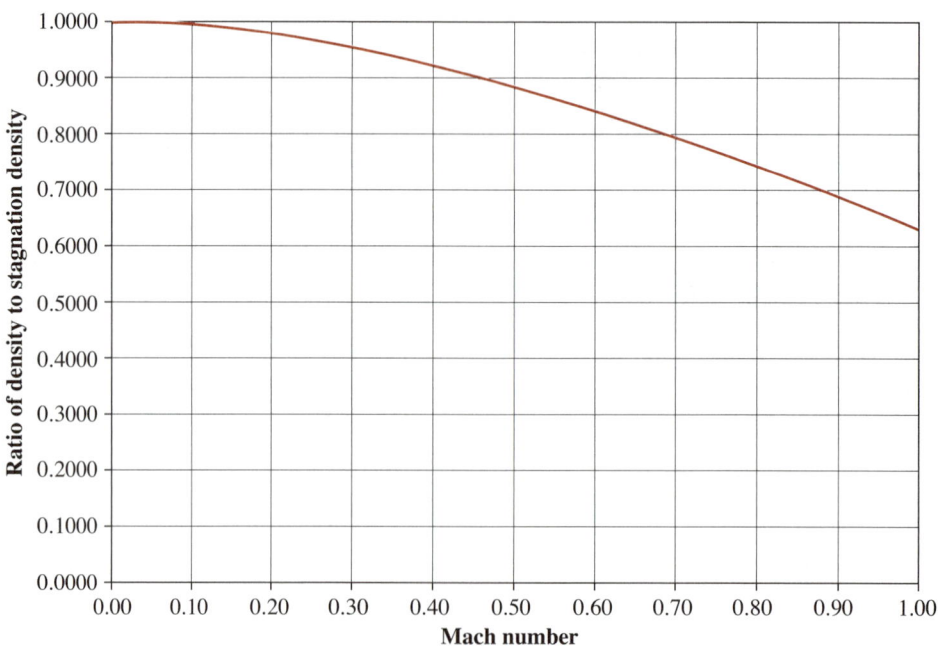

FIGURE B.2 Plot of the density to stagnation density ratio as a function of the Mach number.

Compressible Flow Tables

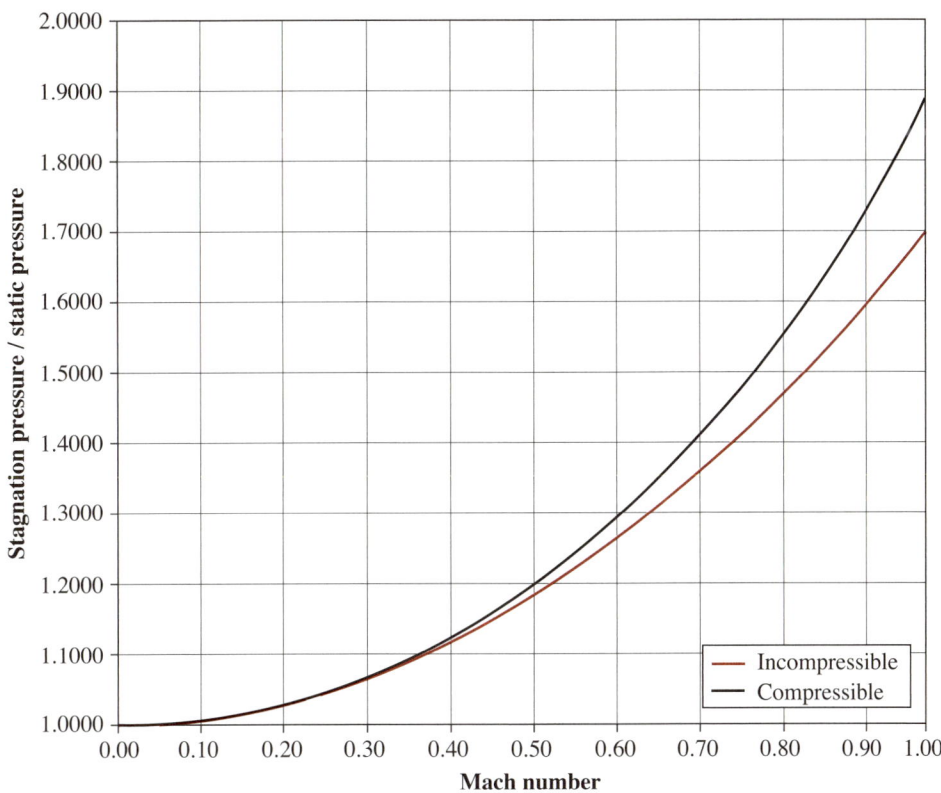

FIGURE B.3 Plot of the stagnation pressure to static pressure ratio as a function of the Mach number for predictions using incompressible and compressible flow theory.

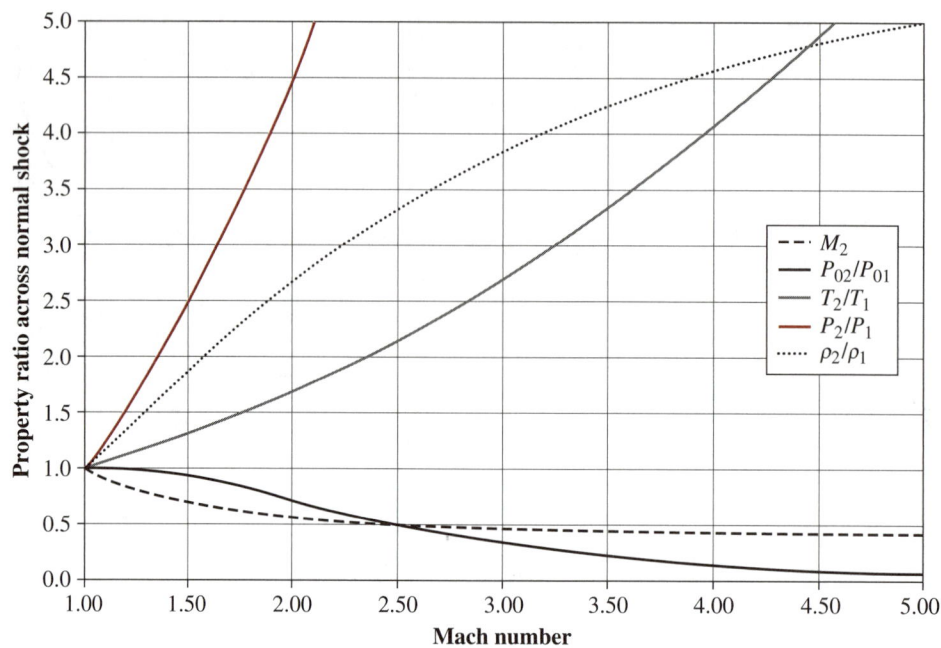

FIGURE B.4 Normal shock relationships as a function of the Mach number.

References

[NACA53] NACA. 1953. *Equations, Tables, and Charts for Compressible Flow.* NACA Report 1135.

C Reynolds Transport Theorem

The Reynolds transport theorem is a general statement about the conservation of certain properties. It can be applied to mass, momentum, and energy, as well as other quantities. This appendix presents the derivation of the Reynolds transport theorem and shows how the other conservation equations of fluid mechanics can be derived from it.

The total amount of any *extensive* property, B, in a system can be written as

$$B = \int_{CV} \beta \, dm = \int_{CV} \beta \rho \, d\forall \tag{C.1}$$

where β is the corresponding *intensive* property. For the study of fluid mechanics, B can take on any of the following values: mass, momentum, angular momentum, energy, or entropy. The relationship between B and β in each of these cases is

If B is:		then β is:	
M	(mass)	1	
$\vec{P}$	(linear momentum vector)	$\vec{V}$	(velocity vector)
$\vec{H}$	(angular momentum vector)	$\vec{r} \times \vec{V}$	(moment vector)
E	(energy)	e	(specific energy)
S	(entropy)	s	(specific entropy)

The general Reynolds transport theorem for B can be written as

$$\frac{dB}{dt} = \frac{\partial}{\partial t}\left(\int_{CV} \beta \rho \, d\forall\right) + \int_{CS} \beta \rho (\vec{V} \cdot d\vec{A}) \tag{C.2}$$

where the left-hand side, dB/dt, represents the total rate of change of B, the first term on the right-hand side is the rate of change of B within the control volume, and the second term on the right-hand side is the rate of flux of B through the surface of the control volume. For each of the following possible values of B, shown in Table C.1, the conditions shown hold true.

Table C.1 Properties in the Reynolds Transport Theorem

Property	Symbol	Equation	Notes
Mass	M	$\frac{dM}{dt} = 0$	Mass cannot be created or destroyed.
Momentum	$\vec{P}$	$\frac{d\vec{P}}{dt} = \vec{F}$	A change in fluid momentum results in a net force exerted by or on the fluid in the control volume.
Angular momentum	$\vec{H}$	$\frac{d\vec{H}}{dt} = \vec{T}$	A change in fluid angular momentum results in a net torque exerted by or on the fluid in the control volume.
Energy	E	$\frac{dE}{dt} = \dot{Q} - \dot{W}$	The first law of thermodynamics must be satisfied; that is, energy cannot be created or destroyed, only transferred from one form to another. If there are no heat transfers across the control volume and no mechanical work is transmitted across the control volume, so that $Q = W = 0$, then $dE/dt = 0$.
Entropy	S	$\frac{dS}{dt} \geq \frac{\dot{Q}}{T}$	The second law of thermodynamics must be satisfied.

D Experimental Uncertainty and Error

■ D.1 Experimental Uncertainty

No measured quantity is ever exact. There are always uncertainties in the numerical value due to the limitations of the measuring equipment. And when these flawed measurements are used to calculate other parameters, those new values also have some error associated with them. So while a calculator can produce eight or more digits, in engineering practice calculated numbers are rarely accurate to more than three digits. And though using three digits in calculations is a good rule of thumb, often the exact magnitude of the uncertainty in calculated quantities needs to be known.

The procedure is to first estimate the uncertainty in each measured quantity (time, distance, mass, temperature, pressure, etc.) and then analyze the propagation of uncertainty into the results calculated from the experimentally obtained data. Uncertainty in measurements can be estimated by looking at the *precision* of the measuring device being used. Alternatively, error can be estimated by having multiple people take a reading at the same flow condition and looking at the range in values measured. If the exact value of a variable is known, the difference between the measured value and the exact value is the *experimental error*. This can be thought of as *bias error* in the system, and if it is known, it can be subtracted off. The previous error, due to differences in measuring by different users, can be thought of as a *random error*. Errors in measurements can be categorized as either *systematic* or *random*. The randomness of such errors leads to *experimental uncertainty*.

A good model for the propagation of errors into a calculated quantity from measured ones is that the error is equal to the square root of the sum of the effects of the square of each of the individual errors, using the chain rule:

$$\Delta f = \sqrt{\sum_i \left(\frac{df}{dx_i} \Delta x_i\right)^2} \qquad (D.1)$$

where each x_i represents an originally measured quantity, and Δx_i is the experimental uncertainty in the measured value of x_i.

EXAMPLE D.1

Find the uncertainty in a density calculation using the ideal gas law, for temperature measured with a thermometer and pressure measured with an electronic barometer. Here the accuracy of the density measurement will be affected by the accuracy of the pressure and temperature measurements:

$$\rho = \frac{PM}{RT}$$

SOLUTION We can assume that M and R are "exact," so that we need only be concerned with uncertainties in the measured values of P and T. (The value of M might not be exact if there are impurities present in the gas mixture, but we will ignore that for now.) If the temperature is measured with a thermometer to an accuracy of $\pm 0.5°C$, and the pressure is read from an electronic barometer with an accuracy of ± 10 Pa, then the uncertainty in the calculated value of the density can be estimated. The formula for the propagation of errors in this case is

$$\Delta\rho = \sqrt{\left(\frac{\partial\rho}{\partial T}\Delta T\right)^2 + \left(\frac{\partial\rho}{\partial P}\Delta P\right)^2}$$

The ideal gas law must be used to evaluate the partial derivatives:

$$\frac{\partial\rho}{\partial T} = \frac{PM}{R}(-T^{-2}) = -\frac{PM}{RT}T^{-1} = -\frac{\rho}{T}$$

$$\frac{\partial\rho}{\partial P} = \frac{M}{RT} = \frac{PM}{RT} \cdot \frac{1}{P} = \frac{\rho}{P}$$

Substituting these relationships into the propagation of errors equation yields

$$\Delta\rho = \sqrt{\left(-\frac{\rho}{T}\Delta T\right)^2 + \left(\frac{\rho}{P}\Delta P\right)^2} = \rho\sqrt{\left(\frac{\Delta T}{T}\right)^2 + \left(\frac{\Delta P}{P}\right)^2}$$

Dividing both sides by the density gives

$$\frac{\Delta\rho}{\rho} = \sqrt{\left(\frac{\Delta T}{T}\right)^2 + \left(\frac{\Delta P}{P}\right)^2}$$

So for the ideal gas law, which is a linear relationship, the *relative error* in the density calculation is equal to the square root of the sum of the relative errors of the pressure and temperature measurements. Plugging in the values above on a day when the measured pressure is 100 kPa and the temperature is 22°C = 295 K, we have

$$\frac{\Delta \rho}{\rho} = \sqrt{\left(\frac{0.5\ \text{K}}{295\ \text{K}}\right)^2 + \left(\frac{10\ \text{Pa}}{100{,}000\ \text{Pa}}\right)^2} = \sqrt{(0.0017)^2 + (0.0001)^2} = 0.0017$$

So for the calculated density of 1.18 kg/m³, the uncertainty is ±0.0017 × 1.18 kg/m³ = ±0.002 kg/m³. For these relatively accurate measurements, it is acceptable to carry three significant digits after the decimal point, but no more than three digits.

EXAMPLE D.2

Consider an apparatus set up to measure the force from a water jet hitting a flat target. The force is measured by recording the displacement of a spring, which is compressed when the water jet hits the target. By Hooke's law the force will be proportional to the change in distance, x, for a linear spring. In this particular apparatus the mass flow rate of the water can also be measured, so that if the density of fluid and the area of the jet are known, then the predicted force can also be calculated. The directly measured force is linearly proportional to the displacement of the balancing weight, so that $F_{\text{dir}} = C\,x$. The uncertainty in measuring x is ±1 mm. Find the uncertainty in the measured and predicted values of force.

SOLUTION The uncertainty in the force is

$$\Delta F = \sqrt{(C \Delta x)^2}$$

So the relative uncertainty in the force is

$$\frac{\Delta F}{F} = \left|\frac{\Delta x}{x}\right|$$

For the indirect, or predicted, force on the flat plate target, the primary uncertainty comes from the measurement of the water flow rate if the density of the fluid is accurately known and the area of the nozzle is precisely measured. Thus the equation for the predicted force can be written as

$$F_{\text{ind}} = \rho A V^2 = \frac{\dot{m}^2}{\rho A}$$

Then the effect of uncertainty in the measurement of the mass flow rate on the calculation of the predicted jet thrust force can be calculated as

$$\Delta F = \sqrt{\left(\frac{\partial F}{\partial \dot{m}} \Delta \dot{m}\right)^2} = \sqrt{\left(\frac{2 \dot{m}}{\rho A} \Delta \dot{m}\right)^2} = \sqrt{\left(\frac{2 \dot{m}^2}{\rho A} \frac{\Delta \dot{m}}{\dot{m}}\right)^2} = \sqrt{\left(2 F \frac{\Delta \dot{m}}{\dot{m}}\right)^2}$$

$$\frac{\Delta F}{F} = \sqrt{2}\left|\frac{\Delta \dot{m}}{\dot{m}}\right|$$

If the uncertainty in the mass flow rate is 5%, then the uncertainty in the indirect force is about 7%.

When the uncertainty is known, these values can be used to create error bars on your graphs. This is easily done in Excel by double-clicking on a data series, selecting error bars, and then selecting the relevant options.

D.2 Experimental Error

Experimental error is defined as the difference between the measured value and the predicted or theoretical value.

$$\% \text{ error} = \frac{\text{measured} - \text{predicted}}{\text{predicted}} \times 100\% \tag{D.2}$$

If the experimental error is less than the experimental uncertainty, than the difference between the predicted and measured values can be attributed wholly to the limited precision of the instruments.

D.3 Data Outliers

Occasionally user or equipment error will result in a data point that does not seem to agree with theory or the rest of the data gathered. There may be times after lab when you are analyzing the data, and the numbers come out looking ridiculous. If every number gives you a ridiculous result, then you know you did the experiment incorrectly or there was an equipment failure. At a research lab, you would have to go back and repeat the experiment (which would be expensive). In class you may not have the opportunity to repeat the experiment.

It may instead be that most of your data looks good, but there are one or two points that do not follow the overall trend. You may feel tempted to simply eliminate these data points from your report, but you may later feel your conscience nagging you that arbitrarily removing data is bad science. To throw out data points you must be able to (a) think of an explanation of how the error occurred (such as user error), (b) define objective criteria by which bad data is identified and eliminated. Fortunately, there is a scientific way of evaluating data statistically to see if a particular data point is an outlier from the overall sample.

Chauvenet's criteria, described below, is a statistically valid method for removing data outliers from a data set. In the example that follows, it is believed that the data (y) should follow a linear fit with the independent variable (x).

Procedure

1. Fit a curve to the (x, y) data set using Excel's *Add Trendline* option. Make sure to print the formula of the trendline to the screen. (The following example uses a linear fit, but the method should work for polynomial or exponential curves, too).
2. Calculate the deviation, d, of each data point from the trendline. For the example, which assumes a linear fit of the form $y = mx + b$, the deviation is

$$\delta_i = y_i - (mx_i + b)$$

for $i = 1, 2, 3, \ldots, N$, where N is the number of x, y data points.

3. Compute the standard deviation, σ, of the data set, using the formula

$$\sigma = \sqrt{\frac{\sum_{i=1}^{N}(\delta_i)^2}{(N-1)}}$$

4. Compute the *deviation ratio*, DR, for each data point.

$$\text{DR} = \frac{\delta_i}{\sigma}$$

5. Then compare the deviation ratio, DR, to the maximum statistically allowable deviation ratio, DR_0, based on the normal distribution and the number of data points used. If $\text{DR} > \text{DR}_0$, the data point can be rejected and omitted from the data set.
6. After dropping the bad data points, recompute a new curve fit (Excel will do this automatically if you just erase the bad data point from the cells the chart is using).

The allowable deviation ratios as a function of number of data points is shown in Table D.1.

Table D.1 Maximum Deviation Ratios for Use with Chauvenet's Criteria

N (number of data points)	DR_0 (max allowed)
5	1.65
10	1.96
15	2.12
20	2.24
25	2.33
50	2.57
100	2.81

Adapted from [Chauvenet 60].

EXAMPLE D.3

Consider the following experimental data set. Plot a trendline and apply Chauvenet's criteria to determine if there are any data outliers. If there are, remove them and plot a new trendline.

x	y
1	1.2
2	1.8
3	3.2
4	5.1
5	5.7

x	y
6	2.2
7	7.5
8	6.5
9	8.1
10	9.9

SOLUTION We used Excel to get the following results. The original data was fit to the linear function $y = mx$.

x	y	Linear fit	Deviation	D-squared	Deviation ratio
1	1.2	0.918	0.282	0.079	0.205
2	1.8	1.837	−0.037	0.001	0.027
3	3.2	2.755	0.445	0.198	0.325
4	5.1	3.674	1.426	2.035	1.041
5	5.7	4.592	1.108	1.228	0.808
6	2.2	5.510	−3.310	10.959	2.415
7	7.5	6.429	1.071	1.147	0.782
8	6.5	7.347	−0.847	0.718	0.618
9	8.1	8.266	−0.166	0.027	0.121
10	9.9	9.184	0.716	0.513	0.522

We compute the deviation ratio, DR, for each data point by taking the absolute value of the deviation and dividing by the standard deviation. In this case we find one outlier, the data point (6, 2.2). After dropping the outlier, we generate the following table of revised data.

x	y	Linear fit	Deviation	D-squared	Deviation ratio
1	1.2	0.918	0.282	0.079	0.331
2	1.8	1.837	−0.037	0.001	0.043
3	3.2	2.755	0.445	0.198	0.523
4	5.1	3.674	1.426	2.035	1.678
5	5.7	4.592	1.108	1.228	1.304
7	7.5	6.429	1.071	1.147	1.260
8	6.5	7.347	−0.847	0.718	0.997
9	8.1	8.266	−0.166	0.027	0.195
10	9.9	9.184	0.716	0.513	0.842

Reference

[Chauvenet60] Chauvenet, W. 1960. *A Manual of Spherical and Practical Astronomy* Vol. II 1863. Reprint of 1891 5th ed. Dover.

Additional Resources

■ E.1 Books

For readers who are interested in or need more information on a particular aspect of fluid mechanics, the following books are recommended. The reading list is organized by subject.

History of Fluid Mechanics

John Anderson. *A History of Aerodynamics*. Cambridge University Press, 1999.

Tom Crouch. *The Bishop's Boys*. Norton, 2003. (Biography of the Wright brothers.)

Hans-Peter Dabrowski. *Horten Flying Wing in World War II*. Schiffer Publishing, 1991.

Ollivier Darrigol. *World of Flow—A History of Hydrodynamics from the Bernoullis to Prandtl*. Oxford University Press, 2005.

Michael Eckert. *The Dawn of Fluid Dynamics*. Wiley, 2006. (Covers era from late 1800s to 1930s.)

Michael H. Gorn. *The Universal Man: Theodore von Karman's Life in Aeronautics*. Smithsonian Institution Press, 1992.

Hunter Rouse. *Hydraulics in the United States, 1776–1976*. University of Iowa, 1976.

G. A. Tokaty. *A History and Philosophy of Fluid Mechanics*. Dover, 1994.

Specialization Areas

Jewel Barlow, William Rae, Alan Pope. *Low Speed Wind Tunnel Testing*. Wiley, 1999.

William Hinds. *Aerosol Technology*. Wiley, 1999.

Sighard Hoerner. *Fluid Dynamic Drag*. Self-published, 1965.

Yukio Hori. *Hydrodynamic Lubrication*. Springer, 2002.

Arthur Lefebvre. *Atomization and Sprays*. CRC Press, 1998.

Nam-Trung Nguyen and Steve Wereley. *Fundamentals and Applications of Microfluidics*. Artech House, 2002.

Howard Shapiro. *Practical Flow Cytometry*. Wiley, 2003.

W. Stepniewski and C. Keys. *Rotary Wing Aerodynamics*. Dover, 1984.

Steven Vogel. *Life in Moving Fluids*. Princeton University Press, 1994.

Turbulence

H. Tennekes and J. Lumley. *A First Course in Turbulence*. MIT Press, 1972.

Arkady Tsinober. *An Informal Introduction to Turbulence*. Kluwer Academic Publishers, 2001.

Birds

David Goodnow. *How Birds Fly*. Periwinkle Books, 1992.

Hank Tennekes. *The Simple Science of Flight*. MIT Press, 1996.

Flow Measurements

Roger Baker. *An Introductory Guide to Flow Measurement*. Professional Engineering Publishing, 2002.

Richard Goldstein. *Fluid Mechanics Measurements*. Taylor & Francis, 1996.

General Reference

R. Blevins. *Applied Fluid Dynamics Handbook*. Krieger, 1992.

Flow Visualization

M. Samimy, K. Breuer, L. G. Leal, and P. Steen. *A Gallery of Fluid Motion*. Cambridge University Press, 2003.

Milton Van Dyke. *An Album of Fluid Motion*. The Parabolic Press, 1982.

The Japan Society of Mechanical Engineers. *Visualized Flow*. Pergamon Press, 1988.

■ E.2 Websites

The following websites may be useful in the study of fluid mechanics. The URLs are current as of the writing of this book, Fall 2008.

Fluid Statics

http://science.howstuffworks.com/submarine1.htm—Submarines

http://science.howstuffworks.com/water1.htm—Water towers

http://auto.howstuffworks.com/pressure-gauge.htm/printable—Tire pressure gauges http://science.howstuffworks.com/question629.htm—How fish float

http://science.howstuffworks.com/helium.htm/printable—Helium balloons

Linear Momentum

http://www.grc.nasa.gov/WWW/K-12/airplane/mflow.html—Jet engine info

http://science.howstuffworks.com/wind-power.htm/printable—Wind turbines

Angular Momentum

http://www.ems.psu.edu/%7Efraser/Bad/BadCoriolis.html—Which way does the water swirl in your toilet?

Energy/Bernoulli Equation

http://entertainment.howstuffworks.com/water-blaster.htm—Squirt guns

http://science.howstuffworks.com/question673.htm—Spray bottles

http://people.howstuffworks.com/hydropower-plant.htm/printable—How hydroelectric power plants work

Aerodynamics

http://www.aeromech.usyd.edu.au/aero/contents.html—General aerodynamics (online textbook)

http://www.paperairplanes.co.uk/—Paper airplanes

http://www.edmunds.com/advice/specialreports/articles/106954/article.html—How aerodynamics affects vehicle mileage

http://www.copters.com/helo_aero.html—Helicopter aerodynamics

Aerodynamics in Sports

http://www.nas.nasa.gov/About/Education/Racecar/physics.html—Introduction to race car aerodynamics

http://www.princeton.edu/~asmits/Bicycle_web/bicycle_aero.html—Bicycle aerodynamics

http://www.eng.vt.edu/fluids/msc/bike.htm—More bicycle aerodynamics

http://www.oddball-mall.com/knuckleball/mego.htm—Why knuckleballs move the way they do in baseball

http://www.billpattersonart.com/dszone.swf—A brief explanation of dynamic soaring—some birds use a form of this to fly long distances

http://units.aps.org/units/dfd/prandtl_vol58no12p42_48.pdf—Ludwig Prandtl and boundary layer theory

http://www.zoo.cam.ac.uk/zoostaff/ellington/aerodynamics.html—Insect flight

Alternative Transportation

http://www.vtol.org/awards/hph.html—Human-powered helicopter competition

http://lancet.mit.edu/decavitator/—Human-powered hydrofoil

http://www.nasm.si.edu/research/aero/aircraft/maccread_condor.htm—Human-powered airplane

Navier-Stokes Equations

http://www.grc.nasa.gov/WWW/K-12/airplane/nseqs.html—As explained by NASA

http://www.claymath.org/millennium/Navier-Stokes_Equations/—Want to make $1,000,000? All you have to do is solve the Navier-Stokes equations.

Computational Fluid Dynamics (CFD)

http://www.fluent.com/about/cfdhistory.htm—History of CFD

General Resources and Informational Links

http://megaconverter.com/Mega2/—Online unit conversion calculator

http://www.denysschen.com/catalogue/density.asp—Air density calculator

http://www.efunda.com/formulae/fluids/overview.cfm—Overview of fluid mechanics theory

Fluid Dynamics Videos and Multimedia

http://modular.mit.edu:8080/reports/index-ifluids.html—Classic fluid dynamics videos. Twenty-seven videos provided by MIT. Most are about half an hour long. The Turbulence video is by far the best, and the ones on Surface Tension and Low Reynolds Number Flows are also good.

http://www.iihr.uiowa.edu/products/dhrm.html—More classic videos from Hunter Rouse at University of Iowa

http://www.colorado.edu/MCEN/flowvis/—Art of fluid flow

http://www.efluids.com/efluids/pages/gallery.htm—Gallery of fluid motion

MATLAB® Codes

This appendix contains MATLAB® codes for the algorithms and some of the examples presented in the text.

F.1 Numerical Integration

```
R = 4;
w = 5;
rho = 850;
g = 9.8;
a = 0.0;
b = R;
n = 100;
% defining step sizes
deltax = (b - a)/n;
% trapezoid method
tsum = 0;
for i =1:n-1
    x = a + i*deltax;
    tsum = tsum + deltax*(rho*g*w*(R-x));
end
tsum = tsum + 0.5*deltax*(rho*g*w*(R-a)+rho*g*w*(R-b))
% example 2
tsum = 0;
for i =1:n-1
    x = a + i*deltax;
    tsum = tsum + deltax*(rho*g*w*(R-sqrt(2*R*x-x^2)));
end
tsum = tsum + 0.5*deltax*(rho*g*w*(R-sqrt(2*R*a-a^2))+rho
*g*w*(R-sqrt(2*R*b-b^2)))
```

F.2 Solution to First-Order Ordinary Differential Equation

```
clear
% problem 1 - dy/dt = 1 - y
time(1) = 0;
answer(1) = 0;
b = 5.0;
n = 500;
dt = b/n;
for i =1:n;
    time(i+1) = time(i) + dt;
    y(i+1)=y(i)+dt*1.0*(1.0-y(i));
% analytical solution
    answer(i+1) = 1 - exp(-time(i+1));
end

answer(1) = 1;
% problem 2 - dy/dt = y
for i =1:n;
    time(i+1) = time(i) + dt;
    y(i+1)=y(i)+dt*1.0*(y(i));
% analytical solution
    answer(i+1) = exp(time(i+1));
end
```

F.3 One-Dimensional Transient Ball Drop

```
% all units are m-k-s system
% 1D ball drop simulation
y = 7.0;
v = 0.0000000001;
m = 0.20;
d = 0.20;
a = pi/4*d^2;
den= 1.2;
g = 9.8;
visc = 1.5d-05;
t = 0.0;
delt = 0.001;
% main program
```

```
for i = 1:100000
  tt(i) = t;
  yy(i) = y;
  t = t + delt;
  Reynum = v*d/visc;
  if Reynum < 1.0;
      cd = 24.0/Reynum;
  else    if Reynum < 1000.0;
          cd = 24.0/Reynum*(1+Reynum^(2/3)/6);
          else
          cd = 0.45;
          end
  end
  cd = 0.45;
  f = 0.5*cd*den*a*v^2;
  v = v+(g-f/m)*delt;
  y = y - v*delt;
  if yy(i) < 0
      yy(i) = 0;
  end
  if y < -0.0000001
      break
  end
end
```

F.4 Two-Dimensional Trajectory of a Projectile

```
% all units are m-k-s system
% 2D baseball trajectory
v0 = 46.9;
m   = 0.150;
diam = 0.073;
a   = pi/4*diam^2;
cd = 0.31;
% could enter atmospheric conditions and use ideal gas law
den= 1.2;
g = 9.8;
angl= 30.0/360.0*(2.0*pi);
% start
x = 0.0;
```

```
y = 1.0;
vx= v0*cos(angl);
vy= v0*sin(angl);
t = 0.0;
delt = 0.001;
% main program
for i = 1:100000
  xx(i) = x;
  yy(i) = y;
  t = t + delt;
  v = sqrt(vx^2 + vy^2);
  f = 0.5*cd*den*a*v^2;
  fx= f*vx/(v);
  fy= f*vy/(v);
  vx=vx-fx/m*delt;
  vy=vy+(-fy/m-g)*delt;
  x = x + vx*delt;
  y = y + vy*delt;
  if yy(i) < 0
      yy(i) = 0;
  end
  if y < 0
      break
  end
end
```

F.5 Two-Dimensional Rocket Trajectory

```
% all units are m-k-s system
% add thrust - 4 N over 0.5 s
ft = 4.0;
tttend = 0.5;
v0 = 0.0000000001;
m  = 0.050;
a  = 0.0001;
cd = 0.75;
% could enter atmospheric conditions and use ideal gas law
den= 1.2;
g = 9.8;
angl= 60.0/360.0*(2.0*pi);
% start
```

```
x = 0.0;
y = 0.0;
vx= v0*cos(angl);
vy= v0*sin(angl);
t = 0.0;
delt = 0.001;
% main program
for i = 1:100000
  xx(i) = x;
  yy(i) = y;
  t = t + delt;
  % check thrust parameters
  % m = m - mburn*delt
  v = sqrt(vx^2 + vy^2);
  f = 0.5*cd*den*a*v^2;
  fx= f*vx/(v);
  fy= f*vy/(v);
  if t < tttend;
     ft = 4.0;
  else
       ft = 0.0;
  end
  ftx=ft*vx/(v);
  fty=ft*vy/(v);
  vx=vx+(ftx-fx)/m*delt;
  vy=vy+((fty-fy)/m-g)*delt;
  x = x + vx*delt;
  y = y + vy*delt;
  if yy(i) < 0
      yy(i) = 0;
  end
  if y < -0.0000001
      break
  end
end
```

F.6   Transient Poiseuille Flow

```
% all units are m-k-s system
clear all;
% this program solves transient Poiseuille flow
```

```matlab
% velocity is u = u(y,t)
% specific geometry
% width of channel (in meters)
H = 0.02 / 2;
% kinematic viscosity of fluid (water) (in m^2/s)
nu = 1E-06;
% density of fluid (water) in (kg/m^3)
rho = 1000;
% axial pressure gradient (Pa/m)
dpdx = 1;
% final time (in seconds)
tmax = 320;
% numerical parameters
% number of time steps
imax = 150;
% number of spatial grid points
n = 10;
% calculate time step and spatial grid size
dt = tmax/imax;
dy = 2*H/n;
% Fourier number (for stability)
Fonum = nu*dt/dy^2
% set initial condition - u(y,0) = 0
for j=1:n+1;
    u(j,1) = 0;
end
% set boundary conditions - u(+H,t) = 0, u(-H,t) = 0;
for i=1:imax+1;
    u(1,i) = 0;
    u(n+1,i) = 0;
end
% main solution loop
for i=1:imax;
    for j=2:n;
        u(j,i+1)=u(j,i)+nu*dt/(dy^2)*(u(j-1,i)
        -2*u(j,i)+u(j+1,i))+dt*dpdx/rho;
    end
end
test2 = u;
% output steady spatial solution
```

```
for j=1:n+1;
    y(j) = -H +(j-1)*dy;
    ufinal(j) = u(j,imax);
end
figure
plot(y,ufinal,'LineWidth',2)
xlabel('Height (m)');
ylabel('Velocity (m/s)');
% check Reynolds number
% use average or centerline velocity
uscale = u(n/2,imax);
Reyn = uscale*H/nu
% out transient centerline velocity
for i=1:imax;
    ucenter(i)=u(n/2,i);
    ttime(i)=(i-1)*dt;
end
figure
plot(ttime,ucenter,'LineWidth',2)
xlabel('Time (s)');
ylabel('Centerline Velocity (m/s)');
```

Glossary

Short definitions of important terms used in this textbook are included here.

Absolute pressure The pressure relative to a perfect vacuum of zero pressure.

Aileron A control surface at the end of a wing used to control an airplane in roll.

Airfoil A shape used in a wing cross-section to produce lift.

Angle of attack The angle between the chord line of a wing and the oncoming fluid velocity vector.

Barometer A device used to measure the local atmospheric pressure.

Bernoulli's equation A simplified form of the energy equation that can be used for steady, frictionless flow with no work or heat transfer.

Boundary layer The thin layer between the surface of a solid object and the free stream in which the velocity rapidly varies from the free-stream value to that of the object.

Buoyancy The upward force on an object immersed in a fluid.

Camber The amount of curvature in an airfoil.

Capillary number The ratio of viscous forces to surface tension forces.

Cavitation The formation of bubbles of vapor in a liquid when the local pressure falls below the vapor pressure.

Chord The width of a wing from front to back.

Compressible Flow traveling at sufficiently high velocities so that a change in velocity can result in a change of gas density.

Compressor A mechanical device that increases the pressure of a flowing gas.

Computational fluid dynamics (*CFD*) The use of numerical methods to find approximate solutions to the Navier-Stokes equations that govern fluid flows.

Convective acceleration The acceleration associated with a change in velocity over distance, even in a steady-state flow.

Drag The resistance force a fluid exerts on an object moving through it. The drag force is always opposite to the direction of motion.

505

Dynamic pressure The pressure energy associated with the kinetic energy of a fluid.

Elevator A control surface at the rear of a plane used to control its motion in roll.

Extensive property A property that depends on the total mass of fluid present.

Fluid A substance without a defined shape, that moves freely when a shear stress is applied.

Fluid dynamics The study of fluid motion.

Fluid mechanics The combined study of fluid dynamics and fluid statics.

Fluid statics The study of fluid at rest.

Form drag The component of drag due to the pressure distribution about an object. The form drag is highly dependent on the shape of the object and whether or not the object is streamlined.

Friction factor A nondimensional measure of the shear stress and pressure loss in internal flows in pipes; also, a nondimensional measure of the resistance to flow for internal flows.

Froude number The ratio of inertial to gravitational forces.

Gage pressure The pressure relative to the local atmospheric pressure.

Head loss The friction loss of a fluid flowing through a pipe, expressed in units of length.

Hydraulic diameter The diameter of a pipe for noncircular ducts that has equivalent flow characteristics to a circular pipe.

Hydrostatic pressure The pressure rise due to depth in a fluid as a result of the gravitational force.

Hypersonic Flow at velocities where the kinetic energy of the molecules in the air is high enough that they can ionize when brought to a stagnation point. Typically $Ma > 5$.

Incompressible Flow in which the density can be safely approximated as being constant.

Induced drag The component of drag resulting from the uneven pressure distribution over wings due to lift. It can been seen in the wingtip vortices shed behind a plane.

Intensive property A property that does not depend on the total mass of the fluid.

Isotropic Uniform in all directions.

Jet A propulsion engine that draws in air, compresses it and uses it to burn fuel, and then forces the hot exhaust out the back.

Kinematic viscosity The absolute viscosity divided by the fluid density. This ratio is used often when computing Reynolds numbers and so is commonly used for convenience.

Laminar The condition of a flow when all the streamlines run parallel and do not cross.

Lift The upward vertical force generated by a fluid flowing over the wings of an object.

Mach number The ratio between a fluid's velocity and the local speed of sound.

Newtonian fluid A fluid in which the strain rate is linearly proportional to the shear stress. In a Newtonian fluid the viscosity is a thermodynamic property.

Non-Newtonian fluid A fluid in which the viscosity varies depending upon the mechanical state in the fluid, as well as the time history of the fluid.

No-slip boundary condition The condition that the velocity of a fluid at a surface be equal to the velocity of the surface at the point of contact.

Pathline The trajectory traced out by an individual particle. It can be obtained by taking a photograph of a reflecting particle and using a long shutter time.

Pitch The motion of an airplane about a horizontal axis aligned with the wings of the plane.

Propeller A spinning airfoil used to produce thrust to propel an airplane forward.

Pump A mechanical device that increases the pressure of a flowing liquid.

Relative roughness The nondimensional ratio of the surface roughness of a pipe to the internal diameter of the pipe.

Relative wind The vector sum of an object's motion minus the wind direction.

Reynolds number A nondimensional number that represents the ratio between inertial and viscous forces in a fluid.

Rheology The study of the properties and characteristics of liquids, usually of non-Newtonian liquids.

Rocket A thrust-producing device that expels fluid without taking in any new fluid. A chemical rocket must carry both its own oxidizer and fuel.

Roll The motion of an airplane about an axis aligned with the body of the plane.

Secondary losses Friction losses in internal flows associated with any fittings in the pipe, including elbows, bends, flanges, expansions, contractions, open valves, and so on.

Span The width of a wing.

Specific gravity The density of a fluid divided by a reference density. For gases the reference is air, and for liquids the reference is water.

Specific weight The weight of a fluid per unit volume. Equal to the product of density and acceleration due to gravity.

Stability A stable system will return to its static equilibrium state when perturbed, while an unstable system will not.

Stagnation pressure The pressure a moving fluid would encounter if it were brought to a complete stop with no friction losses. The sum of the static and dynamic pressure.

Stall The loss of lift on a wing. Usually occurs when the angle of attack is too high and the flow separates from the wing surface.

Static pressure The mechanical pressure of a fluid. Assumed to be equal to the thermodynamic pressure of the fluid.

Streakline A line formed of the locations of all the particles that have passed through a particular point in the flow at any previous time.

Streamline A line that is tangent to the local velocity vector in a fluid all along its length.

Strouhal number The nondimensional frequency of periodic vortex shedding behind a bluff object.

Subsonic Traveling at speeds less than the speed of sound; Ma < 1.

Supersonic Traveling at speeds greater than the speed of sound; Ma > 1.

Surface tension The force arising due to the cohesive attraction between molecules at a liquid surface.

Thrust The force generated by moving fluids so that a net force is produced in the direction of motion. Examples of devices that generate thrust are propellers, jet engines, and rocket engines.

Transonic Flow when the free-stream velocity is subsonic, but there are local pockets of acceleration that have sonic or supersonic flow. The velocity at which transonic flow occurs over a wing is referred to as the *critical Mach number*.

Turbine A mechanical device that extracts energy from a flowing gas.

Turbulent A state of chaotic motion of a fluid, in which individual motions are not repeatable but seem to occur randomly.

Vapor pressure The pressure at which a liquid will boil at a given temperature.

Viscosity The resistance to shearing motion of a fluid.

Vortex shedding The unsteady, cyclic, repeatable motion in the wake behind a blunt object moving through a fluid.

Wake A region of low pressure behind an object moving through a fluid.

Weber number The ratio of inertial forces to surface tension forces in a liquid.

Wing A device used to generate lift in an airplane.

Yaw The rotation of an airplane about its vertical axis.

Greek Alphabet

Due to the large number of variables encountered in science and engineering, coupled with the finite number of letters in the English alphabet, Greek letters are commonly used as symbols for variables. Note that the lowercase Greek letters are more commonly used than uppercase ones. The names and symbols for each letter in the Greek alphabet are shown in Table H.1. A few of these symbols are commonly reserved for mathematical operations, such as Σ for sum, Π for product, and Δ for difference.

Table H.1 Greek Alphabet

Name	Lowercase	Uppercase	(keystroke in Symbol font)
Alpha	α	A	A
Beta	β	B	B
Gamma	γ	Γ	G
Delta	δ	Δ	D
Epsilon	ε	E	E
Zeta	ζ	Z	Z
Eta	η	H	H
Theta	θ	Θ	Q
Iota	ι	I	I
Kappa	κ	K	K
Lambda	λ	Λ	L
Mu	μ	M	M
Nu	ν	N	N
Xi	ξ	Ξ	X
Omicron	o	O	O
Pi	π	Π	P
Rho	ρ	P	R
Sigma	σ	Σ	S
Tau	τ	T	T
Upsilon	ψ	Ψ	U
Phi	ϕ	Φ	F
Chi	χ	X	C
Psi	ψ	Ψ	Y
Omega	ω	Ω	W

Conversion Factors

The following conversion factors are provided for use when converting between English and metric units, and are taken from NASA SP-7012, "Physical Constants and Conversion Factors," published in 1973.

I.1 Metric to English Conversions

Mass

1 kg = 2.205 lbm = 0.0873 slug
1 g = 0.0022 lbm

Length

1 m = 3.28 ft
1 cm = 0.394 in.
1 km = 0.621 mi

Volume

1 m^3 = 1000 L = 35.3 ft^3
1 L = 1000 mL = 0.264 gal
1 mL = 1 cm^3 = 0.061 in^3

Time

1 hr = 3,600 s
1 day = 86,400 s

Velocity

1 m/s = 3.28 ft/s = 2.237 mph
1 km/hr = 0.621 mph

Flowrate

1 m³/s = 35.35 ft³/s
1 L/s = 60 L/min = 15.85 gal/min
1 L/min = 0.264 gal/min

Force/Weight

1 N = 0.225 lbf

Pressure

1 Pa = 0.000145 psi
1 kPa = 1,000 Pa = 0.145 psi
1 bar = 100,000 Pa = 14.5 psi = atm

Torque

1 N-m = 0.737 ft-lbf

Energy

1 J = 0.000948 Btu

Power

1 W = 0.001341 hp
1 kW = 1.341 hp

Absolute Viscosity

1 kg/m-s = 1 Pa-s = 10 poise = 0.672 lbm/ft-s
1 poise = 1 dyne × s/cm² = 1 g/cm-s

Kinematic Viscosity

1 m²/s = 10.76 ft²/s
1 Stoke = 0.0001 m²/s = 0.001076 ft²/s

Temperature

[K] = [°C] + 273
[°C] = ([°F] − 32) × 5/9

I.2 English to Metric Conversions

Mass
1 lbm = 0.4536 kg
1 slug = 32.2 lbm = 14.59 kg

Length
1 in. = 2.54 cm = 0.00254 m
1 ft = 12 in. = 0.3048 m = 30.48 cm
1 mi = 5280 ft = 1.609 km = 1609 m

Volume
1 ft^3 = 0.0283 m^3
1 qt = 0.946 L
1 gal = 4 qt = 231 in^3 = 3.76 L
1 fl oz = 29.57 cm^3

Time
1 hr = 3600 s

Velocity
1 ft/s = 0.3048 m/s
1 mi/hr = 1.609 km/hr = 0.447 m/s

Flowrate
1 ft^3/s = 0.0283 m^3/s
1 gal/min = 3.785 L/min

Force/Weight
1 lbf = 4.45 N

Pressure
1 psi = 6,895 Pa
1 psf = 1/144 psi = 47.88 Pa
1 atm = 14.7 psi = 101,325 Pa

Torque

1 ft-lbf = 1.357 N-m

Energy

1 Btu = 778 ft-lbf = 1055 J

Power

1 hp = 550 ft-lbf/s = 746 W = 0.746 kW

Absolute Viscosity

1 lbm/ft-s = 1.488 kg/m-s = 1.488 N-s/m^2
1 lbf-s/ft^2 = 47.88 kg/m-s

Kinematic Viscosity

1 ft^2/s = 0.0929 m^2/s

Temperature

[°R] = [°F] + 460 = 9/5[K]
[°F] = 9/5[°C] + 32

■ I.3 Other Commonly Used Quantities

g = 9.8 m/s^2 = 32.2 ft/s^2
R = 8314 J/kmol-K = 10.73 ft^3-psi/lbmol-°R = 1545 ft-lbf/lbmol-°R
1 lbf = 32.2 lbm-ft/s^2 = 1 slug-ft/s^2
1 N = 1 kg-m/s^2

Index

A

Absolute pressure, 4, 6, 452
Accelerating systems, 69–75, 104–107. *See also* Rotating systems
Accuracy, 18–19, 112, 193, 451
Adverse flow, 249
Airfoil, 290–293
Anemometer, 431–432, 434
Angle of attack, 290, 295–298
Angular momentum, 113–115, 333–335
Animal flight, 417–422
Area-velocity, 87
Aspect ratio of airplane, 295
Atmosphere of Earth, 30–31, 473
Atmosphere, standard (atm), 4, 5
Atmospheric pressure, 4, 29–31, 474–475
Atomization, 400–402
Automobiles, aerodynamics of, 300–305

B

Balloons. *See* Floating objects
Bar, 4, 5
Barometer, 4
Bearings, lubrication flow past, 184–187
Bend in pipe, 256
Bernoulli equation, 123–141, 162, 212–213, 454
Betz limit, 364
Biological flows, 415–422
Birds, 417–422
Blade advance ratio (J), 350
Blood flow, 416–417
Boats. *See* Floating objects; Water vehicles
Boiling point, 12

Boundary layer, 163, 168–169, 269
 of laminar flow over plate, 167–171
 of turbulent flow, 172
Boussinesq hypothesis, 197
Buckingham pi theorem, 456–462
Buoyant force, 58–68, 279–280
Bypass ratio, 393

C

Camber, 290
Cars, aerodynamics of, 300–305
Cavitation, 12, 342, 343–344
Cavitation number (Ca), 350
Celerity (*c*), 306
Center of pressure, 49–55
Centimeter (cm), 2
Centrifugal pump, 338–342
Centrifugal turbine, 353–354
Centrifuge, 73–75
Centroid method for forces, 55–56
C-g-s system of units, 2
Chauvenet's criteria, 488–489
Chezy's equation, 244–245
Choked flow condition, 378, 384
Chord line, 290
Chord of wing, 290
Circulatory system flows, 416–417
Closed-circuit wind tunnel, 316
Coefficient of performance (C_p), 361
Coefficient of thermal expansion (β), 414
Collision/coalescence of liquid drops, 405–409
Compressible flows, 378–390, 477–482
Compressor/pump, 121, 335–348
Computational fluid dynamics (CFD), 188–199, 316

515

Concorde aircraft, 30
Conditional stability, 193
Conduction of heat, 412
Conservation law(s)
　for energy, 120–122, 123–124, 138, 161, 412
　for mass, 81–93, 138
　for momentum, 94–107, 113–120, 138, 333
　Reynolds transport theorem and, 483–484
Contact angle, 14–15
Continuity equation. *See* Conservation law for mass
Continuum assumption, 16–17
Control volume analysis, 97
　for conservation of energy, 120–121
　for conservation of mass, 81–82
　for conservation of momentum, 94–96
　for hydrostatic pressure distribution, 25–27
　with moving control volume, 103–107, 116–117, 335
　for Navier–Stokes equations, 158–161
　for pressure drop in laminar pipe flow, 210–211
　for uniformly accelerating systems, 69–71
Convection of heat, 413–414
Convective terms of momentum equation, 162, 167
Converging–diverging section, 378, 382–383, 443
Coriolis meter, 451
Courant number, 193
Creeping flow, 175–178
Critical conditions, 381
Critical flow, 249
Critical Mach number, 310
Critical point, 12
Critical Reynolds number, 164, 171, 217, 244, 270
Critical speed of pump, 340
Critical temperature (T_{cr}), 13
Cubic centimeter (cc), 3
Cubic feet (ft^3), 3
Cubic meter (m^3), 3
Cytometry, 21

D

D^2-law, 402, 406
Dams, 38, 50, 250
Darrieus, G.J.M., 359
Darrieus wind turbine, 359
Data outliers, 488–489
Deflagration, 398
Density (ρ), 6–7, 380, 469
Derivatives, approximating, 188–191
Derived units, 2
Design problems, 56
Detached eddy simulation, 197
Detonation, 398
Deviation ratio, 489
Differential equations
　linear, 191
　order of, 191
　ordinary, 108–112
　partial, 160–161, 191
Differential pressure, 33, 429–430
Diffuser, 256
Diffusivity, 11
Dihedral angle, 317
Dimensional consistency/homogeneity, 2, 88, 135, 454
Discharge coefficient (C_d), 398, 448
Discontinuities (shock waves) in flows, 17, 18, 308, 310, 384–387, 389, 477
Displacement pump, 336
Distillation curve, 12
Drag coefficient (C_D), 173, 177, 266, 270, 275, 294–296, 322
Drag divergence Mach number, 310
Drag force, 173, 176, 265–269, 271, 272, 305
　form drag, 273–283
　induced drag, 293–300
　supersonic drag, 307–309
　transient drag, 322–323
　transonic drag, 310–311
　viscous drag, 269–273
　wave drag, 267, 305–307, 307–311
Drag polar, 296
Drops, dynamics of, 400–409
Drop sizing, 435–436
Dynamic pressure (q), 133
Dynamic range of measuring device, 452
Dyne, 2

E

Eddy viscosity, 197
Effective flow area, 399
Efficiency
    of jet engine, 392, 394–395
    of propeller, 350, 394
    of pump, 336, 339, 340
    of rocket, 388–389, 395
    of turbine, 354
    of wind turbine, 361, 364
Eiffel, Gustave, 314
Electronics cooling systems, 411–415
Elevation head, 124
Elliptical lift distribution, 298–300
Elliptic partial differential equation, 191
Emulsion, 15
Energy, 2–3. *See also* Conservation law for energy
English system of units, 3
Enthalpy, 121, 123, 380
Entrance length, 227–229
Erg, 2
Errors, 48, 198, 290, 428, 452, 485–491
Escape velocity of rocket, 112–113
Euler equations, 162–163, 176
Euler method, 108–112
Euler's pump/turbine formula, 339
Evaporation, 12
Explicit numerical method, 193
Extensive property, 483

F

F-5 aircraft, 311
F-14 Tomcat aircraft, 311
F-15 aircraft, 311
Fanno flow, 384
Fans, 348–349
Finite difference approximation, 189–190, 190–191
First law of thermodynamics. *See* Conservation law for energy
Flettner rotor, 359
Flight, dynamics of. *See* Animal flight; Glider aircraft; Lift force; Lifting bodies
Flight regimes for aircraft, 266

Floating objects, 58–59, 65–68
Flow loss in pipe systems, 233
Flow measurement
    flow rate meters, 443–452
    velocity probes, 427–436
    visualization methods, 436–443
Flow rate
    measuring devices, 443–452
    in open channel, 244
    in pipe systems, 233
    over weir, 250–251
Fluid(s)
    incompressible, 7, 27
    mixtures involving, 12, 15
    Newtonian, 8, 142
    non-Newtonian, 8, 9
Fluid flow(s)
    classification of, 19–21
    compressible, 378–390
    creeping, 175–178
    discontinuities (shock waves) in, 17, 18, 308, 310, 384–387, 389, 477
    laminar (*see* Laminar flow)
    lubrication flow, 179–187
    one-dimensional isentropic, 381–384
    in open channel, 240–251
    in pipes (*see* Pipe flow)
    over/between plates, 8, 167–171, 179–182, 191–193, 269–270
    around sphere, 175–177, 273, 275
    steady-state, 85–91, 97, 123, 167–171, 175, 205–213
    transient, 91–93
    turbulent (*see* Turbulent flow)
Fluid properties, 6–18, 469–476
    density, 6–7, 380, 469
    heat of vaporization, 11–12
    speed of sound, 17–18, 307, 379–380
    surface tension, 12, 13–15, 27, 73, 472
    stagnation properties, 380
    vapor pressure, 12
    viscosity, 3, 7–11, 12, 162–163, 170, 197, 469
Foot (ft), 3
Force, 2–3
Forced convection, 414

Form drag, 273–283
Forward difference method, 193
Fourier number (Fo), 193
Fourier's law of conduction, 412
Francis turbine, 351
Free surface, 27, 240
Free-wheeling speed, 353
Friction coefficient (c_f), 173
Friction factor (f), Darcy, 174, 212, 224, 245
Friction losses, 121, 142; *See also* Drag force
    in pipes, 210–213, 219, 254
    for open channel flow, 244
Froude number (Fr), 242, 249–250, 267, 305–306
Fuel injectors, 13, 398–409
Fully developed pipe flow, 227
Fundamental units, 2

G

Gage pressure, 4, 6, 452–453
Gallon (gal), 3
Gas bearings, 187
GE 90 turbofan, 392
Geosynchronous orbit, 112
Glider aircraft, 295–296, 300–302
Grades of oil, 10–11
Gradual expansion, 256
Gradually varying flow in open channels, 245
Gram (g), 2
Grashof number (Gr), 413
Great Flight Diagram of Tennekes, 418
Guggenheim–Katayama surface tension
    formula, 13

H

Head, 124
Head loss, 124, 244
Heat of vaporization, 11–12
Heat transfer, 412–414
Helios aircraft, 31
Hindenburg disaster, 63
HL-10 aircraft, 319–320
Horizontal-axis wind turbine, 359
Horsepower (hp), 3
Hot-wire anemometer, 431–432
Hull speed, 306

Hydraulic depth of open channel, 242
Hydraulic diameter of pipe, 229
Hydraulic jump, 249–250
Hydraulic radius of open channel, 241
Hydraulic turbine/motor, 351–358
Hydrazine, 374, 472
Hydrocarbon fuel, 373
Hydrofoil, 306
Hydrogen, 63, 372–373, 472
Hydrophilic surface, 14–15
Hydrophobic surface, 14–15
Hydrostatic pressure distribution
    and center of pressure, 49–55
    derivation of, 25–27, 141–142
    for gases, 29–30
    and manometer problems, 32–37
    and submerged surfaces, 38–46, 59–61
    and uniformly accelerating systems, 69–75
Hyperbolic partial differential equation, 191

I

Ideal Rankine analysis, 361–364
Immiscible liquids, 13
Impulse turbine, 351–353
Inches of mercury (inch Hg), 5, 32
Inches of water (inch H_2O), 5, 32
Incompressible fluid, 7, 27
Induced drag, 293–300
Inhaler, 409
Inkjet printer, 409
Integral length scale of turbulence, 195
Integration of functions, numerical, 47–49
Intensive property, 483
Interface (liquid–gas/liquid–liquid), 13–15, 27,
    72–73
Interference factor, 363
Internal energy, 121, 123
Inverse problem, 35
Isentropic flow, 381–384, 477
Iterative methods, 235, 252–254, 279

J

J-57 turbojet, 392
Jet-A fuel, 373
Jet engine, 391–398

Joule (J), 2
Journal bearings, flow past, 184–187
JP-8 fuel, 373

K

k-ε model, 196, 197
Kaplan turbine, 353
Kayak, 68
Kerosene, 373
Kilogram (kg), 2
Kinematic viscosity (ν), 11, 170, 469
Kinetic energy correction factor (α), 221–223
Kinetic energy flow rate, 221–223
Knocking, 356
Knudsen number (Kn), 16–17
Kolmogorov scale of turbulence, 195

L

Laminar flow, 20–21
   past bearings, 184–187
   of lubrication fluid, 186
   in open channel, 242–243
   in pipe, 82–85, 205–213, 217–218, 220–221, 222–223
   over/between plates, 167–171, 179–182, 191–193, 269–270
Laminar flow element, 449–450
Lapse rate, 29
Large eddy simulation, 195, 196, 197
Laser Doppler velocimetry, 432–436
Lattice Boltzmann methods, 198
Length, 2, 454
Leonardo da Vinci, 87
Lift coefficient (C_L), 266
Lift distribution, optimal, 298–300
Lift force, 290, 293–300
Lifting bodies, 318–322
Linear differential equation, 191
Linearity of measuring device, 452
Linear momentum, 94–95, 159
Line of action of force, 49–55
Liquid sprays, 398–411
Liter (L), 3

Loss coefficient (K), 254, 256
Low-Earth orbit, 112
Lubrication flow, 179–187

M

M2-F1/M2-F2 aircraft, 319–320
Mach, Ernst, 266
Mach number (Ma), 17, 266, 307, 310
Madaras rotor, 359
Magnus force, 280, 283
Manning's formula, 244–245
Manometer measurements, 32–37
Mass, 2, 454. *See also* Conservation law for mass
Mass-averaged velocity, 84, 205, 221–222
Max q, 133
Mean camber line, 291
Mean free path (λ), 16
Mesosphere, 30–31
Meter (m), 2
Microchannel flow, 16, 17, 73
Milliliter (mL), 3
Millimeters of mercury (mm Hg), 4, 5, 32
Miscible liquids, 15
Mixing length model, 197
Mixtures involving fluids, 12, 15
M-k-s system of units, 2
MOD-1 wind turbine, 365
Model testing, 306–307, 311–317, 459. *See also* Nondimensional numbers and nondimensionalization; Scaling and similarity
Molecular diffusivity (D), 11
Molecular dynamics simulations, 198
Moment. *See* Torque
Momentum; *See also* Conservation law for momentum
   angular, 113–115, 333–335
   linear, 94–95, 159
Momentum flow rate correction factor (β), 223
Monatomic ideal gas, 9
Mono-methyl hydrazine, 374, 472
Monte Carlo methods, 198
Moody chart, 224
Moving control volume, 103–107, 116–117, 335
Multigrade/multiweight oils, 11

Multiphase mixtures, 15, 19–20
Multipipe systems, 251–252, 416
Murray's law, 416

N

NACA airfoil sections, 290–293
Nappe, 250
National Full-Scale Aerodynamics Complex, 315
Natural convection, 413
Navier–Stokes equations, 157
   for boundary layer over flat plate, 167–171, 269–270
   derivation of, 158–161
   for flow around sphere, 175–177, 273, 275
   for flow between plates, 179–182, 192
   for flow past journal bearings, 184–187
   numerical solution of, 191–195
   for pipe flow, 182–184, 206–209
   Reynolds-averaged, 166–167, 195, 196, 197
Net positive suction head, 343
Newton (N), 2
Newtonian fluid, 8, 142
Newton, Isaac, 7
Newton's law of cooling, 413
Newton's law of viscosity, 7–9, 182, 269
Nitric oxides, 30
Noncircular pipes, 229–230
Nondimensional numbers and nondimensionalization, 16, 453–462; *See also* Model testing; Scaling and similarity
   cavitation number, 350
   Courant number, 193
   drag coefficient, 173, 177, 266, 270, 275, 294–296, 322
   Fourier number, 193
   friction coefficient, 173
   friction factor, 174, 212, 224, 245
   Froude number, 242, 249–250, 267, 305–306
   Grashof number, 413
   Knudsen number, 16–17
   lift coefficient, 266
   Mach number, 17, 266, 307, 310
   Nusselt number, 413, 414
   Ohnesorge number, 403
   Prandtl number, 413
   pressure coefficient, 276
   Reynolds number, 164, 171, 196, 205, 217, 243, 244, 270, 290, 454
   spin parameter, 283
   Stokes number, 178, 435
   Strouhal number, 276–278
   Taylor number, 402
   Weber number, 400, 402, 403
Non-Newtonian fluid, 8, 9
Normal shock, 384–387, 477
No-slip boundary condition, 16, 168, 169, 176, 180, 269, 275
Nozzle, 378, 382–383, 389
Nozzle area ratio, 389
Numerical methods
   for approximating derivatives, 188–191
   explicit, 193
   for integration of functions, 47–49
   iterative, 235, 252–254, 279
   for Navier–Stokes equations, 191–195
   for ordinary differential equations, 108–112
   for trajectories, 110–112, 287–288, 376–378
Nusselt number (Nu), 413, 414

O

Object of revolution, volume of, 61
Oblique shock, 384
Ohain, Hans von, 391
Ohnesorge number (Oh), 403
Oil, grades of, 10–11
One-dimensional isentropic flow, 381–384, 477
Open channel flow, 240–251
Open-circuit wind tunnel, 316
Ordinary differential equations, numerical solution of, 108–112
Orifice plate meter, 447–449
Ostwald viscometer, 11
Outliers, 488–489
Oxygen, 372–373, 472, 473
Ozone layer, 30

P

Pappus's theorem, 61
Parabolic partial differential equation, 191

Parabolic velocity profile
　　for laminar pipe flow, 82–85, 206–209
　　for Poiseuille flow, 193
Paraboloid, 73
Parachute, 280
Parallel pipe connection, 251–252
Partial differential equations, 160–161, 191
Particle image velocimetry, 432–436
Pascal (Pa), 4, 5
Passenger cars, aerodynamics of, 300–305
Pathline, 174
Pelton, Lester, 119
Pelton turbine, 351–353
Pelton waterwheel, 119–120, 351, 352
Perfect fluid, 163
Peristaltic pump, 336
Phase Doppler anemometry, 432–436
Pi analysis, 456–462
Piezoelectric effect, 453
Pipe flow, 82–85, 182–184
　　in complex pipe systems, 251–252
　　entrance flow effects, 227–229
　　friction losses in, 210–213
　　fully developed, 227
　　laminar, velocity profile of, 82–85, 206–209
　　in noncircular pipes, 229–230
　　problem-solving methods for, 233–240
　　secondary losses, 254–257
　　turbulent, 216–221, 223
Pitot, Henri, 133
Pitot tube, 133–135, 428–431
Planar flow (over/between plates), 8, 167–171, 179–182, 191–193, 269–270
Poise (P), 3
Poiseuille flow, 179–182, 191–193, 416
Poiseuille's law, 234
Porous media, 178
Pound force (lbf), 3
Pound mass (lbm), 3
Pounds per square foot (psf), 3
Pounds per square inch (psi), 3, 5
　　absolute (psia), 5
　　gage (psig), 4
Power, 2–3, 121–122, 335
　　for bird flight, 418
　　from Pelton turbine, 352–353
　　from Pelton waterwheel, 119, 351
　　for pipe flow, 212, 233
　　from propeller, 349
　　from/to pump, 336, 338, 339
　　for wind tunnel, 316–317
　　from wind turbine, 364
Prandtl, Ludwig, 168, 314
Prandtl number (Pr), 413
Prefixes for units, 2
Pressure
　　absolute, 4, 6, 452
　　atmospheric, 4, 29–31
　　differential, 33, 429–430
　　dynamic, 133
　　gage, 4, 6, 452–453
　　measurement of, 32–37, 452–453
　　stagnation, 133, 380
　　units of, 3, 4–6
　　vacuum, 6, 453
　　vapor, 12
Pressure coefficient (C_P), 276
Pressure drag. *See* Form drag
Pressure drop, 210–211, 219
Pressure head, 124
Pressure transducer, 452–453
Propeller, 349–350, 394
Propeller turbine, 353
Propulsive efficiency, 394–395
Propulsive power, 349
Pulse detonation engine, 398
Pump/compressor, 121, 335–348
Pump head, 124, 338, 340

Q

Quiet Spike, 311

R

Radiation of heat, 412
Rain-X™ glass treatment, 14
Ramjet, 396
Random error, 452
Range of measuring device, 452
Rayleigh breakup, 400–402
Rayleigh flow, 384

Reaction turbine, 351, 352
Repeatability of measurement, 451
Respiratory system flows, 415–416
Reynolds-averaged Navier–Stokes equations, 166, 195, 196, 197
Reynolds number (Re), 164, 196, 205, 243, 270, 290, 454
  critical, 164, 171, 217, 244, 270
Reynolds, Osborne, 157, 164–165, 217, 453
Reynolds stresses, 167, 197
Reynolds transport theorem, 483
Rheology, 9
Rigid body motion, 69–75
Road vehicles, 300–305
Rocket, 371
  efficiency of, 388–389, 395
  escape velocity of, 112–113
  nozzle design, 378, 382–383, 389
  trajectory, 106–108, 110–112, 376–378
Rocket fuel, 372–376
Rotameter, 445–447
Rotary pump, 336
Rotating systems, 71–75, 113–120
  fans, 348–349
  hydraulic turbines, 351–358
  propellers, 349–350
  pumps/compressors, 121, 335–348
  wind turbines, 358–367
Round-off error, 48
Runge-Kutta methods, 109, 290

S

SAE grades of oil, 10–11
Saturn V rocket, 376
Sauter mean diameter, 409
Savonius rotor, 359
Saybolt universal seconds (SUS), 10
Saybolt viscometer, 9–10
Scaling and similarity, 306–307, 454–455; *See also* Model testing, Nondimensional numbers and nondimensionalization
  for fans, 348–349
  for propellers, 350
  for pumps, 342–343
Schlieren photography, 442–443

Scramjet, 396
Second (s), 2, 3
Secondary breakup, 403
Secondary losses, 254–257
Separated wake flow, 273
Series pipe connection, 251–252
Settling velocity, 177
Shadowgram, 442
Shaft power, 335
Shear stress (τ), 7, 8, 171, 172, 182, 210, 225, 269–270
Shock-induced separation, 310
Shock wave, 17, 18, 308, 310, 384–387, 389, 477
Significant digits, 18–19
Similarity. *See* Scaling and similarity
Skin friction. *See* Viscous drag
Slip correction factor (C_C), 275
Slug, 3
Smagorinsky model, 197
Solidity (σ), 364
Solid rocket booster, 374
Specific gravity (SG), 7
Specific impulse, 388
Specific speed of pump/fan, 340, 348
Specific volume (v), 6
Specific weight (γ), 7
Speed for minimum power, 421
Speed of sound (a), 17–18, 307, 379–380
Sphere
  drag on, 273–276
  flow around, 175–177, 273, 275
Spin parameter (S), 283
Sports balls, 280–286
Stability, 59, 67–68, 193, 317–318
Stagnation flow, 273
Stagnation fluid properties, 380
Stagnation point, 133
Stagnation pressure, 133, 380
Standard atmosphere (atm), 4, 5
Standard drag curve, 276
Steady-state systems, 85–91, 97, 123, 167–171, 175, 205–213
Stoichiometric condition, 356
Stokes flow, 175–177, 273, 275

Index

Stokes number (St), 178, 435
Stratosphere, 30–31
Streakline, 174
Streamline, 174
Strouhal number (St), 276–278
Subgrid scale model, 197
Submerged objects, net force on, 38–46, 55, 58–65
Sudden contraction, 256
Sudden expansion, 254
Supercharger, 356–357
Super-soaker™ water gun, 131–132
Supersonic drag, 307–309
Surface roughness, 223–227, 273
Surface tension (σ), 12, 13–15, 27, 73, 472
Surfactant, 15
Suspension, 15
Sutherlands' formula, 469
Systematic error, 452
Swept-back wings, 310–311
Swimming animals, 417–422

T

Taylor analogy breakup model, 402
Taylor number, 402
Taylor series, 109, 188–189
Temperature, 2, 380, 454
Terminal velocity, 177
Theorem of Pappus and Guldinus, 61
Thermal diffusivity (α), 11, 413
Thermal transfer number (B), 402
Thermistor, 453
Thermocouple, 453
Thermosphere, 30–31
Thrust force, 94–96, 97–107, 113–120
Thrust specific fuel consumption, 392
Time, 2, 454
Timeline, 175
Tip speed ratio (X), 364
Torque, 49–55, 113–116, 333
Torque converter, 357–358
Torr, 5
Tractor-trailer, 303–304
Trajectory
  of projectile, 287–288
  of rocket, 110–112, 376–378
Transient drag, 322–323
Transient flow, 91–93
Transonic drag, 310–311
Trapezoid rule, 47–48
Troposphere, 29–31
Trucks, aerodynamics of, 300–305
Turbine, 121, 351–358, 358–367
Turbine flow meter, 450–451
Turbine head, 124
Turbocharger, 355–357
Turbojet/turbofan, 391–394
Turbomachinery. See Rotating systems
Turbulence Reynolds number (Re_t), 196
Turbulent flow, 20–21, 97, 163–167
  boundary layer of, 172, 270–271
  Boussinesq hypothesis for, 197
  characteristics of, 164, 167, 195–197
  computational modeling of, 195–198
  in open channel, 240
  in pipe, 216–221, 223
  and surface roughness, 223–227
  transition to, 163, 164, 171, 205, 217–218, 244, 270
Turndown ratio, 452

U

Ultrasonic flow meter, 451
Uncertainty of measurement, 451, 485–491
Uniform flow in open channel, 242–245
Uniformly accelerating systems, 69–75
Units of measurement, 2–6, 454
Universal gas constant, 5

V

Vacuum pressure, 6, 453
van der Waals force, 13
Vans, aerodynamics of, 300–305
Vaporization, 402–403
Vapor lock, 131
Vapor pressure (P_{vap}), 12
Variable Density Tunnel, 315
Velocity head, 124
Velocity probes, 427–436

Velocity profile
   in circular pipe, 82–85, 206–209
   laminar vs. turbulent, 220, 223
   of uniform open channel flow, 243
Venturi flow meter, 125–127, 443–445
Vertical-axis wind turbine, 359
Viscometer, 9–11
Viscosity, 3, 7–11, 12, 162–163, 197, 469
   kinematic, (ν), 11, 170, 469
   Newton's law of, 7–9, 182, 269
Viscous drag, 269–273
Visualization methods, 436–443
Volume, 3, 61
Volume flow rate (Q), 85, 443–452
Volute of pump, 342
Vortex shedding, 276–278, 293

W

Water sprinkler, 116–119, 335
Water vehicles, 305–307
Waterwheel, 119–120
Watt (W), 2
Wave drag, 267, 305–307, 307–311
Weber number (We), 400, 402, 403
Weight flow rate, 388
Weirs, 250–251
Wenham, Frank, 314
Whittle, Frank, 391
Wind tunnel, 314–317
Wind-tunnel testing. *See* Model testing
Wind turbine, 358–367
Wing, 290
Winter grades of oil, 10–11
Wright Brothers, 300, 314

X

X-15 aircraft, 373
X-24 aircraft, 320
X-29 aircraft, 311
X-38/X-48 aircraft, 321–322
X-43 scramjet, 396